Basic Concepts of
Electrical and Electronics Engineering

Basic Concepts of Electrical and Electronics Engineering

Dr. P. S. Subramanyam

M.Tech., Ph.D.

Professor Emeritus
Formerly HOD, EEE
Vignana Bharathi Institute of Technology,
Aushapur, Hyderabad - 501 301.

Prof. K. Lal Kishore

M.Tech, Ph.D, SMIEEE, FIETE, FIE, FAPAS, LMISTE, LMISHM

Former
Vice-Chancellor
JNTU, Anantapur
Ananthapuramu – 515 002

BSP **BS Publications**

A unit of **BSP Books Pvt., Ltd.**

4-4-309/316, Giriraj Lane, Sultan Bazar,
Hyderabad - 500 095
Phone : 040 - 23445605, 23445688

Basic Electrical and Electronics Engineering *by Dr. P. S. Subramanyam and Prof. Lal Kishore*

Published by :

BSP **BS Publications**

A unit of **BSP Books Pvt., Ltd.**

4-4-309/316, Giriraj Lane, Sultan Bazar,
Hyderabad - 500 095
Phone : 040 - 23445605, 23445688
e-mail : info@bspbooks.net
www.bspbooks.net

ISBN : 978-93-5230-150-8 (Hardback)

Dedicated To

LOTUS FEET

of

Lord

- P. S. Subramanyam

Dedicated To

SRI SARASWATI

the goddess of learning

- K. Lal Kishore

Preface

Syllabus for the subject "Basic Electrical Engineering" has been changed recently and accordingly the contents pertaining to Electronics Engineering are rewritten. These chapters cover the topics to be studied by all Electrical, Electronics, Computer Science and other allied branches.

The contents and topics are changed so that there is almost one to one match between the syllabus and topics given in this text book. This will enable the students and teachers to feel 'at home' while using the book.

A good knowledge of Electrical and Electronic Circuits is necessary for Engineers of all disciplines including Computer Science Engineering and Information Technology. An earnest attempt has been made in this book to elucidate the principles and applications of Electrical and Electronic Engineering and also its importance, so as to evince interest on the topics so that the student gets motivated to study the subject with interest.

The Basic Concepts of Electrical Engineering have been presented in a graded manner in this book, "Basic Concepts of Electrical and Electronics Engineering", with the above objective in mind. For every topic the necessary pre-requisite material has been presented well so as to understand the topic. Also the topics have been rearranged and included in the relevant chapters, so as to make the subject matter coherent. **The book has been revised and topics on Basic Electronics Engineering have been added as regards Electrical Engineering.**

The book deals with the Basic Circuits Analysis of D.C.Circuits and A.C. Circuits including single phase A.C. circuirts. Topics on Resonance and current locus have also been included. Application of different theorems for D.C. and A.C. Circuits are prescribed.

The depth of any particular topic has to be decided by the concerned Faculty since the topics have been designed to cater to the needs of students studying Basic Electrical and Electronics Engineering Course in different Universities. For better understanding and grasp of the subject, a set of Comprehension Questions and Exercises have been provided at the end of each chapter.

By answering all the questions given in the comprehension at the end of each chapter one can answer any objective type of questions easily whatever be the form of the question. Answers for those questions can be obtained by going through the topics in the chapter carefully. It also helps in understanding the subject covered in the chapter thoroughly.

By answering all the questions given in the exercise at the end of each chapter one can answer subjective type of questions easily including problems. The worked examples in the chapter may help in solving these questions. This helps in creating confidence in the student so that he can confidently answer any question subjective or objective covered in the chapter.

My hearty thanks the management of M/s B.S. Publications, Sri Nikhil Shah ji, Mr. Naresh Davergave and his team members, who have helped in bringing out this book.

Comments & Suggestions are Welcome to improve the book in subsequent editions.

I wish to dedicate this book to LORD ALMIGHTY who is the First and Foremost GURU and the Fountainhead of All Knowledge, to my teachers and to my students with whose inquisitiveness I could learn more and develop the subject, in the true tradition of GURU PARAMPARA.

-Authors

About the Author

Dr. Pisupati S. Subramanyam

Dr. Pisupati S. Subramanyam is a holder of Doctorate in Engineering from IIT, Madras has a vast experience in industrial (Electrical Power Generation and Control), academic and administrative areas. He is a fellow of The Institution of Engineers (India), fellow of National Federation of Engineers, senior member of IEEE, member of Computer Society of India, member of Indian Society for Technical Education. He has guided a number of B.Tech, M.Tech and Ph.D. projects in Power Systems, published a number of papers in National and International Journals and authored several textbooks. Basically from Electrical Engineering discipline, he cross migrated to the field of Computer Science and Engineering, working as Head of the Department of Computer Science and Engineering and Information Technology for 20 years. Being Principal of J. B. Institute of Engineering and Technology, Hyderabad, he is instrumental in getting additional branches and additional intake sanctioned. He is also instrumental in getting accreditation by All India Council for Technical Education (AICTE) for J.B.I.E.T. He also got several PG courses like M.Tech., MCA and MBA sanctioned for J.B.I.E.T. He also got the Jawahar Knowledge Centre sanctioned for J.B.I.E.T. and ran it successfully. He got In-campus and Off-campus Placements done by several M.N.Cs for J.B.I.E.T. and students of other colleges also. At present Dr. P. S. Subramanyam is Professor Emeritus in the Department of E.E.E. in Vignana Bharathi Institute of Telchnology, Aushapur, Ghatkesar Mandal at the outskirts of Hyderabad.

Dr. P.S. Subramanyam has been awarded the Best Student Branch Councellor Award twice by IEEE, Hyderabad section.

Professor Kondepudi Lal Kishore

Professor Kondepudi Lal Kishore is a distinguished academician and administrator with over 34 years of teaching experience. A post graduate and Ph.D. degree holder from Indian Institute of Science (IISc), Bangalore. Prof. Kishore began his career in teaching at Jawaharlal Nehru Technological University Hyderabad (JNTUH) in 1977. He is elevated as professor in 1990 in the same University and subsequently held many academic and administrative positions in the University. He has over 190 research publications to his credit till now. So far 24 research scholars obtained Ph.D under his guidance and some more are doing research under him. He wrote Six Text Books in the areas of Electronic Devices, Electronic Circuit Analysis, Linear I.C. Applications, Electronic Measurements and VLSI, which are popular and widely used by engineering students. He has many more distinctions to his credit from National and International organizations. He delivered many invited lectures and chaired Technical sessions.

He has received Best Teacher Award for the year 2004, Sir Mokshagundam Visweswaraya Award, first Bapu Seetharam Memorial Award, Eminent Electronics and Communications Engineer Award, elected fellow of A.P. Akademi of Sciences.

He has held many other administrative positions in JNTU Hyderabad, as Chairman BOS, Head of the Department, Principal of Engineering College, Director SCDE, Director UGC Academic Staff College, Director Academic and Planning etc. He was Convener of State level Competitive entrance examination ECET, (FDH), thrice, during 2000, 2001 and 2005. He was registrar of JNTUH during 2005-08 and Rector of the University during 2008-10. From 2012-15 he was Vice-Chancellor of JNTUA, Anantapur and successfully completed his assignment.

Contents

CHAPTER 1

D.C. CIRCUIT CONCEPTS AND CIRCUIT ELEMENTS

D.C. CIRCUIT CONCEPTS AND CIRCUIT ELEMENTS - I

D.C. CIRCUIT CONCEPTS AND CIRCUIT ELEMENTS - II

CHAPTER 2

D.C. CIRCUIT ANALYSIS

CHAPTER 3

FUNDAMENTALS OF ALTERNATING CURRENT

CHAPTER 4

A.C. CIRCUITS

A.C. CIRCUITS - I - SINGLE PHASE CIRCUITS

CHAPTER 5

NETWORK THEOREMS

CHAPTER 6

SEMICONDUCTOR DIODES, RECTIFIERS AND FILTERS

RECTIFIERS AND FILTERS

CHAPTER 7

BJT AND TRANSISTOR CONFIGURATIONS

CHAPTER 8

JFET AND SPECIAL PURPOSE DEVICES

SPECIAL PURPOSE DEVICES

1

D.C. CIRCUIT CONCEPTS AND CIRCUIT ELEMENTS - I

1.1 INTRODUCTION TO BASICS OF ELECTRICAL ENGINEERING

Electrical Engineering forms the foundation of Electrical, Electronics, Communications, Controls, Computers, Information, Instrumentation, etc. Hence a good grasp of the fundamentals of Electrical Engineering is an absolute necessity to become a good engineer in any discipline.

In this chapter we discuss the basics of Electrical Engineering like sources of electrical energy-voltage and current sources and their conversion, Ohm's law, calculation of electrical power and energy and DC circuit analysis using mesh and nodal analysis.

1.1.1 Circuit Concepts - Concepts of Networks

An Electrical Circuit or an Electrical Nework consists of one or more Electical Energy Sources connected to a number of circuit elements like active or passive elements or both in such a way that there is a connection between the different elements causing a current to flow through the elements. The electrical circuit should be a closed circuit so that current can flow through it. If an Electrical Circuit is open then no current can flow through the circuit.

1.2 CURRENT FLOW

1.2.1 Potential and Potential Difference

An electrically charged particle sets up an electric field around it. If the particle is stationary, then the field set up by it is said to be ***Electrostatic Field***. The electric field lines of a positive charge **+q** will be radial and directed away from the charge. The field set up by the negative charge **–q** will be radial and directed towards the negative charge. Like charges repel and unlike charges attract. The force of attraction or repulsion between two charges q_1 and q_2 will be governed by ***Coulomb's Law*** which states that the force will be proportional to product of the charges $q_1 \times q_2$ and inversely proportional to the square of the distance 'R' between them and depends upon the medium in which the charges are placed.

$$F = \frac{q_1 q_2}{4\pi \in R^2} \qquad \text{.......... (1.2.1)}$$

where F is the force in newtons

q_1 and q_2 are the charges in coulombs

R is the distance in meters

$\in$ is the permittivity of the medium in farads / meter

The Absolute Permittivity for free space or vacuum,

$$\in_o = 8.854 \times 10^{-12} = \frac{1}{36\pi} \times 10^{-9} \text{ Farads / meter}$$

For other media the Absolute Permittivity,

$$\in = \in_o \times \in_r$$

where $\in_r$ is the relative permittivity of the medium which is a mere number.

$\in_r = 1$ for Air and $\in_r = 1$ for Mica

The force will be attractive if q_1 and q_2 are opposite charges and repulsive if q_1 and q_2 are like charges.

If we want to place any charge from one point to another point in the electrostatic field work has to be done against the electrostatic force or coulomb force experienced by that charge.

The work done in bringing a unit positive charge from infinity upto a given point p in an electrostatic field is defined as the potential at that point in the electro-static field. The unit for potential will be Joules/Coulomb. The unit is also called Volts.

The work done in moving a unit positive charge from one point in the electric field to another field in the electric field is known as the potential difference between the two points and is measured in volts. If V_A is the potential at point A and V_B is the potential at point B then the potential difference between the two points A and B will be $V_{AB} = V_A - V_B$. If $V_A > V_B$ then V_{AB} will be positive and is known as potential drop from point A to point B.

If $V_A < V_B$ then V_{AB} will be negative and is known as voltage rise from point B to point A. The voltage rise from point B to point A is generally denoted by the letter E.

$$E_{AB} = (E_A - E_B) = (V_B - V_A) = V_{BA} = -V_{AB} \qquad \text{...... (1.2.2)}$$

1.2.2 Electric Current

An electron placed in an electric field will experience a force and move towards the positive potential of the field since it is negatively charged. Continuous flow of electrons constitute a current flow from negative potential to positive potential of the field. This current is known as ***electron current***. The conventional current flow is opposite to that of the electron current in direction. The conventional current flow which is in opposite direction to electron current flow, will be flowing from a point of higher potential to a point of lower potential.

In metals (conducting materials), a large number of free electrons are available which move from one atom to the other at random when a potential difference is applied between two points of the conducting material and the current starts flowing.

The rate of flow of charges through any cross-section of a conductor is called a ***current*** and is denoted as 'i'. Current is expressed in terms of ***amperes***. Ampere is denoted by A or sometimes by α.

$$i = \frac{dq}{dt} \text{ Amperes} \qquad \text{.......... (1.2.3)}$$

where i is the instantaneous value of the current (value at any particular instant of the current)

The ***steady current*** 'I' is given as,

$$I = \frac{Q}{t} \text{ Amperes} \qquad \text{.......... (1.2.4)}$$

where Q is the charge flowing through the cross section of the conductor in time 't', if the flow of the charges is uniform.

Otherwise,

$$Q = \int i \, dt \text{ coulombs} \qquad \text{.......... (1.2.5)}$$

A wire is said to carry a current of ***one ampere*** when charge flows through it at the rate of one coulomb per second. Hence, one ampere is the current which flows when a charge of one coulomb moves across the cross-section of a conductor in one second.

1.3 ACTIVE AND PASSIVE ELEMENTS

The elements of an electric circuit can be classified into active and passive elements. Active elements are the sources that supply electrical energy to the circuit causing current flow through it. The energy sources can be independent or dependent sources. They may be voltage or current sources.

1.3.1 Sources of Electrical Energy - Voltage Source

The ***voltage source*** is assumed to deliver energy with a specified ***terminal voltage*** V_T, if it is a steady voltage source or $v(t)$ or simply v, if the voltage changes with respect to time. An ideal voltage source is expected to deliver a constant voltage to the outside circuit whatever be the amount of current drawn from the voltage source. The voltage of the source is called the ***Electro-Motive Force*** (**E.M.F.**) and is measured in Volts. It is denoted by the symbol E.

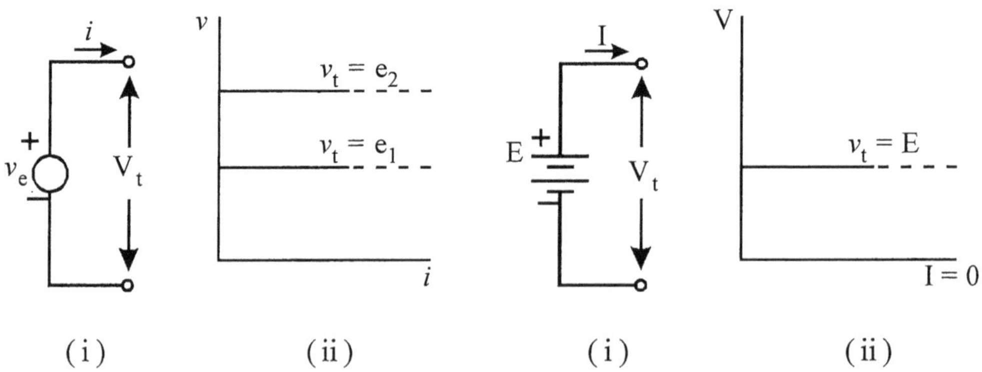

(a) **(b)**

Fig. 1.1 (a) *For a Time-Varying Voltage Source,* ***(b)*** *For a Time-Invariant Voltage Source (represented by a Battery)*

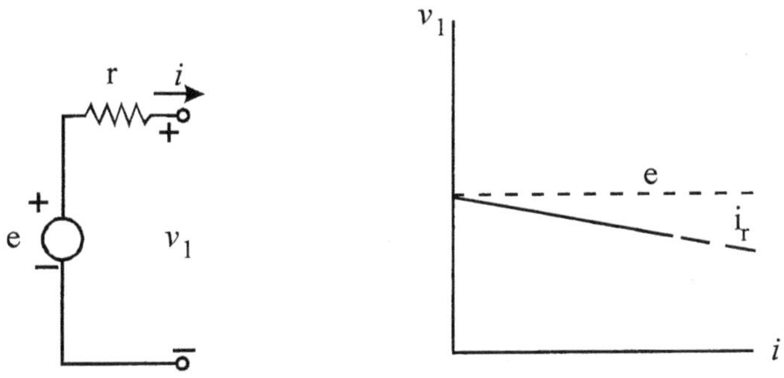

(i) (ii)

(c)

Fig. 1.1 (c)(i) Model for a Voltage Source in which 'r' represents Source Resistance. The internal resistance 'r' will be very low. For the model of (i), the Terminal Voltage depends on source current as shown in (ii) where $v_t = v \;\; ir$

If the source has an internal resistance γ then,

$$V = v(t) - i\,r \qquad \text{.......... (1.3.1)}$$

or

$$V_t = E - I\,r \text{ Volts} \qquad \text{.......... (1.3.1a)}$$

where I is the current drawn from the Voltage source in Amperes (A)

r is the internal resistance of the voltage source in Ohms (Ω).

In practice the terminal voltage of the voltage source will be decreasing as the current drawn from it is increased due to the voltage drop in the internal resistance of the voltage source. The internal resistance has to be very small in order that the voltage drop inside the source will be very small and maximum voltage may be available to the load.

Note : *For any D.C. source, the polarities at the terminals will be same at all instants of time. For Time Varying Sources to polarities indicate the polarities at different terminals at any particular instant of time.*

1.3.2 Sources of Electrical Energy - Current Source

A current source is said to deliver a constant current $i_2 = I$ to the circuit thorugh the terminals, if it is a steady current source or *i*(t) or simply *i*, if the current changes with respect to time. An ideal current source is expected to deliver a constant current to the outside circuit whatever be the circuit. An ideal current source can be represented by the symbol 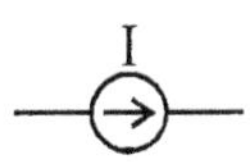

or 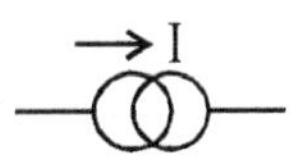

In practice the current supplied by the current source will be decreasing as the voltage across the current source is increasing due to the internal resistance R of the current source, which is assumed to be across the current source. The internal resistance of the current source should be as high as possible so that maximum current will be delivered to the load connected across the current source with the current through the internal resistance being very very small.

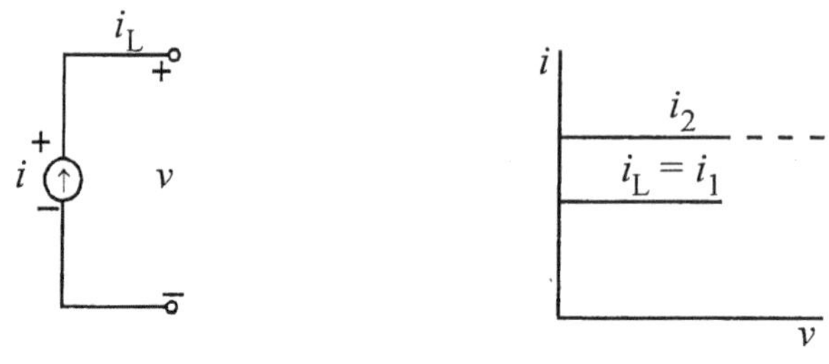

(i) (ii)

Fig. 1.2 (a)(i) Symbol for the Current Source for which i does not depend on v as shown in, (ii) Other lines may be drawn parallel to that shown for a specific current, i_1.

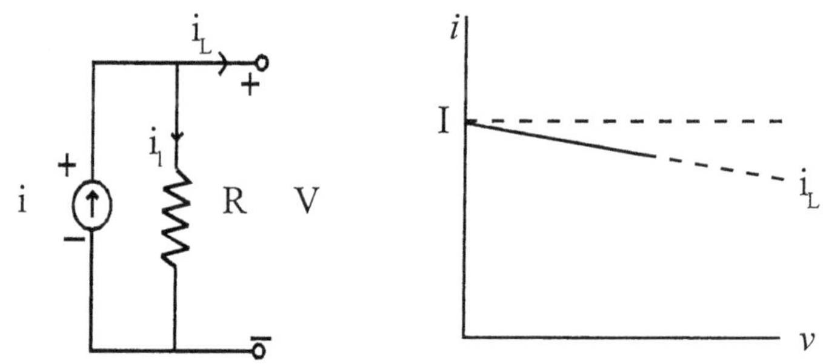

(i) (ii)

Fig. 1.2 (b)(i) *Model for a Current Source in which R represents shunt resistance. For the model of 1.2 (i) the Terminal Current in 1.2 (ii) is given by* $i_L^{\cdot} = i - (1/R)v$ *(ii) Other lines may be drawn parallel to that shown for a specific current,* i_L. The internal resistance 'R' will be very high.

1.3.3 D.C. & A.C. Sources

If the voltage or current supplied by an electrical energy source is constant with respect to time as shown in Fig. 1.3 (a)(i) or Fig. 1.3 (a)(ii) then it is known as ***D.C. Voltage Source*** or ***Direct Current Source*** (or ***Steady Current Source***). D.C. stands for ***Direct Current.***

A D.C. Source has two terminals from which energy is supplied to the outside load. They are known as ***Positive Terminal*** which supplies the positive ions and ***Negative Terminal*** which receives the returning current or which can be assumed as supplying negative charges called electrons in the direction opposite to the conventional current direction. D.C. supply is provided by batteries or D.C. generators. The battery converts chemical energy into electrical energy. The generator converts mechanical energy into electrical energy.

If the voltage or current is varying with respect to time but, has the same polarity as shown in Fig. 1.3(b)(i) then it is known as ***Unidirectional Source***. If the polarity is positive, it is known as ***Positive Source***. If the polarity is negative, it is known as ***Negative Source***.

If the voltage or current supplied by an electrical energy source varies in both magnitude and polarity with respect to time as shown in Fig. 1.3 (b)(ii) then it is known as ***A.C. Voltage Source*** or ***Alternating Current Source***. A.C. stands for ***Alternating Current***.

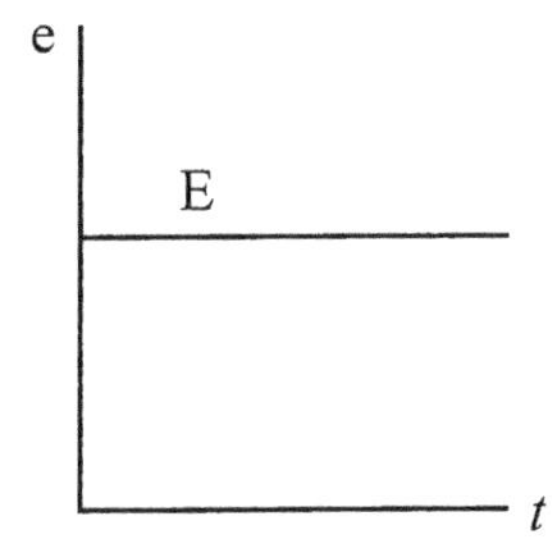

(i) D.C. Voltage Source

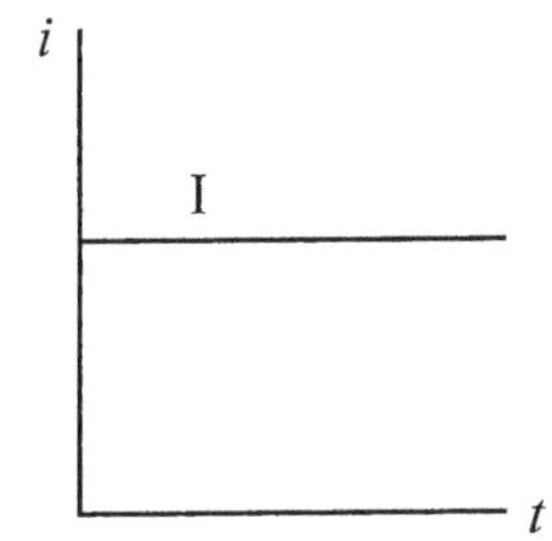

(ii) D.C. Current Source

(a) D.C. Voltage & Current Sources

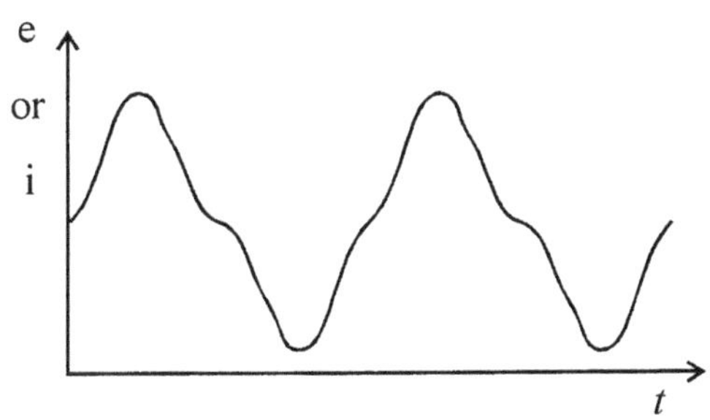

(i) Uni directional voltage or current

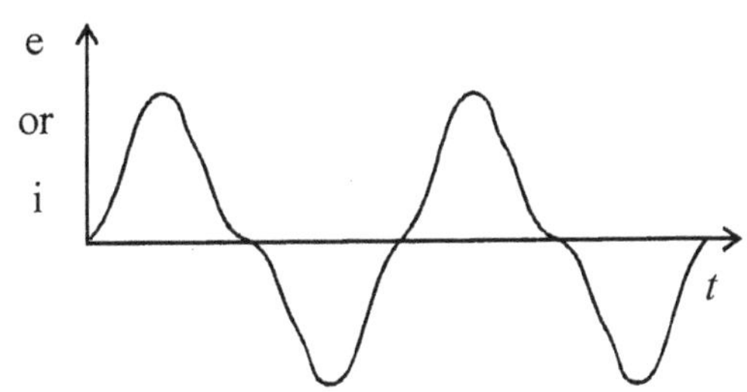

(ii) A.C. Voltage or Current

(b) A.C. Uni-directional Voltage & Current Sources

Fig. 1.3 *A.C and D.C. Voltage & Current Sources*

The value of the voltage or current of an A.C. supply at any instant is called ***Instantaneous Value of Voltage or Current*** and is denoted as $v(t)$ or $i(t)$.

In general, the instantaneous values may also be denoted as v or i. **Generally no polarities will be marked for A.C. voltage or current. If at all polarities are marked for A.C. voltage or current, they mean the polarities of the voltage or current at the marked terminals, at any one particular instant and will be changing from time to time.**

1.3.4 Unilateral and Bilateral Elements

Elements in which current flow is in only one direction are called unilateral elements. Eg. Diode. Elements in which current flow can be in both directions are called bilateral elements. Eg. Resistance, Inductance, Capacitance etc.

i) Passive Elements - Resistance or Resistance Parameter

When a potential difference is applied across a conductor (or wire), the free electrons start moving in a particular direction. While moving through the material, these electrons collide with other atoms and molecules. They oppose this flow of electrons (or current) through it. This opposition is called ***Resistance***. Heat is produced because of the collisions of moving electrons with the other atoms and molecules. Thus whenever a current flows through a conductor, heat is produced in the conductor and this heat has to be dissipated fully. Otherwise, the insulation of the conductor (the Sheath made of insulating material covering the conductor) will get damaged.

The opposition offered to the flow of current (free electrons) is called ***Resistance***. Resistance is denoted by R and is measured in ***ohms*** named after a German mathematician ***George Simon Ohm*** and is represented by the Greek symbol Ω. For very high resistance we use large units such as kilo-ohms (kΩ which is equal to 10^3 Ω) or Mega-ohms (MΩ which is equal to 10^6 Ω) while for small resistances we use smaller units such as milli-ohms (mΩ which is equal to 10^{-3} Ω) or micro-ohms (μΩ which is equal to 10^{-6} Ω).

In electronic circuits, the current will generally be very small in milli-Amperes (mA or 10^{-3}A), micro-Amperes (μA or 10^{-6}A) or nano-Amperes (nA or 10^{-9}A) and hence, the resistance or resistors used will be in kilo-ohms or Mega-ohms and will be denoted simply as k or M (Ω is understood). They will be made of carbon resistors and will have color codes for different digits. They will not be accurate and will have a tolerance limit denoted by another color band and the wattage of the resistors will also be specified as ½ Watt or 1 Watt. For higher wattages of 5 or 8 or 10 Watts etc., wire wound resistors will be used.

If the resistance or resistivity is less the current flowing will be more. The resistance of a material is given by

$$R = \frac{\rho l}{a} \ \Omega \qquad \text{.......... (1.3.2)}$$

where ρ is the specific resistance of the material of the conductor in Ω-meters

l is the length of the conductor in meters

a is the area of cross-section of the conductor in square meters.

Not only conductors, but a coil wound with a conductor or any other electrical equipment offers resistance to current flow. A wire wound coil with two fixed terminals is called a ***resistor*** or ***resistance***. A coil with two fixed terminals and a variable contact terminal which makes contact with the body of the coil is called a ***Rheostat*** or ***Variable Resistor***. A rheostat can be connected in two ways as,

1. ***Series Resistance*** *(as shown in Fig. 1.4 (a))*

If the moving contact is very near the starting terminal the resistance offered by the rheostat will be minimum and if it is nearer to the farthest end terminal, the resistance offered by the rheostat will be maximum. Sometimes the moving contact and one of the end terminals will be connected together in which case, the resistance offered will be the resistance of the remaining part of the winding. As the moving contact is varied the resistance offered by this part will be varying, as the moving contact is moved away from or towards the second terminal. If the moving contact is towards the starting terminal then the resistance offered by the rheostat will be less.

2. ***Potential Divider*** *(as shown in Fig.1.4 (b))*

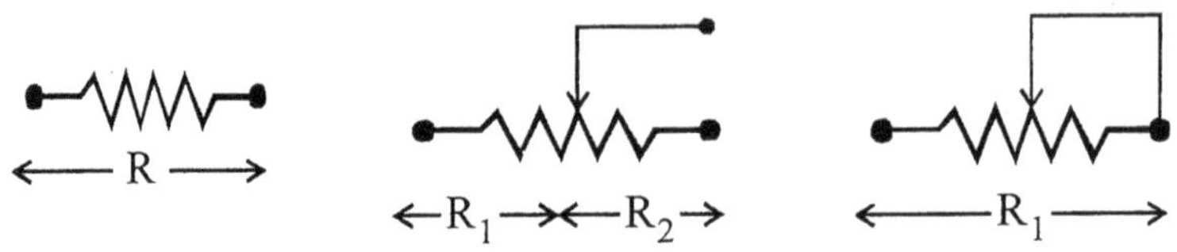

Fig. 1.4 (a) *Series Rheostat*

The two ends of the rheostat are connected across a voltage source which constitutes the input to the potential divider. The output is tapped between the moving contact and one of the end terminals, in which case, part of the input voltage will be the output voltage. The voltage tapped is given by,

$$V_{out} = V_{in} \times \frac{R_t}{R} \text{ Volts} \qquad \text{.......... (1.3.3)}$$

where R is the total resistance of the rheostat and R_t is the reistance of the tapped part of the rheostat winding.

Example 1.1:

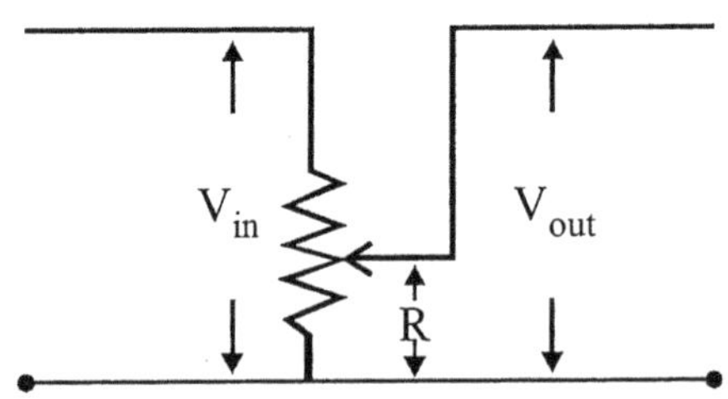

Fig. 1.4 (b) *Rheostat as Potential Divider*

Find the resistance of a coil of mean diameter 4 cm containing 400 turns of manganese wire 0.05 cm in diameter. The resistivity of manganese is 42 $\mu\Omega$ - cm.

Solution:

Here $\rho = 42\ \mu\Omega\text{-cm.}$

$= 42 \times 10^{-6}\ \Omega\text{cm}$

$= 42 \times 10^{-8}\ \Omega\text{ - m}$

$$a = \frac{\pi}{4} d^2$$

$$= \frac{\pi}{4} (\ 0.05\ )^2\ \text{cm}^2$$

$$= \frac{\pi}{4} (\ 0.05\ )^2 \times 10^{-4}\ \text{m}^2.$$

Number of turns (N) of the coil,

$N = 400$

Length per turn of the coil is $\pi \times D$

where D is the diameter of the coil.

Length (l) of the conductor of the coil,

$$l = \pi \times D \times N$$

$$= \pi \times 4 \times 400\ \text{cm}$$

$$= 16\ \pi\ \text{m}$$

Resistance of the coil,

$$R = \frac{42 \times 10^{-8} \times 16\pi \times 4}{\pi\,(0.05)^2 \times 10^{-4}} = 107.52\ \Omega.$$

1.3.5 Effect of Temperature on Resistance

As temperature increases the resistance of most of the conducting materials increase while for some material like carbon, electrolytes, insulators the resistance decreases as the temperature increases. The change in resistance depends upon the temperature-coefficient of resistance, which will be positive if the resistance increases with temperature and negative if the resistance decreases with temperature. The change in resistance per ohm per degree temperature change is called ***temperature-coefficient of resistance*** and its symbol is α.

If a metallic conductor of resistance R_0 at 0°C is heated to a temperature t_1, then the resistance R_1 at temperature t_1 is given by

$$R_1 = R_0\,(\ 1 + \alpha_0 t_1\ ) \qquad \text{.......... (1.3.4)}$$

Since the temperature-coefficient itself varies with temperature, it does not have the same value at all temperatures. Thus if R_1 and R_2 are the resistances of a conductor at temperatures t_1 and t_2, we have

$$R_2 = R_1 (1 + \alpha_1 (t_2 - t_1)) \qquad \text{.......... (1.3.5)}$$

where α_1 is the temperature-coefficient of resistance at t_1°C.

Variation of α is obtained as

$$\frac{1}{\alpha_2} = \frac{1}{\alpha_1} + (t_1 - t_2)$$

or $$\alpha_2 = \frac{\alpha_1}{1+\alpha_1(t_2 - t_1)} \qquad \text{.......... (1.3.6)}$$

Example 1.2:

The resistance of a coil decreases from 70Ω at 75°C to 50Ω at 15°C. Calculate the value of temperature-coefficient of resistance of the material of the coil at 0°C. Find the resistance at 0°C.

Solution:

Let α_0 be the temeperature-coefficient of resistance and R_0 be the resistance at 0°C.

$$70 = R_0 (1 + 75 \alpha_0) \qquad \text{.......... (1)}$$

and $$50 = R_0 (1 + 15 \alpha_0) \qquad \text{.......... (2)}$$

Dividing Eq. (1) by Eq. (2) we get,

$$\frac{70}{50} = 1.4 = \frac{1 + 70\alpha_0}{1 + 15\alpha_0}$$

$$54 \alpha_0 = 0.4$$

or $$\alpha_0 = \frac{0.4}{54} = 0.0074074$$

$$R_0 = \frac{70}{1+75\alpha_0} = \frac{70}{1.5555} = 45\ \Omega$$

1.3.6 Electrical Conductance

The reciprocal of resistance of a conductor is called ***Conductance*** of the conductor and is denoted by 'G' and is expressed in terms of Siemen abbreviated as "S" or mhos (℧).

$$G = \frac{1}{R}$$

$$= \frac{1}{\rho l/a}$$

$$= \frac{a}{\rho l} = \sigma\frac{a}{l} \text{ Siemens} \qquad \text{.......... (1.3.7)}$$

where σ is called specific conductance or conductivity and is measured in Siemens per meter (S/m).

If the conductance or conductivity is less the current flowing will be less.

1.4 V-I RELEATIONSHIP FOR RESISTANCE – OHM'S LAW

The relationship between the current flowing through a conductor and the potential difference across the conductor 47 is given by ***Ohm's Law***.

The *Ohm's Law* states that *the potential difference across a conductor is directly proportional to the current flowing through the conductor, the temperature of the conductor remaining constant.* The constant of proportionality is R, the resistance.

$$V = I \times R \text{ Volts} \qquad \text{.......... (1.4.1)}$$

or

$$V = R \times I \text{ Volts} \qquad \text{.......... (1.4.1a)}$$

Here, V is the voltage drop across the conductor.

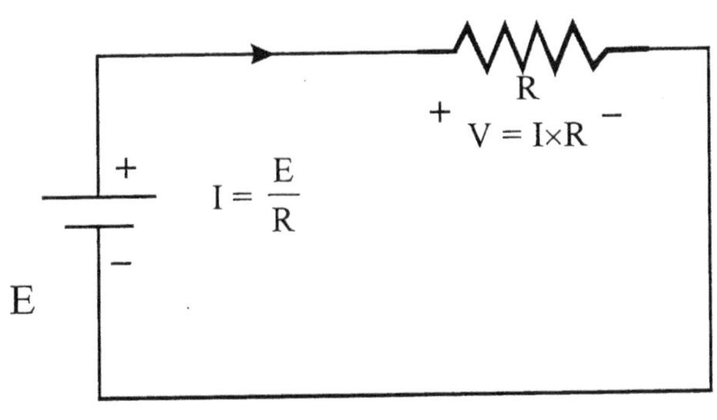

Fig. 1.5 Circuit for Ohm's Law

Note: *In the circuit, the voltage drop caused in the conducting wires connecting the battery and resistance are assumed to be zero as it will be negligible because the wire is a conducting material.*

Note: *While writing the equations for Volt-Ampere relationships in matrix form the second form i.e. Eq. (1.4.1a) will be applicable and can be expressed as,*

$$[\,V\,] = [\,R\,][\,I\,] \qquad \text{.......... (1.4.1b)}$$

Ohm's Law can also be expressed as

$$E = I \times R \text{ Volts} \qquad \text{.......... (1.4.1c)}$$

Here, E is the voltage rise across the conductor.

$$E_{AB} = -V_{AB}$$

The equation for Ohm's Law can also be written as,

$$I = \frac{V}{R} \text{ Amperes} \qquad \text{.......... (1.4.2)}$$

Ohm's Law also gives the Volt-Ampere relationship for an element. Ohm's Law can be applied to a part of a circuit or to the full circuit in which the current flows.

Ohm's Law can also be applied to A.C. Circuits or to circuits with Unidirectional Source in Laplace Transform domain for instantaneous values. For steady state conditions of A.C Ohm's Law using impedances and using RMS values for voltages and currents, all in complex form will be discussed later. However, for resistive circuits consisting of only resistances Ohm's Law can be written as,

$$v = i \times R \text{ Volts} \qquad \text{.......... (1.4.3)}$$

or

$$v = R \times i \text{ Volts} \qquad \text{.......... (1.4.3a)}$$

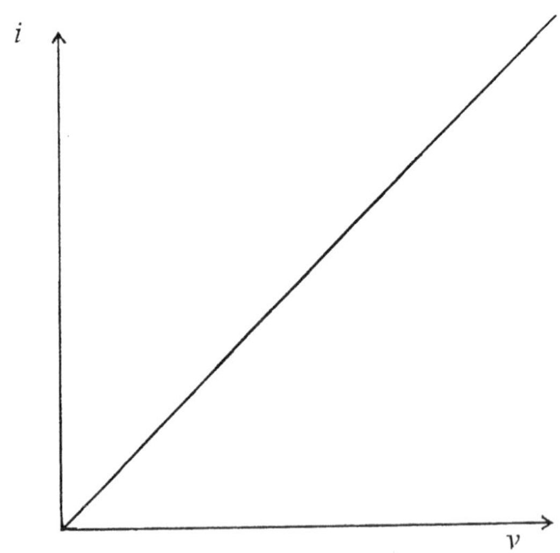

Fig. 1.6 Linear V-I Characteristic for Resistance

or $$i = \frac{v}{R} \text{ Amperes} \qquad \text{.......... (1.4.3b)}$$

where v and i are instantaneous values of Voltage and Current respectively.

1.4.1 Linear & Non-Linear Resistances

Those resistances in which the current flow changes in direct proportion with changes in the voltage applied across them are called ***Linear Resistances***. (curve 1 of fig. 1.6) The v-i characteristics for linear resistances will be current increasing as the voltage across increases. Ohm's Law is applicable as the resistance remains constant.

Those resistances for which the current through them does not vary in direct proportion are called ***Non-Linear Resistances***. For nonlinear resistances the v-i characteristics will be nonlinear. (curves 2 and 3 of fig.1.6)

In certain nonlinear resistances like Thyrite, the current increases more than proportionately with applied voltage with resistance decreasing rapidly like in curve 2 of Fig. 1.7. Hence, it is used in Lightning Arrestors.

In certain other nonlinear resistances like Semiconductors, Thermistors, the current decreases as the voltage across increases like in curve 3 of Fig. 1.6. Hence, thermistor is used in over current protection in Motors, etc.

Example 1.3:

A current of 0.75A is passed through a coil of nichrome wire which has an area of cross-section of 0.01 cm^2. If the resistivity of the nichrome is 108 ´ 10^{-6}W-cm and the potential difference across the ends of the coil is 81V. What is the length of the wire? What is the conductivity and conductance of the wire?

Solution:

Resistance,

$$R = \frac{\rho l}{a}$$

where, $$R = \frac{V}{I}$$

$$= \frac{81}{0.75}$$

$$= 108\ \Omega$$

$$a = 0.01\ \text{cm}^2 = 0.01 \times 10^{-4}\ \text{m}^2$$

$$l = \frac{R \times a}{\rho}$$

$$= \frac{108 \times 0.01 \times 10^{-4}}{108 \times 10^{-8}} = 100\ \text{m}$$

Conductivity,

$$\sigma = \frac{1}{\rho}$$

$$= \frac{1}{108 \times 10^{-8}}$$

$$= 92.59 \times 10^{4}\ \mho/\text{m}$$

Conductance,

$$G = \frac{1}{R}$$

$$= \frac{1}{108}$$

$$= 9.259 \times 10^{-3}\ \mho \text{ or Siemens}$$

1.5 ELECTRICAL POWER

Power is the rate of doing work and is expressed in Joules per second. When one coulomb of electrical charge moves through a potential difference of one volt in one second the work done is one Joule/sec and in electrical engineering it is expressed as one Watt and is denoted by the symbol P.

So Power supplied,

$$P = E \times I \text{ Watts} \qquad \text{.......... (1.5.1)}$$

where E is the source voltage.

Power expended,

$$P = V \times I \text{ Watts} \qquad \text{.......... (1.5.1a)}$$

where V is the voltage drop.

Applying Ohm's Law for V,

$$P = (I \times R) \times I \text{ Watts}$$

$$P = I^2 \times R \text{ Watts} \qquad \text{.......... (1.5.2)}$$

or Applying Ohm's Law for I,

$$P = V \times \left(\frac{V}{R} \right) \text{ Watts}$$

$$P = \frac{V^2}{R} \text{ Watts} \qquad \left(\because I = \frac{V}{R} \right) \qquad \text{.......... (1.5.3)}$$

For A.C. circuits or circuits with unidirectional source, the equations for electrical power can be written using instantaneous values as,

$$p = e \times i \text{ Watts} \qquad \text{.......... (1.5.4)}$$

where e is the source voltage.

Power expended,

$$p = v \times i \text{ Watts} \qquad \text{.......... (1.5.4a)}$$

where v is the voltage drop.

Applying Ohm's Law for v,

$$p = i^2 \times R \text{ Watts} \qquad \text{.......... (1.5.5)}$$

or Applying Ohm's Law for i,

$$p = \frac{v^2}{R} \text{ Watts} \qquad \text{.......... (1.5.6)}$$

The power expended is also known as ***Power Loss*** or ***Copper Loss*** since the power loss takes place in the conductor (generally made of copper). It is called copper loss, even though the conductor material is not made of copper.

The power loss or copper loss appears in the form of heat. This heat has to be dissipated properly or else the insulation of the conductor or the insulation coating (varnish)of the coil will get damaged and there will be short circuits between turns of the coil and the coil may get burnt away in the case of machines and other equipments using coils.

1.6 ENERGY CALCULATIONS

Energy is the work done in a given time to achieve the required state of heating, lighting, lifting weights, moving the objects, etc. As such energy calculations are very important. Of late to have good efficiency in getting the work done and to have good economy Energy Auditing is resorted to in Industries and because energy charges are recurring charges involving expenditure.

According to the *Law of Conservation of Energy*, energy can neither be created nor destroyed. As such energy can atmost be converted from one form of energy into another form like converting mechanical energy into electrical energy and vice versa, converting electrical energy into heat energy and vice versa, etc. In this process the efficiency of the equipment used in conversion plays an important role. Also the constants of conversion are to be considered.

1.7 ELECTRICAL ENERGY

Electrical Energy is the total amount work done and is expressed in Joules or in Watt-seconds in electrical engineering. It is denoted by W. If E is the voltage rise or electromotive force and I is the current then, the energy generated or the energy supplied for a time t seconds is given as,

$$W = P \times t = E \times I \times t \text{ Watt-sec} \quad \text{.......... (1.7.1)}$$

or

$$W = I^2 \times R \times t \text{ Watt-sec} \quad \text{.......... (1.7.2)}$$

or

$$W = \frac{E^2}{R} \times t \text{ Watt-sec} \quad \text{.......... (1.7.3)}$$

For A.C. circuits or circuits with unidirectional source,

$$w = p \times dt = e \times i \times dt \text{ Watt-sec} \quad \text{.......... (1.7.4)}$$

or

$$w = i^2 \times R \times dt \text{ Watt-sec} \quad \text{.......... (1.7.5)}$$

or

$$w = \frac{e^2}{R} \times dt \text{ Watt-sec} \quad \text{.......... (1.7.6)}$$

where p, v, i all stand for instantaneous values and dt is the differential time

So,

$$W = \int p\,dt \text{ Watt-sec} \quad \text{.......... (1.7.7)}$$

or

$$W = \int e \times i\,dt \text{ Watt-sec} \quad \text{.......... (1.7.8)}$$

or $$W = \int \frac{e^2}{R} dt \quad \text{Watt-sec} \qquad \text{.......... (1.7.9)}$$

The energy expended canl be obtained by using the above equations substituting V for E or v for e.

If the power is supplied for time t_1 to t_2 seconds then, the total energy W will be given as,

$$W = \int_{t_1}^{t_2} p\,dt \quad \text{Watt-sec} \qquad \text{.......... (1.7.10)}$$

Since the Watt-sec is a small unit, for practical purposes, energy is expressed in terms of Kilo-Watt-Hour (KWH) or units.

$$1 \text{ unit of energy} = 1 \text{ KWH}$$

$$1 \text{ unit of energy} = \frac{\textit{Power in Watts} \times \textit{Time in seconds}}{1000 \times 60 \times 60}$$

Power distribution companies charge the electrical energy supplied to the consumer in terms of Standard Energy Units (Board of Trade Units) known as Kilo-Watt-Hours.

1.8 KIRCHOFF'S LAWS

There are two more important laws governing the performance of a circuit known as

1. *Kirchoff's Voltage Law (KVL)*
2. *Kirchoff's Current Law (KCL or KIL)*

where I stands for current in KIL.

1.8.1 Kirchoff's Voltage Law (KVL)

Kirchoff's Voltage Law states that, *in a closed electric circuit the algebraic sum of E.M.F.s and Voltage drops is zero.*

By convention, the E.M.F.s or voltage rises are taken to be positive and voltage drops are taken to be negative.

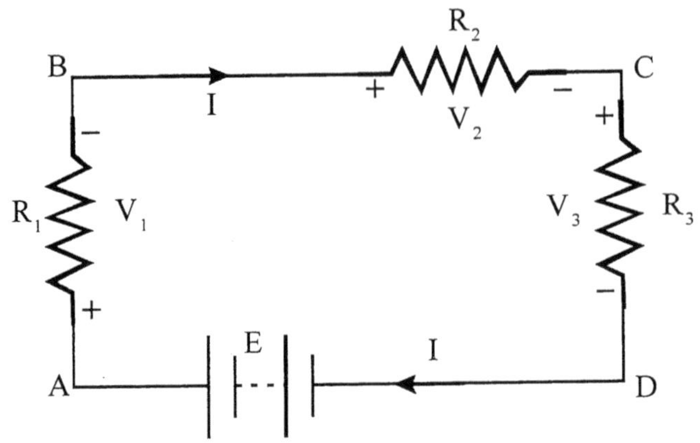

Fig. 1.7 Circuit for Kirchoff's Voltage Law

In the closed circuit ABCDA given in Fig. 1.7, applying Kirchoff's Voltage Law, we have,

$$E - V_1 - V_2 - V_3 = 0 \qquad \text{.......... (1.8.1)}$$

$$E - IR_1 - IR_2 - IR_3 = 0 \qquad \text{.......... (1.8.1a)}$$

or $$IR_1 + IR_2 + IR_3 = E \qquad \text{.......... (1.8.1b)}$$

Sum of voltage drops = Sum of E.M.F.s or Voltage rises

Kirchoff's Voltage Law can be applied to any closed loop (closed circuit) even if there is no voltage source in which case the right hand side of Eq. (1.7.1b) will be zero for several loops. KVL in matrix form is given as,

$$[\,R\,][\,I\,] = [\,E\,] \qquad \text{.......... (1.8.1c)}$$

1.8.2 Kirchoff's Current Law (KCL or KIL)

Kirchoff's Current Law states that, *at any junction (or node) at which different elements are connected, the algebraic sum of the current at the junction is zero A junction or node is the meeting point of more than one element in a circuit. For eg. point B in the circuit given for Example 1.9 is a junction or node.*

By convention, a currents entering the junction are taken to be positive and currents leaving the junction are taken to be negative.

The currents of the current sources entering the junction are positive.

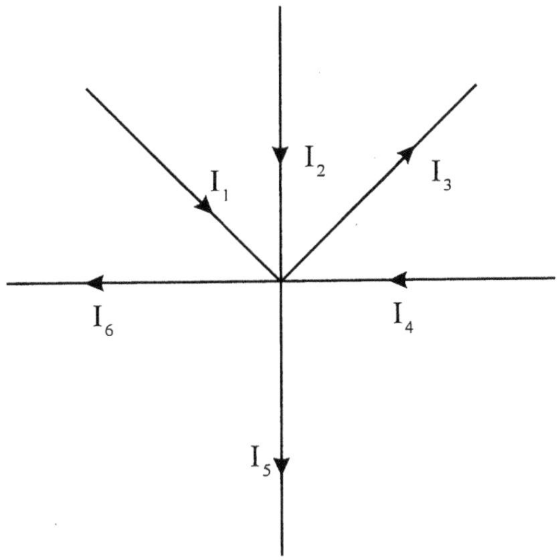

Fig. 1.8 Kirchoff's Current Law

In the Fig.1.8

applying Kirchoff's Current Law to the junction A, we have,

$$I_1 + I_2 + I_4 - I_3 - I_5 - I_6 = 0 \qquad \text{.......... (1.8.2)}$$

or

$$I_3 + I_5 + I_6 = I_1 + I_2 + I_4 \qquad \text{.......... (1.8.2a)}$$

Sum of currents leaving the junctions = Sum of currents entering the junction

Note: Kirchoff's Voltage Law and Kirchoff's Current Law can also be applied to A.C. Circuits or to circuits with Unidirectional Source using instantaneous values for Voltages and currents. In A.C., for steady state values using impedances and using RMS values for voltages and currents, all in complex form will be discussed later. However, for resistive circuits consisting of only resistances Kirchoff's Voltage and Current Laws can be written as,

KVL: $$iR_1 + iR_2 + iR_3 = e \qquad \text{.......... (1.8.2c)}$$

KCL: $$i_3 + i_5 + i_6 = i_1 + i_2 + i_4 \qquad \text{.......... (1.8.2d)}$$

Example 1.4:

Applying KCL and KVL, find the currents in the various elements of the circuit given in Fig. 1.8. Find the power delivered by the battery and the energy supplied by the battery for a period of half an hour. Also calculate the power loss in the 6Ω resistor.

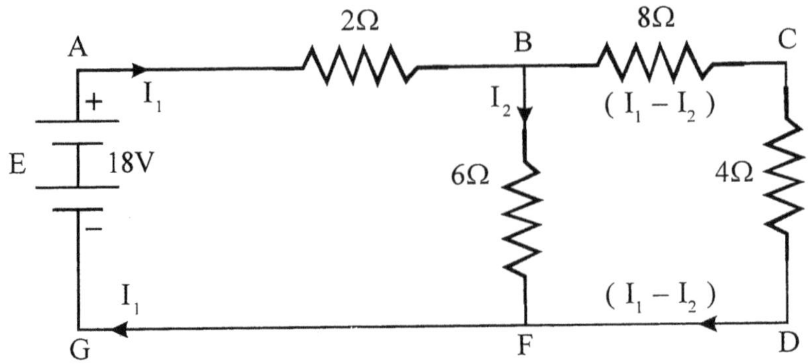

Solution:

Let the current supplied by the battery to junction A be I_1. The same current I_1 flows through the 2Ω resistance towards junction B. At B a part of this current of I_1 flows through the 6Ω resistance towards junction F. Let it be I_2.

Applying Kirchoff's Current Law to junction A, the current 8Ω resistance will be $(I_1 - I_2)$ towards junction C.

Applying Kirchoff's Voltage Law to the loop ABFGA,

$$2 \times I_1 + 6 \times I_2 = 18$$

or $$2I_1 + 6I_2 = 18 \qquad \text{.......... (1)}$$

Applying Kirchoff's Voltage Law to the loop BCDFB,

$$8 \times (I_1 - I_2) + 4 \times (I_1 - I_2) - 6 \times I_2 = 0$$

i.e., $$12I_1 - 18I_2 = 0 \qquad \text{.......... (2)}$$

Solving Eq. (1) and (2) we obtain,

$$I_1 = 3 \text{ A}$$

and $$I_2 = 2 \text{ A}$$

Also, Eq. (1) and (2) can be solved using Cramer's Rule,

$$I_1 = \frac{\begin{vmatrix} 18 & 6 \\ 0 & -18 \end{vmatrix}}{\begin{vmatrix} 2 & 6 \\ 12 & -18 \end{vmatrix}}$$

$$= \frac{18\times(-18) - 6\times 0}{2\times(-18) - 6\times 12}$$

$$= \frac{-324}{-72} = 3\text{ A}$$

Similarly,

$$I_2 = \frac{\begin{vmatrix} 2 & 18 \\ 12 & 0 \end{vmatrix}}{\begin{vmatrix} 2 & 6 \\ 12 & -18 \end{vmatrix}}$$

$$= \frac{2\times 0 - 18\times 12}{2\times(-18) - 6\times 12}$$

$$= \frac{-216}{-108} = 2\text{ A}$$

or $$2I_1 + 6I_2 = 18$$

Substituting the values of $I_1 = 3$ we obtain,

$$I_2 = 2\text{ Amperes}$$

Current in 2Ω resistor,

$$I_1 = 3\text{ A}$$

Current in 6Ω resistor,

$$I_2 = 2\text{ A}$$

Current in 8Ω resistor,

$$I_1 - I_2 = 1\text{ A}$$

Current in 4Ω resistor,

$$I_1 - I_2 = 1\text{ A}$$

Current supplied by the battery,

$$P = EI_1$$

$$= 18 \times 3 = 54\text{ Watts}$$

Energy, $W = P \times t$

Energy supplied by the battery for half an hour (1800 sec),

$$W = 54 \times 1800 = 97200 \text{ Watt-sec.}$$

$$= \frac{97200}{60 \times 60}$$

$$= 27 \text{ Watt-hours}$$

$$= \frac{27}{1000}$$

$$= 0.027 \text{ KWH}$$

Power Loss,

$$P = I^2R \text{ Watts}$$

Power Loss in the 6Ω resistor $= (I_2)^2 \times 6$

$$= 2^2 \times 6$$

$$= 24 \text{ Watts}$$

Example 1.5:

In the circuit of given figure, find the power supplied to the load. Find also the voltage at the load using KCL and KVL Equations. Also find the current through 0.2Ω resistance.

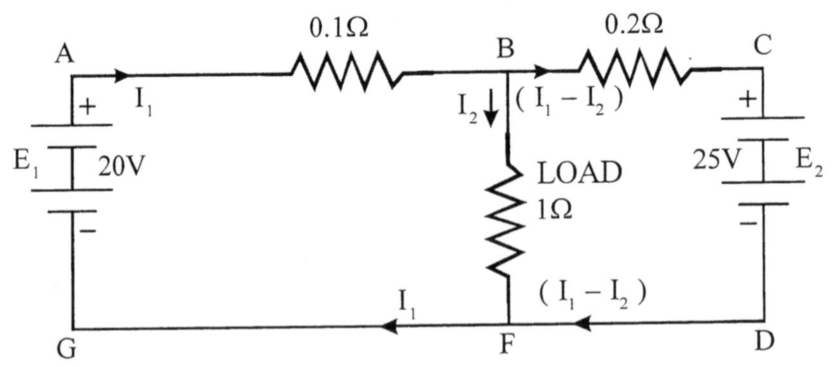

Solution:

Let the current entering node B be I_1 and let the current flowing through the resistance 1Ω resistance be I_2. Applying KCL for node B, the current through 0.2Ω resistance will be $(I_1 - I_2)$.

Writing KVL for loop ABFGA,

$$0.1I_1 + 1 \times I_2 = 20$$

i.e., $$0.1I_1 + I_2 = 20 \qquad \text{.......... (1)}$$

Writing KVL for loop BCDFB,

$$0.2x(I_1 - I_2) + 1x (- I_2) = -25$$

i.e., $$0.2I_1 - 1.2I_2 = -25 \qquad \text{.......... (2)}$$

Solving Eq. (1) and (2) we get,

$$I_1 = -3.125 \text{ A}$$

$$I_2 = -20.3125 \text{ A}$$

Hence, the current in the load is I_2

$$= 20.3125 \text{ A.}$$

Current through 0.2Ω resistance is,

$$(I_1 - I_2) = (-3.125 - 20.3125) = -23.4375 \text{ A}$$

The negative signs for I_1 , I_2, $(I_1 - I_2)$ mean that the direction of the current flow is opposite to the assumed direction i.e., the current flows from B to A and not from A to B as assumed, I_2 from C to B and $(I_1\text{-}I_2)$ from F to B.

Voltage at the load is

$$V = IR$$

$$= 20.3125 \times 1$$

$$= 20.3125 \text{ Volts}$$

Load Power,

$$P = I^2R$$

$$= (20.3125)^2 \times 1$$

$$= 412.598 \text{ Watts.}$$

1.9 RESISTANCES IN SERIES

If the ending terminal of the resistance R_1 is connected to the beginning terminal of the resistance R_2 and the ending terminal of R_2 is connected to the beginning terminal of the resistance R_3 and so on then the resistances R_1, R_2, R_3, etc., are said to be connected in series.

In series circuits, the elements in the series can be connected in any order. For example, R_2, R_3, R_1, etc., instead of R_1, R_2 R_3 etc. In series circuits, the same current will flow through all the elements in series.

In D.C. series circuits, while connecting the elements in series, one should be very careful of the polarities of the meters used to measure the currents or voltages or the polarities of the equipment. Positive polarities of the meters or the equipments should always be connected to the positive of the supply point and the negative terminals should be connected to the negative of the supply point. While two equipments are connected in series, the positive of the first equipment should be connected to the positive terminal of the supply point. Negative terminal of the first equipment should be connected to the positive terminal of the second equipment and the negative terminal of the second equipment should be connected to the positive terminal of the third equipment and so on.

Ammeters are used to measure the currents. The ammeters should always be connected in series in the circuit so that the current to be measured flows through the ammeters. In order that, the voltage drop across the ammeter to be very very small so that full current flows through the circuit, the resistance of the ammeter should be very very small. Hence, if the ammeter is connected across the supply or across two points having large voltage drop, very heavy current will flow through the ammeter and the ammeter will get burnt.

Voltmeters are used to measure the voltage of the supply or voltage drop between two points inorder that the voltmeter does not draw more current so as to allow full current in the circuit, the voltmeters should have very high resistance. If the voltmeter is connected in series, it causes high voltage drop across it and the voltage supplied to the remaining circuit will be less. Hence, voltmeters should be connected only in parallel and not in series.

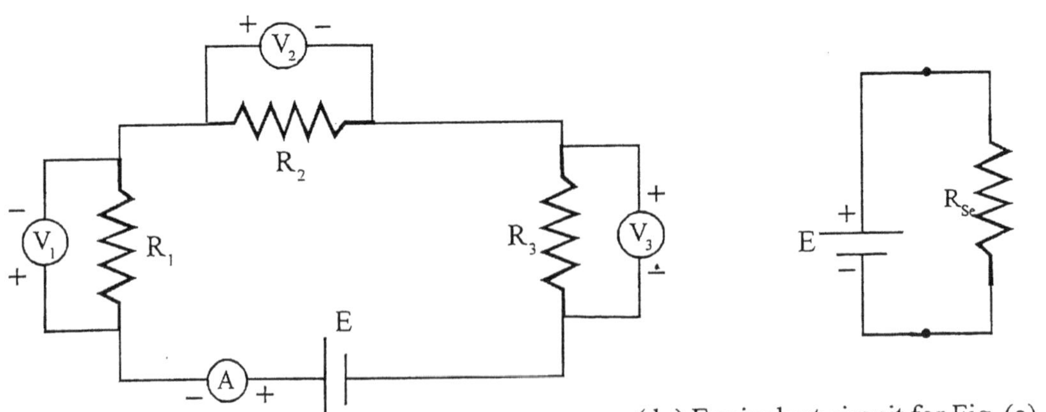

(a) Series Circuit (b) Equivalent circuit for Fig. (a)

Fig. 1.9 Circuit with Resistances in Series

In the closed circuit ABCDA given in Fig. 1.9 (a), applying Kirchoff's Voltage Law, we have,

$$E - V_1 - V_2 - V_3 = 0 \qquad \text{.......... (1.9.1)}$$

$$E - IR_1 - IR_2 - IR_3 = 0 \qquad \text{.......... (1.9.1a)}$$

or

$$IR_1 + IR_2 + IR_3 = E \qquad \text{.......... (1.9.1b)}$$

or

$$I(R_1 + R_2 + R_3) = E \qquad \text{.......... (1.9.1c)}$$

For the equivalent circuit of Fig. 1.9 (b),

$$IR_{Se} = E \qquad \text{.......... (1.9.2)}$$

Comparing Eq. (1.9.1c) and (1.9.2) we have,

$$R_{Se} = (R_1 + R_2 + R_3)$$

In general for n resistances in series,

$$R_{Se} = \sum_{i=1}^{n} R_i \qquad \text{.......... (1.9.3)}$$

In terms of conductances for resistances in series,

$$\frac{1}{G_{Se}} = \sum_{i=1}^{n} \frac{1}{G_i} \qquad \text{.......... (1.9.4)}$$

Example 1.6:

Fig. below shows three resistors R_A, R_B and R_C connected in series to a 250V source; Given $R_C = 50\Omega$, and $V_B = 80$Volts when the current is 2 Amperes, calculate the total resistances, R_A and R_B.

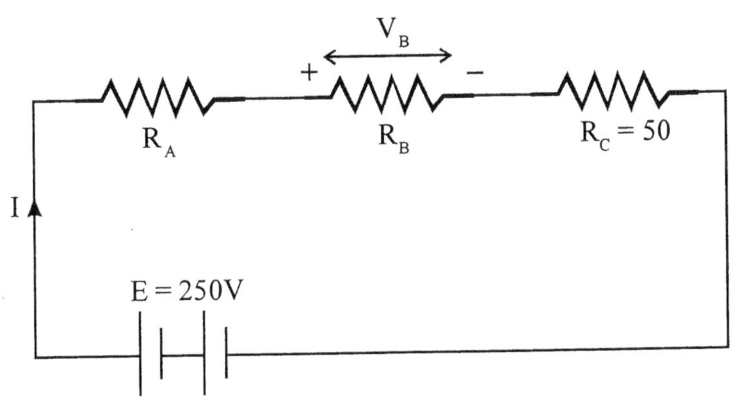

Solution:

Since $I = 2$ Amperes

$$V_B = IR_B$$

$$= 80V$$

$$R_B = 40\Omega$$

Also, $$I = \frac{E}{(R_A + R_B + R_C)}$$

Therefore,

$$R_{Se} = R_A + R_B + R_C = \frac{E}{I} = \frac{250}{2} = 125\Omega$$

Therefore, $R_A = R_{Se} - (R_B + R_C) = 35\Omega.$

Example 1.7:

A lamp rated 500W, 100V is to be operated from 220V supply. Find the value of the resistor to be connected in series with the lamp. What is the power lost in the resistance.

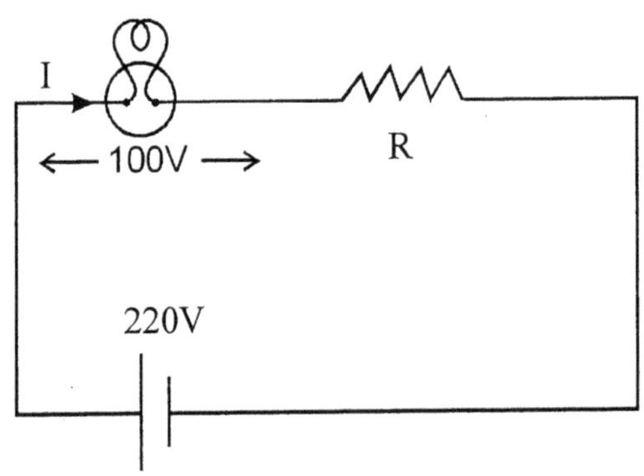

Solution:

$$\text{Current in the lamp} = \frac{W}{E}$$

$$= \frac{500}{100} = 5 \text{ Amperes}$$

Since the Voltage drop across the lamp is 100 V, Voltage to be dropped in the series resistor is 120V.

$$\text{Therefore, Value of the resistor} = \frac{120}{5}$$

$$= 24\ \Omega$$

$$\text{Power Lost in this resistor} = I^2R$$

$$= 5^2 \times 24 = 600\text{W}$$

1.10 RESISTANCES IN PARALLEL

If the starting terminal of two or more elements are connected together and the ending terminals of these elements are connected together then the elements are said to be connected in parallel. A parallel element may also be known as a Shunt Element.

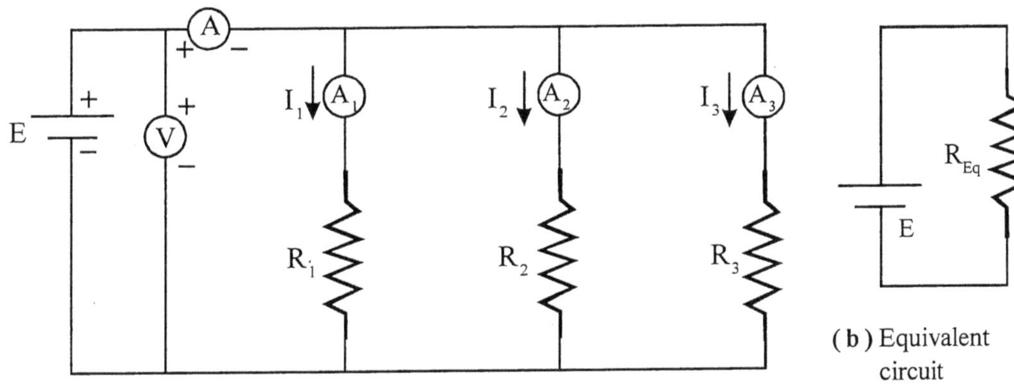

(b) Equivalent circuit

(a) Parallel circuit

***Fig. 1.10** Circuit with Resistances in Parallel*

The voltage across all the elements that are connected in parallel will be same.

In D.C. circuits, if the elements of the meters or the equipments with polarities marked are connected in parallel then terminals of the same polarities should be connected together.

In the parallel circuit given in the Fig. 1.10(a), applying Kirchoff's Current Law to the junction A,

$$I = I_1 + I_2 + I_3 \qquad \text{.......... (1.10.1)}$$

Applying Ohm's Law,

$$I = \frac{E}{R_1} + \frac{E}{R_2} + \frac{E}{R_3} \qquad \text{.......... (1.10.1a)}$$

For the equivalent circuit given in Fig. 1.14b,

$$I = \frac{E}{R_P} \qquad \text{.......... (1.10.2)}$$

Comparing Eq. (1.16.2) and (1.16.1a) we have,

$$\frac{1}{R_P} = \frac{1}{R_1} + \frac{1}{R_2} + \frac{1}{R_3} \qquad \text{.......... (1.10.3)}$$

or In terms of conductances,

$$G_P = G_1 + G_2 + G_3 \qquad \text{.......... (1.10.4)}$$

In general for parallel circuits with n resistances in parallel,

$$\frac{1}{R_P} = \sum_{i=1}^{n} \frac{1}{R_i} \qquad \text{.......... (1.10.5)}$$

or

$$G_P = \sum_{i=1}^{n} G_i \qquad \text{.......... (1.10.6)}$$

If two resistances R_1 and R_2 are in parallel,

$$\frac{1}{R_P} = \frac{1}{R_1} + \frac{1}{R_2} \qquad \text{.......... (1.10.7)}$$

or

$$R_P = \frac{R_1 R_2}{R_1 + R_2} \qquad \text{.......... (1.10.7a)}$$

If the two resistances are equal and in parallel

i.e., $R_1 = R_2 = R$

then,

$$R_P = \frac{R}{2} \qquad \text{.......... (1.10.7b)}$$

If three resistances R_1, R_2 and R_3 are parallel,

$$\frac{1}{R_P} = \frac{1}{R_1} + \frac{1}{R_2} + \frac{1}{R_3} \qquad \text{.......... (1.10.8)}$$

or

$$R_P = \frac{R_1R_2R_3}{R_1R_2 + R_2R_3 + R_3R_1} \qquad \text{.......... (1.10.8a)}$$

If three resistances are equal, i.e., $R_1 = R_2 = R_3 = R$ then,

$$R_P = \frac{R}{3} \qquad \text{.......... (1.10.9)}$$

In general, if n resistances, each of value R are in parallel the,

$$R_P = \frac{R}{n} \qquad \text{.......... (1.10.10)}$$

1.10.1 Division of Currents in Parallel Circuits

If two resistances R_1 and R_2 are connected in parallel and if the total current entering the parallel combination is I then this current I divides into two parts I_1 flowing through R_1 and I_2 flowing through R_2.

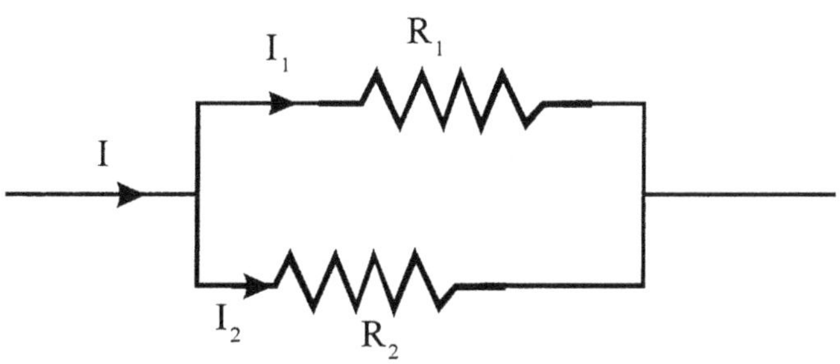

Fig. 1.11 *Division of Currents in Parallel Circuits*

The Voltage Drop across the parallel combination will be

$$V_P = I \times R_P = I \times \left(\frac{R_1R_2}{R_1 + R_2}\right) \text{ Volts}$$

The current I_1 flowing through R_1 will be given as,

$$I_1 = \frac{V_P}{R_1} = \frac{1}{R_1} \times I \times \left(\frac{R_1 R_2}{R_1 + R_2} \right) \text{ Amps}$$

$$I_1 = I \times \left(\frac{R_2}{R_1 + R_2} \right) \text{ Amps} \qquad \text{.......... (1.10.11)}$$

When two resistances R_1 and R_2 are in parallel, Current through R_1 is given as,

$$I_1 = \text{Total Current} \times \frac{\text{Second Resistance}}{\text{Sum of the Two Resistances in Parallel}} \qquad \text{.......... (1.10.12)}$$

This form is used frequently in Electronic Circuits.

Similarly,

$$I_2 = \frac{V_P}{R_2}$$

$$= \frac{1}{R_2} \times I \times \left(\frac{R_1 R_2}{R_1 + R_2} \right) \text{ Amps}$$

$$I_2 = I \times \left(\frac{R_1}{R_1 + R_2} \right) \text{ Amps} \qquad \text{.......... (1.10.13)}$$

Also, $\quad I_2 = (I - I_1)$ Amps

In general, *the current through any parallel path is given as the product of the total current and the parallel equivalent resistance divided by the resistance of that path.*

Example 1.8:

Solve the network shown in the figure for the current through 6Ω resistor.

Solution:

Let the current flowing through various branches be as marked in the figure. Applying Kirchoff's Voltage Law to the following closed circuits,

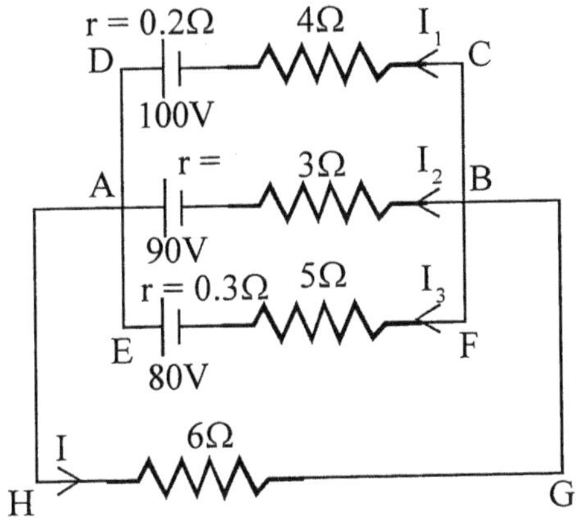

Circuit CDAHGBC,

$$-4I_1 - 0.2I_1 + 100 - 6(I_1 + I_2 + I_3) = 0$$

or $$10.2\ I_1 + 6I_2 + 6I_3 = 100 \qquad \text{.......... (1)}$$

Circuit BAHGB,

$$-3I_2 - 0.25I_2 + 90 - 6(I_1 + I_2 + I_3) = 0$$

or $$6I_1 + 9.25\ I_2 + 6I_3 = 90 \qquad \text{.......... (2)}$$

Circuit FEAHGBF,

$$-5I_3 - 0.3I_3 + 80 - 6(I_1 + I_2 + I_3) = 0$$

or $$6I_1 + 6I_2 + 11.3\ I_3 = 80 \qquad \text{.......... (3)}$$

Subtracting (2) from (1), we get,

$$4.2\ I_1 - 3.25\ I_2 = 10 \qquad \text{.......... (4)}$$

Eqn.(2) x 11.3 - Eqn. (3) x 6 gives

$$31.8\ I_1 + 68.525\ I_2 = 537 \qquad \text{.......... (5)}$$

Eqn.(5) x 4.2 - Eqn.(4) x 31.8 gives

$$391.51I_2 = 1937.4$$

$$I_2 = 4.953\text{A}$$

From Eqn.(5) substituting for I_2

$$I_1 = 6.21\text{A}$$

From Eqn. (3), substituting for I_1 and I_2

$$I_3 = 1.15A$$

Current in 6Ω resistor,

$$I = I_1 + I_2 + I_3$$
$$= 6.21 + 4.953 + 1.15$$
$$= 12.313A$$

Example 1.9:

Find the magnitude and direction of the currents in all branches of the circuit shown in the figure using Kirchoff's Laws. All resistances are in Ohms.

Solution:

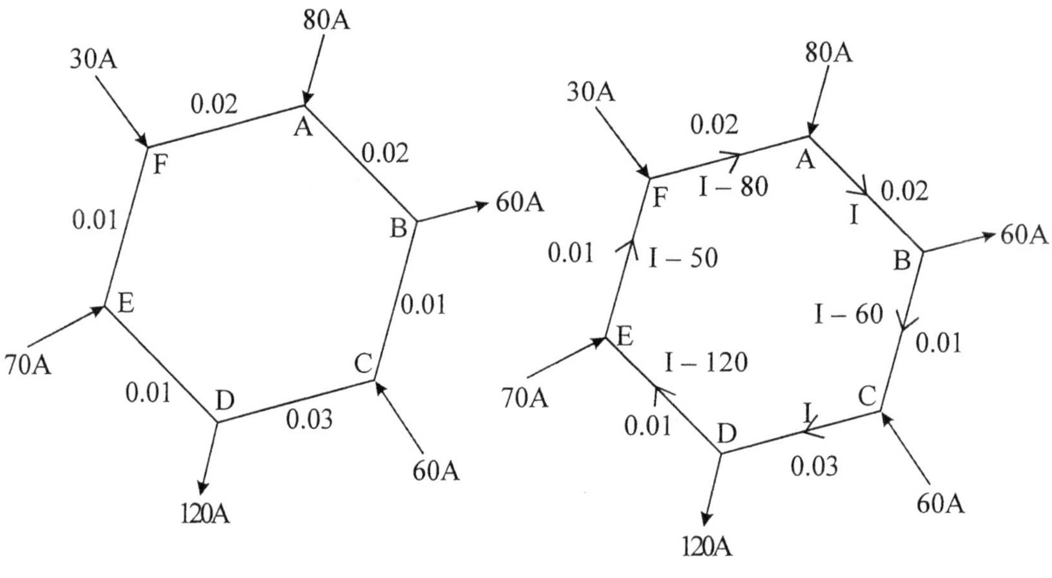

Let current from A to B junctions be I Amps. Applying Kirchoff's First Law, let current flowing through various branches be as shown in the figure.

Applying Kirchoff's Current Law to current in each branch and Kirchoff's Voltage Law to a closed loop ABCDEFA, we get,

$$-0.02I - 0.01(I - 60) - 0.03I - 0.01(I - 120) - 0.01(I - 50)$$
$$- 0.02(I - 80) = 0$$

or $0.02I + 0.01I + 0.03I + 0.01I + 0.01I + 0.02I$

$$= 0.6 + 1.2 + 0.5 + 1.6$$

or $0.1I = 3.9$

or $I = 39A$

Current in various branches is as under :

$I_{AB} = 39A$ (i.e., A to B)

$I_{BC} = I - 60 = -21A$ (i.e., C to B)

$I_{CD} = I = 39A$ (i.e., C to D)

$I_{DE} = I - 120 = -81A$ (i.e., E to D)

$I_{EF} = I - 50 = -11A$ (i.e., F to E)

$I_{FA} = I - 80 = -41A$ (i.e., A to F)

1.11 SERIES-PARALLEL RESISTANCES

In the case of series-parallel resistances, the parallel equivalent of the resistances in parallel are obtained first as a single resistance which will be in series with the other resistances, thus bringing the circuit into a single series circuit. After finding the current flowing through this equivalent series circuit again the parallel equivalent resistance may be replaced with the corresponding parallel circuit and the current in the parallel paths are calculated as given in the Section 1.10, *the current through any parallel path is given as the product of the total current and the parallel equivalent resistance divided by the resistance of that path.*

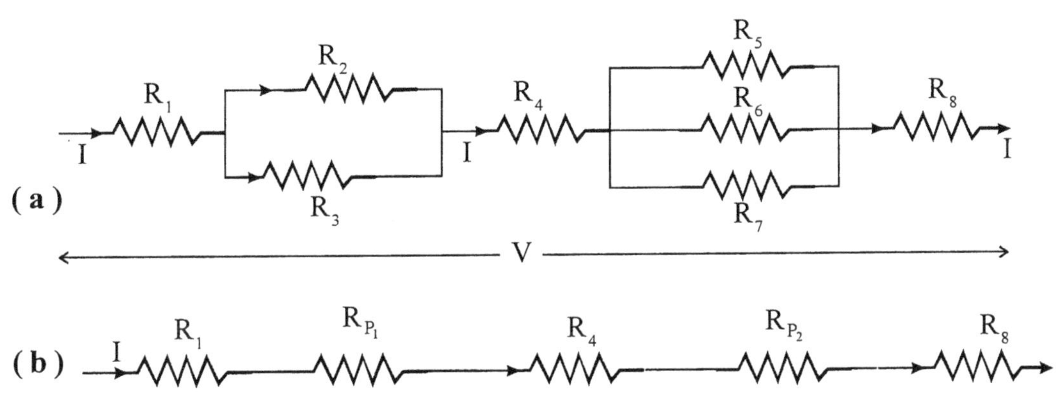

Fig. 1.12 Resistance in Series-Parallel

Typical values of currents are given below for the above circuit,

$$R_{P_1} = \frac{R_2R_3}{(R_2+R_3)}$$

$$R_{P_2} = \frac{R_5R_6R_7}{R_5R_6+R_6R_7+R_7R_5}$$

$$I = \frac{V}{R_1+R_{P_1}+R_4+R_{P_2}+R_8}$$

$$I_2 = I \times \frac{R_3}{(R_2+R_3)}$$

$$I_6 = \frac{I}{R_6}\left(\frac{R_5R_6R_7}{(R_5R_6+R_6R_7+R_7R_5)}\right)$$

Example 1.10:

A Wheatstone Bridge consists of AB = 4Ω, BC = 3Ω, CD = 6Ω and DA = 5Ω. A 2V Cell is connected between B and D and a Galvanometer of 10Ω resistance between A and C. Find the current through the Galvanometer.

Solution:

The circuit is shown in the figure. Applying Kirchoff's Current Law at junction B, A and C, the current in various branches is marked.

Applying Kirchoff's Voltage Law to various closed loops and considering loop BACB, we get,

$$-4I_1 - 10I_3 + 3I_2 = 0$$

or $$4I_1 + 10I_3 - 3I_2 = 0 \qquad \text{.......... (1)}$$

Considering loop ADCA, we get

$$-5(I_1 - I_3) + 6(I_2 + I_3) + 10I_3 = 0$$

or $$5I_1 - 6I_2 - 21I_3 = 0 \qquad \text{.......... (2)}$$

Considering loop BADEB, we get,

$$-4I_1 - 5(I_1 - I_3) + 2 = 0$$

or $$4I_1 - 5I_1 + 5I_3 = -2 \qquad \text{.......... (3)}$$

or $$9I_1 - 5I_3 = 2$$

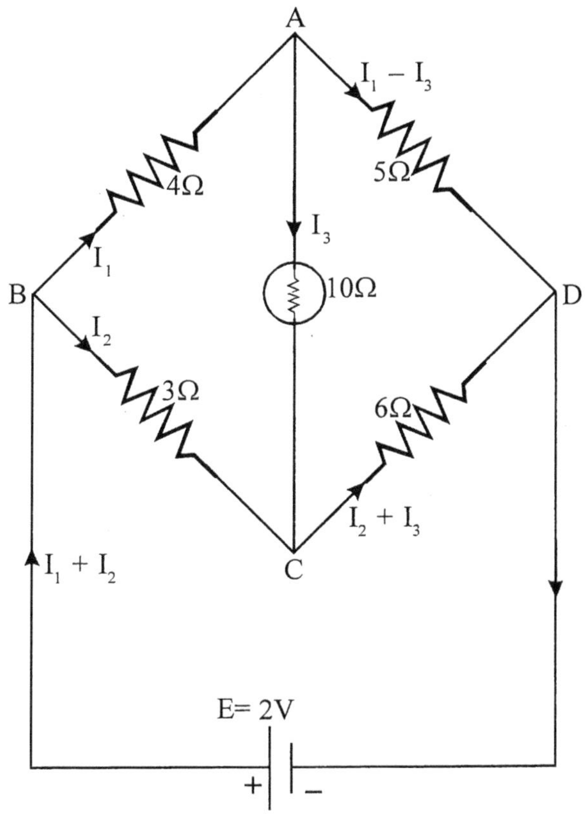

Multiplying Eq. (1) by Eq. (2) and Subtracting from Eq. (2), we get,

$$5I_1 - 6I_2 - 21I_3 = 0$$
$$8I_1 - 6I_2 + 20I_3 = 0$$
$$-3I_1 \quad - 41I_3 = 0$$

or $$I_1 = -\frac{41}{3} I_3$$

Substituting the value of I_1 in Eq. (3), we get,

$$9\left(-\frac{41}{3}I_3\right) - 5I_3 = 0$$

or $$-123I_3 - 5I_3 = 2$$

or $$I_3 = -\frac{1}{64}\text{ A}$$

Example 1.11:

A resistance of 15Ω is connected in series with two resistances each of 30Ω arranged in parallel. A voltage source of 30V is connected to this circuit.

(**a**) What is the current drawn from the source.

(**b**) What resistance should be placed in shunt (parallel) with the parallel combination in order that the current drawn from the source is 1.2 A.

Solution:

The circuit is as shown in the figure given below.

(**a**) R_1 is in series with the parallel combination of R_2 and R_3.

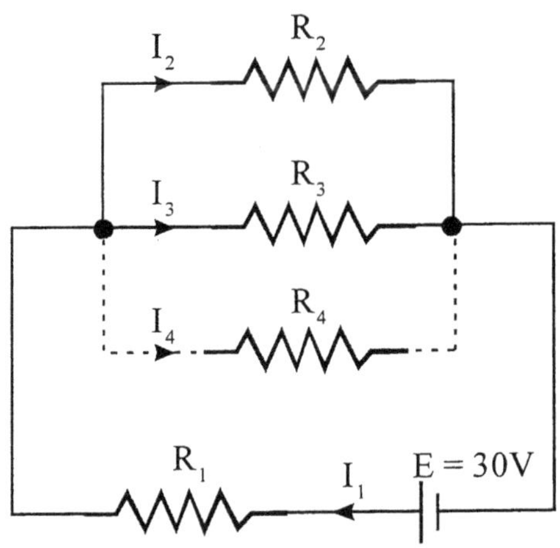

Hence, Total Resistance R_t is,

$$R_t = R_1 + \frac{R_2R_3}{R_2+R_3}$$

$$= 15 + \frac{30\times30}{30+30}$$

$$R_t = 30\Omega$$

Hence current,

$$I_1 = \frac{E}{R_t}$$

$$I_1 = \frac{30}{30} = 1\text{ A}$$

(b) With $I_1 = 1.2\text{A}$, Voltage across R_1

$$= I_1 R_1 = 1.2 \times 15$$

$$= 18\text{ V}$$

Voltage Drop across parallel combination of R_2, R_3 and R_4 is,

$$V_P = 30 - 18$$

$$V_P = 12\text{ V}$$

Hence, Voltage Drop across each resistor R_2, R_3, R_4 is 12 V.

Hence, $$I_2 = \frac{V_P}{R_2} = \frac{12}{30}$$

$$I_2 = 0.4\text{ A}$$

$$I_3 = \frac{V_P}{R_3} = \frac{12}{30}$$

$$I_3 = 0.4\text{ A}$$

$$I_4 = I_1 - (I_2 + I_3)$$

$$= 1.2 - (0.4 + 0.4)$$

$$I_4 = 0.4\text{ A}$$

$$I_4 = \frac{V_P}{R_4}$$

$$0.4 R_4 = 12$$

Hence, $$R_4 = \frac{12}{0.4}$$

$$R_4 = 30\ \Omega$$

1.12 RESISTANCES IN STAR OR DELTA CONNECTIONS

If the end of R_1 is connected to beginning of R_2, the end of R_2 is connected to beginning of R_3 and the end of R_3 is connected to beginning of R_1 then, the resistances R_1, R_2, R_3 are said to be connected in Delta (Δ) and the three points common to any two resistances are connected to the remaining part of the circuit as given in the Fig. 1.13 (a).

If one end of the resistances R_1, R_2, R_3, are all connected together and the other ends of the three resistances are connected to different points of the circuits, as given in the Fig. 1.13 (b), then the resistances R_1, R_2, R_3 are said to be connected in Star or WYE (Y).

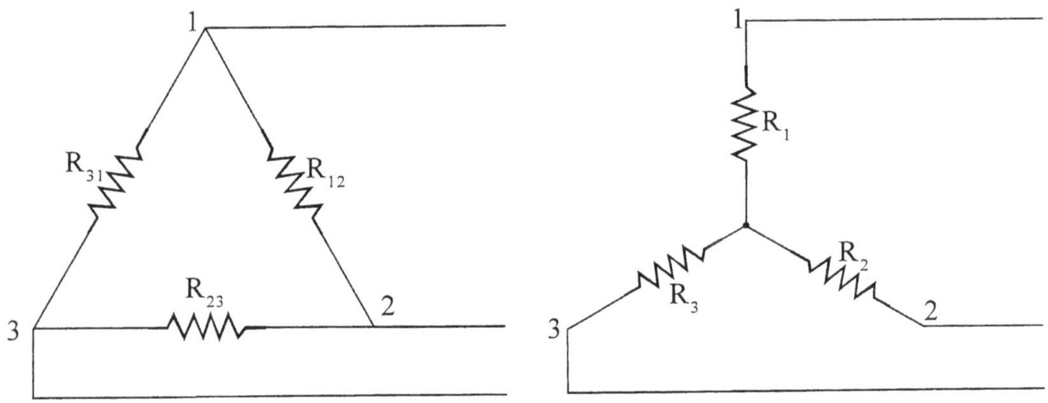

(**a**) Delta Connection (**b**) Star Connection

Fig. 1.13 Resistance in Star-Delta Connection

In general, more than three resistances can be connected in Star.

In general, more than three resistances can also be connected in Mesh.

In some cases, the circuits cannot be solved by means of simple series, parallel or parallel combinations but by finding the equivalent Delta for the given Star connection or equivalent Star connection for the given Delta connection of resistances, the circuit can be resolved into simple Series, Parallel or Series-Parallel combinations.

1.12.1 The Equivalent Star Resistances For Given Delta Connected Resistances

Referring to Fig. 1.13 (a), in Delta Connection, we observe that between the terminals 1 and 2 the resistance R_{12} is in parallel with the series combination of R_{23} and R_{31}.

Referring to Fig. 1.13 (b), in Star Connection, we observe that between the terminals 1 and 2 the resistance R_1 is in series with R_2.

Equating the resistances between terminals 1 and 2 of Star and its equivalent Delta, we have R_{12} in parallel with $(R_{23} + R_{31})$

$$R_1 + R_2 = R_{12} + \frac{R_{23}R_{31}}{R_{23} + R_{31}}$$

Rewriting the above equation we have,

$$R_1 + R_2 = \frac{R_{12}(R_{23} + R_{31})}{R_{12} + R_{23} + R_{31}}$$

$$R_1 + R_2 = \frac{R_{12}(R_{23} + R_{31})}{\sum R_{12}} \qquad \text{.......... (1.12.1)}$$

Similarly equating the resistances between terminals 2 and 3, we have,

$$R_2 + R_3 = \frac{R_{23}(R_{31} + R_{12})}{\sum R_{12}} \qquad \text{.......... (1.12.2)}$$

Similarly equating the resistances between terminals 3 and 1, we have,

$$R_3 + R_1 = \frac{R_{31}(R_{12} + R_{23})}{\sum R_{12}} \qquad \text{.......... (1.12.3)}$$

Subtracting Eq. (1.18.2) from Eq. (1.18.3), we have,

$$R_1 - R_2 = \frac{R_{12}(R_{31} - R_{23})}{\sum R_{12}} \qquad \text{.......... (1.12.4)}$$

Adding Eq. (1.18.1) and Eq. (1.18.4), we have,

$$2R_1 = \frac{R_{12}(2R_{31})}{\sum R_{12}}$$

Hence, $$R_1 = \frac{R_{12}R_{31}}{\sum R_{12}}$$

Rewriting the above equation, we have,

$$R_1 = \frac{R_{12}R_{13}}{\sum R_{12}} \qquad \text{.......... (1.12.5)}$$

Similarly, $$R_2 = \frac{R_{21}R_{23}}{\sum R_{12}} \qquad \text{.......... (1.12.6)}$$

$$R_3 = \frac{R_{31}R_{32}}{\sum R_{12}} \qquad \text{.......... (1.12.7)}$$

Note that from Eq. (1.12.5), Eq. (1.12.6) and Eq. (1.12.7), we observe that, *the equivalent star resistance connected to a given terminal is given by the product of the two Delta resistances that are connected to that terminal divided by the sum of the three Delta connected resistances.*

The same statement can be extended to more than three resistances connected in Mesh and its equivalent Star !! (check up)

1.12.2 The Equivalent Delta Resistances For Given Star Connected Resistances

Multiplying Eq. (1.12.5) and Eq. (1.12.6), we have,

$$R_1R_2 = \frac{R_{12}^2R_{23}R_{31}}{\left(\sum R_{12}\right)^2} \qquad \text{.......... (1.12.8)}$$

Multiplying Eq. (1.12.6) and Eq. (1.12.7), we have,

$$R_2R_3 = \frac{R_{12}R_{23}^2R_{31}}{\left(\sum R_{12}\right)^2} \qquad \text{.......... (1.12.9)}$$

Multiplying Eq. (1.12.7) and Eq. (1.12.5), we have,

$$R_3R_1 = \frac{R_{12}R_{23}R_{31}^2}{\left(\sum R_{12}\right)^2} \qquad \text{.......... (1.12.10)}$$

Adding Eq. (1.12.8), Eq. (1.12.9) and Eq. (1.12.10), we have,

$$R_1R_2 + R_2R_3 + R_3R_1 = \frac{R_{12}R_{23}R_{31}\left(R_{12} + R_{23} + R_{31}\right)}{\left(\sum R_{12}\right)^2}$$

$$= (R_{12}R_{23}R_{31})\frac{\left(\sum R_{12}\right)}{\left(\sum R_{12}\right)^2}$$

$$= R_{12}\left[\frac{R_{23}R_{31}}{\sum R_{12}}\right]$$

$R_1R_2 + R_2R_3 + R_3R_1 = R_{12}R_3$ by using

R_3 from eqn.1.12.7 for the term inside the bracket

Hence, $$R_{12} = \frac{R_1R_2 + R_2R_3 + R_3R_1}{R_3}$$

or $$R_{12} = R_1 + R_2 + \frac{R_1R_2}{R_3} \qquad \text{.......... (1.12.11)}$$

Similarly, $$R_{23} = R_2 + R_3 + \frac{R_2R_3}{R_1} \qquad \text{.......... (1.12.12)}$$

$$R_{31} = R_3 + R_1 + \frac{R_3R_1}{R_2} \qquad \text{.......... (1.12.13)}$$

Note that from Eq. (1.12.11), Eq. (1.12.12) and Eq. (1.12.13), we observe that, *the equivalent Delta resistance connected between any two given terminals is given by the sum of the two Star resistances that are connected between those two terminals plus the product of those two resistances connected between those two terminals divided by the third Star resistance.*

In terms of conductances the equivalent mesh element can be given in terms of the star elements for star or delta having three or more resistances as

$$G_{ij} = \frac{G_i G_j}{\sum G_i} \quad \text{!! (check up)} \qquad \text{.......... (1.12.14)}$$

If $R_{12} = R_{23} = R_{31} = R$ (say), then, the equivalent star resistances will be,

$$R_1 = R_2 = R_3 = \frac{R}{3} \qquad \text{.......... (1.12.15)}$$

If $R_1 = R_2 = R_3 = R$ (say), then, the equivalent delta resistances will be,

$$R_{12} = R_{23} = R_{31} = 3R \qquad \text{.......... (1.12.16)}$$

Example 1.12:

For the network shown in the figure below, find the equivalent resistance between the terminals B and C.

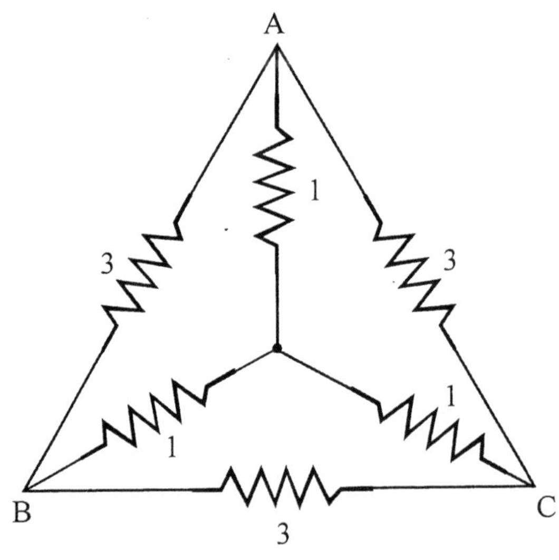

Solution:

The given combination of resistances between the terminals B and C is neither series combinations nor parallel combination. If, the star connection between A, B and C is converted into equivalent delta, we will have a known combination which can be simplified by series parallel simplification.

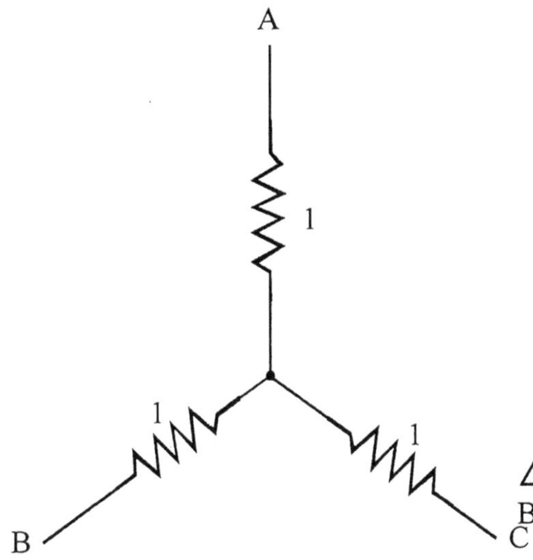

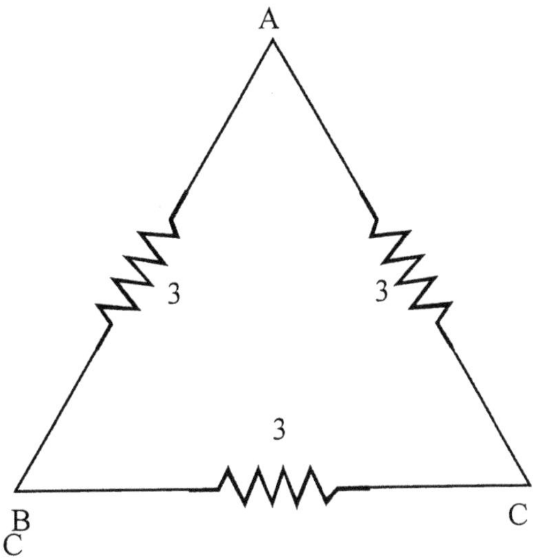

The given figure is redrawn after replacing the star by its equivalent delta.

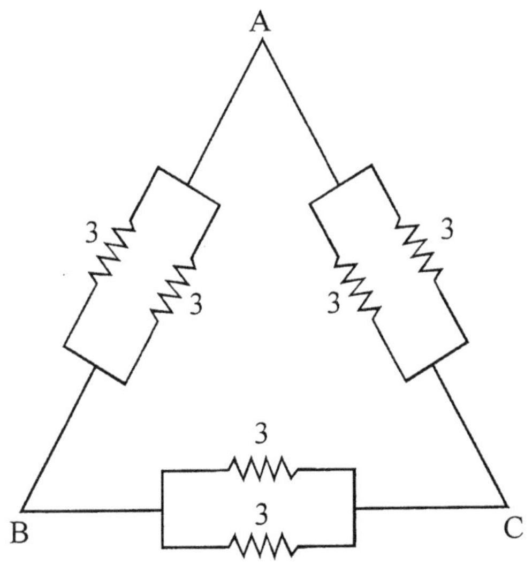

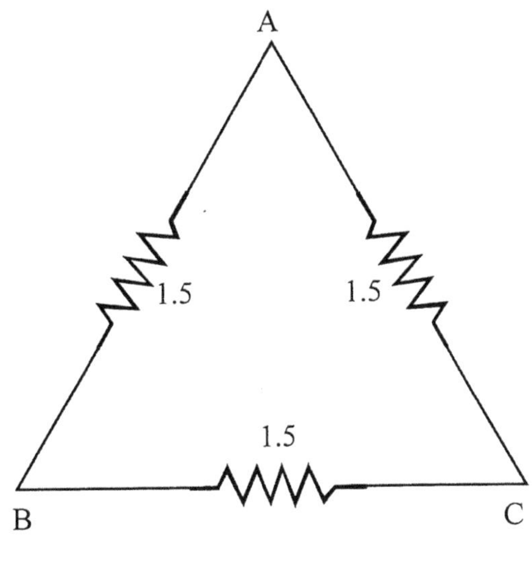

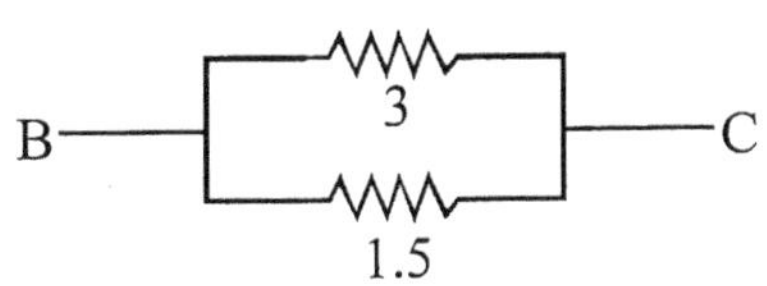

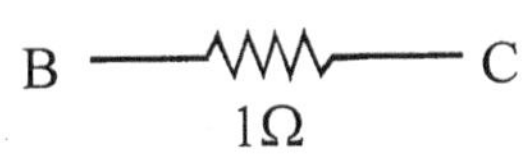

$$R_{BC} = \frac{3 \times 1.5}{3 + 1.5} = 1\Omega.$$

Example 1.13:

Determine the equivalent resistance between A and B.

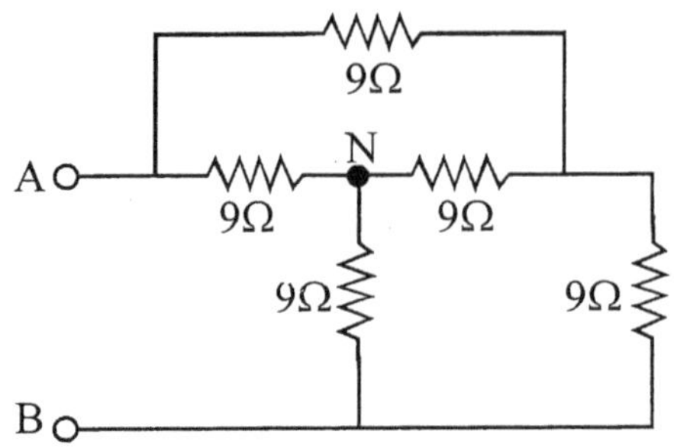

Solution:

The combination is neither series nor parallel. There is Delta and Star connection. Converting the star connection for which N is the star point and redrawing the circuit, we get,

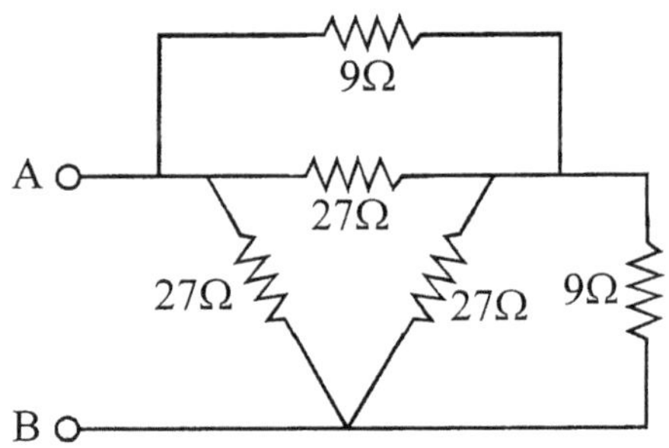

$$\frac{27\times 9}{27+9} = \frac{27\times 9}{36} = 6.75\Omega$$

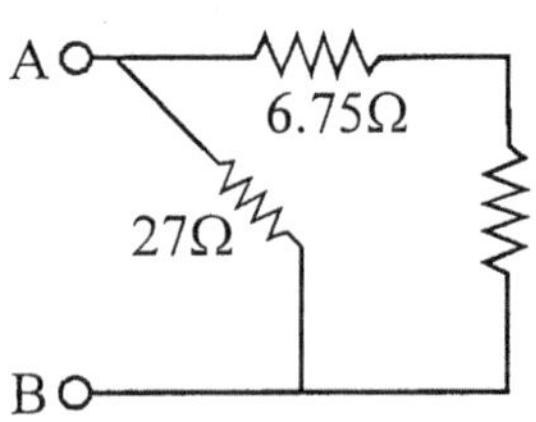

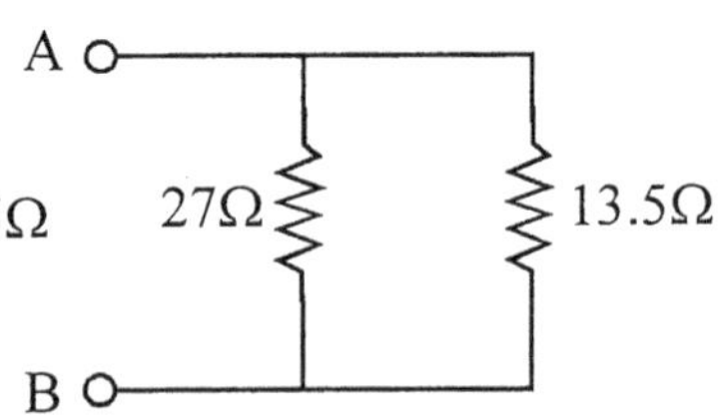

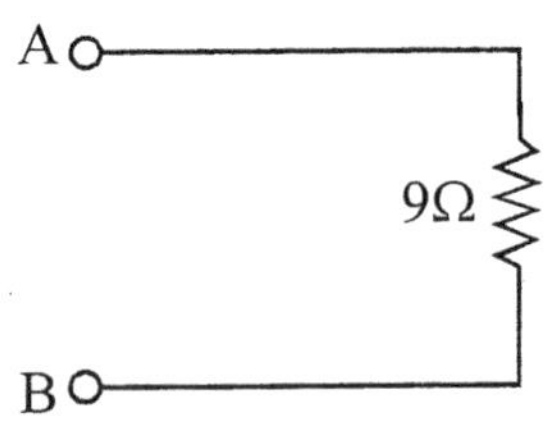

Therefore,

$$R_{AB} = 27 \parallel 13.5$$

$$= \frac{27\times 13.5}{27+13.5} = 9\Omega.$$

Example 1.14:

In the Wheat stone Bridge Circuit of figure (a) given below, find the effective resistance between PQ. Find the current supplied by a 10V Batter connected to PQ.

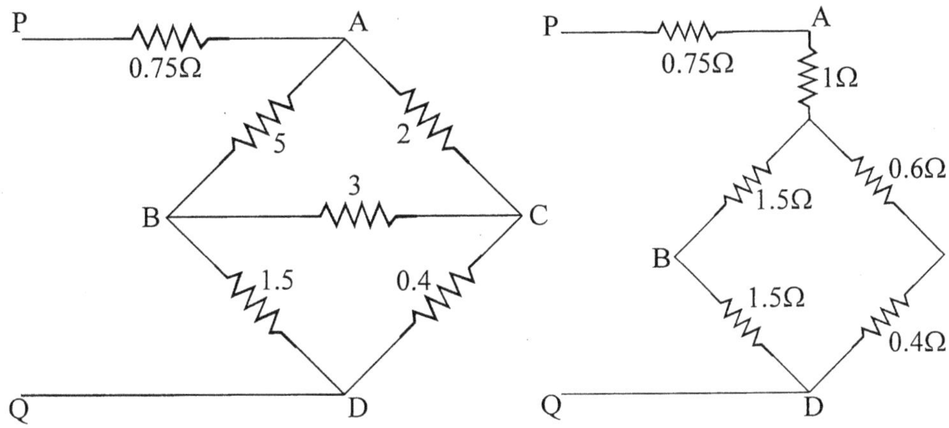

(a) (b)

Solution:

Converting the Δ formed by 5, 2 and 3Ω in a star of figure (b), we have,

$$R_A = \frac{5 \times 2}{5+3+2} = 1\Omega$$

$$R_B = \frac{5 \times 3}{10} = 1.5\Omega$$

$$R_C = \frac{3 \times 2}{10} = 0.6\Omega$$

The circuit then reduces to the following form.

The two 1.5 Ohm resistors are in series and this is in parallel with the 0.6 and 0.4 Ohm resistors which are in series.

Therefore, the total effective resistance between P and Q

$$= 0.75 + 1 + \frac{3 \times 1}{3+1}$$

$$= 2.5\Omega$$

If a 10V battery is attached to PQ, current drawn,

$$= \frac{10}{2.5} = 4A.$$

1.13 USES OF RESISTANCE

1. Resistance is used in any circuit to limit the current.
2. Resistance is used in series resonance circuits to limit the currents at resonance point.
3. Resistance will be used in Wave Shaping Circuits to obtain a different wave form from a given wave form.
4. Resistance can be used in analog circuits, to solve for other systems like Mechanical Systems, Hydraulic System, etc., in terms of Electrical Systems
5. Resistance will be used in Filter Circuits to select or reject certain frequencies.
6. Change of resistances as obtained in ***Resistance Strain Gauges***, which convert physical signals like temperature, strain, load, force etc., into electrical signals are used for measurement of the physical signals.
7. Resistance can also be used in Power Factor changing circuits but because it causes Power Loss and additional real power to be supplied, the power factor changing circuits generally do not employ change of resistances.
8. Resistances are used in quenching the arcs (extinguishing the arcs) in the case of circuit breakers, while breaking heavy current circuits in electrical transmission and distribution.
9. Resistance can be used in Measurement Circuits. For example, Variable Resistance is used to balance the bridge networks. Resistance shunt is used to extend the range of an ammeter or galvanometer, by diverting a major part of the circuit current through the shunt resistance so that only the allowable current flows through the ammeter or galvanometer. The actual current is calculated from the reading of the ammeter or galvanometer using a factor in terms of resistance of the galvanometer and the resistance of the shunt as in the case of parallel circuit. *A shunt is a very small pure resistance of the order 1Ω or less.*
10. Resistance shunt can also be used as a series resistance causing a small voltage drop across it, so that this voltage drop can be used in CRO (Cathode Ray Oscilloscope) to trace the current waveform which will be proportional to the voltage waveform across the resistance shunt.

D.C. CIRCUIT CONCEPTS AND CIRCUIT ELEMENTS - II

1.14 INTRODUCTION TO ENERGY STORING ELEMENTS

Earlier we have studied ahout Resistance Parameter. Resistance is a bilateral element and it can be linear or non-linear. Generally linear resistances are widely used. Resistance is an element which will be used in both D.C. and A.C. circuits.

Now we discuss about Capacitance and Inductance Parameters which are known as ***Energy Storing Elements***. Capacitance and Inductance elements are generally used in A.C. Circuits. They are used in D.C. circuits only for the purpose of making use of transients (transient voltage or transient current) for special purposes like wave shaping, differentiation, integration circuits. They are also used in filter circuits for allowing or rejecting currents or voltages of certain frequencies.

Inductance and capacitance can be used in resonance circuits to select or reject the particular frequency voltage or current. They can also be used in narrow band filters based on resonance to select or reject a narrow band of frequencies.

In the case of resistance parameter, the resistance reduces the current and at the same time causes *Power Loss* in the resistance.

Just like resistance, the capacitance and also the inductance reduce the current, but they store energy in them. The energy stored in the case of capacitance will be in the form of Electrostatic Energy and the energy stored in the case of inductance will be in the form of Electromagnetic Energy.

For an ideal capacitor and for an ideal inductor there will be no power loss. Just like the resistance, the capacitance and also the inductance depend upon the dimensions and geometry.

Capacitance also depends on the dielectric medium between the capacitance plates and the inductance depends up the core material of the inductor.

1.15 CAPACITANCE PARAMETER

Two conducting plates of equal and opposite charges, separated by a distance in air or any other dielectric material constitutes a capacitor and there will be force of attraction between the plates given by Coulombs Law.

The electric field between the plates will be parallel and having a value,

$$E = \frac{F}{Q} \qquad \text{.......... (1.15.1)}$$

where Q is the total charge of the *capacitance* or *capacitor*.

Capacitor is also called ***Condenser***.

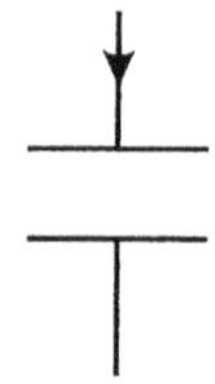

Fig. 1.14 Parallel Plate Capacitor

The electric field E will exists between charged conductors of any arbitrary shape also. But the parallel plate capacitors will have a charge density of Q/A, where A is the area of the parallel plate. According to ***Gauss's Law***, the total normal electric flux emanating from a closed surface enclosing a point charge Q or distributed charge of total value Q will be equal to Q.

The electric flux,

$$Q = D \times A \qquad \text{.......... (1.15.2)}$$

where D is the electric flux density and

A is the area of the capacitor plate.

$$D = \varepsilon E \qquad \text{.......... (1.15.3)}$$

where E is the electric field intensity and the permittivity, $\varepsilon = \varepsilon_0 \varepsilon_r$

The *Voltage Gradient*,

$$E = \frac{V}{d} \qquad \text{.......... (1.15.4)}$$

where d is the distance between the two capacitor plates and

V is the voltage across the capacitor plates.

$$V = Ed = \left(\frac{d}{\varepsilon A}\right) Q$$

From Eq. (1.15.2) and Eq. (1.15.3), we have,

$$E = \frac{D}{\varepsilon}$$

and the electric flux density,

$$D = \frac{Q}{A}$$

$$Q = V\frac{\varepsilon A}{d} \qquad \text{.......... (1.15.5)}$$

$$Q = C \times V$$

where C is the capacitance of the parallel plate capacitor

and d is the distance between the two capacitor plates

$$C = \frac{\varepsilon A}{d} \qquad \text{.......... (1.15.6)}$$

1.15.1 V-I Relationship for Capacitor – The Current Flowing through a Capacitor

Since, the dielectric medium between the parallel plates is an insulator, the steady state current flowing through the capacitor, when connected in a D.C. circuit will be equal to zero. If the capacitor is connected in an A.C. circuit, the instantaneous current flowing through the capacitor

$$i = \frac{dQ}{dt} \qquad \text{.......... (1.15.7)}$$

Hence, $$Q = \int i\,dt = \int_{-\infty}^{t} i\,dt$$

This equation may be written in a different form by separating the integral into two parts.

$$Q = \int_{-\infty}^{0} i\,dt + \int_{0}^{t} i\,dt$$

$$= Q_0 + \int_{0}^{t} i\,dt$$

$$i = \frac{dQ}{dt}$$

$$= \frac{d}{dt}(CV)$$

$$i = C\frac{dV}{dt} \qquad \text{if C is constant.} \qquad \text{.......... (1.15.8)}$$

If C is also varying with respect to time,

$$i = \frac{dQ}{dt}$$

$$= \frac{d}{dt}(CV)$$

$$= C\frac{dV}{dt} + V\frac{dC}{dt} \qquad \text{.......... (1.15.9)}$$

Hence, the voltage across the capacitor or capacitance,

$$V = \frac{1}{C}\int i\,dt$$

$$= \frac{1}{C}\int_{-\infty}^{t} i\,dt$$

$$= \frac{1}{C}\int_{-\infty}^{0} i\,dt + \frac{1}{C}\int_{0}^{t} i\,dt$$

$$= v_0 + \frac{1}{C}\int_{0}^{t} i\,dt \qquad \text{.......... (1.15.10)}$$

where v_0 is the initial voltage across the capacitor.

$\frac{1}{C}$ is also known as ***Elastance*** and is denoted by D and is measured in Daraf (Farad written in reverse direction).

Note: *Current can flow through the capacitance only so long as the charge flow is varying with respect to the time since,*

$$i = \frac{dQ}{dt}$$

This is possible only in A.C. circuits since current varies with respect to time. In D.C., when the voltage across the capacitance is steady the capacitor gets charged fully to the supply voltage and there will be no flow of current in the capacitor since,

$$I = \frac{dQ}{dt} = 0$$

1.15.2 Energy Stored in a Charged Capacitance

Capacitance stores energy between its plates to the form of Electrostatic field since electrostatic field is established between the two charged plates

The instantaneous value of power to the capacitor,

$$p = V \times i \text{ Watts}$$

$$= V \times C\frac{dV}{dt} \text{ Watts}$$

The total energy supplied to the capacitor when potential difference across the capacitor is increased from 0 to V Volts is given as,

$$W_C = \int p\,dt = \int_0^V vC\frac{dv}{dt}dt$$

From Eq. (1.15.8), we have,

$$W_C = \int_0^V Cv\,dv$$

$$= \tfrac{1}{2}C\,v^2\Big|_0^V$$

$$= \tfrac{1}{2}CV^2 \text{ Joules or Watt-sec.}$$

also
$$W = \tfrac{1}{2}C\left[\frac{Q}{C}\right]^2$$

$$= \tfrac{1}{2}\frac{Q^2}{C} \qquad \because\ v = \frac{Q}{C} \qquad \text{.......... (1.15.11)}$$

For a capacitor with dielectric of thickness d meters and area A square meters, energy per cubic meter is,

$$\frac{1}{2}\frac{CV^2}{Ad} = \frac{1}{2}\frac{\in A}{d}\frac{V^2}{Ad}$$

$$= \frac{1}{2}\in\left(\frac{V}{d}\right)^2 = \frac{1}{2}\in E^2 \text{ Watt-sec} \qquad \text{........ (1.15.12)}$$

$$W_C = \frac{1}{2}DE$$

$$= \frac{1}{2}\frac{D^2}{\in} \text{ Joules or Watt-sec.} \qquad \text{.......... (1.15.13)}$$

where D is the Flux Density and

E is the Electric Field Intensity.

Note :

1. *In a system altered instantaneously by the closing or opening of a switch, the voltage across the capacitance cannot change immediately and is equal to v_0 the initial voltage across the capacitance since,*

$$V_C = V_o + \int_0^t i\,dt$$

Hence, the voltage across a capacitance cannot change all on a sudden. This principle is made use in designing of Pulse and Digital Circuits.

2. *The Capacitors used in electronic circuits will have Paper or Ceramic or an electrolyte as the dielectric medium. The Ceramic capacitors will be of the order micro-Farads ($\mu F = 10^{-6}$ Farads) or pico-Farads ($pF = 10^{-12}$ Farads)and will have a color coding painted on them to indicate their values. The electrolytic capacitors will have polarities marked and they have to be connected to the same polarities of the supply else the capacitor will burst. The symbol for the electrolytic capacitor is given in Fig. 1.15.*

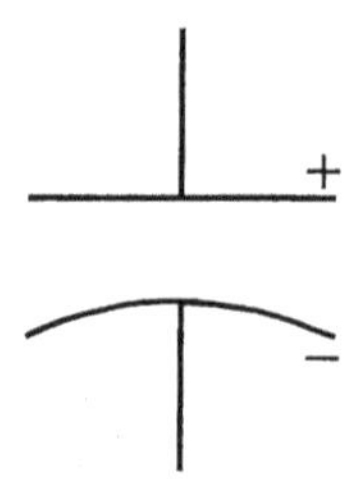

Fig. 1.15 *Capacitor*

3. *A pure capacitance will have no resistance. The capacitance gets charged to its full value and the voltage across the capacitance remains the same when the capacitor circuit is opened, since, the charge of the capacitance cannot be discharged. Hence the capacitance gives an electric shock when touched after the circuit is opened. To prevent the shock, the capacitor has to be discharged by connecting a wire across the two terminals of the capacitor. In some capacitors like power capacitors used in electrical power circuits, a high resistance will be connected across the capacitor to discharge the capacitor when the circuit is opened.*

4. *The unit of capacitance is Farads and is denoted by F. Farad is a large unit and hence, capacitance have values of micro-Farads. The capacitor used in electronic circuits may be denoted by the manufacturer as MF. It should not be mistaken as Mega-Farads, since, Farad itself is a big unit and it should be taken as micro-Farads only.* ***A Farad may be defined as the capacitance of a capacitor between the plates of which there appears a potential difference of one volt when it is charged by one coulomb of electricity.***

5. *The electric power lines have charged conductors and hence, there will be capacitance between any two conductors and between a conductor and earth. Since, it is a long line the capacitance will be distributed capacitance and expressed in micro-Farads per kilometer length of line.*

6. *For precision work, the capacitance between the two ends of a coil and also the capacitance between two turns of the coil should be considered.*

Example 1.15:

A 1μF parallel plate capacitor with air as the dielectric medium is charged to a potential of 200V. The distance between the two plates is 1 cm, Calculate

(i) *The electric stress with in the dielectric*

(ii) *The electric stress on the plate surface*

(iii) *The charges on the plates*

Calculate the values (i), (ii) and (iii) if air dielectric is replaced by paper of relative permittivity 4.

Solution:

Capacitance of capacitor,

$$C = 1\mu F = 1 \times 10^{-6} F$$

Distance between plates,

$$d = 1 \text{ cm} = 0.01\text{m}$$

Potential Difference across pates,

$$V = 200V$$

(i) Electric Stress of Intensity in the Dielectric,

$$E = \frac{V}{d}$$

$$= \frac{200}{0.01}$$

$$= 20 \times 10^3 \text{ V/m}$$

$$= 20\text{KV/m}$$

(ii) Electric Stress on the plate surface is same as in the Dielectric Medium, i.e.,

$$E = 20\text{KV/m}$$

(iii) Charge on the plates

$$Q = CV$$

$$= 1 \times 10^{-6} \times 200$$

$$= 2 \times 10^{-4}C$$

When air is replaced by paper of relative permittivity,

$$\varepsilon_r = 4$$

Electric stress in the medium = Electric Stress on the plates,

$$\frac{V}{d} = \frac{200}{0.01}$$

$$= 20\text{KV/m}$$

Note :

Capacitance of capacitor,

$$C_P = \varepsilon_r C$$

$$= 4 \times 1 = 4\mu F = 4 \times 10^{-6}F$$

Therefore, Charge on the plates,

$$Q = C_P V$$

$$= 4 \times 10^{-6} \times 200$$

$$= 8 \times 10^{-4} C$$

Example 1.16:

Calculate the capacitance and energy stored in a Parallel Plate Capacitor which consists of two metal plates each $60 cm^2$ separated by a dielectric of 1.5mm thickness and $\varepsilon_r = 3.5$ if Potential Difference of 1000V is applied across it.

Given $\varepsilon_0 = 8.854 \times 10^{-12}$ F/m.

Solution:

Area of each plate,

$$A = 60 cm^2 = 60 \times 10^{-4} m^2$$

Thickness of Dielectric,

$$d = 1.5 mm = 1.5 \times 10^{-3} m$$

Relative Permittivity,

$$\varepsilon_r = 3.5$$

Potential Difference across Capacitor,

$$V = 1000V$$

Capacitance of the capacitor,

$$C = \frac{\varepsilon_0 \varepsilon_r A}{d}$$

$$= \frac{8.854 \times 10^{-12} \times 3.5 \times 60 \times 10^{-4}}{1.5 \times 10^{-3}}$$

$$= 123.96 \text{ pF}$$

Energy Stored $= \frac{1}{2} CV^2$

$$= \frac{1}{2} \times 123.96 \times 10^{-12} \times (1000)^2$$

$$= 61.98 \times 10^{-6} \text{ Joules}$$

1.15.3 Parallel Plate Capacitor with Guard Ring

One difficulty encountered in practice with a parallel plate capacitor is due to the fringing of flux which occurs at the edges. This difficulty is overcome by the provision of a ***Guard Ring*** around one electrode, as shown in the Fig. 1.16

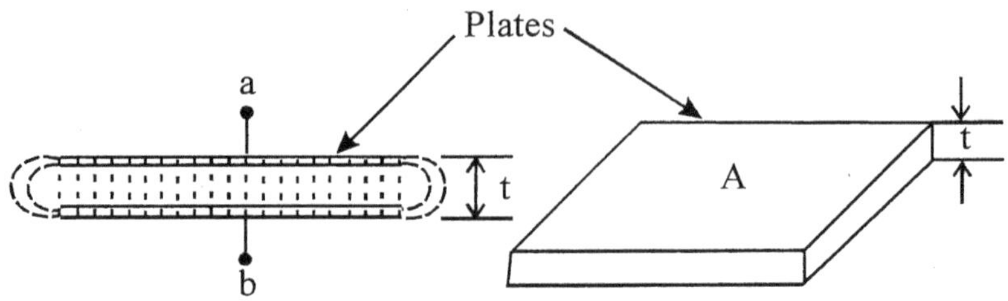

Fig. 1.16 *Parallel - Plate Capacitor*

The potential of the guard ring is maintained at the same level as that of the central electrode, but from a separate source. Fringing of flux is now confined to the outer edge of the guard ring, while the flux density over the central electrode is uniform. That is what we require as then the charge that actually reaches the electrodes through the lead L is accurately given by the equation charge density,

$$D = \sigma = \frac{Q}{A} \quad \text{Coulombs/m}^2$$

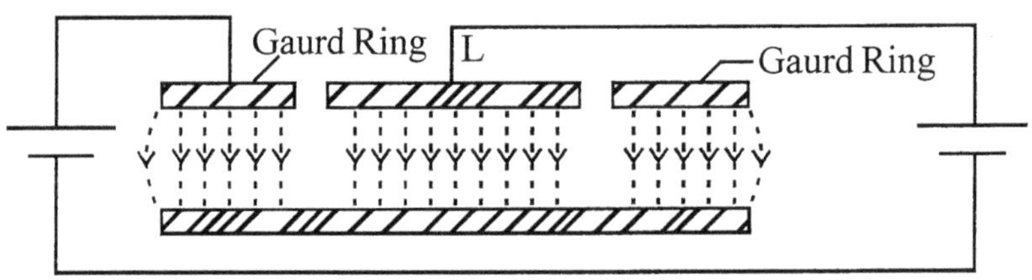

Fig. 1.17 *Parallel - Plate Capacitor with Gaurd Ring*

1.15.4 Capacitances in Series or Series Capacitors

Applying Kirchoff's Voltage Law,

$$V = V_1 + V_2 + V_3$$

$$= \frac{1}{C_1}\int i\,dt + \frac{1}{C_2}\int i\,dt + \frac{1}{C_3}\int i\,dt$$

$$V = \left(\frac{1}{C_1} + \frac{1}{C_2} + \frac{1}{C_3}\right)\int i\, dt \qquad \text{.......... (1.15.14)}$$

For an equivalent capacitance,

$$V = \left(\frac{1}{C_{Se}}\right)\int i\, dt \qquad \text{.......... (1.15.15)}$$

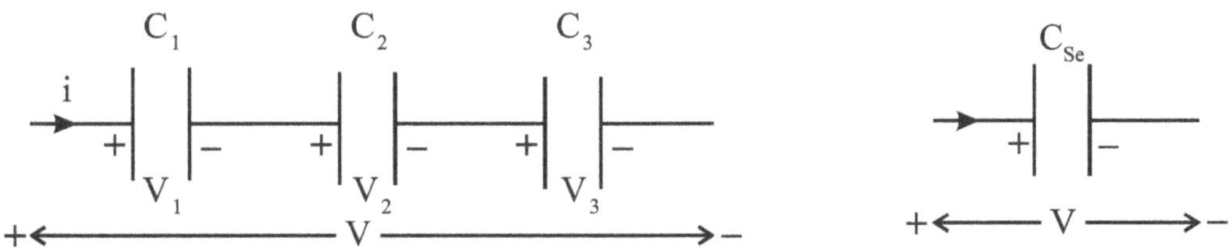

Fig. 1.18 *Capacitance in Series*

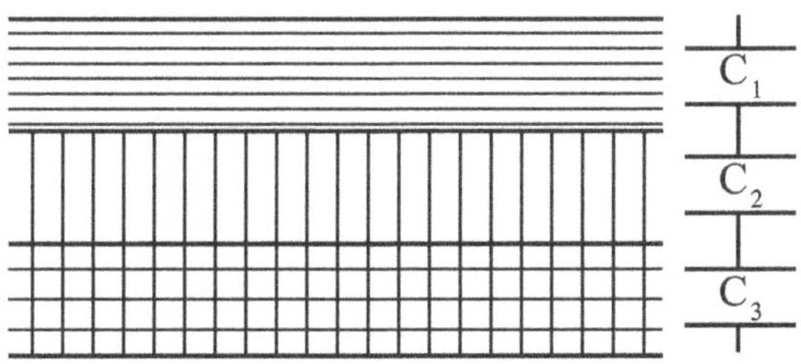

Fig. 1.19 *Composite Dielectric Capacitors*

Comparing Eq. (1.15.14) and Eq. (1.15.15), we get,

$$\frac{1}{C_{Se}} = \frac{1}{C_1} + \frac{1}{C_2} + \frac{1}{C_3} \qquad \text{.......... (1.15.16)}$$

For the Capacitance in Series,

$$\frac{1}{C_{Se}} = \sum_{i=1}^{n} \frac{1}{C_i} \qquad \text{.......... (1.15.17)}$$

Example 1.17 :

Deduce an expression for the capacitance of a parallel-plate capacitor having two dielectric media.

A parallel plate capacitor has a plate separation t. The capacitance with air only between the plates is C. When a slab of thickness t′ and relative permittivity ε_r is placed on one of the plates, the capacitance is C′. Show that,

$$\frac{C'}{C} = \frac{\varepsilon_r t}{t' + \varepsilon_r (t - t')}$$

Solution :

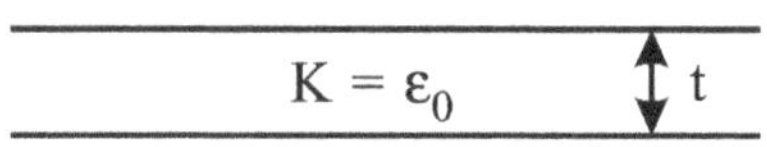

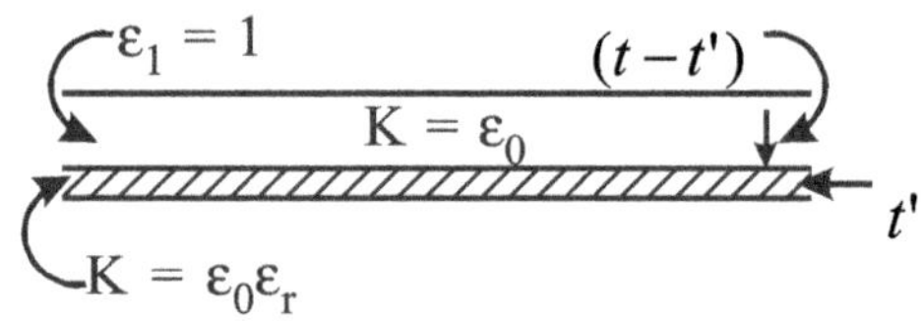

With air alone as dielectric,

$$C = \varepsilon_0 \qquad (\text{ as } \varepsilon_r = 1) \qquad \text{.......... (1)}$$

Now, with the introduction of a slab of thickness t', the thickness of air-film is reduced to $(t - t')$ Let E_1 and E_2 be the field intensities in the air-film and slab, respectively and V_1 and V_2 the potential differences across them.

If V is the total potential difference across the capacitor, then

$$V = V_1 + V_2 \qquad \text{.......... (2)}$$

$$V_1 = E_1 (t - t') \qquad \text{.......... (3)}$$

$$V_2 = E_2 t' \qquad \text{.......... (3a)}$$

If Q is the charge on the plates, then, from the boundary relations, the normal components of electric flux densities being equal,

$$D = \frac{Q}{A}$$

we have,
$$E_1 = \frac{D}{\varepsilon_0}$$

and $$E_2 = \frac{D}{\varepsilon_0 \varepsilon_r}$$

Thus substituting for E_1 and E_2, we obtain from Eq. (3) and Eq. (3a),

$$V = \frac{Q}{A}\left[\frac{1}{\varepsilon_0}(t - t') + \frac{1}{\varepsilon_0 \varepsilon_r} t'\right]$$

$$= \frac{Q}{A\varepsilon_0}\left[\frac{1}{\varepsilon_r} t' + (t - t')\right] \qquad \text{.......... (4)}$$

The new value of capacitance is then,

$$C' = \frac{Q}{V}$$

$$= \frac{A\varepsilon_0}{\frac{t'}{\varepsilon_r} + (t - t')}$$

$$= \varepsilon_0 \frac{A\varepsilon_r}{t' + \varepsilon_r (t - t')} \qquad \text{.......... (5)}$$

From Eq. (1) and Eq. (5), we get,

$$\frac{C'}{C} = \frac{\varepsilon_r t}{t' + \varepsilon_r (t - t')} \qquad \text{.......... (6)}$$

Example 1.18:

A parallel plate paper capacitor has 17 plates each having an effective area of $5 cm^2$ and each separated by a paper sheet of 0.005mm thick. Find the capacitance. Take the relative permittivity of paper as 4.

Solution:

Number of parallel plates,

$$n = 17$$

Area of each plate,

$$A = 5\ cm^2 = 5 \times 10^{-4} m^2$$

Thickness of paper sheet,

$$d = 0.005 \text{ mm} = 5 \times 10^{-6} \text{ m}$$

Relative Permittivity of Paper,

$$\varepsilon_r = 4$$

Capacitance of Multi-plate Capacitor (Series Capacitors),

$$C = (n-1)\frac{\varepsilon_0 \varepsilon_r A}{d}$$

$$= (17-1)\frac{8.854 \times 10^{-12} \times 4 \times 5 \times 10^{-4}}{5 \times 10^{-6}}$$

$$= 56.66 \times 10^{-9} \text{F}$$

Example 1.19:

Consider that three capacitors having capacitances 5, 10, 12 microfarads are connected in series across 600V D.C. main. Determine :

(i) *Equivalent Capacitance of the Combination*
(ii) *Charge on each capacitor*
(iii) *Potential across each capacitor*
(iv) *Stored energy in each of the capacitor*

Solution :

Capacitance of each capacitor connected in series,

$$C_1 = 5\mu F; \qquad C_2 = 10\mu F; \qquad C_3 = 12\mu F;$$

Voltage across the grouping,

$$V = 600V$$

(i) Let C be the total or effective capacitance of the grouping, then,

$$\frac{1}{C} = \frac{1}{C_1} + \frac{1}{C_2} + \frac{1}{C_3}$$

$$= \frac{1}{5} + \frac{1}{10} + \frac{1}{12}$$

$$= \frac{23}{60}$$

(ii) Charge on each capacitor,

$$Q = CV = 2.608 \times 600$$
$$= 1565\mu C$$

(iii) Potential across each capacitor,

$$V_1 = \frac{Q}{C_1}$$
$$= \frac{1565 \times 10^{-6}}{5 \times 10^{-6}}$$
$$= 313V$$
$$V_2 = \frac{Q}{C_2}$$
$$= \frac{1565 \times 10^{-6}}{10 \times 10^{-6}}$$
$$= 156.5V$$
$$V_3 = \frac{Q}{C_3}$$
$$= \frac{1565 \times 10^{-6}}{12 \times 10^{-6}}$$
$$= 130.5V$$

(iv) Energy stored in each capacitor,

$$W_1 = \tfrac{1}{2} C_1 V_1^2$$
$$= \tfrac{1}{2} \times 5 \times 10^{-6} - (313)^2$$
$$= 0.245J$$
$$W_2 = \tfrac{1}{2} C_2 V_2^2$$
$$= \tfrac{1}{2} \times 10 \times 10^{-6} - (156.5)^2$$
$$= 0.122J$$
$$W_3 = \tfrac{1}{2} C_3 V_3^2$$
$$= \tfrac{1}{2} \times 12 \times 10^{-6} - (130.5)^2$$
$$= 0.102J$$

1.15.5 Capacitances in Parallel or Parallel Capacitors

Parallel plate capacitance is obtained in the case of composite dielectric capacitors.

Applying Kirchoff's Current Law,

$$I = i_1 + i_2 + i_3$$

$$= C_1 \frac{dV}{dt} + C_2 \frac{dV}{dt} + C_3 \frac{dV}{dt}$$

$$= (C_1 + C_2 + C_3) \frac{dV}{dt} \qquad \text{.......... (1.15.18)}$$

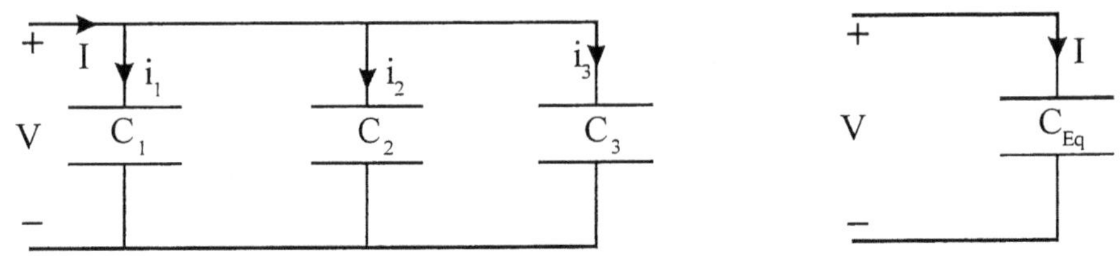

Fig. 1.20 Capacitance in Parallel

For the Equivalent Capacitance,

$$I = C_{Eq} \frac{dV}{dt}$$

Hence, $$C_{Eq} = C_1 + C_2 + C_3 \qquad \text{.......... (1.15.19)}$$

For the Capacitance in Parallel,

$$C_{Eq} = \sum_{i=1}^{n} C_i \qquad \text{.......... (1.15.20)}$$

Table 1.1 *Important Insulating Materials*

Material	Relative Permittivity
Vacuum	1.0
Air	1.0006
Paper (dry)	2.0 to 2.5
Polythene	2.0 to 2.5
Insulating Oil	3.0 to 4.0

Bakelite	4.5 to 5.5
Glass	5.0 to 10.0
Rubber	2.0 to 3.5
Mica	3.0 to 7.0
Porcelain	6.0 to 7.0
Distilled Water	80.0
Barium Titanate	6000+

Table 1.2 *Dielectric Strengths (MV/m)*

Material	**Thickness (mm)**	**Dielectric Strength (MV/m)**
Air (at normal pressure and temperature)	0.2	5.75
	0.6	4.92
	1.0	4.46
	6.0	3.27
	10.0	2.98
Mica	0.01	200
	0.1	176
	1.0	61
Glass (density 2.5)	1.0	28.5
	5.0	18.3
Ebonite	1.0	50
Paraffin-waxed paper	0.1	40 to 60
Transformer Oil	1.0	200
Ceramics	1.0	50

Example 1.20:

Three capacitors have capacitances of $C_1 = 4\mu F$, $C_2 = 10\mu F$ and $C_3 = 20\mu F$. Find the total capacitance and the charge on each capacitor when connected in parallel to 200V supply. Calculate the total capacitance and voltage across each capacitor when connected in series to the same 200V supply.

Solution:

Fig. (a), shows the parallel circuit connection.

Then,

$$Q_1 = C_1 V$$

$$= 4\mu F \times 200V = 800\mu C$$

$$Q_2 = C_2 V$$

$$= 10\mu F \times 200V = 2000\mu C$$

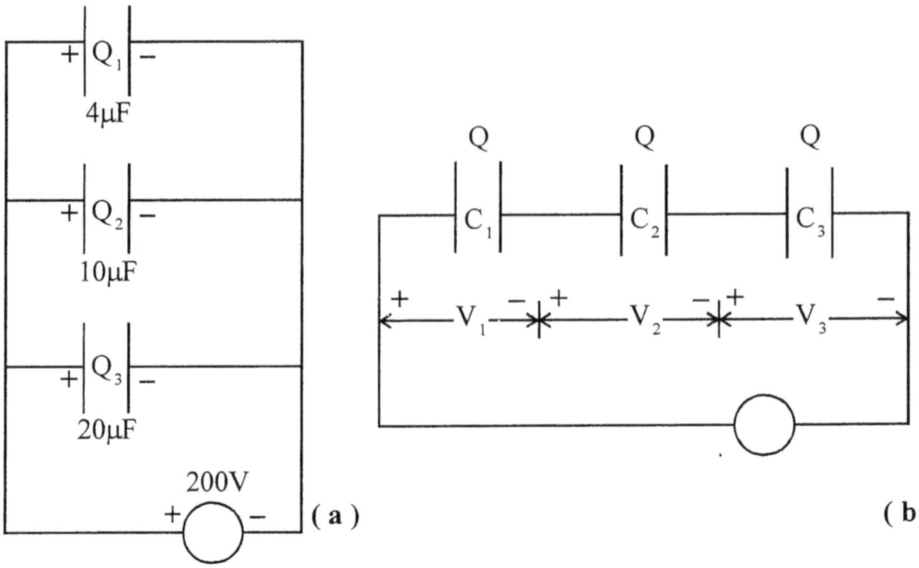

(a) (b)

$$Q_1 = C_3V$$

$$= 20\mu F \times 200V = 4000\mu C$$

$$C = C_1 + C_2 + C_3 = 4 + 10 + 20 = 34\mu F$$

Fig. (b), shows the series circuit connection. The charge on each capacitor is the same.

$$\frac{1}{C} = \frac{1}{C_1} + \frac{1}{C_2} + \frac{1}{C_3}$$

Hence, $$C = \frac{C_1C_2C_3}{C_1C_2 + C_2C_3 + C_3C_1}$$

$$= \frac{4 \times 10 \times 20}{(4 \times 10) + (10 \times 20) + (20 \times 4)}$$

$$= 2.5\mu F$$

$$Q = CV$$

$$= 2.5 \times 10^{-6} \times 200C = 500\mu C$$

$$Q = C_1V_1 = C_2V_2 = C_3V_3$$

Hence,

$$V_1 = \frac{Q}{C_1}$$

$$= \frac{500\mu C}{4\mu F} = 125V$$

$$V_2 = \frac{Q}{C_2}$$

$$= \frac{500\mu C}{10\mu F} = 50V$$

$$V_3 = \frac{Q}{C_3}$$

$$= \frac{500\mu C}{20\mu F} = 25V$$

Example 1.21:

A variable air capacitor has 15 movable plates and 16 fixed plates. The area of each plate is 15 cm^2 and separation between opposite plates is 0.2mm. Determine

(i) *The maximum capacitance of this variable capacitor*

(ii) *The capacitance when ¼ area of movable plates is overlapping the fixed plates*

Solution:

Number of movable plates,

$$n_1 = 15$$

Number of fixed plates,

$$n_2 = 16$$

Area of each plate,

$$A = 15cm^2 = 15 \times 10^{-4}m^2$$

Distance between opposite plates

$$d = 0.2\text{mm} = 2 \times 10^{-4}\text{m}$$

Total Number of plates,

$$n = n_1 + n_2$$
$$= 15 + 16 = 31$$

(i) Capacitance will be maximum when movable plates are rotated so that two sets of plates are completely overlapping each other.

$$C_m = (31-1)\frac{8.854\times10^{-12}\times1\times15\times10^{-4}}{2\times10^{-4}}$$

$$= 19.92 \times 10^{-2}\text{F}$$

(ii) Capacitance of capacitor, when ¼ area of movable plates is overlapping the fixed plates.

$$C_1 = (31-1)\frac{8.854\times10^{-12}\times1\times15\times10^{-4}}{2\times10^{-4}\times4}$$

$$= 4.08 \times 10^{-10}\text{F}$$

1.15.6 Capacitances in Series-Parallel

The equivalent capacitances for the parallel combinations of capacitances are found. These equivalent capacitances are treated to be in series with the remaining capacitances. After solving the circuit, the current in the parallel branches are calculated from the total current.

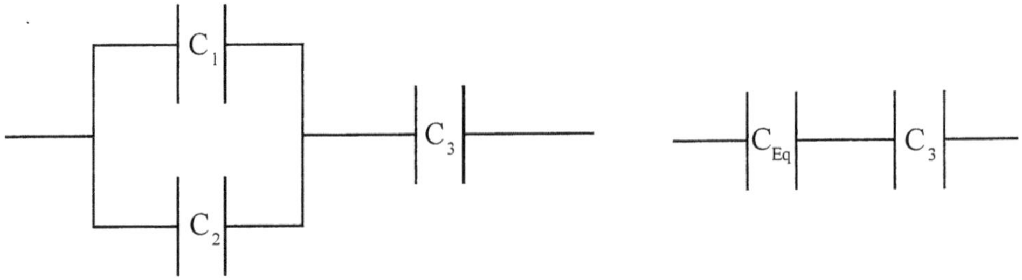

Fig. 1.21 Capacitance in Series-Parallel

Example 1.22:

Calculate the charge and potential difference across each capacitor in the given circuit.

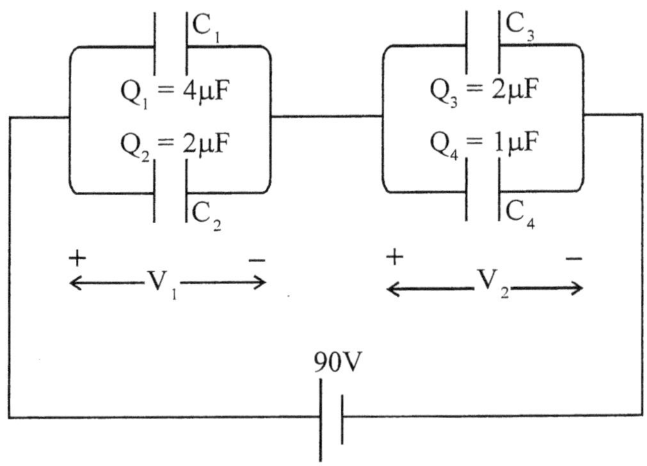

Solution:

C_1 and C_2 in parallel are equivalent to

$$C_{12} = C_1 + C_2 = (4 + 2) = 6\mu F.$$

C_3 and C_4 in parallel are equivalent to

$$C_{34} = C_3 + C_4 = (2 + 1) = 3\mu F$$

Capacitance of the whole circuit consisting of C_{12} and C_{34} in series is equivalent to C_t given by,

$$C_t = \frac{C_{12}C_{34}}{C_{12} + C_{34}}$$

$$= \frac{6 \times 3}{6+3}\mu F \quad = \quad 2\mu F$$

Charging on the combination is

$$Q_t = C_t V$$

$$= 2\mu F \times 90V$$

$$= 180\mu C.$$

$$V_1 = \frac{Q_t}{C_{12}}$$

$$= \frac{180\mu C}{6\mu F} = 30V$$

$$V_2 = 90 - 30 = 60V$$

$$Q_1 = C_1V_1$$

$$= 4\mu F \times 30V = 120\mu C$$

$$Q_2 = C_2V_2$$

$$= 2\mu F \times 30V = 60\mu C$$

$$Q_3 = C_3V_3$$

$$= 2\mu F \times 60V = 120\mu C$$

$$Q_4 = C_4V_4$$

$$= 1\mu F \times 60V = 60\mu C$$

1.15.7 Dielectric Strength and Breakdown

One of the important properties of a dielectric is its capability to withstand the electric stress (viz., the potential gradient) without rupture when the potential difference is raised. The material will cease to be a nonconductor or dielectric once the field produced exceeds a certain value. The maximum value of field-intensity or potential-gradient that the material can withstand without disruption may be referred to as the Dielectric Strength of the material. According to Maximum Stress Theory, if the potential difference across the material is raised to such an extent that the electric stress exceeds the above limit, breakdown is sure to occur. Breakdown may also occur due to presence of impurities, lack of homogeneity, surface-irregularities, etc.

The high electric field intensity surrounding high voltage power lines accounts for an additional energy loss in the transmission of power. The high voltage gradient at the surface of a wire sometimes accelerates electrons in the air sufficiently to ionize air molecules by collision. If the voltage gradient at the wire exceeds a certain critical value, the process of ionization becomes cumulative and results in appreciable loss of energy. The ionization is characterized by a faint glow surrounding the wire and is called ***Corona***.

The dielectric strength of air at barometric pressure of 76cm at 25°C is 30KV/Cm (Peak Value) or 21.1KV RMS value / Cm.

1.15.8 Electrostatic Induction

A positively charged conductor induces negative charges on a nearby conductor. This is called ***Electrostatic Induction***. The charges tend to get distributed at the outer surface of the conductor. The charges get concentrated at sharp points of the conductor. Hence, in the case of lighting arrestors the charged clouds induce opposite charges in the sharp end of arrestor rod which passes the charges to ground, thus, saving the damage to the structure.

1.15.9 Uses of Capacitors

1. In resonance circuits to tune and select particular frequency signals as in Radio and TV., etc.
2. In Filter Circuits to select or reject certain frequencies
3. In Operational Amplifiers which is used for integration, differentiation, etc.
4. In Wave Shaping Circuits to obtain a desired wave form from a given waveform
5. In analog circuits, to solve for other systems like Mechanical Systems, Hydraulic System, etc., in terms of Electrical Systems
6. In ***Transducers***, which convert physical signals into electrical signals as in microphone.
7. In Power Factor Changing Circuits
8. To improve the Power Factor of Load.
9. To improve the voltage profile of transmission lines.
10. In High Voltage Impulse Generators.
11. In Measurement Circuits.

1.16 INDUCTANCE

Inductance is also an energy storing element like capacitor. The energy stored in the inductance is in the form of Electromagnetic Field. Both Inductance can also be used in wave shaping circuits. Inductance is also called as Inductor.

Inductance exhibits delay in the rise and fall of currents through it. Hence, it is used to represent a mass possessing inertia in Electrical Analog Circuits used to represent a Mechanical System.

A current flowing through a conductor sets up an electromagnetic field around the conductor. The direction of the magnetic field setup is given by, ***Maxwell's Right Hand Cork-Screw Rule*** or ***Right Hand Threaded Screw Rule*** which states that, *the direction of the magnetic field setup by a current flowing through a conductor will be given by the direction of rotation of a right hand cork-screw if the screw is to advance in the direction of the current flow.* Hence, it will be circular around the conductor as given in the Fig 1.22

Note: *The dot inside the conductor of Fig. (1.22) represents the arrow head and indicates that the current is directed towards us i.e., away from the plane of the paper. The cross inside the conductor of Fig. (1.22) represents the tail of an arrow and indicates that the current is directed away from us i.e., into the plane of the paper.*

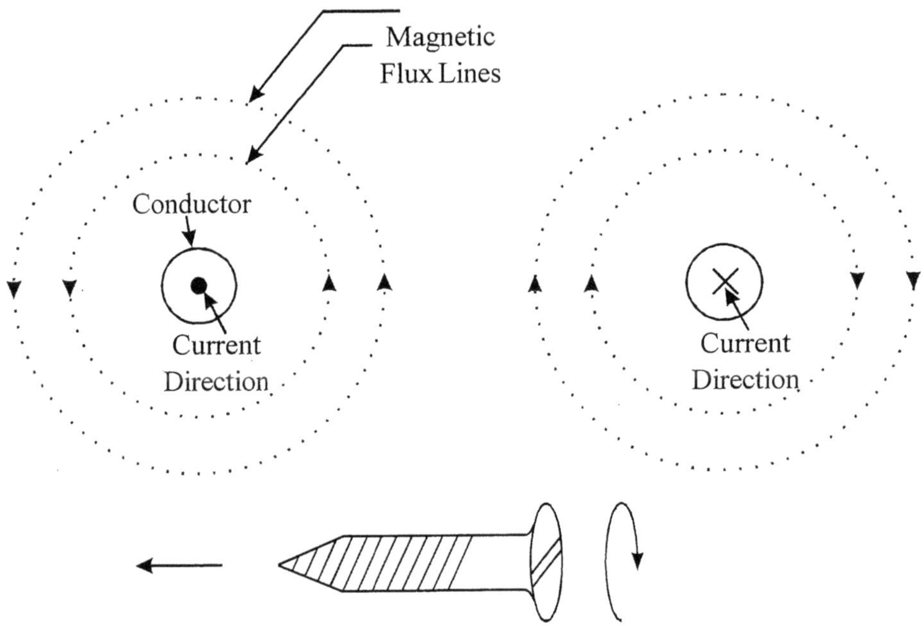

Fig. 1.22 *Magnetic Field Setup by a Current Carrying Conductor*

1.16.1 Magnetic Field of a Solenoid

In the Fig 1.23 (b) and Fig. 1.23 (c), the conductor is shown by circular cross-section with a dot or cross inside to indicate the direction of the current. The circle around the conductor along with its direction indicates the direction of the magnetic field setup by the current flowing through this conductor.

If a coil is wound on a soft iron rod, as in Fig. 1.23 and connected to a battery, the steel becomes magnetized and behaves like a temporary magnet. This magnetic field of the electromagnet is represented by the dotted lines and its direction by the arrow heads.

The direction of the magnetic field produced by a current in a solenoid may be deduced by applying either the Right Hand Threaded Screw or the Right Hand Grip Rule.

If the axis of the screw is placed along that of the solenoid and if the screw is turned in the direction of the current, it travels in the direction of the magnetic field inside the solenoid, namely towards the right in Fig. 1.24

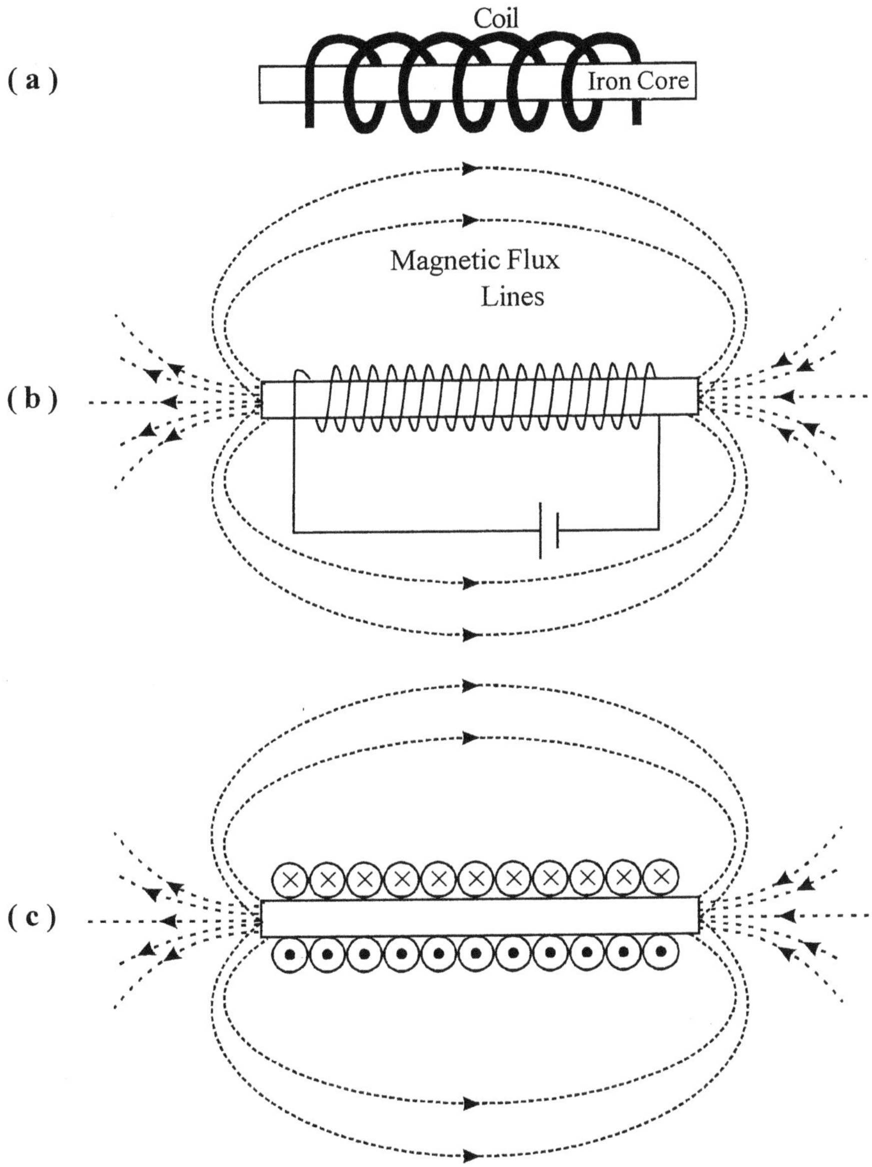

Fig. 1.23 Solenoid and its Magnetic Field Setup by the Solenoid

The ***Right Hand Grip Rule***, can be expressed thus : *if the solenoid is gripped with the right hand, with the fingers pointing in the direction of the current, i.e., conventional current, then the thumb outstretched parallel to the axis of the solenoid points in the direction of the magnetic field inside the solenoid.*

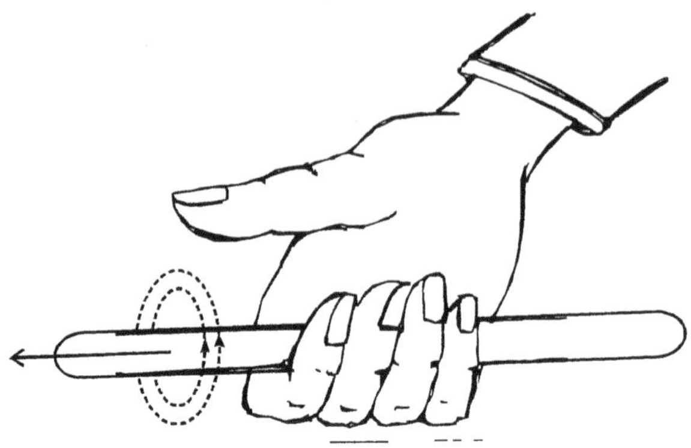

Fig. 1.24 *Right Hand Grip Rule*

1.16.2 Self Inductance of a Coil

If the current flowing through a coil is A.C. current then, the electromagnetic field setup by the current will be time varying i.e., varying with respect to time. According to Faraday's Law, this time varying magnetic field will induce an alternating voltage called induced E.M.F. (E) in the coil or conductor.

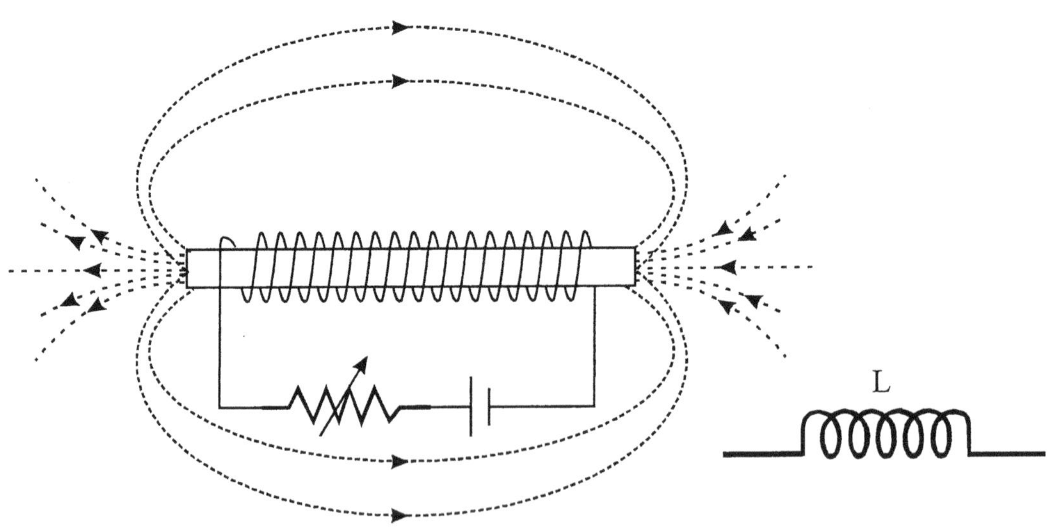

(a) Setting up of Self Inductance **(b)** Symbol for Self Inductance

Fig. 1.25 *Self Inductance of a Coil*

According to Lenz's Law, the E.M.F. induced in the conductor will be opposing the cause that produces it. That means it will try to reduce the current flowing through the conductor which sets up the magnetic field. This reduction in current can be treated as reduction in current due to a parameter called ***Inductance Parameter*** (L) just as the resistance (R) reduces the current. The inductance is denoted by L and is measured in Henry.

A circuit has an inductance of one henry (1 H) if an E.M.F. of one volt is induced in the circuit when the current varies uniformly at the rate of one ampere per second. Henry is also a big unit and hence, the coils will generally have inductance of the order of milli-Henrys.

In this case, the E.M.F. induced in the coil is due to the magnetic field setup by the current flowing in the same conductor. Hence is known as ***self induced E.M.F.*** and also the inductance is known as ***self inductance*** or ***coefficient of self induction***.

The self inductance of a coil is defined as the flux linkages linking that coil for unit current flowing through that coil. If the flux produced by the current i flowing through that coil is is ϕ webers and if the coil has N turns, then the flux linking that coil will be NΦ weber-turns. Then the self inductance,

$$L = \frac{N\Phi}{i} \text{ henrys or Weber-turns/ampere.} \qquad \text{.......... (1.16.1)}$$

This is one way of calculating the Self Inductance.

Hence, a coil is said to have a self-inductance of one henry if a current of one ampere flowing through it produces flux linkages of one wb-turns in it.

Alternatively, for a Solenoid or Toroid of N turns, the magnetization H is given as

$$H = \frac{NI}{l} \text{ AT / m} \qquad \text{.......... (1.16.2)}$$

where l is the length of the magnetic path = $2\pi r$ meters

r is the radius of the toroidal ring of the magnetic material in meters

a is the area of cross-section of the magnetic ring in square meters.

The magnetic flux,

$$\Phi = B \times a \text{ Wb} = \mu_o \mu_r Ha \text{ Wb}$$

$$\Phi = \frac{Ni}{l/\mu_o \mu_r a} \text{ wb} \qquad (\because H = \frac{NI}{l}) \qquad \text{........ (1.16.3)}$$

where B is Flux density in wb/m^2 or Tesla 'T', a μ_o is the permeability of free space with μ_o equals $4\pi 10^{-7} H/m$, μ_r is relative permeability with μ_r=3000 for Iron, a is area of cross section of the magnetic path in square meters

$$\Phi = \frac{\text{Magnetomotive Force}}{\text{Reluctance}}$$

$$\Phi = \frac{\mathscr{F}}{\mathscr{R}} \text{ or } \frac{F}{S} \text{ wb} \qquad \text{.......... (1.16.4)}$$

The magnetomotive force,

$\mathscr{F}$ or F = NI amp-turns .... (1.16.4a)

Reluctance, $\mathscr{R}$ or $S = \frac{1}{a\mu_0\mu_r} AT/wb$

The *self inductance* of a coil is the flux linkages with that coil per unit current flowing through that coil.

Hence,

$$L = \frac{N\Phi}{i} \text{ Wb Turns / Amp. or H} \qquad \text{.......... (1.16.5)}$$

This is another way of calculating the Self Inductance.

Also, $$L = \frac{N\Phi}{i} = \frac{N}{i} \frac{N \times i}{l/\mu_o \mu_r a}$$

$$L = \frac{N^2}{l/\mu_o \mu_r a} \text{ henry} \qquad \text{.......... (1.16.6)}$$

$$L = \frac{N^2}{S} \qquad \text{.......... (1.16.7)}$$

This is another way of calculating the Self Inductance and is in terms of the geometry of the coil.

Note: *Henry is denoted by symbol H* ***and should not be confused with magnetization force H.***

1.16.3 Voltage Drop across the Inductance

According to Faraday's Law and Lenz's Law, the self induced E.M.F. in a coil is the negative of the rate of change of flux linkages in that coil.

$$e_l(t) = -\frac{d\lambda}{dt} \text{ or } -\frac{d\psi}{dt} = -\frac{d}{dt}(N\Phi)$$

$$= -\frac{d}{dt}(Li) \qquad \because \; L = \frac{N\Phi}{i}$$

$$e_l(t) = -L\frac{di}{dt} \qquad \text{.......... (1.16.8)}$$

The voltage drop across the inductance,

$$v_l(t) = -e_l(t)$$

$$= L\frac{di}{dt} \text{ Volts} \qquad \text{.......... (1.16.8a)}$$

Hence, $$L = \frac{v_l}{di/dt} \text{ Henry}$$

or $$L = -\frac{e_l}{di/dt} \text{ Henry} \qquad \text{.......... (1.16.9)}$$

This is another way of calculating the Self Inductance.

Hence, we can also say that a coil or a circuit has an inductance of one henry (1 H) if an E.M.F. of one volt is induced in the coil or circuit when the current varies uniformly at the rate of one ampere (1 A) per second.

From Eq. (1.16.8a), the Volt-Ampere relationship of the inductance parameter is given as

$$v_l(t) = L\frac{di(t)}{dt}$$

or $$v_l = L\frac{di}{dt}$$

where v_l and i stand for instantaneous values or values at any particular instant of voltage and current respectively.

$$i = \frac{1}{L} \int v \, dt$$

$$= \frac{1}{L} \int_{-\infty}^{t} v \, dt$$

$$i = \Gamma \int_{-\infty}^{t} v \, dt \qquad \text{.......... (1.16.10)}$$

where Γ is the Greek Capital Letter Gamma $= \frac{1}{L}$

$$= \frac{1}{L} \int_{-\infty}^{0} v \, dt + \frac{1}{L} \int_{0}^{t} v \, dt$$

$$i = i_l(0^+) + \frac{1}{L} \int_{0}^{t} v \, dt \qquad \text{.......... (1.16.10a)}$$

1.16.4 Energy Stored in an Inductor or Coil or Inductance

The instantaneous power p(t) of a coil is

$$p(t) = v(t) \times i(t) \text{ Watts}$$

or $$p = v \times i \text{ Watts}$$

$$= L \frac{di}{dt} i$$

The instantaneous energy w(t) stored in a coil is given as

$$w(t) = p(t) \, dt$$

or $$w = p \, dt$$

$$w = L \frac{di}{dt} i \, dt$$

$$\therefore \quad w = L i \, di \qquad \text{.......... (1.16.11)}$$

The total energy absorbed by the magnetic field when the current increases from 0 to I Amperes is

$$W_L = L\int_0^I i\,di$$

$$= L\,\frac{1}{2}i^2\Big|_0^I$$

$$\therefore \quad W_L = \frac{1}{2}LI^2 \text{ Joules or Watt-sec.} \qquad \text{.......... (1.16.12)}$$

From Eq. (1.16.5), the Self Inductance,

$$L = \frac{N^2}{l/a\mu}$$

$$= N^2\mu\frac{a}{l}$$

for a homogeneous magnetic circuit of uniform cross-sectional area a and length *l*.

Therefore energy per cubic meter ω_t is

$$\omega_t = \frac{1}{2}\,N^2\mu\,\frac{a}{l}\,I^2 \text{ x } \left(\frac{1}{al}\right) = \frac{1}{2}N^2I^2\frac{\mu}{l^2}$$

Since, $\quad H = \dfrac{NI}{l}$,

$$\omega_t = \frac{1}{2}\mu H^2 \text{ Joules or Watt-sec.} \qquad \text{.......... (1.16.13)}$$

Since the magnetic flux density,

$$B = \mu H \text{ wb/m}^2$$

$$\therefore \quad \omega_t = \frac{1}{2}HB \quad \text{Joules or Watt-sec.}$$

$$\therefore \quad \omega_t = \frac{1}{2}\,\frac{B^2}{\mu_o\mu_r} \text{ Joules / cubic meter} \qquad \text{.......... (1.16.14)}$$

Example 1.23:

The iron core of a choke has mean length of 25cm with an air gap of 1mm. The choke is designed for an inductance of 15H when operating at a flux density of $1 Wb/m^2$. The iron core has a relative permeability of 3000 and $8 cm^2$ area of cross-section. Determine the required number of turns of the coil.

Solution:

From Eq. (1.16.5), Inductance of the coil,

$$L = \frac{N^2}{S_T}$$

where S_T is the total reluctance of the magnetic circuit.

S_T = Reluctance of Iron Core + Reluctance of Air Gap

$$S_T = \frac{l_i}{a\mu_0\mu_r} + \frac{l_g}{a\mu_0}$$

$$= \frac{0.25}{8\times10^{-4}\times4\pi\times10^{-7}\times3000} + \frac{1\times10^{-3}}{8\times10^{-4}\times4\pi\times10^{-7}}$$

$$= 1077612 \text{ Ampere-Turns/Wb}$$

Now,

$$N = \sqrt{LS_T}$$

$$= \sqrt{15\times1077612} = 4020.5 \text{ turns.}$$

Example 1.24:

A Solenoid of 1m in length and 10cm in diameter has 5000 turns. Calculate the energy in the magnetic field when a current of 2A flows in the Solenoid.

Solution:

Inductance of the Solenoid,

$$L = \frac{N^2}{l/\mu_o\mu_r a}$$

$$= \frac{N^2\mu_o\mu_r a}{l} = N^2\mu_o\mu_r a \qquad (\text{ Since } l = 1\text{m })$$

where $N = 5000,\qquad a = \pi\dfrac{d^2}{4} = 25\pi \times 10^{-4} m^2$

$l = 1m;\qquad \mu_r = 1,\qquad \mu_0 = 4\pi \times 10^{-7}$ Henry/m

$$L = (5000)^2 \times 25\pi \times 10^{-4} \times 4\pi \times 10^{-7} \times 1 = 0.2467H$$

Energy Stored $= \frac{1}{2}LI^2$

$$= \tfrac{1}{2} \times 0.2467 \times (2)^2 = 0.4934 \text{ Joules}$$

Example 1.25:

A coil of 2000 turns is wound on a toroidal magnetic core having reluctance of 5×10^5 Ampere-Turns/Wb. If the current in the coil is 8A and is increasing at the rate of 300A/sec. Determine

(i) *Energy stored in the magnetic circuit*
(ii) *Voltage applied across the coil.*

Repeat the above calculation if the coil resistance is 4Ω.

Solution:

From Eq. (2.3.5),

$$L = \frac{N^2}{S}$$

$$= \frac{(2000)^2}{5\times10^5} = 8 \text{ Henry}$$

(i) Energy stored $= \frac{1}{2}LI^2 = \frac{1}{2} \times 8 \times (8)^2$

$$= 256 \text{ Joules}$$

(ii) Voltage applied across the coil = Self Induced E.M.F. in the coil.

Hence, $V = L\dfrac{dI}{dt}$

$$= 8\times\frac{300}{1} = 2400 \text{ V}$$

If the coil resistance is 4Ω, there results additional loss of $(8)^2 \times 4 = 256J$ in the coil resistance. But the energy stored in the coil remains the same. However the Voltage across the coil will increase by $8A \times 4\Omega = 32V$ and will now becomes 2432V.

1.16.5 Inductances in Series or Series Inductors without Mutual Induction

Applying Kirchoff's Voltage Law,

$$V = V_1 + V_2 + V_3$$

$$= L_1 \frac{di}{dt} + L_2 \frac{di}{dt} + L_3 \frac{di}{dt}$$

$$V = (L_1 + L_2 + L_3) \frac{di}{dt} \qquad \text{.......... (1.16.15)}$$

For the Equivalent Series Inductance,

$$V = L_{Se} \frac{di}{dt} \qquad \text{.......... (1.16.16)}$$

Comparing Eq. (1.16.15) and Eq. (1.16.16), we get,

$$L_{Se} = L_1 + L_2 + L_3 \qquad \text{.......... (1.16.17)}$$

In general,

$$L_{Se} = \sum_{i=1}^{n} L_i \qquad \text{.......... (1.16.17a)}$$

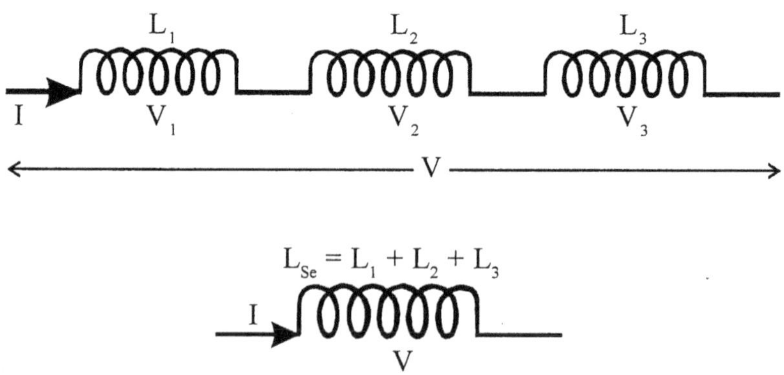

Fig. 1.26 Inductance in Series without Mutual Induction

Example 1.26:

A coil of 250 turns is wound on a magnetic circuit of reluctance 1,00,000 Ampere-Turns/Wb. If a current of 2A flowing in the coil is reversed in 5milli-sec, find the average E.M.F. induced in the coil.

Solution:

$$\phi = \frac{\text{M.M.F.}}{\text{Reluctance}} = \frac{NI}{S}$$

where, $N = 250, I = 2A, S = 1,00,000$ Ampere-Turns/Wb

$$\phi = \frac{250 \times 2}{100000} = 5 \text{ m-Wb}$$

Average Induced E.M.F.,

$$e = N\frac{d\phi}{dt}$$

where $d\phi = 5 - (-5) = 10$ m-Wb

$$e = \frac{250 \times 10 \times 10^{-3}}{5 \times 10^{-3}} = 500 \text{ V}$$

1.16.6 Inductances in Parallel or Parallel Inductors without Mutual Induction

Applying Kirchoff's Current Law,

$$i = i_1 + i_2 + i_3$$

$$= \frac{1}{L_1}\int v\,dt + \frac{1}{L_2}\int v\,dt + \frac{1}{L_3}\int v\,dt$$

$$i = \left(\frac{1}{L_1} + \frac{1}{L_2} + \frac{1}{L_3}\right)\int v\,dt \qquad \text{.......... (1.16.18)}$$

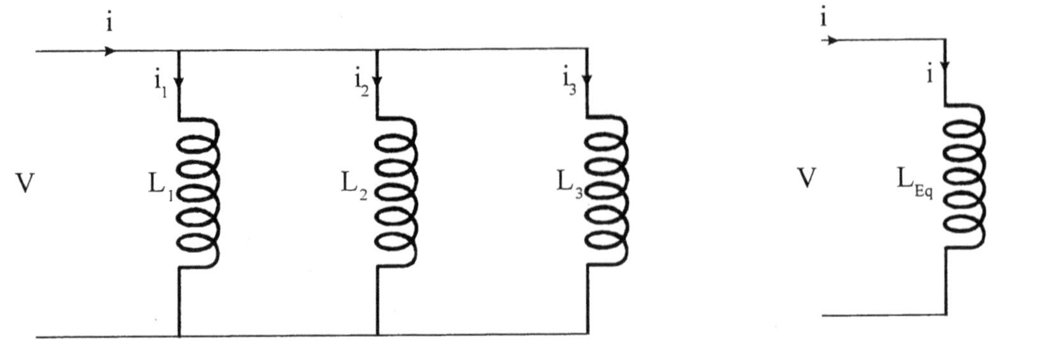

Fig. 1.27 Inductance in Parallel without Mutual Induction

For the Equivalent Parallel Inductance,

$$i = \frac{1}{L_{Eq}} \int v \, dt \qquad \text{.......... (1.16.19)}$$

Comparing Eq. (1.16.18) and Eq. (1.16.19), we get,

$$\frac{1}{L_{Eq}} = \frac{1}{L_1} + \frac{1}{L_2} + \frac{1}{L_3} \qquad \text{.......... (1.16.20)}$$

In general,

$$\frac{1}{L_{Eq}} = \sum_{i=1}^{n} \frac{1}{L_i} \qquad \text{.......... (1.16.20a)}$$

1.16.7 Inductances in Series-Parallel Without Mutual Induction

The equivalent inductances for the parallel combinations of inductances are found. These equivalent inductances are treated to be in series with the remaining series inductances. After solving the circuit, the current in the parallel branches are calculated from the total current. Fig. 1.28 gives two inductances in parallel and one inductance in series, non of them having any mutual induction.

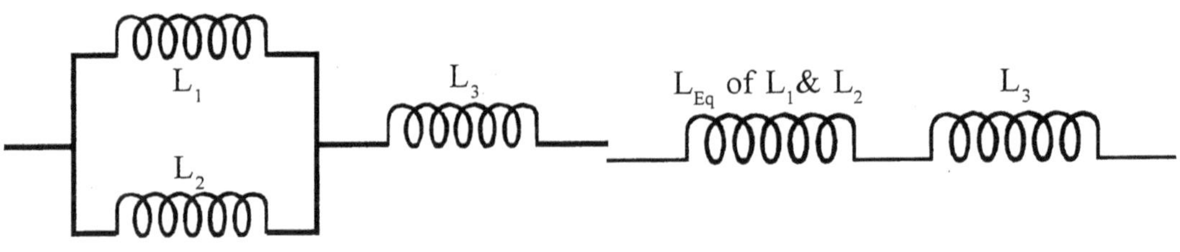

Fig. 1.28 Inductance in Series-Parallel without Mutual Induction

1.16.8 Mutual Induction Between Coils

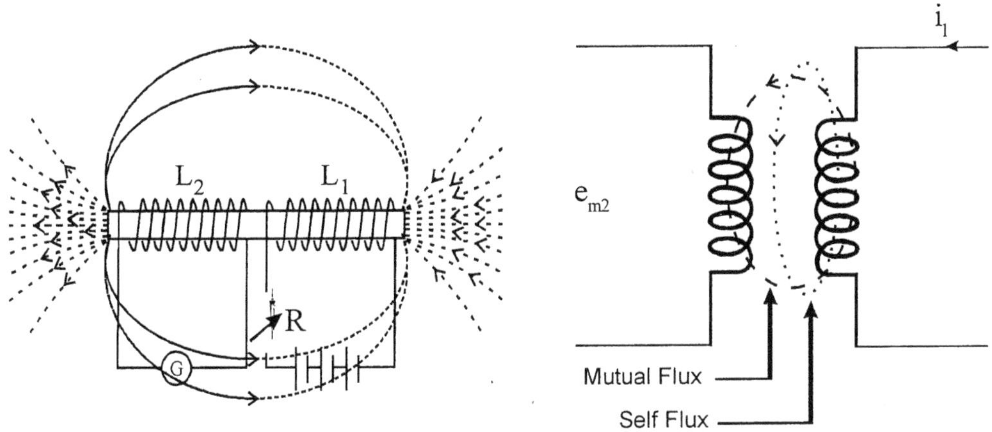

Fig. 1.29 Mutual Inductance

If the E.M.F. is induced in a coil due to the magnetic field setup by another nearby coil then the E.M.F. induced in that coil is called ***Mutual E.M.F.*** and the inductance induced in the first coil gets changed and this additional change in inductance is called ***Mutual Inductance***.

The direction of the induced E.M.F. depends upon the direction of the current in the other coil or flux setup by the other coil. If the mutual E.M.F. adds to the self induced E.M.F. of the first coil, then the mutual inductance adds to the self inductance of the first coil. If the induced E.M.F. opposes self induced E.M.F. of the first coil, then the mutual inductance reduces the self inductance of the first coil.

Mutual inductance may be defined as the ability of one coil (or circuit) to produce an E.M.F. in a nearby coil by induction when the current in the first coil changes. This action being reciprocal, the second coil can also induce an E.M.F. in the first when current in the second coil changes. This ability of reciprocal induction is measured in terms of the ***coefficient of mutual induction*** M.

If a current i_1 flowing through a coil 1 of turns N_1 produces a flux Φ_1 wb and if the flux Φ_1 links with coil 2 of turns N_2 then the mutual flux linking the coil 2 will be $N_2 \times \Phi_1$ and hence, the mutual inductance of coil 2 due to current in coil 1 is

$$M_{21} = \frac{N_2 \Phi_1}{i_1} \qquad \text{.......... (1.16.21)}$$

The first subscript of M indicates response and the second subscript indicates the excitation of source. M_{21} *denotes the mutual inductance of coil 2 due to current* i_1 *in coil 1.*

Similarly, if a current i_2 flowing through a coil 2 of turns N_2 produces a flux Φ_2 wb and if the flux Φ_2 links with coil 1 of turns N_1 then the mutual flux linking the coil 1 will be $N_1 \times \Phi_2$ and hence, the mutual inductance of coil 1 due to current in coil 2 is

$$M_{12} = \frac{N_1\Phi_2}{i_2} \qquad \text{.......... (1.16.22)}$$

In general,

$$M_{12} = M_{21} = M \qquad \text{.......... (1.16.23)}$$

where M is called the mutual inductance or coefficient of mutual inductance.

Alternatively,

The *mutual inductance* of a coil of N_1 turns is the flux linkages with that coil per unit current flowing through the neighboring coil of N_2 turns. Hence,

$$M_{12} = \frac{N_1\Phi_2}{i_2}$$

Since, $\Phi_2 = B_2 a = \mu_0\mu_r H_2 a = \mu_0\mu_r\left(\frac{N_2 i_2}{l}\right)a$

We have,

$$M_{12} = \frac{N_1 N_2}{l/\mu_0\mu_r\, a}$$

$$M_{12} = \frac{N_1 N_2}{S} \text{ henry} \qquad \text{.......... (1.16.24)}$$

Alternatively,

$$M = \frac{\mu_0\mu_r\, a\, N_1 N_2}{l} \text{ Henry} \qquad \text{.......... (1.16.25)}$$

The mutually induced E.M.F. e_{m2} in coil 2 due to current i_1 flowing in coil 1 is given as,

$$e_{m2} = L_{21} \times \frac{di_1}{dt}$$

$$= M_{21} \times \frac{di_1}{dt}$$

$$e_{m2} = M \times \frac{di_1}{dt} \qquad \text{.......... (1.16.26)}$$

where M_{12} is the mutual inductance between the coils 1 and 2 when i_1 flows through the coil 1.

The mutually induced E.M.F. e_{m1} in coil 1 due to current i_2 flowing in coil 2 is given as,

$$e_{m1} = L_{12} \times \frac{di_2}{dt}$$

$$= M_{12} \times \frac{di_2}{dt}$$

$$e_{m1} = M \times \frac{di_2}{dt} \qquad \text{.......... (1.16.27)}$$

The self induced E.M.F. e_{s2} in coil 2 due to current i_2 flowing in coil 2 is given as,

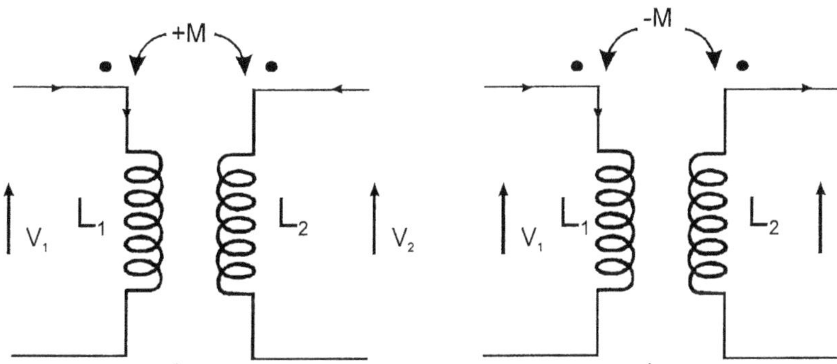

Fig 1.30 Mutual Inductances

$$e_{s2} = L_{22} \times \frac{di_2}{dt} \qquad \text{.......... (1.16.28)}$$

The self induced E.M.F. e_{s1} in coil 1 due to current i_1 flowing in coil 1 is given as,

$$e_{s1} = L_{11} \times \frac{di_1}{dt} \qquad \text{.......... (1.16.29)}$$

$$V_1 = e_{s1} \pm e_{m1}$$

$$V_1 = L_1 \frac{di_1}{dt} \pm M \frac{di_2}{dt} \qquad \text{.......... (1.16.30)}$$

$$V_2 = e_{s2} \pm e_{m2}$$

$$V_2 = L_2 \frac{di_2}{dt} \pm M \frac{di_1}{dt} \qquad \text{.......... (1.16.31)}$$

1.16.9 Dot Convention for Mutual Inductances

When two current carrying coils are located nearby, if the magnetic flux setup by the coil 1 which links coil 2 aids the magnetic flux setup by coil 2 then, the net flux in coil 2 will be increased and the total E.M.F. induced in coil 2 (sum of self and mutually induced E.M.F) will increase. Then the mutual inductance M is considered to be positive. Similarly, when two current carrying coils are located nearby, if the magnetic flux setup by the coil 2 which links coil 1 aids the magnetic flux setup by coil

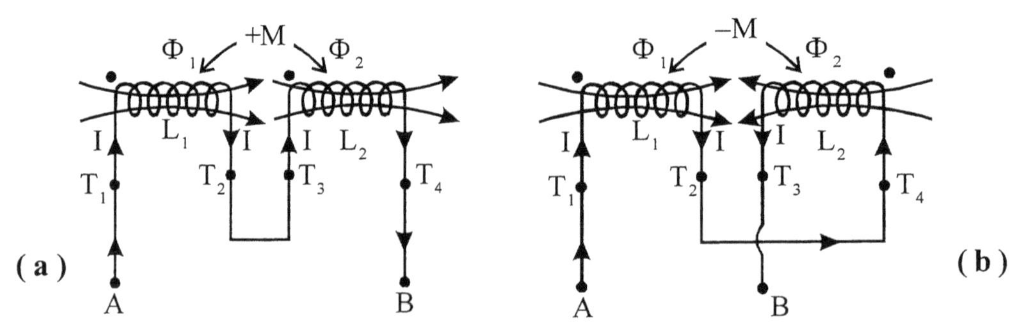

Fig. 1.31 Dot Convention of Mutual Inductance

1 then, the net flux in coil 1 will be increased and the total E.M.F. induced in coil 1 (sum of self and mutually induced E.M.F.) will increase. Then the mutual inductance M is considered to be positive.

When two current carrying coils are located nearby, if the magnetic flux setup by the coil 1 which links coil 2 opposes the magnetic flux setup by coil 2 then, the net flux in coil 2 will be decrease and the total E.M.F. induced in coil 2 (sum of self and mutually induced E.M.F). will decrease. Then the mutual inductance M is considered to be negative. Similarly, when two current carrying coils are located nearby, if the magnetic flux setup by the coil 2 which links coil 1 aids the magnetic flux setup by coil 1 then, the net flux in coil 1 will be decrease and the total E.M.F. induced in coil 1 (sum of self and mutually induced E.M.F.) will decrease. Then the mutual inductance M is considered to be negative.

This phenomenon is represented by dot convention as shown in Fig. 1.31.

Referring to Fig. 1.31 in dot convention, if the current in the two coils enter the terminals marked with a dot then it is assumed that the self flux of a coil and the mutual flux due to the other coil will aid and hence, the total E.M.F. induced in that coil will increase.

$$e_1 = L_1 \frac{di_1}{dt} + M_{12} \frac{di_2}{dt}$$

$$e_1 = L_1 \frac{di_1}{dt} + M \frac{di_2}{dt} \qquad \text{.......... (1.16.32)}$$

$$e_2 = L_2 \frac{di_2}{dt} + M_{21} \frac{di_1}{dt}$$

$$e_2 = L_2 \frac{di_2}{dt} + M \frac{di_1}{dt} \qquad \text{.......... (1.16.33)}$$

$$\because \quad M_{12} = M_{21} = M$$

Referring to Fig. 1.31, in dot convention, if the current in one coil enters the dotted terminal marked with a dot and the current in the other coil leaves the dotted terminal marked with a dot then it is assumed that the self flux of a coil and the mutual flux due to the other coil will be opposing and hence, the total E.M.F. induced in that coil will decrease.

$$e_1 = L_1 \frac{di_1}{dt} - M_{12} \frac{di_2}{dt}$$

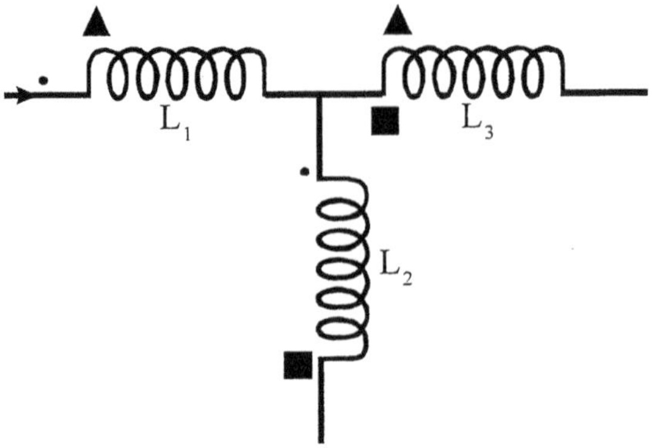

Fig. 1.32 Dot Convention of Mutual Inductance

$$e_1 = L_1 \frac{di_1}{dt} - M \frac{di_2}{dt} \quad .... (1.16.34)$$

$$e_2 = L_2 \frac{di_2}{dt} - M_{21} \frac{di_1}{dt}$$

$$e_2 = L_2 \frac{di_2}{dt} - M \frac{di_1}{dt} \quad .... (1.16.35)$$

$$\because \quad M_{12} = M_{21} = M$$

If there are more than two coils which have mutual inductance then, for a set of two coils with mutual inductance, dots may be indicated for their mutuals. For the second set of coils with mutuals to avoid confusion, instead of dots small triangles or small squares may be used to indicate the mutual inductance as given in Fig. 1.32.

In general, when two coils mutual coupling are connected in series, the total E.M.F. induced across the two coils which are in series with current i flowing through them will be,

$$e_t = L_1 \frac{di}{dt} + L_2 \frac{di}{dt} \pm 2M \frac{di}{dt} \quad .... (1.16.36)$$

The positive sign is used for 2M (+2M), if the current enters into both the coils and the negative sign is used for 2M (–2M), if the current enters into one coil and leaves away from the second coil as given by the dot convention.

In the Fig. 1.32.

Between L_1 & L_2, the Mutual Induction is Positive

Between L_2 & L_3, the Mutual Induction is Negative

Between L_3 & L_1, the Mutual Induction is Positive

1.16.10 Expression for Mutual Inductance M in Terms of Dimensions of Two Coils

The flux in the first coil (primary coil) is,

$$\Phi_1 = \frac{N_1 I_1}{l/\mu_o \mu_r a} \text{ Wb} \qquad \text{.......... (1.16.37)}$$

$$L_{11} = L_1 = \text{Flux/Ampere} = \frac{\Phi_1}{I_1}$$

$$L_{11} = L_1 = \frac{N_1}{l/\mu_o \mu_r a} \text{ Henry} \qquad \text{.......... (1.16.38)}$$

We assume that the whole of this flux F_1 links with the other coil having N_2 turns. Then the Wb-Turns in the second coil due to this flux/ampere in the first coil are:

$$M = \frac{N_2 \Phi_1}{I_1}$$

$$= \frac{N_1 N_2}{l/\mu_o \mu_r a} \text{ Henry}$$

Thus, $$M = \frac{N_1 N_2}{S} \text{ Henry}$$

where S is the reluctance and equals $l/\mu_o\mu_r a$.

Alternatively,

$$M = \frac{\mu_o \mu_r a N_1 N_2}{l} \text{ Henry} \qquad \text{.......... (1.16.39)}$$

1.16.11 Coefficient of Magnetic Coupling or Magnetic Coupling between two Coils

Normally only a part of the flux produced by one coil will be linking with the other coil due to leakage flux which will not be linking with the other coil. Two coils are said to be magnetically coupled if full or part of the flux produced by one links with the other.

Let L_1 and L_2 be the self inductances of the two coils, M their mutual inductance, and k be the coefficient of mutual coupling between the two coils. ϕ_1 and ϕ_2 are the flux setup by coils 1 and 2 respectively. N_1 and N_2 are the number of turns of coils 1 and 2 respectively. S is the reluctance of the magnetic path.

The mutual inductance of coil 2 due to flux setup by coil 1,

$$M_{21} = N_2 \frac{k\phi_1}{I_1}$$

$$= kN_2 \frac{N_1}{S}$$

Similarly, the mutual inductance of coil 1 due to flux setup by coil 2,

$$M_{12} = N_1 \frac{k\phi_2}{I_2}$$

$$= kN_1 \frac{N_2}{S}$$

$$M_{12} = M_{21} = M$$

Hence, $$M^2 = k^2 \frac{N_2^2}{S} \times \frac{N_1^2}{S}$$

$$= k^2 L_1 L_2$$

$$k^2 = \frac{M^2}{L_1 L_2}$$

Therefore,

$$k = \frac{M}{\sqrt{L_1 L_2}} \qquad \text{.......... (1.16.40)}$$

When all the flux produced by one coil links with the other, then mutual inductance between the two coils is maximum and with k = 1, M is given by

$$M = \sqrt{L_1 L_2} \qquad \text{.......... (1.16.41)}$$

When there is no common flux between the two coils, they are said to be magnetically isolated. Since, in that case, M = 0, k = 0 also.

Hence, *coefficient of coupling may be defined as the ratio of actual mutual inductance present between the two coils to the maximum possible value of the mutual inductance between the two coils.*

Since the A.C. current passing through a coil sets up an inductance in addition to the resistance, in order to have a pure resistance offered by the coil, the coil is split up into two parts and each part is wound in the opposite direction in which case, the magnetic flux of the two parts cancel each other and the inductance offered by the two parts of the coil will be come zero, resulting only in ***pure resistance***. A pure resistance is denoted by the symbol 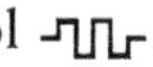.

Example 1.27:

A wooden ring has mean diameter of 150mm and a cross-sectional area of 250mm^2. It is wound with 1500 turns of insulated wire. A second coil of 900 turns in wound on the top of the first. Assuming that all flux produced by the first coil links with the second, calculate the Mutual Inductance and the coefficient of coupling.

Solution:

Mutual Inductance,

$$M = \frac{N_1 N_2}{l} a\mu_o\mu_r$$

where $N_1 = 1500, N_2 = 900, \ l = \pi D = 0.15\pi$–meter

$a = 250 \times 10^{-6} m^2, \quad \mu_r = 1$

$$M = \frac{1500 \times 900}{0.15\pi} \times 250 \times 10^{-6} \times 4\pi \times 10^{-7} \times 1 \quad = \quad 0.9 \text{ mH}$$

$$L_1 = \frac{N_1^2}{l} \mu_0 \mu_r a$$

$$= \frac{1500^2}{0.1 \times \pi} \times 250 \times 10^{-6} \times 4 \times \pi \times 10^{-7} \times 1$$

$$L_1 = 1.5 \text{ mH}$$

$$L_1 = \frac{N_2^2}{l} \mu_0 \mu_r a$$

$$= \frac{900^2}{0.1 \times \pi} \times 250 \times 10^{-6} \times 4 \times \pi \times 10^{-7} \times 1$$

$$= 0.54 \text{ mH}$$

Coefficient of Coupling,

$$k = \frac{M}{\sqrt{L_1 L_2}} = \frac{0.9}{\sqrt{1.5 \times 0.54}}$$

$$= \frac{0.9}{\sqrt{0.81}} = \frac{0.9}{0.9} = 1$$

Note: *Here there is perfect coupling between the two coils since one coil is wound fully over the other coil.*

Example 1.28:

Two coils have Mutual Inductance of 0.6H. If current in one coil is varied from 4A to 1A in 0.2sec, calculate

(i) *The average E.M.F. induced in the other coil*

(ii) *The change of flux linking the later assuming that it is wound with 150 turns.*

Solution:

Mutually Induced E.M.F.,

$$e_m = M \frac{dI_1}{dt}$$

where $M = 0.6H$, $dI_1 = 4 - 1 = 3A$, $dt = 0.2$ sec.

$$e_m = 0.6 \times \frac{3}{0.2}$$

$$= 9V$$

Now, $$e_m = N_2 \frac{d\Phi_{12}}{dt}$$

Change of flux with second coil,

$$d\Phi_{12} = \frac{e_m \times dt}{N_2}$$

$$= \frac{9 \times 0.2}{150} = 12\text{m-Wb}$$

1.16.12 Variable Inductance

It is known that the inductance of a coil depends upon the flux linkages linking with that coil. If the core material is a magnetic material of high permeability then, the flux linking with the coil will be maximum and hence, the inductance (self or mutual) will be maximum. For that purpose, an iron core is used instead of air core for the coil. As we draw the core inside into the coil or away from the coil by means of linear motion or screwed motion by means of a dial or wheel, the inductance of the coil will increase or decrease accordingly. Now the inductance consists of two parts which are equivalent two partial coils, one with air core and the other with magnetic core. By varying the length of the magnetic core, the inductance is varied.

1.16.13 Magnetic Hum

The movement of the magnetic core or the vibrations of the laminated parts of the core of the electrical machinery, if they are not bound tightly due to the forces on them setup by the magnetic field, a humming noise will be produced. To reduce the noise the magnetic core should be fixed tightly.

1.17 USES OF INDUCTORS

1. Inductors are used in the form of Solenoids for operating Gate Valves to control the fluid flow etc.,
2. The inductance coils are used in Electromagnetic relays for opening and closing the control circuits.
3. The principle of Mutual Induction is used in Transformers in which Electrical Energy at certain voltage in one circuit can be transferred magnetically to another circuit at a different voltage with out actual connection between the two circuits. Transformers are used in Electric Power Transmission and Distribution.
4. Extra High Voltage (EHV) Lines will have large capacitance due to its long length and will cause higher receiving and voltage than the sending end voltage of the transmission line. It will also cause High Voltage Oscillations when the line is switched off. To overcome this, Large Inductors are used between the transmission lines and the Earth to compensate for the line capacitance.
5. Inductors can also be used in tuning. But since the inductance will be heavy, instead of inductors, capacitors are used to tune and select particular frequency signals as in Radio and TV., etc. The coils used in Electronic Circuits (Ferrite Rods) for tuning will have resistance also.

6. Inductors can also be used in Filter Circuits to select or reject certain frequencies.

7. Inductors can also be used in Wave Shaping Circuits to obtain a desired wave form from a given waveform

8. Because of its inertia, Inductors can be used in analog circuits, to solve for other systems like Mechanical Systems, Hydraulic System, etc., in terms of Electrical Systems

9. In ***Transducers***, which convert physical signals like displacement, etc., into electrical signals as in the case of Linear Variable Differential Transformer (LVDT).

10. Inductors can also be used in Power Factor Changing Circuits

11. Inductance can be used in Measurement Circuits.

* * * * * *

COMPREHENSION - 1

Note: By answering all the questions given in this comprehension one can answer any objective type of questions easily whatever be the form of the question. Answers for these questions can be obtained by going through the topics in this chapter carefully.

1. Define the terms
   (a) Electric Field Intensity
   (b) Electric Flux Density
   (c) Potential
   (d) Potential Difference
   and mention their units.
2. Define the term Current and mention its unit.
3. What is Steady State Current?
4. What is the relationship between the Current and the Charge?
5. Define the unit of Current.
6. What is meant by Resistance Parameter? What are its Units?
7. How are Mega-Ohms, Kilo-Ohms, Milli-Ohms and Micro-Ohms denoted?
8. Explain the Color Code for Resistors used in Electronic Circuits.
9. What are the factors that affect a Resistance?
10. Given an expression for the Resistance?
11. Explain the different ways of connecting a Rheostat.
12. What will be the output in the case of a Potential Divider connection?
13. What is meant by Temperature Coefficient of Resistance?
14. Discuss the effect of Temperature on Resistance. Mention materials having negative temperature coefficients. What is meant by Negative Temperature Coefficient?
15. What is meant by Electrical Conductance? What are its Units?
16. Explain the differences between Ideal and Practical Sources of Voltage and Current.
17. What should be the nature of the Internal Resistance for
   (a) Voltage Source?
   (b) Current Source?

18. Distinguish between D.C. and A.C. Sources giving their waveforms with respect to time.
19. State and Explain Ohm's Law.
20. Explain the difference between Potential Rise and Potential Drop.
21. What is the difference between Linear and Non-Linear Resistances? Draw their *v-i* Characteristics. Give examples.
22. What is meant by Electrical Power? Give different forms of expression for Electrical Power. Give its Units.
23. Give expressions for Instantaneous Value of Power Loss and Power Supplied.
24. State and Explain Kirchoff's Voltage Law (KVL).
25. State and Explain Kirchoff's Current Law (KCL).
26. Give KVL and KCL in Matrix Form.
27. Explain the different methods of connecting Resistances.
28. Derive the Equivalent Series Resistance for a number of Resistances which are connected in Series.
29. Derive the Equivalent Parallel Resistance for a number of Resistances which are connected in Parallel.
30. Explain the nature of the Division of Currents in Parallel Circuits.
31. What are the Salient Features of Series and Parallel Connections of Resistances?
32. How do you reduce Series-Parallel Resistances into a Single Equivalent Resistance?
33. How are the Voltages and Currents measured in a circuit?
34. What should be the Resistance of a Voltmeter and an Ammeter? Why? How should they be connected in a circuit?
35. In D.C. circuits, What should be the nature & Polarities of the Voltmeter and Ammeter while connecting?
36. Explain Star and Delta Connection of Resistances.
37. Give the connection diagram for Mesh or Star connection using three and more than three resistances. How are their equivalent resistances obtained?
38. Give expressions for the Equivalent Star Resistances for a given Delta connected Resistances.

39. Give expressions for the Equivalent Delta Resistances for a given Star connected Resistances.
40. Explain the features of resistances.
41. Explain the uses of resistances.
42. Explain the differences between Electrical and Electronic Circuits.
43. What is meant by Energy Storing Elements? Name them.
44. What is the salient difference between the Resistance and Capacitance or Inductance Parameters?
45. Compare the Resistance Parameter and Capacitance or Inductance Parameter.
46. What is a Capacitance? What is its other name? What is its Unit?
47. Derive an expression for the value of a Parallel Plate Capacitance.
48. Obtain the relationship between the charge Q, the voltage V and the capacitance C.
49. Can a Capacitor pass DC current through it? Why?
50. What are the expressions for the current flowing through a Capacitance and the voltage?
51. Give an expression for the current through a capacitance if both voltage across the capacitance and the value of the capacitance change with respect to time.
52. Obtain the expression for the voltage across a capcitance including initial voltage.
53. What is meant by Elastance? What is its Unit?
54. Obtain different forms of expressions for the energy stored in a capacitance. What will be its nature?
55. Why is it that the voltage across a capacitance does not change suddenly?
56. What is the Electric Stress?
57. Why should a capacitor be discharged prior to its use?
58. Define a Farad.

59. What will be the nature of the capacitance of Electrical Transmission Lines?

60. What is a Guard Ring? Why should a Guard Ring be provided for a parallel plate capacitor?

61. Obtain an expression for the Effective Capacitance for capacitances connected in Series.

62. If a capacitance consists of a multiple dielectric media placed between the plates, what will be nature of the capacitance?

63. Obtain an expression for the energy stored in a capacitance.

64. Obtain an expression for the equivalent capacitance when the capacitors are connected in parallel.

65. Obtain an expression for the total capacitance in the case of series-parallel connection of capacitors.

66. Obtain an expression for the force on a dielectric slab placed between the plates of a capacitor.

67. What is a Gang Condensor? Where is it used?

68. Give a brief note on Dielectric Strength and Breakdown.

69. What is meant by electrostatic induction?

70. What is the principle of lightning arrestor used to prevent damage due to lightning stroke?

71. Enumerate the uses of a Capacitance.

72. What is meant by an Inductance? What is its Units?

73. What is the nature of the energy stored in an inductor?

74. What is the average power in an inductance?

75. State and Explain Maxwell's Right Hand Cork-Screw Rule. When is it used?

76. Draw the magnetic field lines in the case of a current carrying conductor with current flowing into the plane of the paper, with current flowing away from the plane of the paper indicating both the direction of current and the magnetic flux.

77. Draw the magnetic fields set up by a coil or a solenoid.
78. State and Explain Right Hand Grip Rule.
79. What is meant by Self Induced EMF in a coil?
80. Give expressions for the voltage across an inductance and the current through the inductance.
81. Obtain an expression for the self inductance of a coil.
82. Give an expression for the magnetisation force H.
83. Derive different forms of expressions for the self inductance of a coil.
84. Obtain different forms of expression for the energy stored in an inductance in terms of the current or the flux density B or magnetisation force H or both B and H.
85. Obtain an expression for the total inductance when two or more inductances are connected in series.
86. Obtain an expression for the total inductance when two or more inductances are connected in parallel.
87. Obtain an expression for the total inductance when many inductances are connected in series-parallel.
88. What is meant by Mutual Induction and Coefficient of Mutual Induction?
89. Explain the dot conventions for Mutual Inductance.
90. Obtain expressions for mutual inductance between two coils - M_{12} and M_{21}.
91. Obtain an expression for the total inductance when two inductances with mutual induction are connected in series.
92. Obtain an expression for the total inductance when two inductances with mutual induction are connected in parallel.
93. Explain the dot convention when more than two coils with mutual induction are present.
94. Obtain an expression for the total inductance when many inductances with mutual induction are connected in series-parallel.
95. Obtain an expression for mutual inductance in terms of the dimensions of two coils.

96. Derive an expression for the Coefficient of Magnetic Coupling between two coils.
97. State the Principle of Constant Flux Linkages.
98. Why is it that the current through an inductance does not change suddenly?
99. What is a Variable Inductance? Where is it used?
100. What is meant by Magnetic Hum? How do you prevent it?

* * * * * *

EXERCISE - 1

Note: By answering all the questions given in this exercise one can answer subjective type of questions easily including problems. The worked examples in the chapter may help in solving these questions.

1. (a) Distinguish between potential difference between two points and the potential gradient.

   (b) Find the potential difference and the potential gradient between two points A and B when the potential of point A is 30V and the potential of point B is –20V if they are separated by a distance of 2mts.

   [Ans: V_{AB} = 50Volts; E = 25V/mts]

2. (a) What is meant by electric current? What is its units?

   (b) If a charge of 25C passes a given point in a circuit in a time of 125ms, determine the current in the circuit.

   [Ans: I = 200A]

3. (a) What is meant by resistance? Give an expression for resistance in terms of its dimensions.

   (b) An aluminium wire 400mts long has a resistance of 0.25Ω. Find its area of cross-section. Find the area of cross-section required if the wire is of copper (specific resistances of copper and aluminium are 1.73 ′ 10^{-8} and 2.83 ′ 10^{-8}Ω-m)

   [Ans: a = 0.4528Sq-cm; 0.2768Sq-cm]

4. (a) What are the factors that effect a resistance and How?

   (b) Give the color code for the resistors used in electronic circuits.

   (c) A conductor of 10mts long and with 2mm^2 cross-sectional area has resistance of 0.4Ω. Calculate the specific resistance of the conductor material. What is its conductance?

   [Ans: $\rho = 0.8 \times 10^{-7}$ Ω-m; G = 2.5Siemens]

5. (a) What is the effect of temperature on resistance?

   (b) Define temperature coefficient of resistance.

   (c) The resistance of the armature of a D.C. generator is 0.92Ω at 25°C. After working for 6Hrs on full load, the resistance increases to 1Ω. Calculate the temperature rise of the armature, if the temperature coefficient of resistance is 0.0040 at 0°C.

   [Ans: 23.9°C]

6. (a) How does the temperature coefficient of a resistance vary with respect to temperature?
   (b) What is meant by Negative Temperature Coefficient of Resistance? Mention some materials having Negative Temperature Coefficient of Resistance.
   (c) A coil has resistance of 20Ω when its mean temperature is 30°C and 24Ω when its mean temperature is 50°C. Find the temperature coefficient of resistance at 0°C. Also find the mean temperature rise when its resistance is 26Ω and the surrounding temperature is 20°C.

[Ans: α_0 = 0.0143/°C; 40°C]

7. (a) What is meant by EMF of a source?
   (b) Distinguish between ideal and practical voltage sources drawing the V-I characteristics.
   (c) Distinguish between ideal and practical current sources drawing the V-I characteristics.
   (d) The internal resistance of a 12V battery is 0.9Ω. What will be its terminal voltage when the current drawn from the battery is 2A?

[Ans: V = 10.2Volts]

8. (a) Distinguish between DC and AC sources of voltage and current drawing the waveforms with respect to time.
   (b) What is meant by Instantaneous Voltage and Instantaneous Current? How are they represented?
   (c) What is meant by Polarity of a source? When do you generally use it? Can you use the polarities for AC sources?

9. (a) State and explain Ohm's Law.
   (b) State Ohm's Law in Point Form.
   (c) A lamp takes 2.5A at 230V when switched on at room temperature (25°C). At the normal working temperature the current drops to 0.2A. What is the cold resistance and hot resistance of the filament at the operating temperature?

[Ans: 92Ω;1150Ω]

10. (a) What is meant by electrical power and electrical energy?
    (b) Twenty lamps of 60W are used each for 4Hrs in a building. Calculate the current drawn when all the lamps are working and the monthly electricity charge at Rs. 1.20 per unit. Assume supply at 240V.

[Ans: 0.25A; Rs. 172.8]

11. A factory has a 220V supply and has the following loads:

Lighting:	Five Hundred 100W and Five Hundred 60W lamps: 5Hrs per day.
Heating:	120KW: 10Hrs per day.
Motors:	48KW with average efficiency of 80%: 4Hrs per day.
Others:	15KW: 4Hrs per day

Assume 6 day week, calculate the daily and weekly consumption of the factory. Calculate the current drawn with only the lighting load, lighting and heating.

[Ans: 1900KWH; 11400KWH; 363.6A; 909A]

12. (a) Explain Kirchoff's Voltage Law (giving an example of a circuit with atleast two sources of different polarities and two or three resistances in series)

(b) Explain Kirchoff's Current Law (for a junction with two or three source currents of opposite directions and two or three load currents)

(c) Using Kirchoff's Laws, find the magnitude and direction of current flowing through the 15Ω resistance in the given circuit.

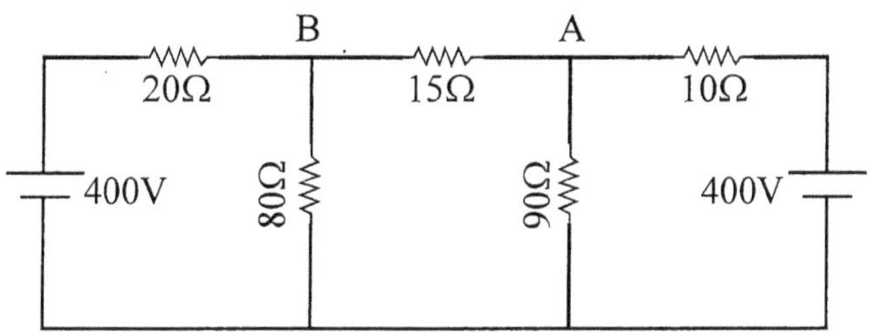

[Ans: 1A from A to B]

13. (a) Explain how the total resistance of a circuit having more than one resistance in series is calculated.

(b) A 60W, 240V lamp is connected in series with a 40W, 200V lamp across 250V supply. Calculate the current taken, voltage across each lamp and power given by each lamp. Assume that the resistance of the lamps remain constant.

[Ans: 0.12755A; 127.45V; 15.62W, 16.27W]

(c) A voltmeter has a resistance of 20,000Ω. When connected in series with an external resistance across a 230V supply, the instrument reads 160V. What is the value of external resistance?

[Ans: R = 8750Ω]

14. (a) Explain how the parallel equivalent resistance of a circuit having more than one resistance in parallel is calculated. Also explain in terms of conductances.

    (b) A resistor R is connected in series with a parallel circuit comprising of two resistances 12 and 8Ω respectively. The total power dissipated in the circuit is 96W when the applied voltage is 24V. Draw the circuit diagram and calculate the value of R.

    [Ans: R = 1.2Ω]

15. (a) Explain the division of current in the parallel branches.

    (b) A circuit consists of three resistances of 12Ω, 18Ω and 36Ω respectively joined in parallel, and the parallel combination is connected in series with a fourth resistance R_4. The whole circuit is supplied at 60V and it is found that power dissipated in 12Ω resistance is 36W. Draw the circuit diagram and determine the value of fourth resistance R_4 and the total power dissipated in the circuit.

    [Ans: R_4 = 11.32Ω; 207.84W]

16. Three loads A, B and C are connected in parallel to a 250V source. Load A takes 50A. Load B is a resistance of 10W and load C takes 6.25KW. Calculate

    (i) R_A and R_C (ii) I_B and I_C

    (iii) Power in Loads A and B (iv) Total Current

    (v) Total Power

    (vi) Total Effective Resistance

    [Ans: R_A = 5Ω; R_C = 10Ω; I_B = I_C = 25A;

    P = 25KW; R_{Eq} = 2.5Ω]

17. (a) How do you reduce the circuit and obtain the total resistance in the case of a circuit with series and parallel resistances?

    (b) A 10Ω resistor is connected in series with a group of two resistances of 15Ω each connected in parallel. Draw the circuit diagram and find out what resistance R must be shunted across this parallel combination so that the total current taken from 20V supply is 1.5A.

    [Ans: R = 6Ω]

18. Determine the effective resistance between terminals A and B in the circuit of the given figure. If the current drawn at A is 9A, find the current in and the voltage drop across each element.

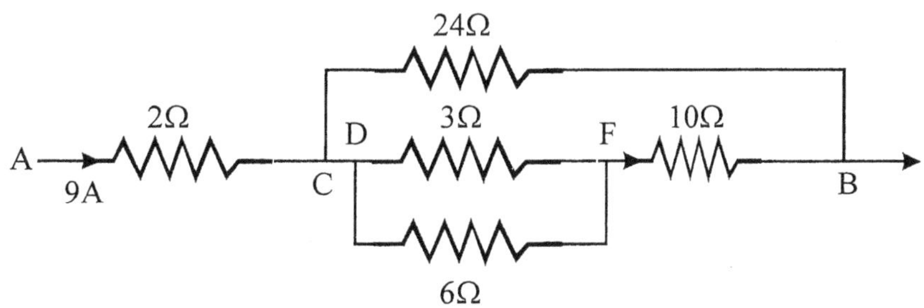

[Ans: 9A, 18V; 3A, 72V; 4A, 12V; 2A, 12V; 6A, 60V]

19. (a) Explain the Potential Divider connection of a Rheostat.

(b) A direct voltage of 240V is applied to a uniform 200Ω resistor connected between A and C. A load 40Ω resistance is connected between the terminal A and the tapping B (between the terminals A and C) Draw the circuit diagram and find the resistance AB in order that 1A flows in the 40Ω resistor.

[Ans: R = 71.65Ω]

20. (a) Explain how three resistances can be connected in Star and Delta.

(b) How do you extend Star-Mesh connection to more than three resistances?

(c) Derive expressions for equivalent star resistances for given delta and equivalent delta resistances for given star.

(d) How do you extend the finding out the equivalent resistances in the case of Star-Mesh connections having more than three resistances?

21. (a) Enumerate the features and uses of resistance.

(b) Enumerate the differences between Electrical Circuits and Electronic Circuits.

22. In the given figure, 160V are applied to the terminals AB. Determine the resistance between the terminals A & B and the current.

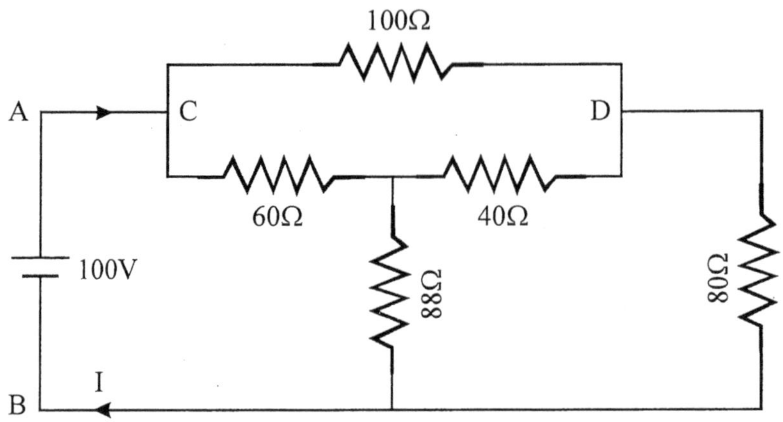

[Ans: R_{AB} = 80Ω; I = 2A]

23. (a) What is a capacitor?

(b) A capacitor having a capacitance of 80μF is connected across a 500V DC supply. Calculate the charge.

[Ans: 40mC]

(c) A capacitor consists of two similar square aluminium plates, each 10cm × 10cm mounted parallel and opposite each other. What is their capacitance in μμF (pF or 10^{-12} Farads) when the distance between them is 1cm and the dielectric air? If the capacitor is given a charge of 500 μμC, What will be the difference of potential between the plates? How will be the difference of potential between plates? How will this be affected if the space between the plates is filled with wax which has a relative permitivity of 4?

[Ans: 8.854pF; 56.5V; 14.1V]

24. (a) A potential difference of 400V is maintained across a capacitor of value 25μF. Calculate

(i) the Charge

(ii) the Electric Field Strength

(iii) the Electric Flux Density

in the dielectric, if the distance between plates of capacitor is 0.5mm and area of the plates is 1.2m².

Given $\varepsilon_0 = 8.854 \times 10^{-12}$ F/m.

[Ans: (i) 0.01C; (ii) 0.8 × 10^6V/m; (iii) 0.0083C/m²]

(b) Capacitance of a capacitor formed by two parallel metal plates each $100cm^2$ in area seperated by a dielectric 5mm thick is 20pF. A potential difference of 200V is applied. Calculate

(i) Total Charge on the plates

(ii) Potential Gradient in the dielectric medium

(iii) Electric Flux Density

(iv) Relative Permitivity of the dielectric.

[Ans: (i) 0.4μC; (ii) 4×10^4V/m; (iii) 4×10^{-6}C/m^2; (iv) 11.29]

25. (a) Derive an expression for the energy stored in a capacitance in terms of Capacitance and Voltage, Charge and Capacitance, Electric Field Intensity and Electric Flux Density

(b) Calculate the capacitance and energy stored in a parallel plate capacitor which consists of two metal plates each $60cm^2$ seperated by a dielectric of 1.5mm thickness and of $e_r = 3.5$ if a potential difference of 1000V is applied across it.

[Ans: 124pF; 62μJ]

26. (a) Find an expression for the effective capacitance when two or more capacitances are connected in series.

(b) In a parallel plate capacitor, the plates are spaced 6mm apart in air. Now two slabs of materials having dielectric constants 3 and 6 and thickness 2 and 4mm respectively are inserted between the plates. By what distance should one plate be further moved so as to restore the capacitance of the capacitor to its original value.

[Ans: 4.666mm]

27. (a) If two capacitors having capacitances of 6 and 10μF respectively are connected in series across a 200V supply. Find

(i) The Potential Difference across each capacitor

(ii) The charge on each capacitor

[Ans: (i) 125V, 75V; (ii) 750μC]

(b) Each plate in a parallel plate capacitor has area of $15cm^2$ and are seperated by 4mm thck sheet of mica of relative permitivity $\varepsilon_r = 6$. Calculate

(i) the capacitance of this capacitor

(ii) Next one plate is moved away permitting on air gap of 1mm between it and the mica sheet. Calculate the value of this capacitor.

(iii) If potential difference of 10V is maintaine across the plates, find the charge on the plates in two cases.

[Ans: (i) 19.92pF; (ii) 7.97pF; (iii) 79.7pC, 79.7pC]

28. (a) Deduce an expression for the equivalent capacitance when the capacitances are connected in parallel.

(b) Three capacitors have capacitances of 2, 4 and 8μF respectively are connected in

(i) Parallel

(ii) Series

Find the total capacitance in each case.

[Ans: (i) 14μF; (ii) 1.413μF]

29. Determine the capacitance of a capacitor consisting of two parallel metal plates 30cm × 30cm surface area, seperated by 5mm in air. What is the total energy stored by the capacitor if the capacitor is charged to a Potential Difference of 500V? What is the Energy Density?

[Ans: 159.2pF; 19.90 × 10^{-6}J; 4.427 × 10^{-6}J/m^3]

30. Deduce an expression for the joint capacitance of two capaictors C_1 and C_2

(i) in series

(ii) in parallel

If C_1 = 100μF, C_2 = 50μF, Calculate the joint capacitance and the total energy stored with a steady applied potential difference of 1000V in each case.

[Ans: C = 33.3μF; 150μF; 16.66J; 75J]

31. (a) Derive an expression for the self inductance of a coil in terms of its dimensions.

(b) A solenoid consists of 1500 turns of wire wound on a length of 60cm. A search coil of 500 turns enclosing a mean area of 20cm^2 is placed centrally in the solenoid. Find the Mutual Inductance.

[Ans: 3.14mH]

32. (a) Derive an expression for the energy stored in an inductance.

(b) A magnetic field is produced by a coil of 300 turns which is wound on a closed iron ring. The ring has a cross-section of 20cm^2 and mean length of 120cm. The permeability of the iron is 800. If the current in the coil is 10A. Find the energy stored in the magnetic field.

[Ans: 7.55J]

33. (a) Obtain an expression for the voltage drop across an inductance.

(b) An air cored toroidal coil has 450 turns and a mean diameter of 30cm and a cross-sectional area of 5cm^2. Calculate the inductance of the coil and the average induced EMF if a current of 4A is reversed in 60 milli-seconds.

[Ans: 135μH; 18mV]

34. (a) If the current through a coil having an inductance of 0.5H is reduced from 5A to 2A in 0.05sec, calculate the mean value of the EMF induced in the coil

[Ans: –30V]

(b) A coil of 300 turns, wound on a core of non-magnetic material, has an inductance of 10mH. Calculate

(i) the flux produced by current of 5A

(ii) the average value of the EMF induced when a current of 5A is reversed in 8milli-sec.

[Ans: (i) 167μWb; (ii) 12.5V]

35. An iron rod 2cm in diameter and 20cm long is bent into a closed ring and is wound with 3000 turns of wire. It is found that when a current of 0.5A is passed through this coil, the flux density in the coil is 0.5Wb/m^2. Assuming that all the flux links with every turn of the coil, what is the B-H ratio for the iron and what is the inductance of the coil. What voltage would be developed across the coil if the current through the coil is interrupted and the flux in the iron falls to 10% of its former value in 0.001second.

[Ans: 6.67×10^{-5}H/m; 0.94H; 424V]

36. (a) Define Mutual Inductance.

(b) Explain dot convention for mutual inductance between two coils and also for three coils.

(c) Derive an expression for coefficient of mutual coupling.

(d) Two identical coils X and Y of 1000 turns each lie in parallel planes such that 80% of flux produced by one coil links with the other. If a current of 5A flowing in X produces a flux of 0.05mWb in it, find the mutual inductance between X and Y.

[Ans: 8×10^{-3}H]

37. (a) Obtain expressions for the effective inductance of two mutually coupled inductances connected in

(i) series (ii) parallel

(b) When two coils are connected in series, their effective inductance is found to 10.0H. However, when the connections to one coil are reversed, the effective inductance is 6.0H. If the coefficient of coupling is 0.6, calculate the self-inductance of each coil and the mutual inductance.

[Ans: L_1 = 7.63H or 0.37H; L_2 = 0.37H or 7.63H; M = 1.0H]

38. Two coils A of 5000 turns and B of 3000 turns lie in parallel planes. A current of 6A in coil A produces a flux of 0.1mWb. If 60% of the flux produced by coil A links with the turns of coil B, calculate the EMF induced in coil B when the current in coil A changes from 5A to –5A in 0.01second.

[Ans: 30V]

* * * * *

2

D.C. CIRCUIT ANALYSIS

2.1 INTRODUCTION

In the first charpter we have studied D.C. circuit analysis for circuits with using combination of resistances in series, parallel, series parallel combination. We have also studied the simplification of D.C. circuit with resistances which are star connected or delta connected and also conversion of star connected resistances into delta connected resistances and vice-versa.

In this chapter we will be studying the analysis of D.C. circuits using mesh or loop analysis and nodal analysis. Mesh and nodal analysis provide an elegant way of finding out the currents through, voltage acrosss and power consumed in different components of resistances and also different parts of the circuits. **Mesh analysis is also known as loop analysis.** A mesh is a part of the circuit which forms a closed part for currents to flow through. A mesh is a sub-set of a circuit as the circuit may contain more than one mesh. Nodal analysis provide ways of finding out the voltages at different nodes or junctions of the circuit from which the currents, voltage and power in any resistance element of the circuit.

Before studying the loop and nodal analysis we will go in for a study of conversion of energy sources i.e., voltage and current sources and also shifting of these sources from one point in the circuit to another point in the circuit so as to make the circuit analysis easier. We also study loop and nodal analysis with independent and dependent sources.

2.1.1 Controlled or Dependent Voltage and Current Sources

Earlier we have studied use of independen voltage and current sources. The ideal and practical voltage & current sources are called ***independent sources*** because the vlaue of the source quantity is not affected in any way by activities in the remainder of the circuit. This is in contrast with yet another kind of ideal source, the dependent or controlled source in which the source quantity is determined by a voltage or current existing at some other location in the electrical system under examination. To distinguish between independent and dependent source, we introduce the diamond symbols shown in the Fig. 2.1 (a).

Sources such asthese will appear in the equivalent electrical models for many electronic devices, such as transistors, operational amplifiers and integrated circuits. Dependent and independent voltage and current sources are active elements, they are capable of delivering power to some external device.

2.1.2 Controlled Source or Dependent Source

In a controlled source, the source voltage or current (depending on the type of source is not fixed, but is dependent on a voltage or current at some other location in the network.

2.1.3 Controlled Voltage Source

The *Controlled Voltage Source*, for example, may be visualized as a voltage source for which the magnitude -of the voltage of the source is determined by the setting of a dial which, in turn, is controlled by some measured votage or current. One kind of controlled source is shown in Fig. 2.1 (a) and (b).

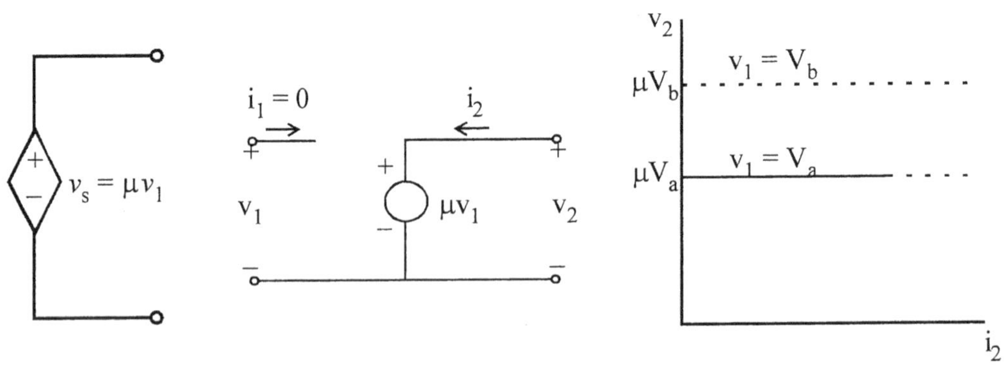

(a) Symbol (b) Circuit (c) Characteristics

Fig. 2.1 *The voltage-controlled voltage source for which v_2 depends on v_1 but not on i_2*

Here the control variable is the voltage v_1 which controls the source voltage μv_1, where μ is a constant. For such a controlled source, the voltage-current characteristics become a set of curves, as shown in Fig. 2.1 (c).

The controlled source of Fig. 2.1 is a three terminal model. The three terminals are paired, with one common terminal, and one pair is described as the input, the other pair as the output. Observe also that the controlled source is unilateral in the sense that the input variable v_1 in Fig. 2.1 (b) controls the output v_2, but that conditions at the output, such as the magnitude of the current i_2 have no influence on the input.

The equivalent circuit for controlled voltage source given in Fig. 2.1 (b) can be used to represent a vaccum tude, a voltage amplifier, etc., which is a voltage controlled device with its output voltage controlled by the input voltage.

2.1.4 Controlled Current Source

The controlled source of Fig. 2.1 is only one of several kinds of such models, depending on wheter the control variable is voltage or current, and whether the source which is controlled is a voltage source or a current source.

Another combination is shown in Fig. 2.2 (a) for which the control variable is the current i_1 and the controlled source is a current source. An equivalent and simplified representation of this model is shown in Fig. 2.2 (b).

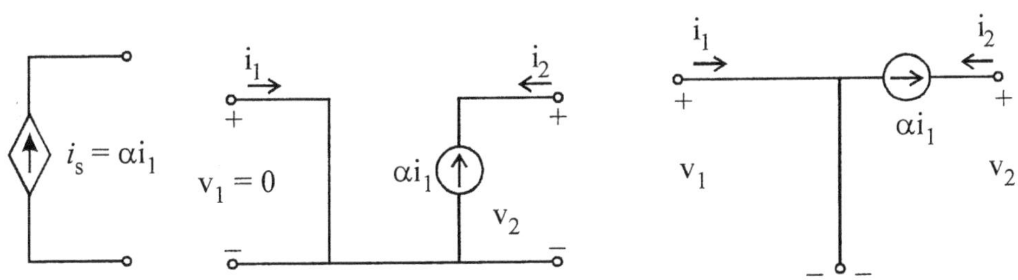

(a) Symbol (b) Equivalent Circuit (c) Characteristics

Fig. 2.2 The current-controlled current source with two equivalent identical representations

Our need for the controlled source model arises from applications in electronic systems that use either the transistor or vaccum tube. The equivalent circuit given in Fig. 2.2(b) and Fig. 2.2 (c) can be used to represent a transistor which is a current controlled device with its output current controlled by the input current.

2.1.5 Combination of Energy Sources

The Voltage Sources that are in Series can be combined into a single equivalent Voltage Source as shown below.

2.1.6 Voltage Sources in Series

Refer Fig. 2.3 (a), where 3 voltage sources are in series between the terminals A and B. Its equivalent single source is shown in Fig. 2.3 (b), where,

$$E = E_1 + E_2 - E_3$$

and

$$R = R_1 + R_2 + R_3$$

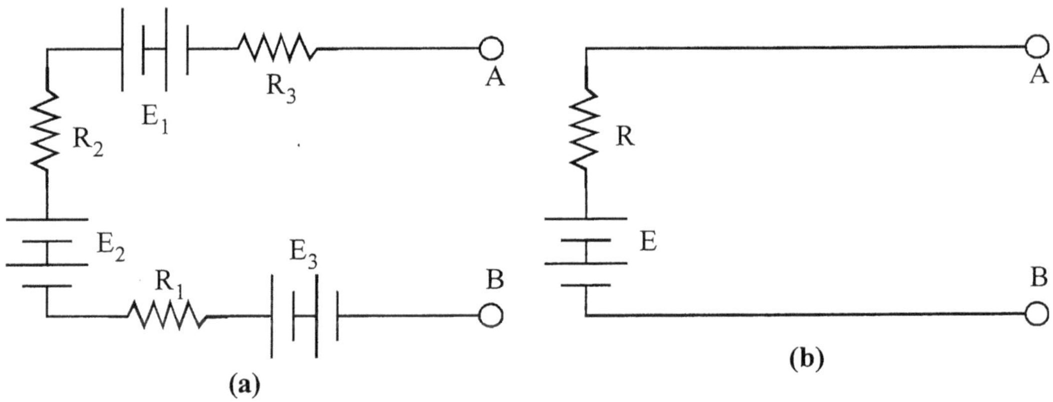

Fig. 2.3 Equivalent Voltage Source for Three Voltage Sources in Series

2.1.7 Current Sources in Parallel

The Current Sources that are in parallel can be combined into a single equivalent Current Source as shown below in Figures 2.4 (a) and (b)

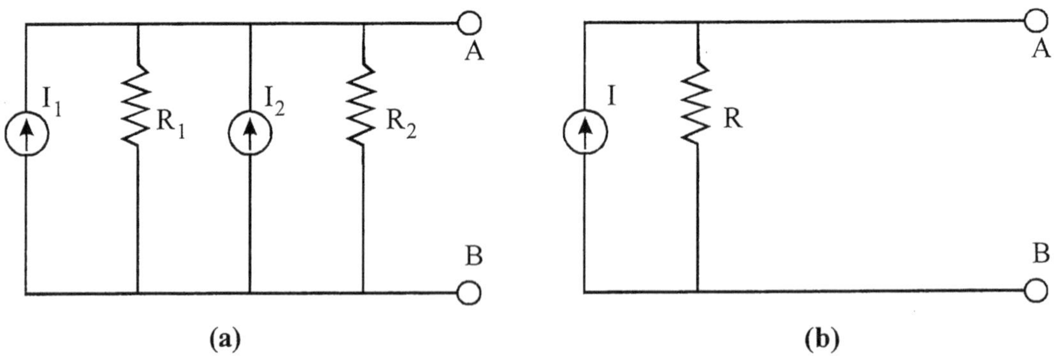

Fig. 2.4 Equivalent Current Source for Two Current Sources in Parallel

$$I = I_1 + I_2$$

$$R = \frac{R_1 R_2}{(R_1 + R_2)}$$

where R_1 and R_2 being in parallel.

If the directions of I_1 and I_2 are opposite to each other then,

$$I = I_1 \sim I_2$$

where the sign ~ indicates the difference.

The arrow for I will be in the direction of that of I_1 or I_2 whichever is larger.

2.1.8 Voltage Sources in Parallel and Current Sources in Series & Parallel

The *conditions for paralleling* of two voltage sources as in Fig. 2.5 are that the magnitudes and polarities of the two voltage sources should be the same and for D.C. the same polarities of two voltage sources should be connected together.

The maximum total current that can be supplied by the parallel combination of the voltage sources will be equal to the sum of the maximum current capacity of each source. When more load than the capacity of a given generator has to be supplied then another voltage source has to be connected in parallel. There is no advantage of connecting the current sources in series. Paralleling of generators with non-similar voltage waveforms, for example, results in heavy currents and equipment damage.

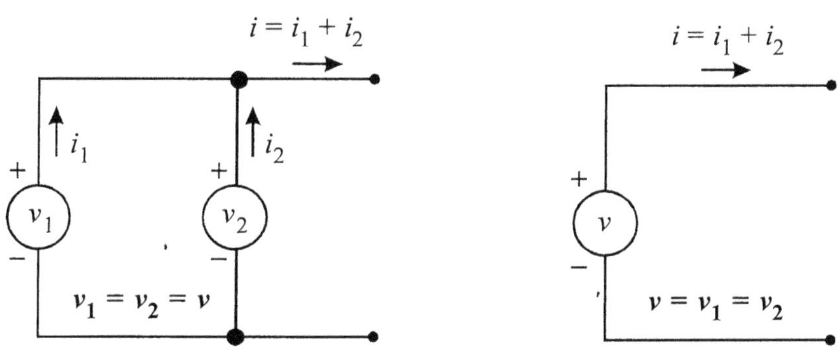

(a) **(b) Eqv.ckt. for Fig. a**

Fig. 2.5 *Parallel Voltage Sources*

There is no advantage of connecting the current sources in series since the same current has to pass through the 2 series current sources. Connection of two current sources with currents I_1 and I_2 in parallel will lead to a supply of total current of $I_1 + I_2$ to the load as shown in Fig,.2.6 (a) and (b)

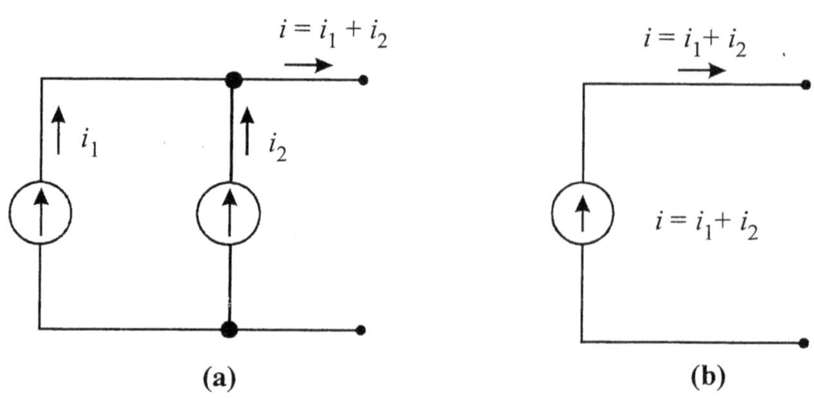

Fig. 2.6 Parallel Current Sources

2.1.9 Source Transformations

An voltage source can be transformed into an equivalent current source or a current source can be transformed into a equivalent voltage source as follows:

Let v be the voltage of the source and v_t the voltage at the node located between resistor R_1 and the rest of the network. Kirchoff's Voltage Law for the circuit of Fig. 2.7(a) is,

$$v = R_1 i + v_t \qquad \text{.......... (2.1.1)}$$

Solving Eq. (2.1.2) for i gives,

$$i = \frac{v}{R_1} - \frac{v_t}{R_1} \qquad \text{.......... (2.1.2)}$$

In this current equation, we will identify each of the individual terms in relationship to the network of Fig. 2.7 (b). The term v/R_1 represents the current from the current source, the term v_t/R_1 represents the current in the resistor R_1 connected in parallel with the current source. The difference of these two currents is the current to the external network, i. Note that i and v_t are the same for the two networks. With respect to the remainder of the network, the source-resistor combinations of Fig. 2.7(a) and Fig. 2.7 (b) are described by the same equations, Eq. (2.1.1) and Eq. (2.1.2), and so are equivalent. *Note that, the voltage and the current sources shown in the figure are ideal sources and R_1 is external resistance.*

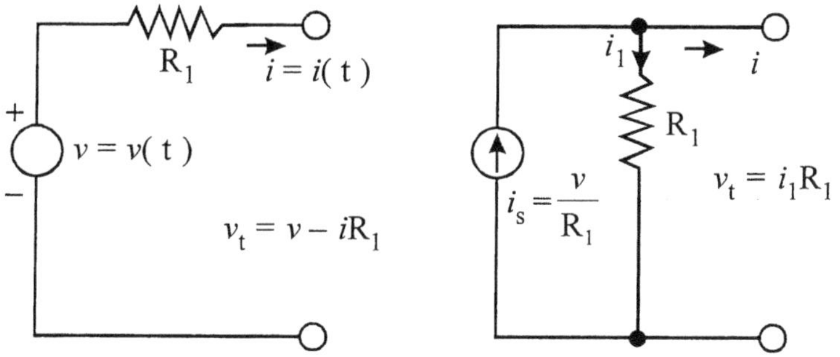

(*a*) *Transformation of Voltage Source into Equivalent Current Source*

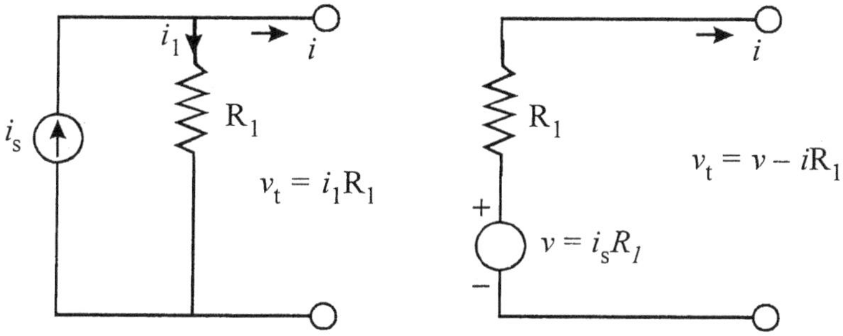

(*b*) *Transformation of Current Source into Equivalent Voltage Source*

Fig. 2.7 *Source Transformations using a Single Resistance*

2.1.10 Source Shifting

Some times it may become necessary to shift a given voltage source into the elements connected to the node to which the voltage source is connected as given in Fig. 2.8 and Fig. 2.9.

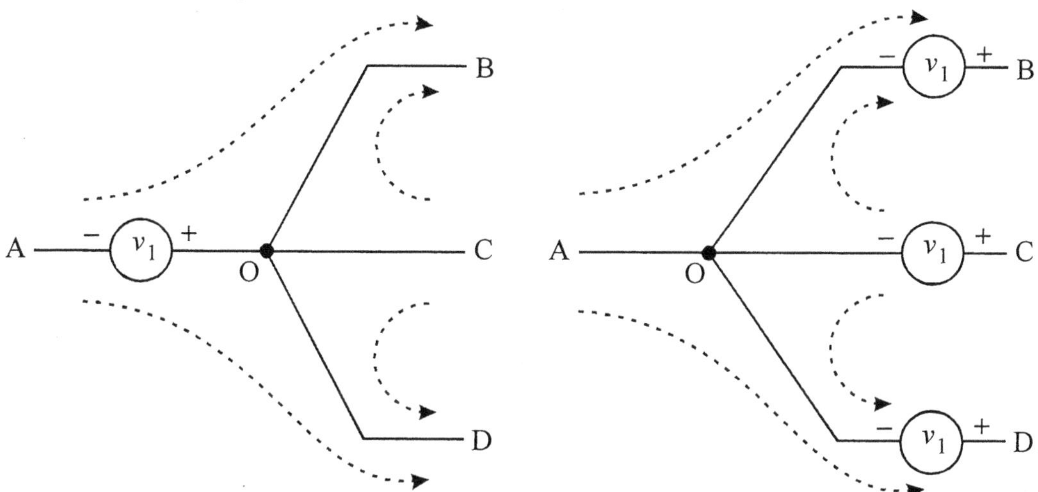

Fig. 2.8 *Shifting a Voltage Source through a node in a network*

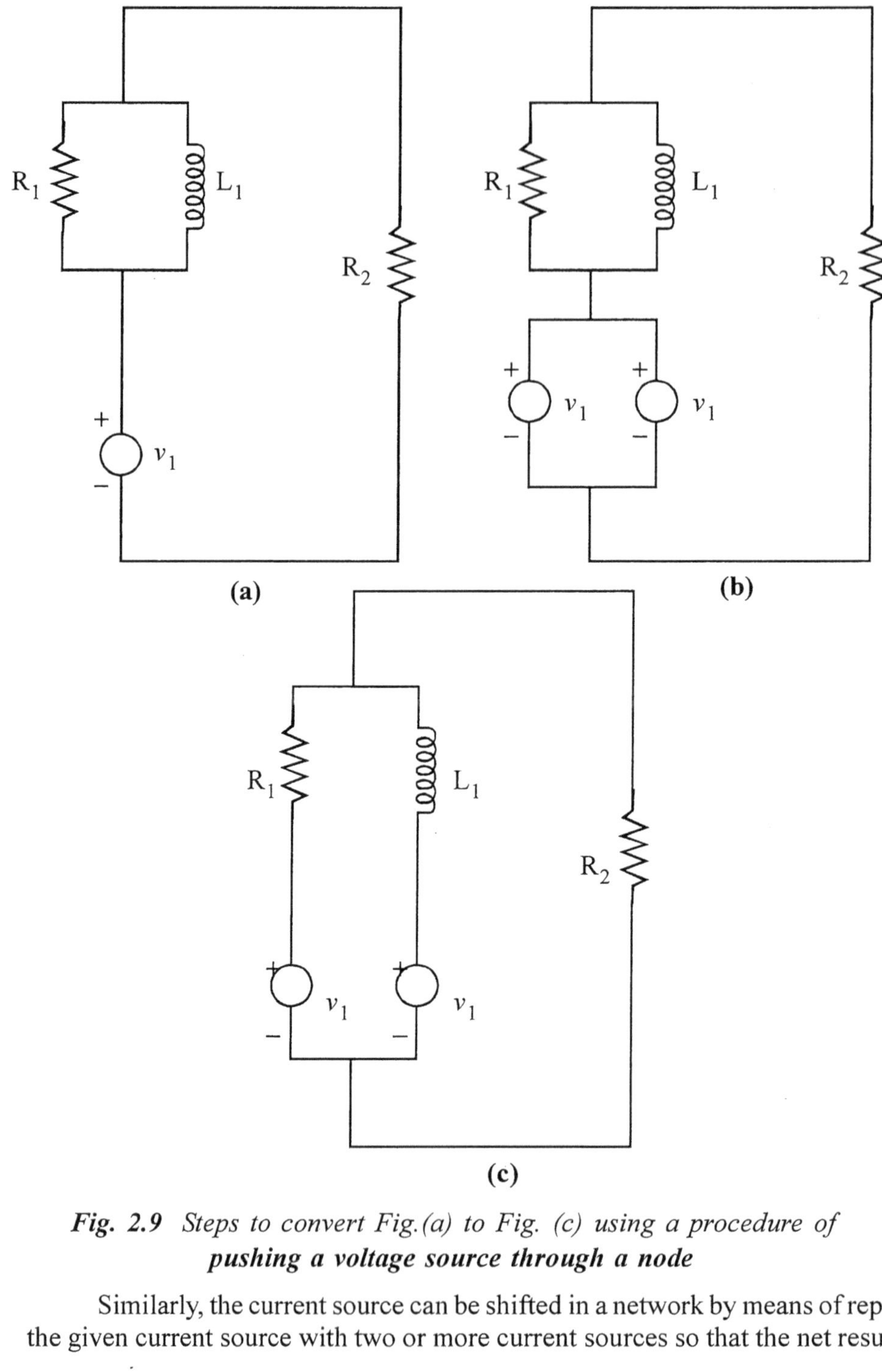

Fig. 2.9 *Steps to convert Fig.(a) to Fig. (c) using a procedure of* ***pushing a voltage source through a node***

Similarly, the current source can be shifted in a network by means of replacing the given current source with two or more current sources so that the net result will be the said current of the given current source will be leaving the same node and entering the same node as in the original circuit. Fig. 2.10 and Fig. 2.11.

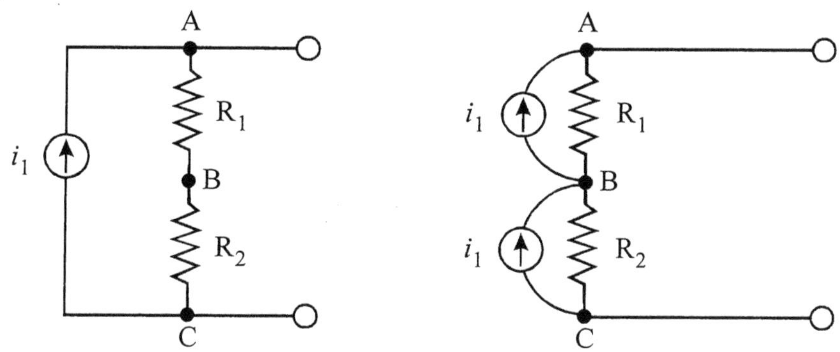

Fig. 2.10 *Shifting of Current Source for Two Branches*

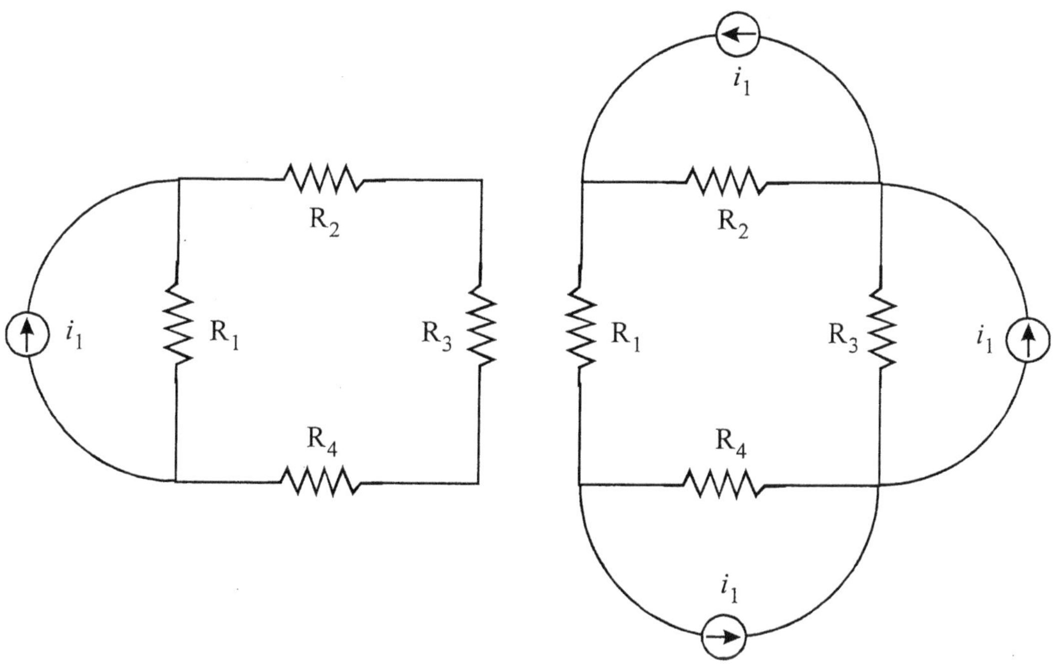

Fig. 2.11 *Shifting of Current Source for Three Branches*

2.2 LOOP OR MESH ANALYSIS AND NODAL ANALYSIS

Earlier we have studied the analysis of circuits having a single loop or two or three loops, using Kirchoff's Voltage Law and Kirchoff's Current Law. This method, no doubt gives a solution directly, but is difficult to apply in circuits having more loops. Applying Kirchoff's Voltage Law for each loop will have many equations in terms of loop currents, some of which will be a linear combination of other equations indicating that they are not independent loops.

For successful analysis of loop analysis we have to identify independent loops to have a set of independent equations. One way of obtaining independent loops using ***Network Topology Method***, is to identify a chosen tree and the links for the tree and form the basic loop incidence matrix to obtain fundamental loops. This process also involves some efforts as had been discussed earlier.

Another way of identifying independent loop is the ***Window Pane Method***, by which the loops that appear like window panes are considered to be independent loops. Loop analysis can be had for circuits having independent voltage sources, independent current sources, a combination of them or using controlled voltage or current sources or a combination of independent and controlled sources.

Basic Loop Impedance Matrix can also be formed by inspection for small networks.

Nodal Analysis is another elegant way of analysing the circuits, wherein the voltages of the nodes are obtained applying KCL to the nodes. Here again we have to identify the nodes that are independent and the nodes whose voltage can be known by some method which will give equations that are in excess of the independent equations.

One way of identifying the independent nodal equations by *Network Topology* method, by identifying a chosen tree and obtaining the basic cutset incidence matrix for that tree branches leading to the formation of nodal admitance matrix.

Nodal Admitance matrix can also be formed by inspection for small networks. Nodal Analysis can be had for circuits having independent current sources, voltage sources, etc.

2.3 LOOP ANALYSIS FOR D.C. CIRCUITS

For any given network, we can identify the loops or meshes which form a closed path for a current to flow through using Kirchoff's Voltage Law, we can write the loop equation of that loop.

A set of loop equations can be expressed in matrix form as

$$[R][I] = [E]$$

where R is the resistance matrix of the network

I is the column vector of the loop currents or mesh currents

E is the column vector of the source voltages in the loops

If the loops identified are fundamental loops, then we get, a set of independent loop equations.

In general, the diagonal elements R_{ii} of the R matrix will be equal to the sum of all the resistances that are connected in that loop. The off diagonal elements R_{ij} of the R matrix will be equal to the sum of all the resistances that are common to the loop i and loop j. This property helps us in writing the loop equation by inspection in the case of small networks having a few loops.

The solution of the loop equations obtained by applying either Cramer's Rule or Matrix Inversion or any other method of solving the equations gives the loop currents. The current through any branch in the loop can be obtained in terms of the loop currents and hence the voltage across the element or power loss in the element, etc., can be calculated.

2.3.1 Redundant Elements

If an element is connected across a voltage source then the voltage across the element is already known. Hence, this element can be treated as a redundant element and it can be removed in the loop analysis.

If an element is connected in series with a current source, then the current flowing through that element is already known. Hence, this element can be treated as a redundant element and it can be removed in nodal analysis.

2.3.2 Identification of independant loops

In any network, there will be many loops. In a planar network, where there are no cross over of elements, the independant loops can be identified by ***Window-Panes Method*** where in each independant loop looks like a pane of a window as identified in Example 6.1.

Alternatively, one of the trees of the graph for the given network can be identified and the fundamental or basic circuits are formed by connecting only one link at a time to create a closed loop. These fundamental loops form the independent loops.

If the network is very large an automatic identification of a particular tree using a computer program can be made and the fundamental loops of the tree can automatically be identified by the computer program.

If the network is very large, identification of independent loops either by window-pane method or by identifying a tree will be very difficult. In such cases, the node element incidence matrix and the primitive or elementary admittance matrix may be made use of to obtain the node admittance matrix. Or with the use of basic loop incidence matrix and primitive or elementary incidence matrix the loop imepdance matrix can be obtained.

2.3.3 Loop Analysis Using Independent Voltage Sources

Consider a three basic loop D.C. circuit with two Independent Voltage Sources E_1 and E_2 as shown in Fig. 2.12. The basic loops are identified by ***Window Pane Method*** and are marked with loop currents I_1, I_2 and I_3. Note that many more loops can be identified in the circuit, but they will turn out to be a linear combination of these three basic loops. I_1 is the loop current for loop 1 and so on.

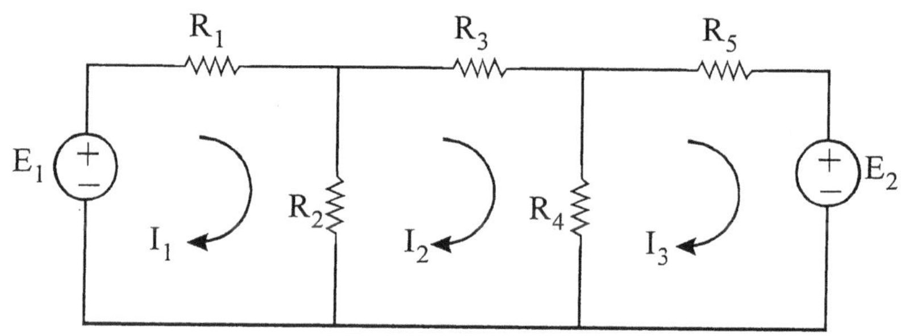

Fig. 2.12 *Mesh Analysis for a Three Basic Loop D.C. Circuit*

Applying KVL to Loop 1, we have,

$$E_1 - I_1R_1 - (I_1 - I_2)R_2 = 0$$

Rearranging the above equation,

$$(R_1 + R_2)I_1 - R_2I_2 = E_1 \quad \text{.......... (2.3.1)}$$

Applying KVL to Loop 2, we have,

$$(I_2 - I_1)R_2 + I_2R_3 + (I_2 - I_3)R_4 = 0$$

Rearranging the above equation,

$$-R_2I_1 + (R_2 + R_3 + R_4)I_2 - R_4I_3 = 0 \quad \text{.......... (2.3.2)}$$

Applying KVL to Loop 3, we have,

$$(I_3 - I_2)R_4 + I_3R_5 + E_2 = 0$$

Rearranging the above equation,

$$R_4 I_2 + (R_4 + R_5) I_3 = -E_2 \qquad \text{.......... (2.3.3)}$$

Writing the above three rearranged equations in matrix form, we get,

$$\begin{bmatrix} (R_1+R_2) & -R_2 & 0 \\ -R_2 & (R_2+R_3+R_4) & -R_4 \\ 0 & -R_4 & (R_4+R_5) \end{bmatrix} \begin{bmatrix} I_1 \\ I_2 \\ I_3 \end{bmatrix} = \begin{bmatrix} E_1 \\ 0 \\ -E_2 \end{bmatrix}$$

.... (2.3.4)

From Eq. (2.3.4), it can be observed that,

1. *The self term R_{11} for loop 1 equals the sum of the resistances in loop 1, the self term R_{22} for loop 2 equals the sum of the resistances in loop 2 and the self term R_{33} for loop 3 equals the sum of the resistances in loop 3.*

2. *The mutual term R_{12} is equal to the negative of the resistance which is common to loop 1 and loop 2. The mutual term R_{13} is equal to 0 because, there is no resistance which is common to loop 1 and loop 3. The mutual term R_{23} is the negative of the resistance which is common to loop 2 and loop 3.*

3. *R_{21} is equal to R_{12}, R_{13} is equal to R_{31} and R_{23} is equal to R_{32}. Hence, this resistance matrix is a symmetrical matrix.*

4. *In general, the diagonal elements R_{ii} of the R matrix will be equal to the sum of all the resistances that are connected in that loop. The off diagonal elements R_{ij} of the R matrix will be equal to the sum of all the resistances that are common to the loop i and loop j.*

5. *The above observations, help us in writing the loop equations by inspections.*

Example 2.1:

Use the mesh analysis to determine the three unknown mesh currents in the circuit of the given figure.

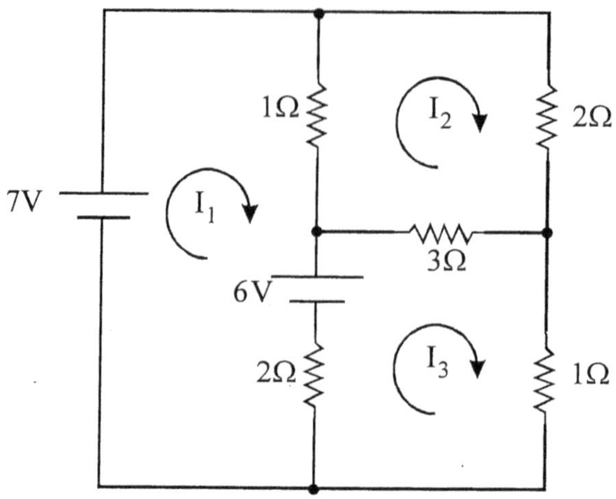

Solution:

The three required mesh currents are assigned as indicated in the figure, and we methodically apply KVL about each mesh :

$$-7 + 1(i_1 - i_2) + 6 + 2(i_1 - i_3) = 0$$

$$1(i_2 - i_1) + 2i_2 + 3(i_2 - i_3) = 0$$

$$2(i_3 - i_1) - 6 + 3(i_3 - i_2) + 1i_3 = 0$$

Simplifying,

$$3i_1 - i_2 - 2i_3 = 1$$

$$-i_1 + 6i_2 - 3i_3 = 0$$

$$-2i_1 - 3i_2 + 6i_3 = 6$$

and Cramer's rule leads to the formulation for i_3 :

$$i_3 = \frac{\begin{vmatrix} 3 & -1 & 1 \\ -1 & 6 & 0 \\ -2 & -3 & 6 \end{vmatrix}}{\begin{vmatrix} 3 & -1 & -2 \\ -1 & 6 & -3 \\ -2 & -3 & 6 \end{vmatrix}} = \frac{117}{39} = 3\text{A}$$

The other mesh currents are $i_1 = 3A$ and $i_2 = 2A$.

Again we notice that we have a denominator determinant that is symmetrical about the major diagonal and has positive terms on the major diagonal and zero or negative terms off it. This occurs for circuits that contain only independent voltage sources when clockwise mesh currents are assigned; in such circuits the elements appearing in the first row of the determinant are the ordered coefficients of $i_1, i_2, \ldots, i_M$ in the KVL equation about the first mesh, the second row corresponds to the second mesh, and so on. This symmetrical array appearing in the denominator is the determinant of the resistance matrix of the network.

$$[R] = \begin{bmatrix} 3 & -1 & -2 \\ -1 & 6 & -3 \\ -2 & -3 & 6 \end{bmatrix}$$

2.3.4 Loop Analysis Using Independent Current Sources (Super Mesh)

The loop containing current source is identified seperately and its loop current is already known and is equal to the current of the current source. Such a loop or mesh having independent current source is known as super mesh.

Alternatively, the current source can be converted to a equivalent voltage source and the loop analysis is carried on using this equivalent voltage source in the loop containing the current source.

Example 2.2:

Find the voltage across the 2Ω resistance in the given circuit.

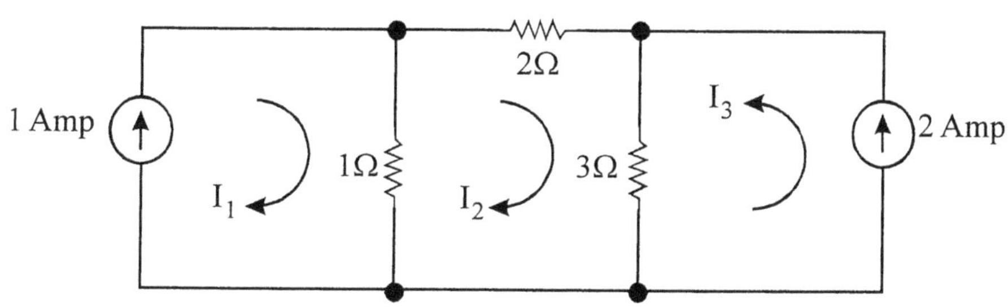

Solution:

There are three loops. Loop currents I_1 and I_3 are known. Hence, writing the loop equations for loop 2, we have,

$$(I_2 - I_1) \times 1 + I_2 \times 2 + (I_2 + I_3) \times 3 = 0$$

$$(I_2 - 1) \times 1 + I_2 \times 2 + (I_2 + 2) \times 3 = 0$$

$$I_2 - 1 + 2I_2 + 3I_2 + 6 = 0$$

$$6I_2 + 5 = 0$$

$$I_2 = -5/6 \text{ Amp.}$$

The voltage across the 2Ω resistor is,

$$V_2 = I_2 \times 2$$

$$= \left(-\frac{5}{6}\right) \times 2$$

$$= -1.66 \text{ Volts}$$

The negative sign for I_2 indicates that the current through the 2Ω resistance is in the direction opposite to that of the assumed loop current I_2. Hence the voltage drop is also in the opposite direction.

2.3.5 Loop Analysis using Independent Voltage and Current Sources

As already stated, the loop current of the loop containing independent current source is treated as a seperate loop and its current is already known. Hence the loop analysis is carried out with the remaining basic loops taking into account the current of the current source also.

Example 2.3:

Find the current flowing through the 3Ω resistor of the given circuit.

Solution:

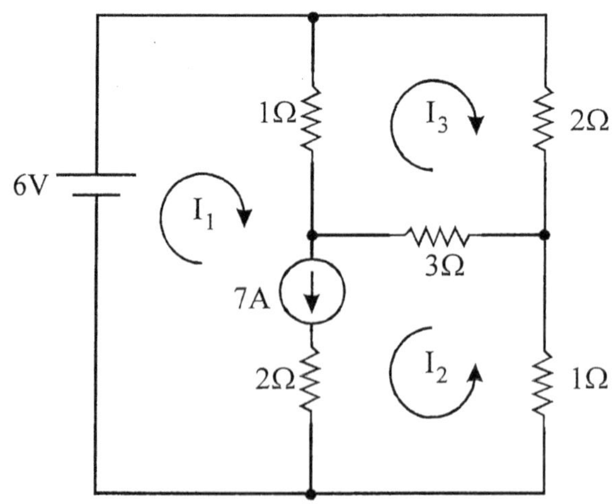

The current in the loop 2 is the current of the current source. Therefore,

$$I_2 = 7 \text{ A.}$$

Writing KVL for Loop 1,

$$(I_1 - I_3) \times 1 + (I_1 + I_2) \times 2 = 6$$

$$3I_1 + 2I_2 - I_3 = 6$$

We know that,

$$I_2 = 7 \text{ A}$$

$$3I_1 - I_3 + 14 = 6$$

or $$3I_1 - I_3 = -8 \qquad \text{.......... (1)}$$

writing KVL for Loop 3,

$$(I_3 - I_1) \times 1 + 2I_3 + (I_2 + I_3) \times 3 = 0$$

$$-I_1 + 3I_2 + 6I_3 = 0$$

$$-I_1 + 6I_3 + 21 = 0$$

or $$-I_1 + 6I_3 = -21 \qquad \text{.......... (2)}$$

$$\begin{bmatrix} 3 & -1 \\ -1 & 6 \end{bmatrix} \begin{bmatrix} I_1 \\ I_3 \end{bmatrix} = \begin{bmatrix} -8 \\ -21 \end{bmatrix}$$

$$I_3 = \frac{\begin{vmatrix} 3 & -8 \\ -1 & -21 \end{vmatrix}}{\begin{vmatrix} 3 & -1 \\ -1 & 6 \end{vmatrix}}$$

$$= \frac{-71}{-17} = 4.176 \text{ A}$$

The current through the 3Ω resistance is

$$(I_2 + I_3) = 7 + 4.176 = 11.176 \text{ A}$$

Example 2.4:

Convert the current sources into equivalent voltage sources and write the loop equations for the given circuit.

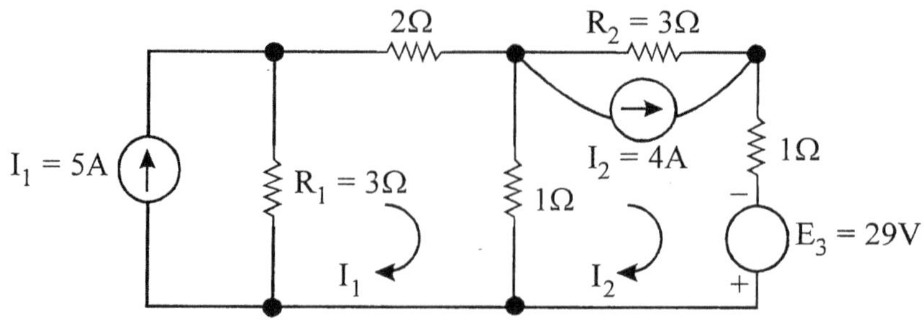

Solution:

For solving the problems on the mesh current (loop current) method, the preferable method is to convert the practical current sources into practical voltage sources.

The current source I_1 in parallel with resistance R_1 is transformed into an equivalent voltage source E_1 of 15V in series resistance R_1. The current source I_2 in parallel with resistance R_2 is transformed into an equivalent voltage source E_2 of 12V in series with resistance R_2.

The original circuit is redrawn as in the figure below after transforming the current sources into voltage sources.

By inspection, writting the above circuit equations in the matrix form,

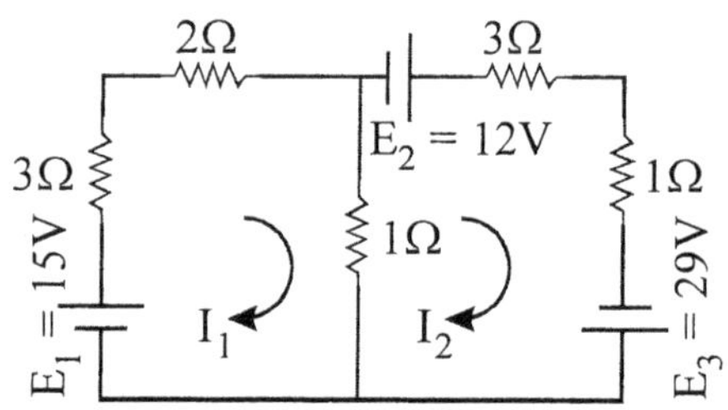

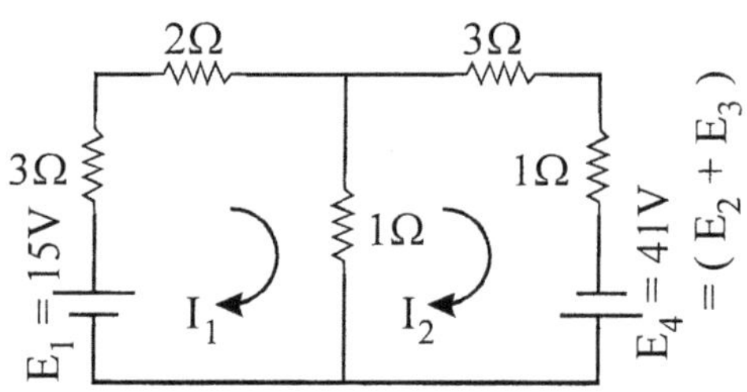

$$\begin{bmatrix} 6 & -1 \\ -1 & 5 \end{bmatrix} \begin{bmatrix} I_1 \\ I_2 \end{bmatrix} = \begin{bmatrix} 15 \\ 41 \end{bmatrix}$$

$$\Delta = \begin{vmatrix} 6 & -1 \\ -1 & 5 \end{vmatrix}$$

$$= 30 - 1 = 29$$

$$\Delta_1 = \begin{vmatrix} 15 & -1 \\ 41 & 5 \end{vmatrix} = 75 + 41 = 116$$

$$\Delta_2 = \begin{vmatrix} 6 & 15 \\ -1 & 41 \end{vmatrix} = 246 + 15 = 261$$

Therefore, $I_1 = \dfrac{\Delta_1}{\Delta} = \dfrac{116}{29} = 4A$

$$I_2 = \frac{\Delta_2}{\Delta} = \frac{261}{29} = 9A$$

2.4. NODAL ANALYSIS

In Nodal Analysis, generally one of the nodes is taken as a reference node, preferably earth whose potential is taken to be zero. The voltages of the independent nodes are obtained by solving the equations formed by applying KCL to the independent nodes.

By convention, the source currents or the branch currents entering a node are taken to be positive and the source currents or branch currents leaving the node are taken to be negative.

The voltage of the node to which a voltage source is connected is known, and hence, it can be eliminated in nodal analysis. Such a node is known as ***Super Node*** or ***Redundant Node***.

Nodal Admittance Matrix can also be formed by inspection for small networks. The performance equation using Nodal Admittance Matrix Y is given as,

$$[Y][V] = [I] \qquad \text{.......... (2.4.1)}$$

where [V] is the column vector of the nodal voltages or junction voltages

[I] is the column vector of the source currents entering the nodes

For D.C. networks, we use the term ***Conductance Matrix*** instead of ***Nodal Admittance Matrix***. A set of nodal equations can be expressed in matrix form as

$$[G][V] = [I] \qquad \text{.......... (2.4.2)}$$

where [G] is the nodal admittance matrix of the network

The nodal equation for any node for any node, is written by using KCL at that node. The nodal equations can also be written as, algebraic sum of branch currents connected to a node is equal to the algebraic sum of the source currents connected to that node.

By convention, the branch currents leaving a node are taken to be positive and the source currents entering a node are taken to be positive and the branch currents entering a node are taken to be negative and the source currents leaving a node are taken to be negative.

The number of nodal equations is equal to the number of independent nodes in the network. In writing nodal equations, we can transform voltage source into current source or vice versa, remove the redundant elements and remove the super nodes or redundant nodes whose voltages are known.

2.4.1 Identification of Independant Nodes

As already stated, the reference node and the super node or redundant nodes whose voltages are already known can be eliminated and the remaining nodes will be independent nodes. This method will be helpful for small networks.

Alternatively, one of the trees of the graph for the given network can be identified and the fundamental or basic cutsets are formed by cutting one tree branch at a time. The voltages of the nodes forming the tree branches will be the independent nodes.

If the network is very large an automatic identification of a particular tree using a computer program can be made and independent nodes can be identified automatically as the nodes which are connected by tree branches.

If the network is very large, identification of a tree by inspection will be difficult. In such cases, the node element incidence matrix and the primitive or element admittance matrix may be made use of to obtain the nodal admittance matrix.

From the nodal admittance matrix, the nodal impedance matrix can be obtained by inversion of the nodal admittance matrix.

$$Z_{Nodal} = Y^{-1}$$

For this the Y matrix should not be singular. Only if both Y and Z have a common reference node namely earth or ground then it is possible to obtain the nodal impedance matrix by inverting the nodal admittance matrix.

2.5 NODAL ANALYSIS FOR D.C. CIRCUITS

2.5.1 Nodal Analysis using Independent Current Sources

The nodal analysis using independent current sources is straight forward.

Example 2.5:

Find the node voltages V_A and V_B using nodal analysis.

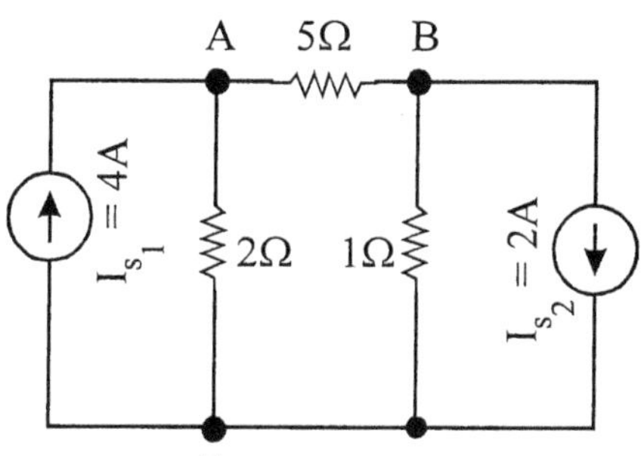

(a) Given Circuit

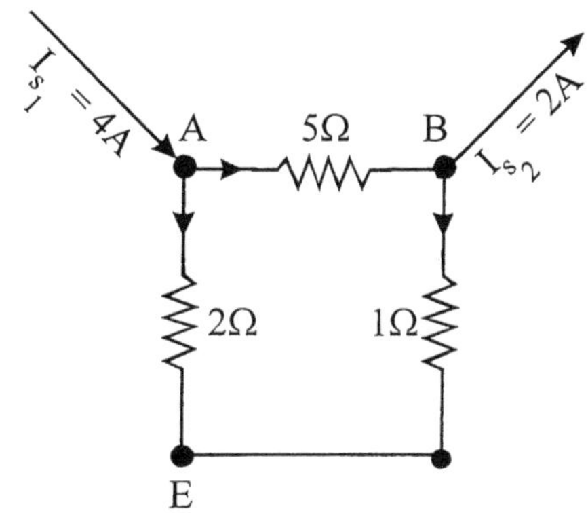

(b) Given Circuit in different form

Solution:

There are three nodes A, B and E in the given circuit. The node E is taken as reference node as is assumed to be connected to earth with zero potential. The independent nodes are A and B. The voltages of the independent nodes are V_{AE} and V_{BE} and are measured with reference to the node E and hence can simply be represented as V_A and V_B.

The branch currents leaving the node A are,

$$G_{AB}(V_A - V_B) = \frac{1}{5}(V_A - V_B) = (0.2V_A - 0.2V_B)$$

and
$$G_{AE}V_A = \frac{1}{2}V_A = 0.5V_A$$

The source current entering the node A is,

$$I_{s_1} = 4A$$

Applying KCL to node A or stating that algebraic sum of the branch currents leaving the node is equal to the algebriac sum of the source currents entering the node, we have,

$$(0.2V_A - 0.2V_B) + 0.5V_A = 4$$

$$0.7V_A - 0.2V_B = 4 \qquad \text{.......... (1)}$$

The branch currents leaving the node B are,

$$G_{AB}(V_B - V_A) = \frac{1}{5}(V_B - V_A) = (0.2V_B - 0.2V_A)$$

and $$G_{BE}V_B = \frac{1}{1} V_B = V_B$$

The source current entering the node B is,

$$I_{s_2} = -2A$$

because I_{s_2} is leaving the node.

Applying KCL to node B, or stating that algebraic sum of the branch currents leaving the node is equal to the algebriac sum of the source currents entering the node, we have,

$$(0.2V_B - 0.2V_A) + V_B = -2$$

$$-0.2V_A + 1.2V_B = -2 \qquad \text{.......... (2)}$$

From Eq. (1) and Eq. (2), we get,

$$V_A = \frac{\begin{vmatrix} 4 & -0.2 \\ -2 & 1.2 \end{vmatrix}}{\begin{vmatrix} 0.7 & -0.2 \\ -0.2 & 1.2 \end{vmatrix}}$$

$$= \frac{4.8 - 0.4}{0.84 - 0.04} = \frac{4.4}{0.8}$$

$$V_A = 5.5V$$

$$V_B = \frac{\begin{vmatrix} 0.7 & 4 \\ -0.2 & -2 \end{vmatrix}}{\begin{vmatrix} 0.7 & -0.2 \\ -0.2 & 1.2 \end{vmatrix}} = \frac{-0.6}{0.8}$$

$$V_B = -0.75V$$

2.5.2 Nodal analysis using Independent Voltage Sources (Super Nodes)

The voltages of the nodes to which the independent voltage sources are connected will be equal to the source voltages. Hence, those nodes are *Super Nodes* which are not independent. One node will be reference node. Only the remaining nodes will be the independent nodes.

For the circuit given in Fig. 2.13, A and D are the super nodes and E is the reference node. Hence, B and C are the independent nodes.

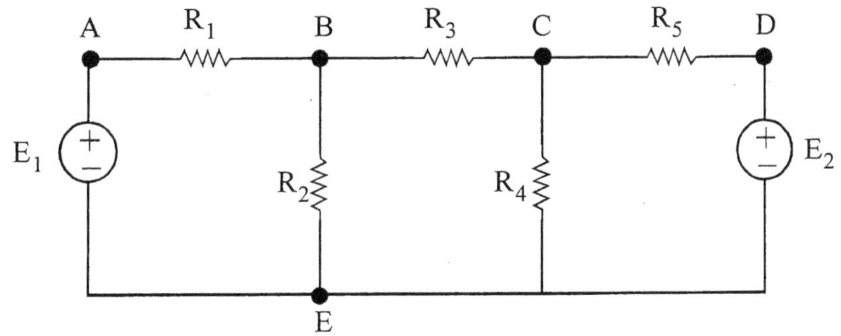

Fig. 2.13 *Nodal Analysis using Independent Voltage Sources*

Writing nodal equation for node B we have,

$$G_1(V_B - V_A) + G_2V_B + G_3(V_B - V_C) = 0$$

i.e.,
$$\frac{1}{R_1}(V_B - E_1) + \frac{1}{R_2}V_B + \frac{1}{R_3}(V_B - V_C) = 0 \quad \text{.......... (2.5.1)}$$

Writing nodal equation for node C we have,

$$G_5(V_C - V_D) + G_4V_C + G_3(V_C - V_B) = 0$$

i.e.,
$$\frac{1}{R_5}(V_C - E_2) + \frac{1}{R_4}V_C + \frac{1}{R_3}(V_C - V_B) = 0 \quad \text{.......... (2.52)}$$

Here E_1 and E_2 are the voltages of the nodes A and D respectively.

Example 2.6:

Find Node Voltages of Nodes A, B, C and D of the given network using Nodal Analysis.

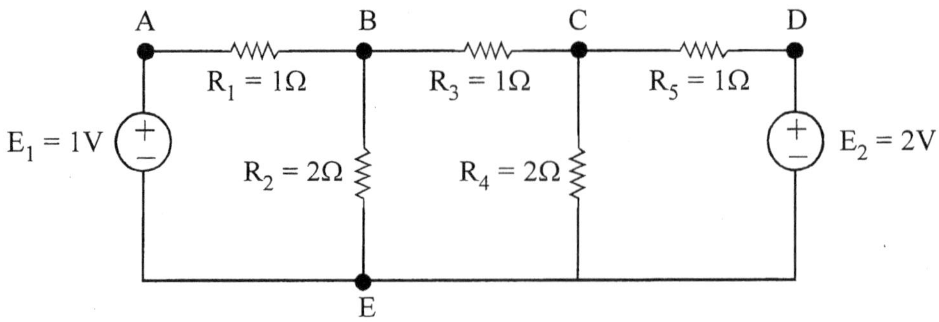

Solution:

The node voltages V_A, V_B, V_C and V_D are measured with respect to node E whose voltage is assumed to be 0.

Voltage source E_1 is connected to Node A, node A is a super node and its voltage will be 1V. Similarly, voltage source E_2 is connected to Node D, node D is a super node and its voltage will be 2V.

Nodes B and C are the independent nodes.

Applying KCL to node B, we have,

$$\frac{1}{R_1}(V_B - E_1) + \frac{1}{R_2}V_B + \frac{1}{R_3}(V_B - V_C) = 0$$

$$\frac{1}{1}(V_B - 1) + \frac{1}{2}V_B + \frac{1}{1}(V_B - V_C) = 0$$

$$V_B + 0.5V_B + V_B - V_C = 1$$

$$2.5V_B - V_C = 1 \qquad \text{.......... (1)}$$

Applying KCL for Node C, we have,

$$\frac{1}{R_5}(V_C - E_2) + \frac{1}{R_4}V_C + \frac{1}{R_3}(V_C - V_B) = 0$$

$$\frac{1}{1}(V_C - 2) + \frac{1}{2}V_C + \frac{1}{1}(V_C - V_B) = 0$$

$$V_C + 0.5V_C + V_C - V_B = 2$$

$$-V_B + 2.5V_C = 2 \qquad \text{.......... (2)}$$

Therefore,

$$V_B = \frac{\begin{vmatrix} 1 & -1 \\ 2 & 2.5 \end{vmatrix}}{\begin{vmatrix} 2.5 & -1 \\ -1 & 2.5 \end{vmatrix}} = \frac{4.5}{5.25} = 0.86 \text{ V}$$

$$V_C = \frac{\begin{vmatrix} 2.5 & 1 \\ -1 & 2 \end{vmatrix}}{\begin{vmatrix} 2.5 & -1 \\ -1 & 2.5 \end{vmatrix}} = \frac{6.0}{5.25} = 1.14 \text{ V}$$

2.5.3 Nodal Analysis using Independent Voltage and Current Sources

Since the voltage of the node to which the independent voltage source is connected, is already known, that node can be eliminated and the voltage difference between that node and any other node will be the voltage across a given element from which the current through it can be calculated or estimated and used in nodal analysis.

Alternatively, the voltage source can be converted into an equivalent current source and the nodal analysis can be carried out.

The treatement of independent voltage source, taking the node to which it is connected as a super node has already been explained in the previous section. Now we will take up the method of converting the voltage source into its equivalent current source and carrying out the nodal analysis by means of an example. Convertion of the voltage source into its equivalent current source has already been explained.

Example 2.7:

Find the currents I_1, I_2 and I_3 and the voltages V_B and V_C for the circuit given below using nodal analysis.

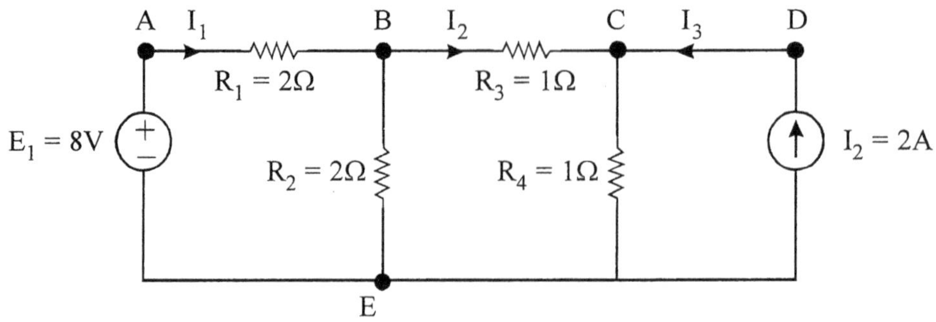

Solution:

The voltage source E_1 in series with resistance R_1 can be converted to an equivalent current source of current

$$I_1 = \frac{E_1}{R_1} = \frac{8}{2} = 4A$$

in parallel with resistance R_1 of 2Ω.

The node voltages V_B and V_C are measured with respect to node E whose voltage is assumed to be 0.

Applying KCL to Node B, we have,

$$\frac{V_B}{R_1} + \frac{V_B}{R_2} + \frac{1}{R_4}(V_B - V_C) = 4$$

$$\frac{V_B}{2} + \frac{V_B}{2} + \frac{1}{1}(V_B - V_C) = 4$$

$$2V_B - V_C = 4 \qquad \text{.......... (1)}$$

Applying KCL to Node C, we have,

$$\frac{1}{R_3}(V_C - V_B) + \frac{V_C}{R_4} = 2$$

$$\frac{1}{1}(V_C - V_B) + \frac{V_C}{1} = 2$$

$$-V_B + 2V_C = 2 \qquad \text{.......... (2)}$$

Therefore,

$$V_B = \frac{\begin{vmatrix} 4 & -1 \\ 2 & 2 \end{vmatrix}}{\begin{vmatrix} 2 & -1 \\ -1 & 2 \end{vmatrix}} = \frac{10}{3} = 3.33 \text{ V}$$

$$V_C = \frac{\begin{vmatrix} 2 & 4 \\ -1 & 2 \end{vmatrix}}{\begin{vmatrix} 2 & -1 \\ -1 & 2 \end{vmatrix}} = \frac{8}{3} = 2.66 \text{ V}$$

$$I_1 = \frac{E_1 - V_B}{2} = \frac{8 - 3.33}{2} = 2.335 \text{ A}$$

$$I_2 = \frac{1}{R_3}(V_B - V_C)$$

$$= \frac{1}{1}(3.33 - 2.66) = 0.67 \text{ A}$$

$$I_3 = 2 \text{ A}$$

The nodal equations can be written in matrix form,

$$[G][V] = [I] \qquad \text{.......... (2.5.3)}$$

where G is the nodal admittance matrix.
V is the column vector of the node votlages
I is the column vector of the source currents

Here,

$$[G] = \begin{bmatrix} G_{BB} & G_{BC} \\ G_{CB} & G_{CC} \end{bmatrix} \qquad \text{.......... (2.5.4)}$$

In general, it may be noted that from the above analysis we can observe that for the independent nodes we can write the ***Nodal Admittance Matrix*** (nodal conductance matrix for D.C. Circuits) ***by Inspection*** as follows:

The values of self-terms and mutual terms of the nodal admittance matrix are,

Y_{ii} = the sum of all the admittances connected to the node i

Y_{ij} = the negative of the sum of all the admittances connected between the node i and node j.

Y_{ji} = Y_{ij}, since nodal admittance matrix will be symmetrical.

Example: 2.8

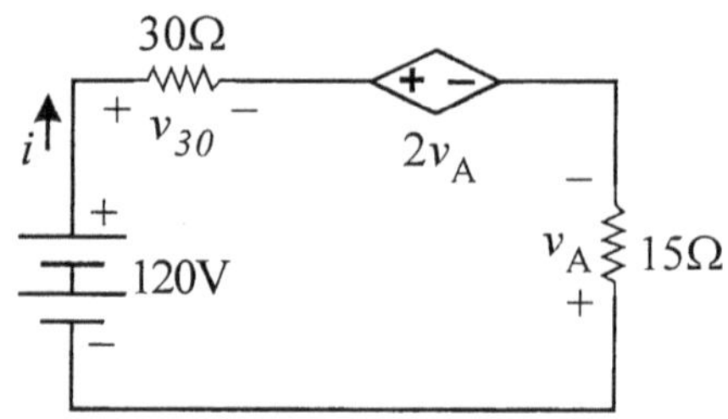

For the given circuit, with a controlled voltage source, find the total power dissipated in the circuit.

Solution:

We again assign a reference direction for the current i and a reference polarity for the voltage v_{30}. There is no need to assign a voltage to the 15Ω resistor, since the controlling voltage v_A for the dependent source is already available. It is worth noting, however that the reference signs for v_A are reversed from those we would have assigned, and that Ohm's Law for this element must thus be expressed as $v_A = -15i$. we apply Kirchoff's Voltage Law around the loop,

$$-120 + v_{30} + 2v_A - v_A = 0 \quad \text{.......... (1)}$$

Using Ohm's Law for the two resistances,

$$v_{30} = 30i$$

$$v_A = -15i$$

Using the value of v_{30} and v_A in Eq. (1)

$$-120 + 30i - 30i + 15i = 0$$

$$\therefore \quad i = 8\text{A}$$

The power relationship show that the 120V battery supplies 960W, the dependent source supplies 1920W, and the two resistors together dissipate 2880W.

Example 2.9:

Determine the values of v and the power absorbed by the independent current source in the given figure.

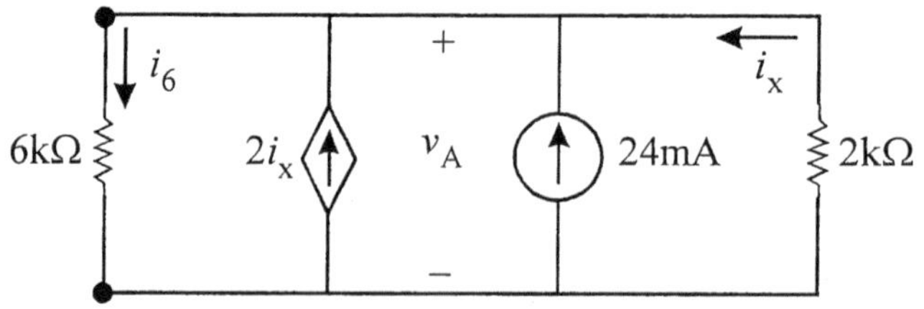

Solution:

We have defined the voltage v, arbitrarily with the positive reference at the top, and a current i_6 in the 6kΩ resistor.

Applying KCL to the upper node, we have,

$$i_6 - 2i_x - 0.024 - i_x = 0 \qquad \text{.......... (1)}$$

Applying Ohm's Law to each resistor, we have,

$$i_6 = \frac{v}{6000}$$

and $$i_x = \frac{-v}{2000}$$

Hence, $$\frac{v}{6000} - 2\left(\frac{-v}{2000}\right) - 0.024 - \left(\frac{-v}{2000}\right) = 0$$

and $$v = 600 \times 0.024 = 14.4 \text{ V}$$

Any other information we may want to find for this circuit is now easily obtained, usually in a single step. For example, the power supplied by the independent source is $p_{24} = 14.4 - 0.024 = 0.346$W, and the current flowing to the right in the central conductor at the top is $i = -0.024 + (14.4/2000) = -0.0168$A or -16.8mA.

It may be noted that, if voltage v is in Volts and the resistance R is in kilo-ohms (kΩ or k) then the current i will be in milli-Amperes (mA).

Example 2.10:

From the given figure, find the power supplied and voltage of the dependent current source.

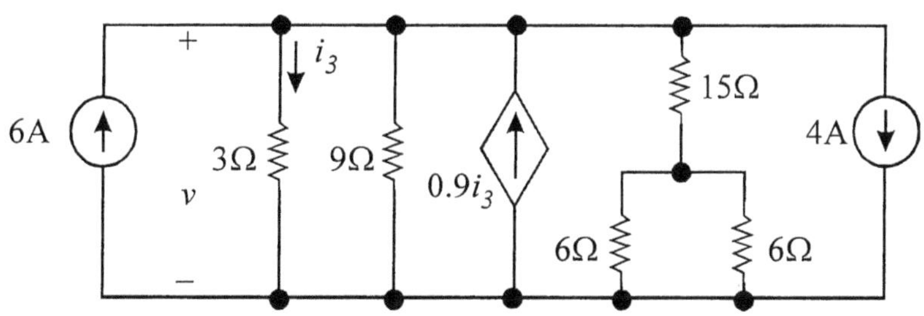

(**a**) Given Circuit with a controlled current source

Solution:

The independent current sources 6A and 4A a single current source of 2A as shown in the equivalent circuit. The dependent source is left as it is. The two resistances in parallel are combined into an equivalent 3Ω resistor, which will be in series with the 15Ω resistor giving a total of 18Ω resistance. This 18Ω resistance in parallel with 9Ω resistance will given an equivalent resistance of 6Ω.

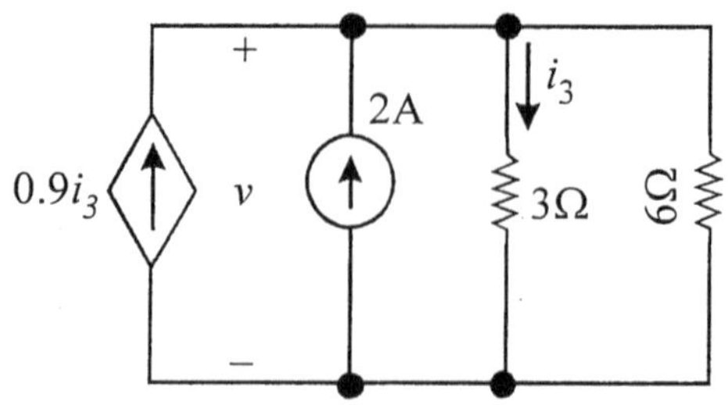

(**b**) Equivalent circuit for the given circuit

The 3Ω resistance is not combined with the 6Ω resistance in parallel because it is the source resistance of the current source.

From the equivalent circuit in the figure, we have,

$$-0.9i_3 - 2 + i_3 + \frac{v}{6} = 0$$

and $$v = 3i_3$$

yielding $$i_3 = \frac{10}{3} \text{ A}$$

and $$v = 10\text{V}$$

Thus the dependent source furnishes a power of,

$$v \times 0.9i_3 = 10 \times 0.9 \times \frac{10}{3} = 30\text{W}$$

to the remainder of the circuit.

If the power dissipated in the 15Ω resistor is asked, then, the 15Ω resistance cannot be removed in simplification. This resistor is in series with the equivalent resistance of 3Ω which is obtained by combining the two parallel resistances of 6Ω each. The total resistance will now be 18Ω and the voltage across it will be 10V. Hence, the current flowing through the 15Ω resistance,

$$\frac{10}{18} \text{ or } \frac{5}{9} \text{A}$$

The power absorbed by the 15Ω resistance is given as,

$$\left(\frac{5}{9}\right)^2 (15) \text{ or } 4.63\text{W}.$$

Example 2.11:

Use mesh analysis to evaluate the three unknwon currents in the circuit given in the figure.

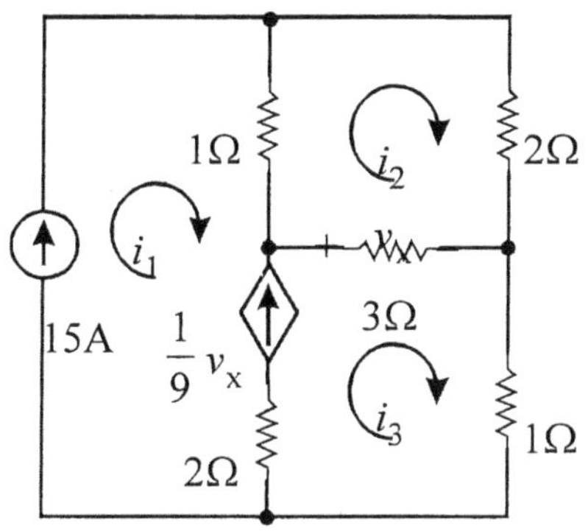

Solution:

We first assign three mesh currents and then apply KVL around mesh 2.

$$1(i_2 - i_1) + 2i_2 + 3(i_2 - i_3) = 0$$

The current sources appear in meshes 1 and 3. Since the 15A source is located on the perimeter of the circuit, we may eliminate mesh 1 from consideration. Then the dependent current source is on the perimeter of the modified network, and thus we avoid any equation writing for mesh 3.

Only mesh 2 remains, and we already have an equation for it. We therefore, turn our attention to the source quantities, obtaining

$$i_1 = 15$$

and $$\frac{1}{9}v_x = i_3 - i_1 = \frac{1}{9}[3(i_3 - i_2)]$$

Thus,

$$-i_1 + 6i_2 - 3i_3 = 0$$

$$i_1 = 15$$

$$-i_1 + \frac{1}{3}i_2 + \frac{2}{3}i_3 = 0$$

from which we have $i_1 = 15$, $i_2 = 11$ and $i_3 = 17$A.

Alternatively it may be noted the mesh current i_1 could have been given immediately as 15A because its value is already known and mesh 1 is a super mesh.

2.6 NODAL ANALYSIS WITH DEPENDENT SOURCES

Voltage Amplifier, Operational Amplifier, Transistors, etc., are solved by considering their equivalent circuits. Voltage Amplifiers, Operational Amplifiers, etc., are represented by means of Voltage Controlled Dependent Sources in their equivalent circuits. The transistors are represented by means of Current Controlled Dependent Sources in their equivalent circuits.

Example 2.12:

Find the value of v_x in the circuit given in figure (a) below.

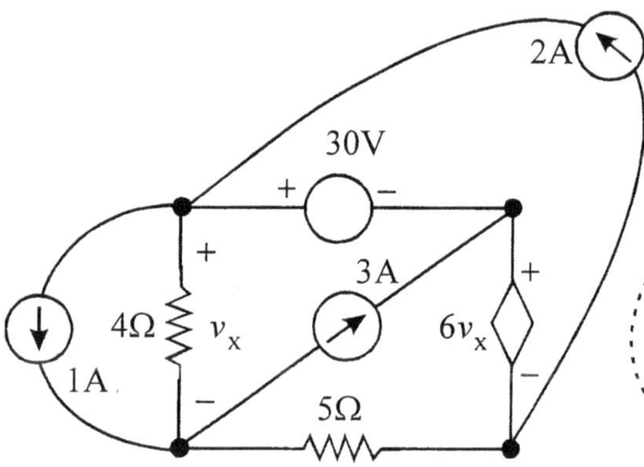

(a) Circuit

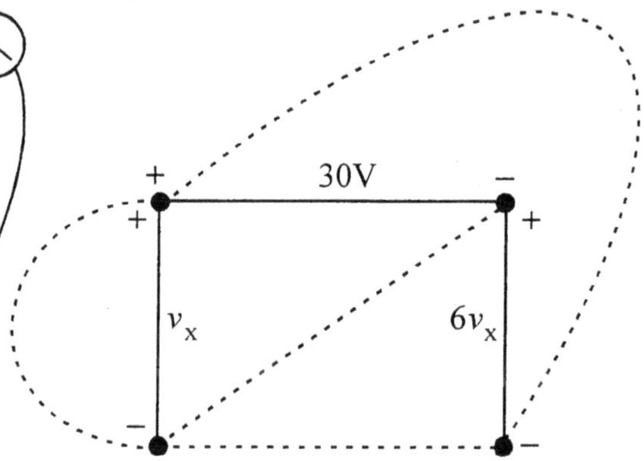

(b) Graph

Solution:

The two voltage sources and the control voltage establish the three branch tree shown in figure (b). Since the two upper nodes and the lower right node all join to form one supernode, we need write only one KCL equation. Selecting the lower left node, we have,

$$-1 - \frac{v_x}{4} + 3 + \frac{-v_x + 30 + 6v_x}{5} = 0$$

Hence, $\quad v_x = 10.666\text{V}$

Employing mesh currents or node-to-reference voltages would require more equations and more effort.

COMPREHENSION - 2

Note: **By answering all the questions given in this comprehension one can answer any objective type of questions easily whatever be the form of the question. Answers for these questions can be obtained by going through the topics in this chapter carefully.**

1. What is meant by dependent sources? Explain the symbol and characteristics of Controlled Voltages Sources and Controlled Current Sources?
2. What is meant by Source Transformation? How do you transfer the Current and Voltages Sources into Voltage and Current Sources respectively?
3. What is meant by Source Shifting? How do you shift a voltage source through a mode and current sources?
4. What is the need for writing a set of independent equations for Loop and Nodal Analysis?
5. What should be the number of independent equations required for Loop and Nodal Analysis?
6. How do you identify the independent equations in Loop Analysis and in Nodal Analysis?
7. How do you establish that a particular equation is not an independent equation?
8. Give the performance equations in the matrix from for Loop or Mesh Analysis in D.C. Circuits.
9. What is meant by redundant elements? How do you remove them?
10. What is meant by Window-Panes Method?
11. How do you obtain the self and mutual terms of the resistance matrix in Loop Analysis by inspection method?
12. How do you treat the independent current sources in Loop Analysis when you have both the independent voltage and current sources?
13. How do you treat the voltage and current sources that are in series and parallel in writing Loop Equations?
14. Give the performance equation for Nodal Analysis in D.C. circuits.

15. How do you identify the Independent Nodes?
16. How do you treat the Independent Current Sources in Nodal Analysis?
17. What do you do if you have both the Independent Voltage Sources and Independent Current Sources in writing Nodal Equations?
18. How do you obtain the self and Mutual terms of the Nodal Conductance Matrix or Nodal Admittance Matrix?
19. How do you deal with Dependent Sources in writing the Nodal Equations?

* * * * *

EXERCISE - 2

Note: **By answering all the questions given in this exercise one can answer subjective type of questions easily including problems. The worked examples in the chapter may help in solving these questions.**

1. (a) Explain the method of solving a network using Loop and Mesh Analysis.

   (b) How do you treat the current sources in Loop Analysis.

   (c) Explain the concept of Super Mesh Analysis.

2. (a) Explain the method of solving a network using Nodal Analysis.

   (b) How do you treat the voltage sources in Ndoal Analysis.

   (c) Explain the concept of Super Node Analysis.

3. (a) Write the Loop or Mesh Equations for the given circuit and solve for the individual branch currents.

[Ans : I_{AB} = 16/21 A ; I_{BC} = 3/21 A ; I_{cd} = 2/21 A ; I_{BG} = 13/21 A; I_{CF} = 31/1A]

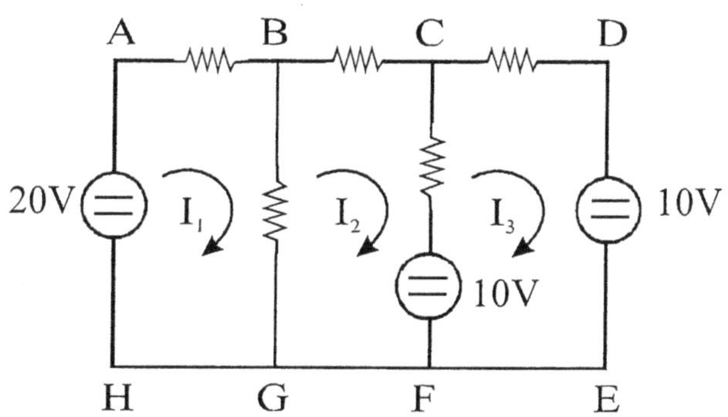

(b) In the given circuit, determine the current and energy supplied by each battery.

[Ans : B_1 = 0.6A, 6W; B_2 = 0.2A, 1W; B_2 = 0.8A, SW; B_4= 2.2A, 11W; B_5 = 1.4A, 28W]

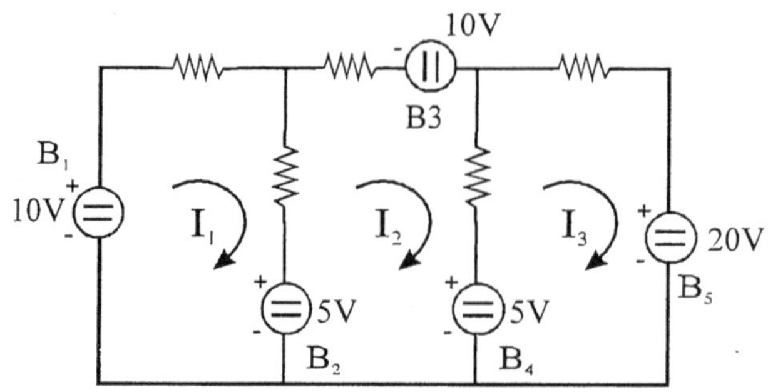

4. (a) A Wheatstone bridge ABCD is arranged as follows : Resistances between AB, BC, CD, DA and BD are $10\,\Omega$, $2\,\Omega$, $8\,\Omega$, $4\,\Omega$ and $5\,\Omega$ respectively. A 100V supply is connected between terminals A and C. Determine the currents in branches AB and AD of the circuit and the total current taken from the supply using Loop Analysis.

[Ans : AB = 250/21A; AD = 1625/84 A and Total Current = 625/84A]

(b) Find the value of R and the current through it in the circuit shown in the figure below when the current is zero in branch O.A. using Mesh Analysis. Verify it by Bridge Balancing Method (Wheatstone Bridge - under balanced conditions, products of the two arms are equal.)

[Ans : R = $6\,\Omega$]

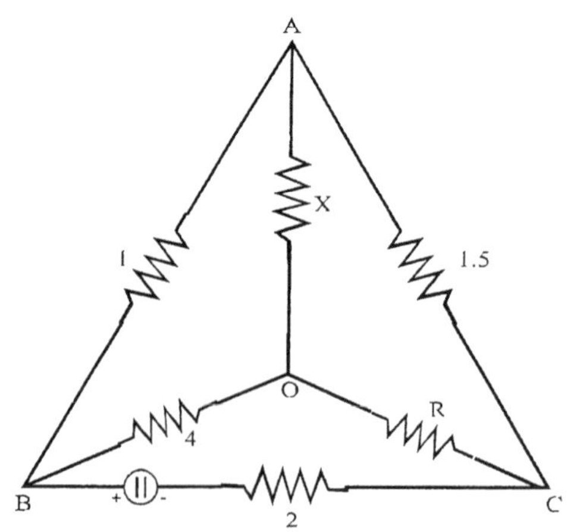

5. (a) Determine the currents i_1, i_2, and i_3 and the Voltage V_1 and V_2 in the network given in 5 (a).

[Ans: i_1 = 40A, i_2 = 10A, i_3 = -10A; V_1 = 112.5V, V_2 = 109 V]

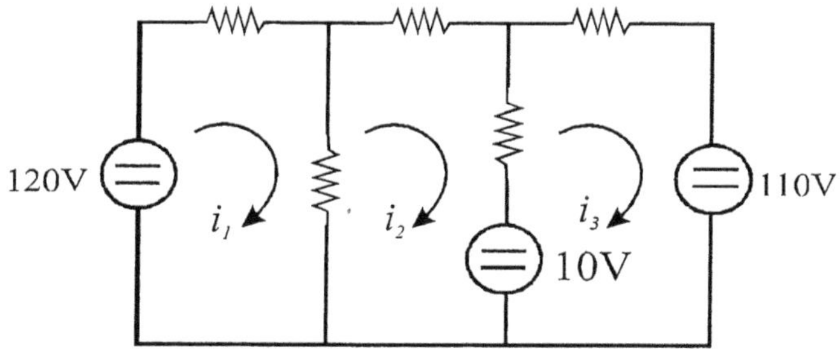

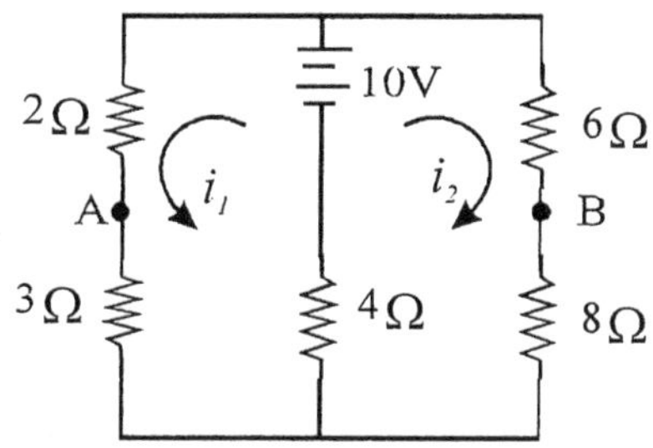

Fig. 5 (b)

(b) Determine the current in the battery, the current in each branch and the potential difference (PD) across AB in network given in 5 (b).

[Ans: 1.303A; 0.96A, 0.343A;
V_{aC} = 2.88 V$_9$ V_{aC} = 2.744V$_9$ V_{AB} = 0.14 V]

6. (a) Determine the current in the 5 Ω resistor in the network given below.

[Ans : 4.66A]

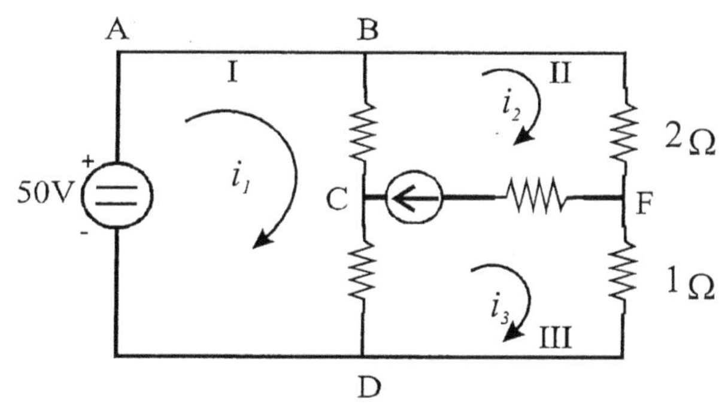

(b) Write the Mesh Equations for the network shown below by Loop Analysis and verify by converting the current source into voltage source. Find the power absorbed by the 3 Ω resistor.

[Ans: 12W]

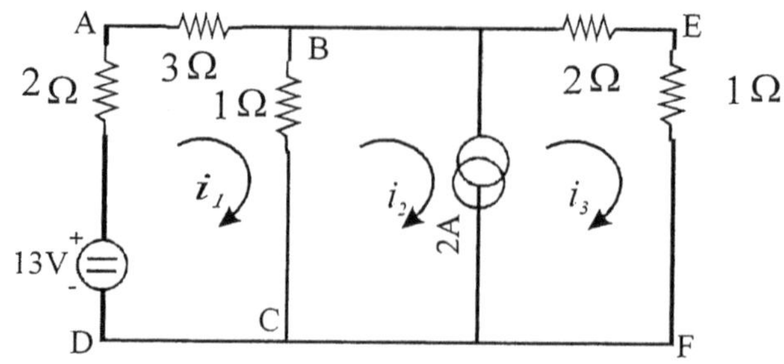

7 Using Node Voltage Method, find the currents in the different branches of the given net.

[Ans : (–19/9)A; (10/9)A; (17/9)A; (–26/9)A]

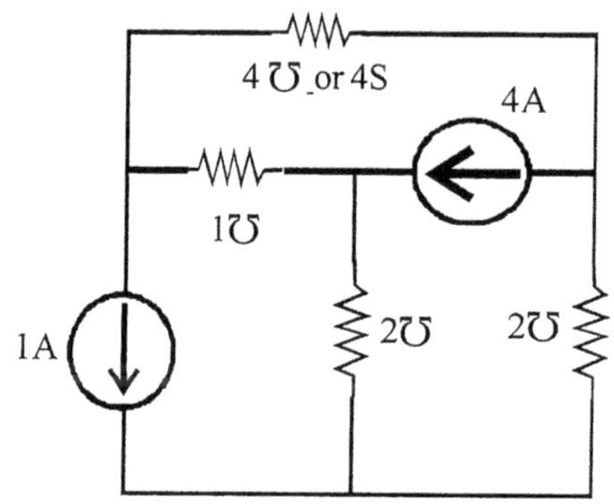

* * * * *

3

FUNDAMENTALS OF ALTERNATING CURRENT

3.1 INTRODUCTION

Most of the electrical energy supplied is in the form of alternating currents, because generation, transmission and distribution of electric supply in the form of AC current will be economical and easy.

In this chapter, we introduce fundamentals of *Alternating Currents* like *periodic* and *aperiodic waveforms*, representation of AC quantities in different forms, Power & Power Factor and Energy.

3.2 GENERATION OF ALTERNATING VOLTAGES & CURRENTS

A *D.C. Voltage* or *D.C. Current* will have a steady or constant value with respect to time. In other words, the values of voltage or current in D.C. will be constant at all instants of time. On the other hand, the *Alternating Voltage* or *Alternating Current* will be varying in value and also in direction with respect to time. In other words, the values of A.C. Voltages and Currents will be changing in *magnitude and direction* from time to time.

D.C. Voltages are generated by ***Batteries*** or ***D.C. Generators***. Batteries convert *chemical energy into electrical energy*. D.C. Generators convert *mechanical energy into* D.C. *Electrical Energy*.

A.C. Voltages are generated by ***A.C. Generators*** or ***Alternators*** which convert *mechanical energy into* A.C. *Electrical Energy*. Use of alternating voltages are very popular in view of easy generation, transmission and distribution.

A D.C. Generator also generates A.C. Voltages inherently, but provision is made to convert the A.C. Voltage into D.C. Voltage for use by means of ***Commutators*** and ***Brushes***. The value of the waveform at any particular instant of time is known as the ***Instantaneous Value*** of the quantity. It is denoted by small case letters like v, i or $v(t)$, $i(t)$ etc.

3.2.1 D.C. Waveform or Steady Waveform

The pictorial representation of the value of a steady quantity like D.C. Voltage or Current with respect to time is known as a ***Steady Waveform***.

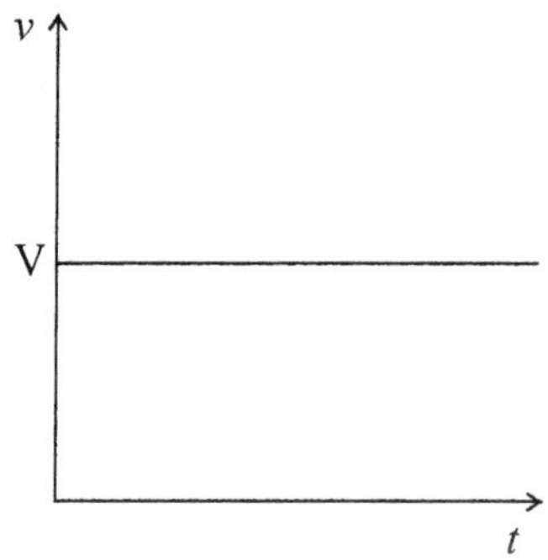

Fig. 3.1 *D.C. Waveform*

3.2.2 A.C. Waveforms

The pictorial representation of the value of any quantity like A.C. Voltage or Current which varies with respect to time is known as an ***Alternating Waveform*** of that quantity.

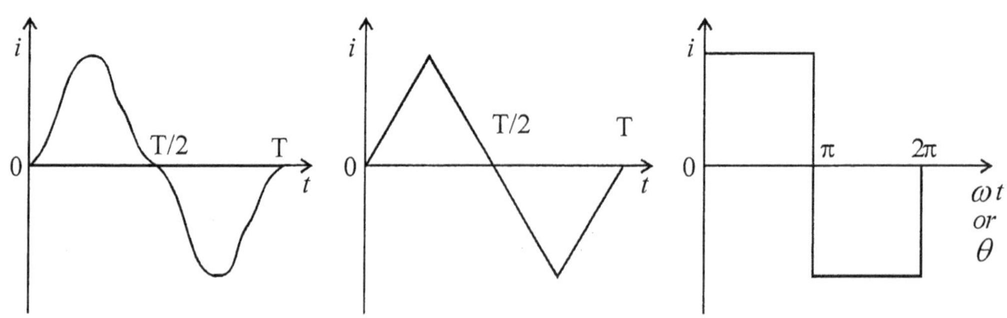

(**a**) Irregular Waveform (**b**) Triangular Waveform (**c**) Square Waveform

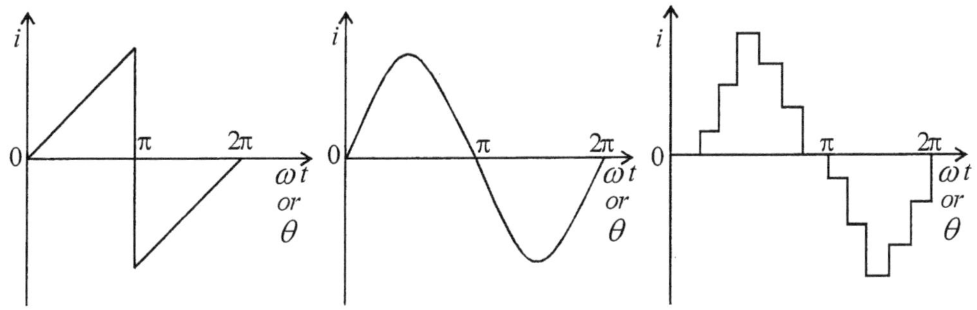

(d) Sawtooth Waveform (e) Sinusoidal Waveform (f) Stepped Waveform

Fig. 3.2 A.C. Waveforms

3.2.3 Periodic and Aperiodic Waveforms

If a pattern of a waveform over a certain period of time goes on repeating successively over intervals of time equal to the interval for the given pattern then, the waveform is said to be a ***Periodic Waveform***. If the wave form does not repeat in time, it is an ***Aperiodic Waveform***. Some periodic and aperiodic waveforms are given below.

(a) Irregular Waveform

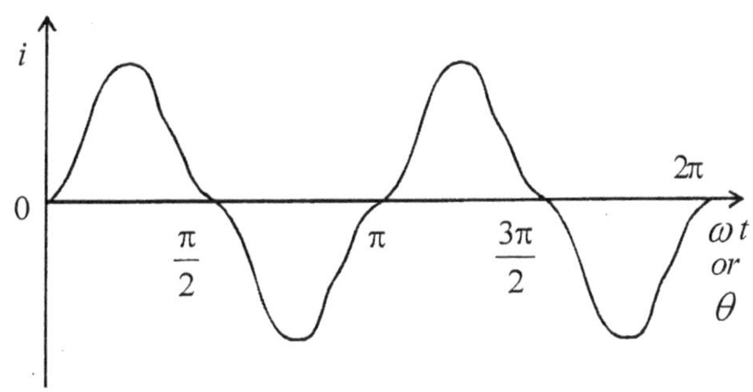

(b) Triangular Waveform

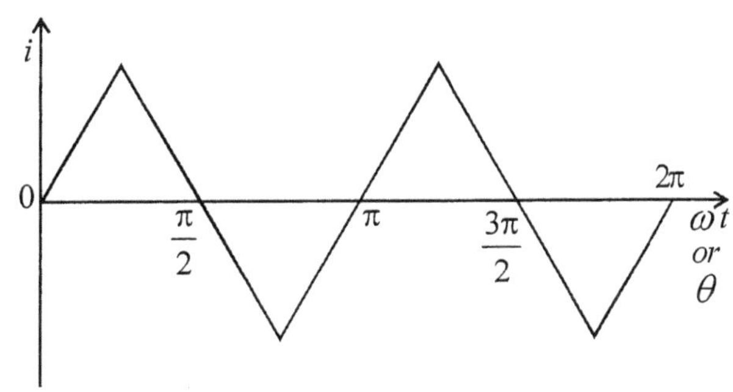

(c) Square Waveform

(d) Saw Tooth Waveform

(e) Sinusoidal Waveform

(f) Stepped Waveform

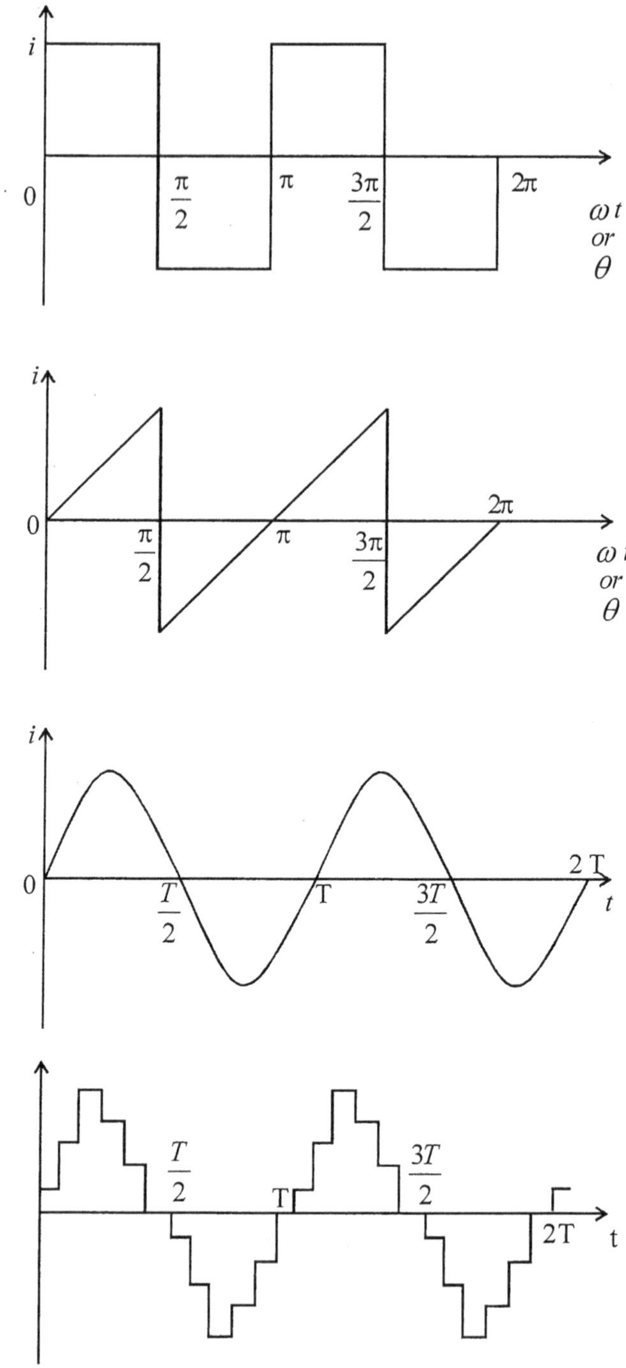

Fig. 3.3 Periodic A.C. Waveforms

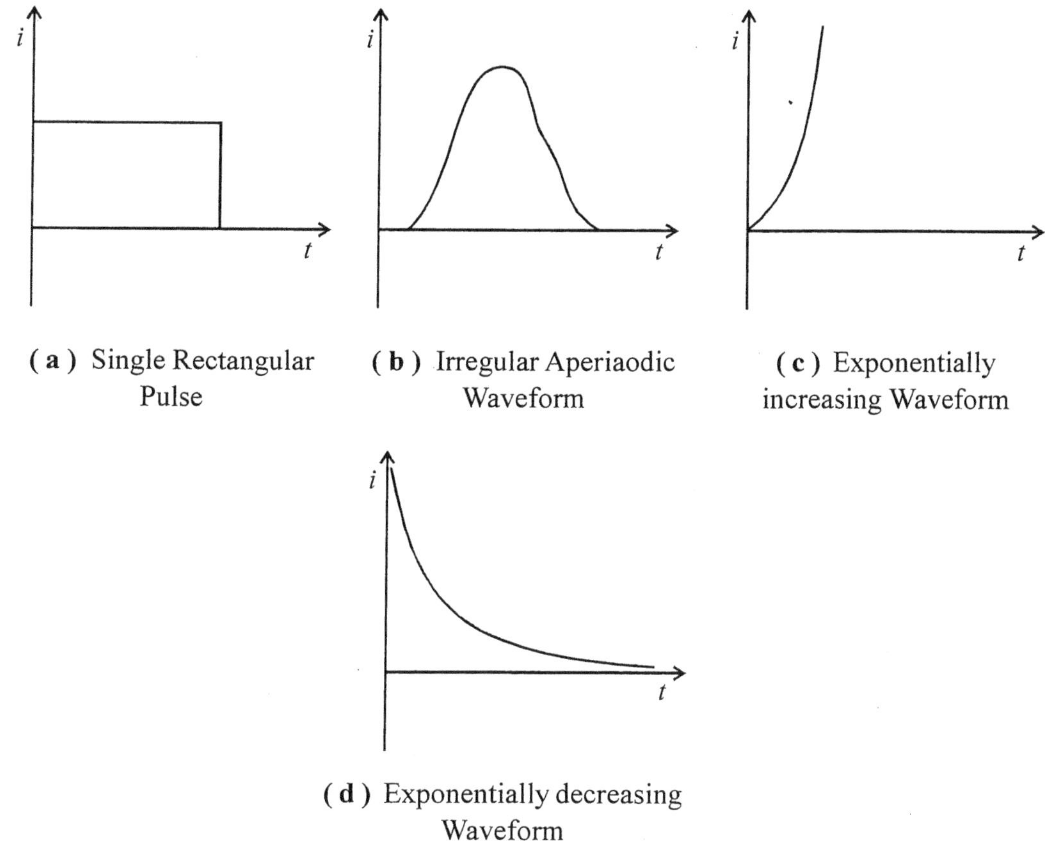

Fig. 3.4 *Aperiodic Waveforms*

3.2.4 Sinusoidal Waveforms

Among the periodic waveforms, ***Sinusoidal Waveforms*** are very popular because

1. ***Sinusoidal Voltages*** *whose values vary sinusoidally with respect to time can be generated easily using A.C. Generators.*
2. ***Non-sinusoidal*** *periodic waveform can be expressed as a sum of a constant term and a number of sine and cosine waveforms of frequencies equal to multiples of the fundamental frequency* ω, *using Fourier Series Expansion. The constant value is known as D.C. term.*

The sine and cosine waveforms of angular frequency ω are known as the fundamental waves and the waves of frequencies 2ω, 3ω, etc., are known as **Second Harmonic**, **Third Harmonic Waves**, *etc. Harmonics are undesirable since they increase heating effect. Methods have to be taken to suppress the harmonics and have as smooth a sinusoidal waveform as possible.*

3. *A cosine waveform is a sine waveform displaced by an angle of $\pi/2$ radians or 90° or by a time period $T/4$ where T is the time period of the waveform which is the time taken for the repeating part of the wave to occur.*

A Cosine or Sine wave form can be expressed in terms of Exponential Waveforms.

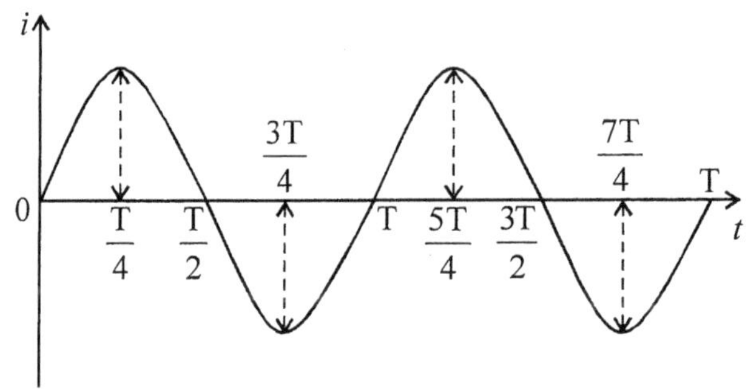

Fig. 3.5 *Sine Waveform*

3.2.5 Generation of Sinusoidal Waveforms

When a current carrying conductor placed in a magnetic field exhibits a motion then the magnetic flux linkages linking the conductor changes with respect to time and an E.M.F. (e) is induced in the conductor. This is called ***Motional E.M.F.*** or ***Dynamically Induced E.M.F.***, which states that, whenever there is a relative motion between a magnetic flux and a conductor an E.M.F. is induced in the conductor and the direction of induced E.M.F. is given by ***Lenz's Law***. This induced E.M.F. will oppose the cause that produces it. The instantaneous value of the voltage is given as

$$e = Blv \sin\theta$$

where B is the flux density in Wb/m^2

l is the length of the conductor in meters

v is the velocity of the motion of the conductor in m/sec.

θ is the angle between the motion of the conductor with the direction of the magnetic field

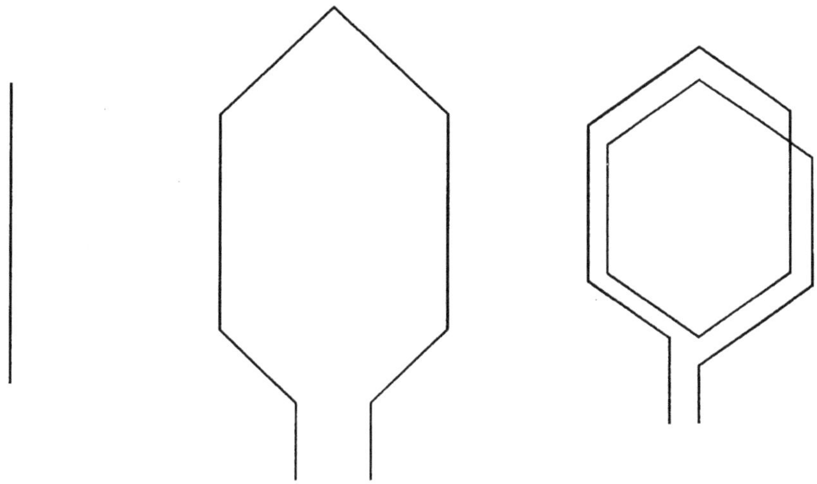

(**a**) Conductor

(**b**) a turn of two conductors or two coil sides

(**c**) a coil of two turns

Fig. 3.6 *A conductor and a coil*

A turn consists of two conductors which are connected at one end. A coil consists of number of turns connected in series.

In an alternator (A.C. Generator), a coil of two sides is placed on the armature which rotates in a magnetic field setup by a pair of electro-magnets called ***Poles***. One side of the coil will be under a particular pole called ***North Pole*** from which the magnetic flux emanates and the other side of the coil called ***South Pole*** will be in the opposite pole into which the flux enters as shown in Fig. 3.7.

Hence, the E.M.F. induced in the conductors of one coil side will be having a particular polarity say positive and the conductors on the other side of the coil will have E.M.F. of opposite sign say negative induced in them. Because of the connection between the two sides of the coil, the E.M.F. on the two sides of the coil get added up. Since the number of coil turns are in series, the total E.M.F. induced in the coil will be equal to the coil per turn multiplied by number of turns per coil. The direction of E.M.F. is given by Fleming's Right Hand Rule.

Note :

$$e = -N\frac{d\phi}{dt} \qquad \text{as in Fig. 3.7 (b).}$$

Hence, if the coil sides are directly under the pole, maximum E.M.F. will be induced in the conductors, even though the flux linkages with the coil is zero because the rate of change of flux linkages with the coil will be maximum at that instant. When the coils sides as in Fig. 3.7 (a), the E.M.F. induced in the conductors will be zero, even though maximum flux links with the coil because the rate of change of flux linkages at that instants will be zero.

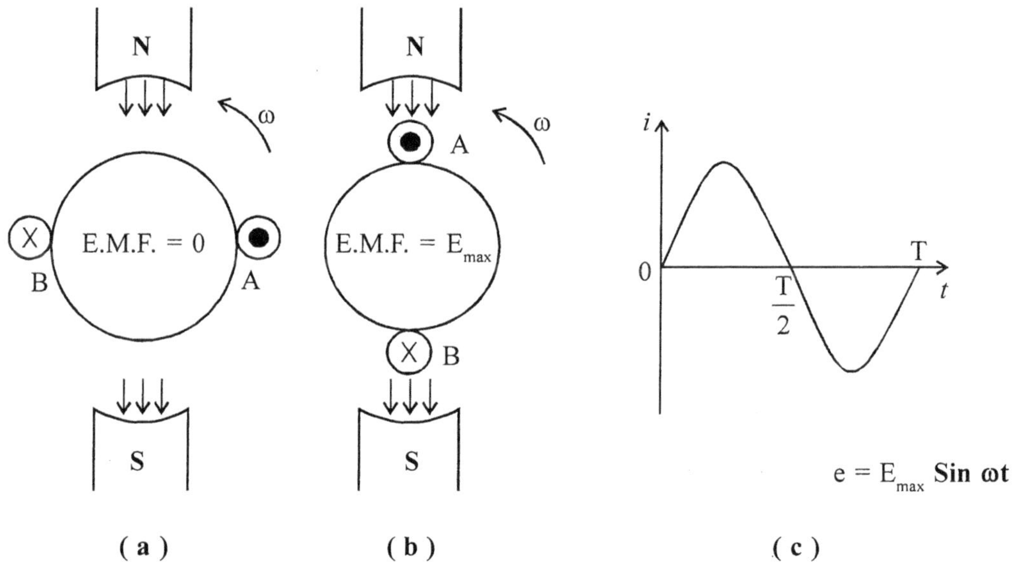

Fig. 3.7 Two - Pole Generator

3.3 EXPRESSION FOR THE E.M.F. GENERATED BY A SINGLE PHASE ALTERNATOR

The expression for e.m.f. generated in a coil of N turns each turn having two sides rotating in the magnetic field of maximum flux Φ_m Wb at an angular velocity of is ω radians per second, is given by the expression

$$e = \omega N\Phi_m \mathbf{Sin}\ \theta \text{ Volts}$$

$$e = E_m \mathbf{Sin}\ \theta \text{ Volts}$$

where $\theta = \omega t$ radians, ... t in seconds

or $$\theta = \frac{\omega t}{\pi} \times 180^\circ$$

e = instantaneous value of the voltage induced in Volts.

E_m = Max. value of the E.M.F. induced.

$$E_m = \omega N \Phi_m$$
$$= 2 \times \pi \times f \times N \times B_m \times A \text{ Volts}$$

where B_m is the maximum flux density in Wb/m^2

A is the area of the coil in $m^2 = b \times l$

The wave form of voltage generated will be sinusoidal in nature with respect to time or sinusoidal in respect to the angle ωt covered during rotation of the coil.

The e.m.f. can also be expressed as

$$e = E_m \mathbf{Sin}\ \theta \text{ Volts}$$

where θ is the angle covered in time 't' seconds. θ will be in radians. θ can also be expressed in degrees (electrical degrees different from mechanical degrees).

Electrical degrees equals mechanical degrees into p where p is the number of pairs of poles = P/2.

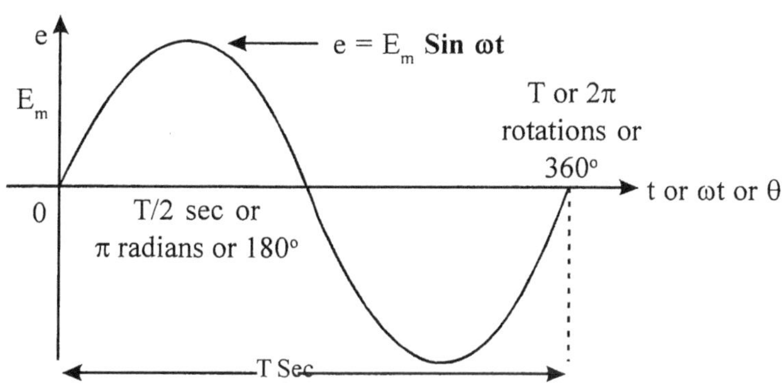

***Fig. 3.8** One cycle of e.m.f induced by a single phase alternator*

3.3.1 Time Period (T)

Time Period is the period of time over which the given wave exhibits a pattern which repeats periodically over the same intervals of time period. In the case of sinusoidal waveform, the time taken by the waveform to start from zero value, increase sinusoidally to a positive maximum value, start decreasing sinusoidally to zero value and then increase negatively from zero to negative maximum and returns to zero value, sinusoidally.

3.3.2 Amplitude

The maximum value positive or negative attained by an alternating quantity in one cycle is called its ***Amplitude*** or ***Peak Value*** or ***Maximum Value*** and is denoted by E_m (or V_m) and I_m respectively for sinusoidal voltage and current wave forms.

3.3.3 Cycle

If an alternating quantity undergoes a set pattern of variation over a period of time T then, it is said to have undergone one full ***Cycle***.

3.3.4 Frequency

Frequency of a given alternating waveform is the number of periodic variations, the waveform undergoes in one second which is measured in Hertz (Hz). The normal frequency of the A.C. supply in India and in UK will be 50Hz, whereas in USA, the normal frequency of operation is 60Hz. It was earlier measured in cycles per second.

3.3.5 Relation between Frequency and Time Period

If the frequency of an alternating quantity is f Hertz or cycles/sec (c/s). Hence, the time taken to complete one full cycle is $T = \frac{1}{f}$ sec, then the frequency of the waveform f is given as

$$f = \frac{1}{T} \text{ Hertz} \qquad \text{.......... (3.3.1)}$$

or $$f = \frac{1}{T} \text{ Cycles/sec}$$

Hertz is also denoted as Hz and is named after the scientist Hertz.

The instantaneous value of a sinusoidal voltage is given as

$$e = E_m \mathbf{Sin}\ \omega t \qquad \text{.......... (3.3.2)}$$

or $$e = E_m \mathbf{Sin}\ \theta \qquad \text{.......... (3.3.2 a)}$$

where E_m is the maximum value of the voltage

t is the time

θ is the angle in radians; $\theta = \omega t$ radians

ω is the angular velocity in radians/sec.

Similarly, the instantaneous value of a sinusoidal current is given as

$$i = I_m \mathbf{Sin}\ \omega t \qquad \text{.......... (3.3.3)}$$

or $$i = I_m \mathbf{Sin}\ \theta \qquad \text{.......... (3.3.3a)}$$

where I_m is the maximum value of the current.

3.3.6 Relation between Frequency and Angular Velocity

If the alternating quantity has a frequency of f c/s, the angular distance covered in one full cycle is 2π radians. Hence the angular distance covered per second in f cycles is $2\pi f$ radians. Hence the angular velocity,

$$\omega = 2\pi f \text{ radians} \qquad \text{.......... (3.3.4)}$$

where f is the frequency in Hz.

3.3.7 Relationship Between Frequency, Speed And Number of Pole Pairs

The frequency

f = The number of cycles per second

= The number of cycles per revolution × number of revolutions per second

$$f = p \times n \text{ Hz.} \qquad \text{.......... (3.3.5)}$$

where p is the number of pairs of poles and

n is the speed in number of revolutions per second

or $$f = \frac{P}{2} \times \frac{N}{60}$$

$\therefore$ $$f = \frac{PN}{120} \qquad \text{.......... (3.3.6)}$$

where P is the number of poles

N is the speed in number of revolutions per minute

Since it is not possible to have lesser than two poles, the highest speed at which a 50Hz A.C. Generator can be operated is

$$N = \frac{120f}{P}$$

$$= \frac{120 \times 50}{2}$$

$$= 3000 \text{ rpm}$$

For a frequency of 60Hz used in USA, the maximum speed will be 3,600 rpm.

3.3.8 Average Value of An Alternating Current

Most of the electrical energy is provided by rotating A.C. Generators. The E.M.F.s and the resulting voltages and currents are for the most part sinusoidal which is the waveform on which we have concentrated. However, the use of electronic switching has resulted in many circuits operating with waveforms which are anything but Sinusoidal Square Waveforms are especially common in communication circuits. Hence, a term called ***Average Value*** is used in the case of A.C. waveforms of current, voltage or any other quantity.

*The **Average Value** of an A.C. Current is defined as that value of the steady D.C. Current which when flowing through an electric circuit for a given time T transfers the same amount of electric charges as the charges transferred by the given A.C. Current which when flowing through the same electric circuit for the same amount of time.*

Mathematically, the average value of the A.C. Current,

$$I_{av} = \frac{1}{T}\int_0^T i(t)dt \text{ Amps.} \qquad \text{.......... (3.3.7)}$$

or $$I_{av} = \frac{1}{T}\int_0^T i\,dt \text{ Amps.} \qquad \text{.......... (3.3.8)}$$

where T is the time period of the given A.C. Current

$i(t)$ or i is the instantaneous value of the A.C. Current

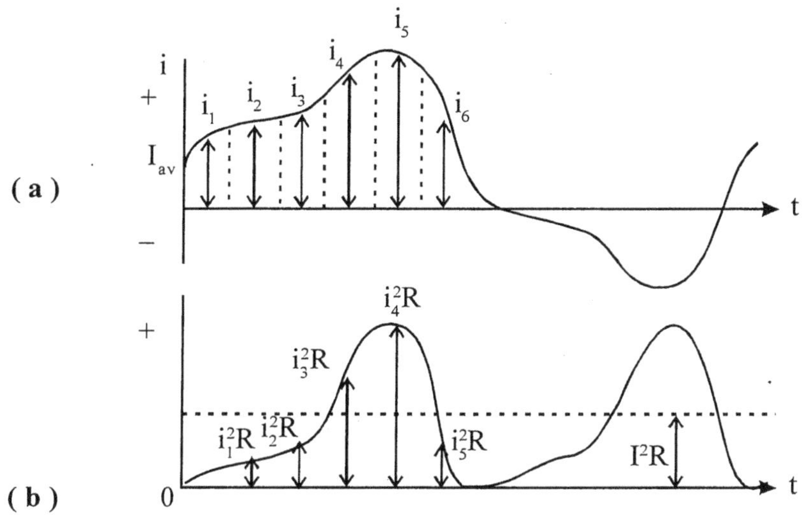

Fig. 3.9 *Average and R.M.S. Values*

Graphically for a symmetrical waveform which cannot be represented by a simple mathematical expression, by considering n equidistant mid-ordinates, i_1, i_2, i_3, etc., are taken over either the positive or the negative half-cycle for a symmetrical waveform, ***Average Value*** of current over half a cycle is

$$I_{av} = \frac{i_1 + i_2 + i_3 + \ldots + i_n}{n} \qquad \text{.......... (3.3.9)}$$

or alternatively, average value of current is,

$$\frac{\text{Area enclosed over half - cycle}}{\text{Length of base over half - cycle}} \qquad \text{.......... (3.3.10)}$$

But in the case of a unsymmetrical alternating current (like half-wave rectified current) the average value must always be taken over the whole cycle.

Note: A Symmetrical Waveform ***will have the same form, but with opposite signs in the positive and negative half-cycles.***

For a sinusoidal waveform or any symmetrical waveform, the expression,

$$I_{av} = \frac{1}{T}\int_0^T i(t)dt \text{ Amps} = 0 \qquad \text{.......... (3.3.11)}$$

For Sinusoidal Waveform,

One period or One cycle

$$I_{av} = \frac{1}{2\pi}\int_0^{2\pi} I_m \text{Sin}\,\theta\, d\theta$$

$$= \frac{1}{2\pi} I_m \left[-\textbf{Cos }\theta\right]_0^{2\pi}$$

$$= 0$$

Hence, ***in such cases, as a sinusoidal or any symmetrical alternating waveform the averaging is done over one half-cycle instead of over a full-cycle.*** For a sinusoidal alternating current

$$I_{av} = \frac{1}{T/2}\int_0^{T/2} i(t)dt \text{ Amps.}$$

or $$I_{av} = \frac{1}{\pi}\int_0^{\pi} I_m \text{Sin}\,\theta\, d\theta$$

$$I_{av} = \frac{1}{\pi} I_m \left[-\textbf{Cos }\theta\right]_0^{\pi}$$

$$= \frac{I_m}{\pi/2}$$

$$= 0.637\, I_m \qquad \text{.......... (3.3.12)}$$

or $$= \frac{2I_m}{\pi} \qquad \text{.......... (3.3.12a)}$$

Note: *The above integral can also be expressed as*

$$I_{av} = \frac{1}{\pi}\int_{\omega t=0}^{\pi} I_m \, \mathbf{Sin}\, \omega t \, d(\omega t) \qquad \text{.......... (3.3.13)}$$

$$= \frac{1}{\pi}\int_{0}^{\pi} I_m \text{Sin}\,\theta\, d\theta \qquad \text{.......... (3.3.13a)}$$

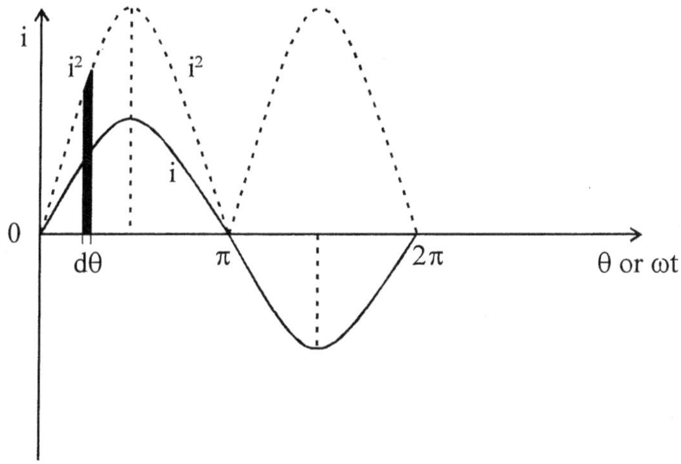

Fig. 3.10 Wave forms to find R.M.S. and average values of current

3.3.9 Root Mean Square (RMS) Value Of An Alternating Current

*The **RMS Value** of an A.C. Current is defined as that value of the steady D.C. Current which when flowing through an electric circuit for a given time T produces the same amount of heat as the heat produced by the given A.C. Current which when flowing through the same electric circuit for the same amount of time.*

The RMS Value is also known as ***Effective Value***. The RMS Value of the current is denoted by I_{rms} or simply I. In all A.C. Circuits unless otherwise stated, the voltages and currents given will be RMS Values and denoted by V or I.The Voltmeters and Ammeters used for measuring the voltages and currents will always read the RMS values of the voltages and currents.Generally, moving iron meters will be used for measuring the A.C. Voltages and A.C. Currents.

$$I_{rms} = \sqrt{\frac{1}{T}\int_{0}^{T} i^2(t)dt} \text{ Amps.} \qquad \text{.......... (3.3.14)}$$

or $$I_{rms} = \sqrt{\frac{1}{T}\int_0^T i^2\, dt} \text{ Amps.} \qquad \text{.......... (3.3.14a)}$$

where T is the time period of the given A.C. Current

$i(t)$ or i is the instantaneous value of the A.C. Current

Graphically for a waveform which cannot be represented by a simple mathematical expression, by considering n equidistant mid-ordinates, i_1, i_2, i_3, etc., are taken over either the positive or the negative half-cycle for a symmetrical waveform, ***Average Value*** of current over half a cycle is

$$I_{rms}^2 = \frac{i_1^2 + i_2^2 + i_3^2 + \ldots + i_n^2}{n}$$

$$I_{rms} = \sqrt{\frac{i_1^2 + i_2^2 + i_3^2 + \ldots + i_n^2}{n}} \qquad \text{.......... (3.3.15)}$$

$$I_{rms} = \sqrt{\text{Mean of Squares of Instantaneous Values}}$$

$$\text{.......... (3.3.15a)}$$

For a sinusoidal waveform,

$$I_{rms} = \sqrt{\frac{1}{T}\int_0^T i^2(t)\, dt} \text{ Amps.} \qquad \text{.......... (3.3.16)}$$

$$= \sqrt{\frac{1}{T}\int_0^T (I_m \textbf{Sin}\, \omega t)^2\, dt} \text{ Amps.} \qquad \text{.......... (3.3.17)}$$

or $$I_{rms} = \sqrt{\frac{1}{2\pi}\int_{\omega t = 0}^{2\pi} (I_m \textbf{Sin}\, \omega t)^2\, d(\omega t)}$$

or $$I_{rms} = \sqrt{\frac{1}{2\pi}\int_{\theta = 0}^{2\pi} (I_m \textbf{Sin}\, \theta)^2\, d\theta} \qquad \text{.......... (3.3.18)}$$

Since, the waveform for the square of the current will be the same for each half cycle, we can also calculate I_{rms} as,

$$I_{rms} = \sqrt{\frac{1}{T/2}\int_0^{T/2} i^2(t)\,dt} \text{ Amps.} \qquad \text{.......... (3.3.19)}$$

or

$$I_{rms} = \sqrt{\frac{1}{\pi}\int_{\theta=0}^{\pi} (I_m \mathbf{Sin}\,\theta)^2\, d\theta} \qquad \text{.......... (3.3.19a)}$$

$$= \sqrt{\frac{I_m^2}{\pi}\int_0^{\pi} \mathbf{Sin}^2\,\theta\, d\theta}$$

$$= \sqrt{\frac{I_m^2}{\pi}\int_0^{\pi}\left(\frac{(1-\mathbf{Cos}\,2\theta)}{2}\right) d\theta}$$

$$= \sqrt{\frac{I_m^2}{2\pi}\int_0^{\pi}(1-\mathbf{Cos}\,2\theta)\, d\theta}$$

$$= \sqrt{\frac{I_m^2}{2\pi}\left(\theta - \frac{\mathbf{Sin}\,2\theta}{2}\right)_0^{\pi}}$$

$$= \sqrt{\frac{I_m^2}{2\pi}\left((\pi-0) - \frac{\mathbf{Sin}\,2\pi - \mathbf{Sin}\,0}{2}\right)}$$

$$= \sqrt{\frac{I_m^2}{2\pi}[(\pi-0)-(0-0)]}$$

$$= \sqrt{\frac{I_m^2}{2}}$$

$$= \frac{I_m}{\sqrt{2}} = 0.707\, I_m \qquad \text{.......... (3.3.19b)}$$

Example 3.1:

An alternating voltage has the equation,

$$v = 141.4 \textbf{ Sin 377t}$$

What are the values of :

(a) *R.M.S. Voltage*
(b) *Frequency*
(c) *The instantaneous Voltage when t = 3 ms*

Solution:

The relation is of the form,

$$v = V_m \textbf{Sin } \boldsymbol{\omega} \textbf{t}$$

and by comparison,

$$V_m = 141.4V = \sqrt{2}\, V_{rms}$$

Hence,

$$V = \frac{141.4}{\sqrt{2}} = 100V$$

Also by comparison,

$$\omega = 377 \text{ rad/s} = 2\pi f$$

Hence,

$$f = \frac{377}{2\pi} = 60\text{Hz.}$$

Finally,

$$v = 141.4\textbf{Sin 377t}$$

When, $t = 3 \times 10^{-3}\text{sec}$

$$v = 141.4 \textbf{ Sin (377} \times \textbf{3} \times \textbf{10}^{-3} \textbf{)}$$

$$= 141.4 \textbf{ Sin 1.131}$$

$$= 141.4 \times 0.904$$

$$= 127.8V$$

Note that, in this example, it was necessary to determine the sine of 1.131rad, which could either be obtained from suitable tables, or from a calculator. Alternatively, 1.131rad may be converted into degree measurement, i.e.,

$$1.131 \text{ rad} \cong 1.131 \times \frac{180}{\pi}$$

$$= 64.8^{\circ}$$

3.3.10 Form Factor And Peak Factor

There exists a definite relation among the average value, RMS Value and Peak Value of an alternating quantity. The relationship is expressed by the two factors namely, form factor and peak factor.

(i) *Form Factor :*

The ratio of RMS Value to Average Value of an alternating quantity is called ***Form Factor***.

Mathematically,

$$\text{Form Factor } F = \frac{\text{RMS Value}}{\text{Average Value}}$$

$$F = \frac{I_{rms}}{I_{av}} \text{ or } \frac{E_{rms}}{E_{av}} \qquad \text{.......... (3.3.20)}$$

For the current varying sinusoidally,

$$\text{Form Factor} = \frac{I_{rms}}{I_{av}}$$

$$= \frac{\frac{I_m}{\sqrt{2}}}{\frac{2I_m}{\pi}}$$

$$= \frac{\pi I_m}{2\sqrt{2} I_m}$$

$$= 1.11 \qquad \text{.......... (3.3.21)}$$

(ii) *Peak Factor :*

The ratio of maximum value to RMS Value of an alternating quantity is called ***Peak Factor***. Peak Factor is also known as ***Crest Factor*** or ***Amplitude Factor***.(k_m or k_p or k_c.

Mathematically,

$$\text{Peak Factor } k_a = \frac{\text{Maximum Value}}{\text{RMS Value}}$$

$$= \frac{I_m}{I_{rms}} \text{ or } \frac{E_m}{E_{rms}} \quad \text{.......... (3.3.22)}$$

For current varying sinusoidally,

$$\text{Peak Factor} = \frac{I_m}{I_{rms}} = \frac{I_m}{\frac{I_m}{\sqrt{2}}}$$

$$= \sqrt{2} = 1.4142 \quad \text{.......... (3.3.23)}$$

Example 3.2:

Calculate the R.M.S. value, average value, form factor and crest factor for the given square wave voltage.

Solution:

For period of $0 < t < 0.2$ second,

$v = 10$ V

For period of $0.2 < t < 0.5$ second,

$v = 0$

The periodic time of the voltage is

T = 0.5 second

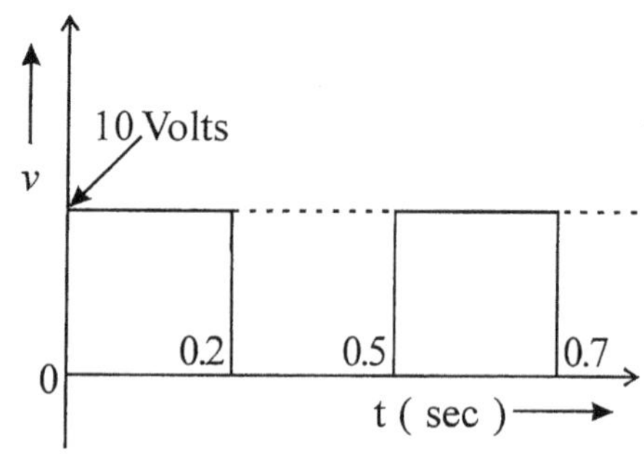

$$V^2 = \frac{1}{T}\int_0^T v^2\,dt \quad \frac{1}{T} = \left[\int_0^{0.2} 10^2\,dt + \int_{0.2}^{0.5} 0\,dt\right]$$

$$= \frac{1}{0.5}\left[(10)^2 t\right]_0^{0.2}$$

$$= \frac{100 \times 0.2}{0.5} = 40$$

Hence,

$$V = V_{rms} = \sqrt{40} = 6.234 \text{ Volts}$$

Average Voltage,

$$V_{av} = \frac{1}{T}\int_0^T v \, dt = \frac{1}{T}\left[\int_0^{0.2} 10 dt + \int_{0.2}^{0.5} 0 dt\right]$$

$$= \frac{1}{0.5}\int_0^{0.2} 10 \, dt$$

$$= \frac{10 \times 0.2}{0.5}$$

$$= 4 \text{ Volts}$$

Form Factor,

$$F = \frac{V}{V_{av}}$$

$$= \frac{6.234}{4}$$

$$= 1.58$$

Crest Factor or Peak Factor or Amplitude Factor

$$k_a = \frac{10}{6.234}$$

$$= 1.6$$

Examples 3.3:

An alternating voltage has an amplitude of 100V. Find its (i) R.M.S. value (ii) average value.

Solution:

We have, for an alternating voltage :

(i) R.M.S. Value $E = \frac{E_m}{\sqrt{2}}$

$$= 0.707\ E_m$$

Given $E_m = 100$ V.

$$\therefore\ E = 0.707 \times 100 =$$

$$= \mathbf{70.7\ Volts.}$$

(ii) Average value $E_{uv} = \frac{E_m}{(\pi/2)}$

$$\therefore\ E_{uv} = 0.637\ E_m$$

$$= 0.637 \times 100$$

$$= 63.7\ \text{Volts.}$$

***Answer:* 70.7V, 63.7 V**

Example 3.4:

An alternating current has an effective value of 200A. If its frequency is 50 Hz, find its average value and write down the expression for the current.

Solution:

Given, RMS value, I = 200A and f = 25 Hz.

We have, $I = 1.11\ I_{av}$ for a pure sine wave.

$$\therefore\ \text{Average value} = \frac{I}{1.11} = \frac{200}{1.11}$$

$$= \mathbf{180.18\ Amps.}$$

Also, Maximum value = $\sqrt{2}$ (R.M.S. Value)

$$\therefore\ I_m = \sqrt{2} \times 200$$

$$= 282.84\ A$$

$\therefore$ The mathematical expression for the current is :

$$i = I_m \sin 2\pi ft$$

$$= 282.84 \sin\ (2\pi \times 50 \times t)$$

$$i = 282.84 \sin 314\ t.$$

3.4 NON-SINUSOIDAL WAVEFORMS

3.4.1 R.M.S. & Average Values of Halfwave Rectifier Output

A halfwave rectifier uses a diode to allow transmission of one half cycle (positive half cycle) of A.C. Voltage (or current) and disallow transmission of the other half. of one cycle. Fig. 3.11 shows a simple halfwave rectifier driven by a sinusoidal voltage and its output voltage waveform using load resistor R.

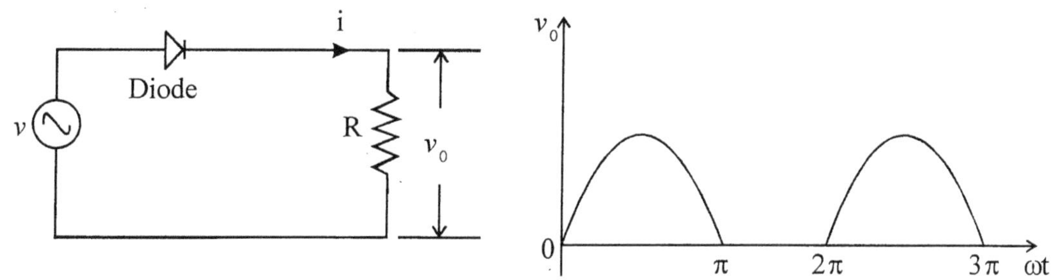

(**a**) Basic Circuit (**b**) Halfwave Rectifier Voltage

Fig. 3.11 *Basic Circuit of a Halfwave Rectifier*

R.M.S. Value of Half Rectified Sine Wave:

$$I_m = \frac{V_m}{(R + R_f)} \qquad \text{.......... (3.4.1)}$$

where R_f is the diode forward (conducting) resistance.

To find the R.M.S. value, summation of the square of the current is carried out over the period for which the current actually flows through the load resistor R i.e., for the period 0 to π radians. But this sum is averaged for the whole cycle i.e., from 0 to 2π radians. Let I be the R.M.S. value of current. Then,

$$I^2 = \frac{1}{2\pi}\int_0^{\pi} i^2 \, d\theta$$

$$I^2 = \frac{I_m^2}{2\pi}\int_0^{\pi} \mathbf{Sin}^2\,\theta \, d\theta \qquad \text{.......... (3.4.2)}$$

$$= \frac{I_m^2}{4\pi}\int_0^{\pi} (1 - \mathbf{Cos}\,2\theta) \, d\theta$$

$$= \frac{I_m^2}{4\pi}\left[\theta - \frac{\mathbf{Sin}\,2\theta}{2}\right]_0^{\pi} = \frac{I_m^2}{4} \qquad \text{.......... (3.4.3)}$$

Hence,

$$I = \frac{I_m}{2} = 0.5 I_m \qquad \text{.......... (3.4.4)}$$

Average Value of Half Rectified Sine Wave :

$$I_{av} = \frac{1}{2\pi}\int_0^{\pi} i \, d\theta$$

$$= \frac{I_m}{2\pi}\int_0^{\pi} \mathbf{Sin}\,\theta \, d\theta \qquad \text{.......... (3.4.5)}$$

$$= \frac{I_m}{2\pi}\left[-\mathbf{Cos}\,\theta\right]_0^{\pi} = \frac{I_m}{\pi} \qquad \text{.......... (3.4.6)}$$

Form Factor of Half Rectified Sine Wave :

$$F = \frac{\text{R.M.S. Value}}{\text{Average Value}}$$

$$= \frac{I_m}{2} \Big/ \frac{I_m}{\pi} = \frac{\pi}{2} = 1.57 \qquad \text{......... (3.4.7)}$$

3.4.2 R.M.S. & Average Values of Fullwave Rectifier Output

A fullwave rectifier uses two diodes each acting as a halfwave rectifier. The circuit is so arranged that one diode conducts during say positive half of the sine wave input while the other diode conducts during the other half i.e., the negative half of the input voltage. Thus Fig. 3.12 shows the output current where I_m is the amplitude of the output current.

$$I_m = \frac{V_m}{(R + R_f)} \qquad \text{.......... (3.4.8)}$$

where R is the load resistance and R_f is the diode forward (conducting) resistance.

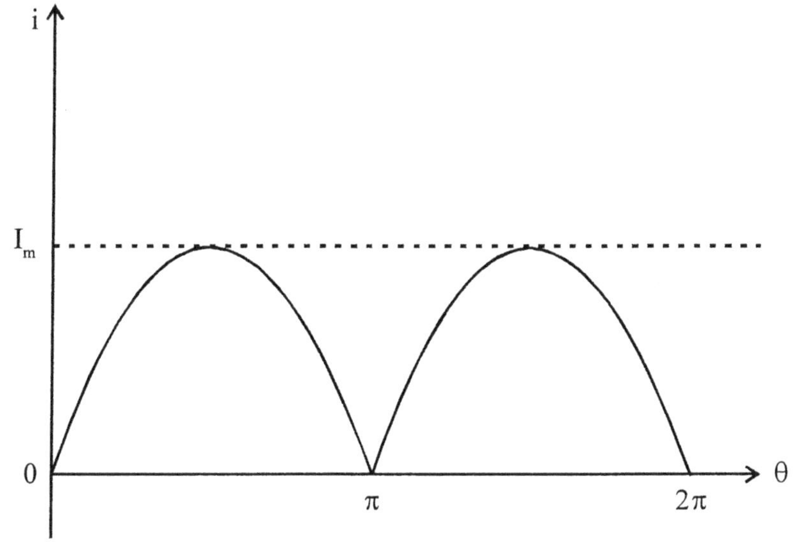

Fig. 3.12 *Output Current of a Fullwave Rectifier*

R.M.S. Value of Full Rectified Sine Wave :

The two half waves are identical. Hence,

$$I^2 = \frac{1}{2\pi}\int_0^{2\pi} 2i^2 \, d\theta$$

$$= \frac{I_m^2}{\pi}\int_0^{\pi} \mathbf{Sin}^2\,\theta \, d\theta \qquad \text{.......... (3.4.9)}$$

$$= \frac{I_m^2}{\pi}\int_0^{\pi}\frac{(1-\mathbf{Cos2\theta})}{2}\,d\theta$$

$$= \frac{I_m^2}{2\pi}\left[\theta - \frac{\mathbf{Sin\,2\theta}}{2}\right]_0^{\pi} = \frac{I_m^2}{2} \qquad \text{.......... (3.4.10)}$$

Hence,

$$I = \frac{I_m}{\sqrt{2}} \qquad \text{.......... (3.4.11)}$$

Average Value of Full Rectified Sine Wave :

$$I_{av} = \frac{1}{2\pi}\int_0^{\pi} 2i\,d\theta$$

$$= \frac{I_m}{\pi}\int_0^{\pi}\mathbf{Sin\,\theta}\,d\theta \qquad \text{.......... (3.4.12)}$$

$$= \frac{I_m}{\pi}[-\mathbf{Cos\,\theta}]_0^{\pi}$$

$$= \frac{2I_m}{\pi} \qquad \text{.......... (3.4.13)}$$

Note:

The averaging can be done over 2π radians or π radians since the waveform is the same in both the half cycles.

Form Factor of Full Rectified Sine Wave :

$$F = \frac{\text{R.M.S. Value}}{\text{Average Value}}$$

$$= \frac{I_m}{\sqrt{2}}\Big/\frac{2I_m}{\pi}$$

$$= \frac{\pi}{2\sqrt{2}} = 1.11 \qquad \text{.......... (3.4.14)}$$

Example 3.5:

A sinusoidal voltage v = 100 **Sin 314t** Volts is applied to a diode which offers resistance to the flow of current in one direction and infinite resistance to the flow of current in the opposite direction. A resistor of 200Ω is connected in series with the A.C. source and the diode. Calculate the R.M.S. and Average Values of the current over one full cycle and the Form Factor.

Solution:

This is a half wave rectifier circuit

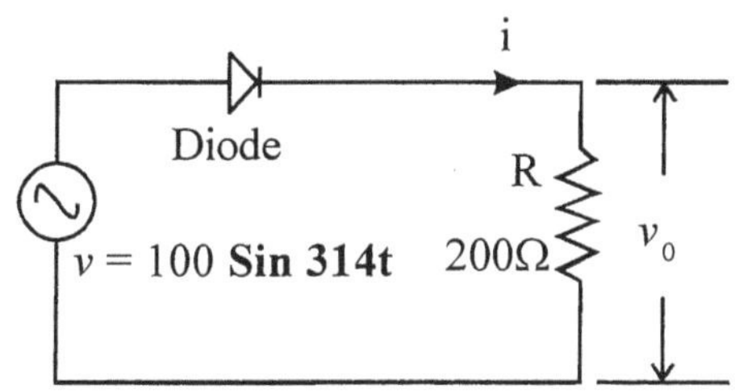

Current, $i = \frac{V}{200\Omega}$

$$= \frac{100 \textbf{ Sin 314t}}{200\Omega}$$

$$= 0.5 \textbf{ Sin 314t} \text{ Amp}$$

$$I_m = 0.5\text{Ampere}$$

R.M.S. Value for the half wave rectifier

$$I = \frac{I_m}{2}$$

$$= \frac{0.5}{2} = 0.25 \text{ Amp}$$

Average Value for the half wave rectifier

$$I_{av} = \frac{I_m}{\pi}$$

$$= \frac{0.5}{\pi} = 0.159\text{Amp}$$

Form Factor for the half wave rectifier

$$F = \frac{I}{I_{av}}$$

$$= \frac{0.25}{0.159} = 1.57$$

Example 3.6:

A halfwave rectifier in series with a Moving Iron ammeter for AC and a permanent magnet moving coil (PMMC) ammeter for D.C is fed from a sinusoidal supply. The A.C. ammeter reads 16Amperes. What will be the reading of the PMMC ammeter? What will be the readings of these ammeters if half wave rectifier is replaced by a fullwave rectifier i.e., the rectifier which conducts during both half cycles.

Solution:

A.C. ammeter reads R.M.S. value while PMMC meter reads average value of the rectifier output current. For halfwave rectifier,

$$I_{rms} = \frac{I_m}{2}$$

while
$$I_{av} = \frac{I_m}{\pi}$$

But A.C. ammeter reads 16Amperes i.e., I = 16A. Hence,

$$I_m = 2I$$
$$= 32A.$$

Hence,
$$I_{av} = \frac{I_m}{\pi}$$
$$= \frac{32}{\pi} = 18.18A.$$

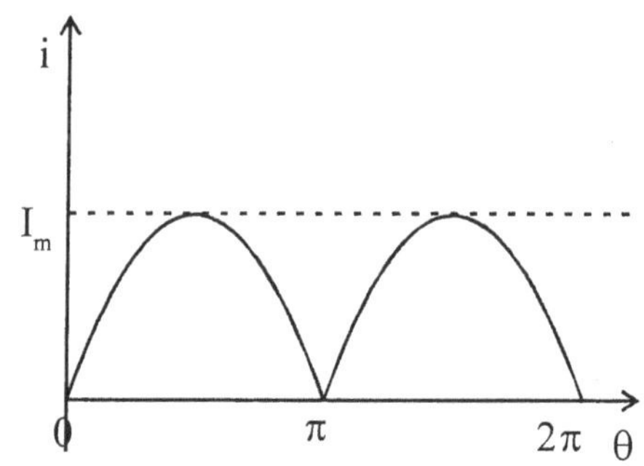

Thus, PMMC ammeter will read 18.18Amperes.

In case of full wave rectifier, current flows during both half cycles

$$I = \frac{I_m}{\sqrt{2}}$$
$$= \frac{32}{\sqrt{2}} = 22.63\text{Amperes}$$

Hence, A.C. Ammeter will read 22.63 Amperes. The Average Current over the complete cycle is given by,

$$I_{av} = \frac{1}{2\pi}\int_0^{\pi} 2i\, d\theta = \frac{I_m}{\pi}\int_0^{\pi} \textbf{Sin}\ \theta\, d\theta$$

$$I_{av} = \frac{I_m}{\pi}\left[-\textbf{Cos}\,\theta\right]_0^{\pi}$$

$$= \frac{2I_m}{\pi}$$

For the given circuit,

$$I_{av} = \frac{2\times 32}{\pi}$$

$$= 20.37\text{Ampere}$$

Hence, D.C. Ammeter will read 20.37Amperes

Having demonstrated the determination of average and R.M.S. values for sinusoidal currents and voltages, it is a relatively short step to consider non-sinusoidal quantities.

Example 3.7:

Calculate the form factor for each of the waveforms shown in the following figure.

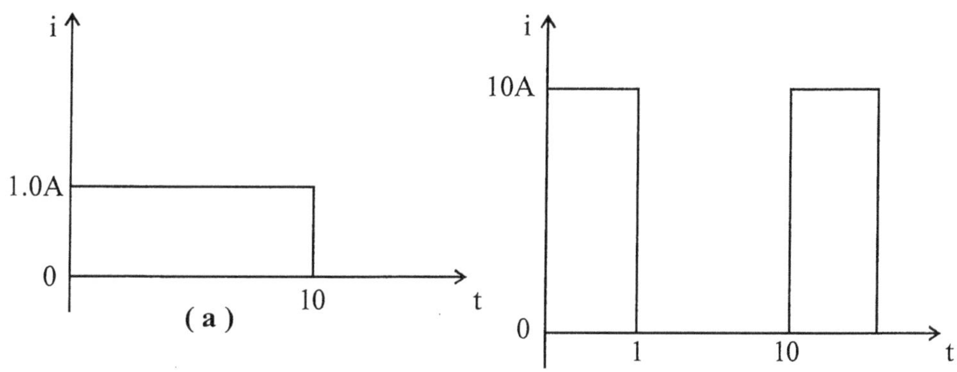

(a) (b)

Solution:

For Fig. **(a)** :

$$I_{av} = \frac{1(10-0)}{10-0} = 1.0\ \text{Amp.}$$

$$\text{Alternatively } I_{av} = \frac{1}{10}\int_0^{10} 1dt = \frac{1}{10}[t]_0^{10} = 1.0\text{A}$$

$$I_{rms} = \left(\frac{1^2(10-0)^2}{10-0}\right)^{½} = 1.0 \text{ Amps.}$$

$$\text{Alternatively } I = I_{rms} = \sqrt{\frac{1}{10}\int_0^{10} 1^2 dt} = \sqrt{\frac{1}{10}[t]_0^{10}} = 1.0 \text{ Amp.}$$

$$k_f = \frac{I_{rms}}{I_{av}} = \frac{1.0}{1.0} = 1.0$$

For Fig. **(b)** :

$$I_{av} = \frac{10(1-0)+0(10-1)}{10-0} == 1.0 \text{ Amp.}$$

$$\text{Alternately } I_{av} = \frac{1}{10}\left[\int_0^1 10dt + \int_1^{10} 0dt\right] = 1.0\text{A}$$

$$I_{rms} = \left(\frac{10^2(1-0)^2 + 0^2(10-1)^2}{10-0}\right)^{½} = 3.16 \text{ Amp.}$$

$$\text{Alternately } I_{rms} = \sqrt{\left[\frac{1}{10}\int_0^1 10^2 dt + \int_1^{10} 0dt\right]}$$

$$= \sqrt{\frac{1}{10}[100t]_0^1} = 3.16\text{A}$$

$$k_f = \frac{I}{I_{av}} = \frac{3.16}{1.0} = 3.16$$

It will be noted that the first waveform is that of direct current in which the R.M.S. current and the mean current have the same value. It is for this reason that the R.M.S. value of an alternating current may be equated to the mean value of a direct current both giving the same heating effect.

Example 3.8:

Average value, R.M.S. value and form factor of periodic wave forms

(a) Triangular wave:

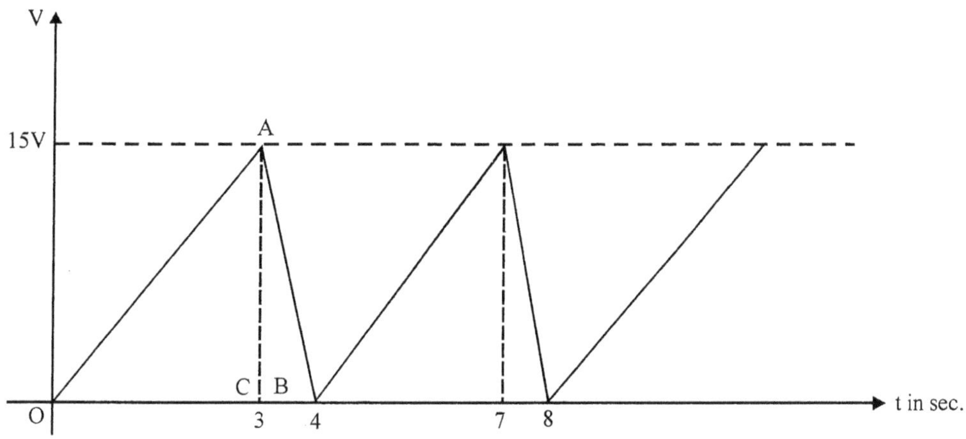

Consider the triangular wave shown in the figure.

Peak value AC = 15 V ; period T = 4 sec.

The wave repeats after every 4 sec.

From a study of the wave form it is eident that there are two distinct parts of the wave OA and AB.

In the interval t = 0 to t = 3 sec, the magnitude increases linearly from 0 to 15 V.

$$\text{Slope of OA} = \frac{AC}{OC} = \frac{15\text{Volts}}{3\text{sec}} = 5\,V/s$$

$$\text{But slope} = \frac{v}{t} \quad \therefore \frac{v}{t} = 5 \text{ or } v = 5t$$

In the interval t = 3 sec to t = 4 sec. the magnitude decreases linearly from 15 V to 0.

$$\text{Slope of } \frac{AC}{CB} = -\frac{15}{(4-3)} = -15\,V/\text{sec}.$$

$$\therefore \frac{v}{t} = -15 \text{ or v=-15t}$$

At t = 3 sec, v = 15 V and at t = 4 sec, v = 0.

The equation which satisfies there values is **v = -15t + 60**

$\therefore$ v = 5 t in the interval 0 < t < 3, and

v = -15t + 60 in the interval 3 < t < 4.

In general, average value $\frac{I}{T}\int_0^T v\,dt$

Average value of the given triangular wave is given by

$$V_{av} = \frac{1}{4}\left[\int_0^3 5t\,dt + \int_3^4 (-15t + 60)dt\right]$$

$$= \frac{1}{4}\left[5\left|\frac{t^2}{2}\right|_0^3 + (-15)\left|\frac{t^2}{2}\right|_0^4 + 60|t|_3^4\right]$$

$$= \frac{1}{4}\left[\frac{45}{2} - 15\left(8 - \frac{9}{2}\right) + 60(4-3)\right]$$

= 7.5 Volts.

$$\text{R.M.S. value} = \sqrt{\frac{1}{T}\int_0^T v^2 dt}$$

$$= \sqrt{\frac{1}{4}\left[\int_0^3 (5t^2)dt + \int_5^4 (-15t + 60)^2 dt\right]}$$

$$= \sqrt{\frac{1}{4}\left[25\int_0^3 t^3 dt + 15^2\int_5^4 (t^2 - 8t + 16)dt\right]}$$

$$= \sqrt{\frac{1}{4}\left[25\left|\frac{t^3}{3}\right|_0^3 + 225\left|\frac{t^3}{3} - 8\frac{t^2}{2} + 16t\right|_3^4\right]}$$

= 8.66 V on simplification

$$\text{Form factor} = \frac{\text{RMS value}}{\text{Average value}}$$

$$= \frac{8.66}{7.5} = 1.1547$$

(b) Square wave:

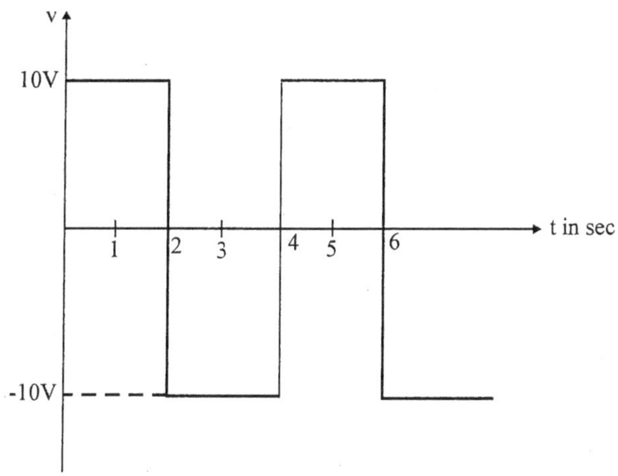

From the figure, it is obvious that :

v = 10V in the interval t = 0 to t = 2, and

v = -10 V in the interval t= 2 to t = 4

$\therefore$ v = 10V for 2 > t> 0, and v = -10V for 4 > t> 2.

Time period T = 4 sec.

We have (i) Average value $V_{av} = \frac{I}{T}\int_0^T v\,dt$

$$\therefore \quad v_{av} = \frac{1}{4}\left[\int_0^4 10dt + \int_2^4 (-10dt)\right]$$

$$= \frac{10}{4}\left[\int_0^2 dt - \int_2^4 dt\right]$$

(Even otherwise it is obvious. Since the positive and negative half-cycles are identical, the average value over one complete cycle is zero.

If however, for such symmetrical waves (symmetrical positive and negative parts), the average value over only a half-cycle is to be evaluated, we have :

$$V_{av} = \frac{2}{T}\int_0^{T/2} v\,dt$$

$$= \frac{2}{4}\int_0^{2} 10\,dt$$

$$= 5|t|_0^2 = \textbf{10 Volts.}$$

[Even this is obvious, since v = 10V throughout the half cycle]

(ii) R.M.S. value $= \sqrt{\frac{I}{T}\int_0^T v^2 dt}$

$$\therefore V_{RMS} = \sqrt{\frac{1}{4}\left[\int_0^2 10^2 dt + \int_2^4 (-10)^2 dt\right]}$$

$$= \sqrt{\frac{100}{4}\left[|t|_0^2 + |t|_2^4\right]}$$

$$= 10 \text{ Volts.}$$

For RMS value of such symmetrical waves integration over the entire period including positive and negative values has to be done.

(iii) Form factor $= \frac{V_{RMS}}{V_{av}} = \frac{10}{10} = 1$

(c) Saw tooth wave:

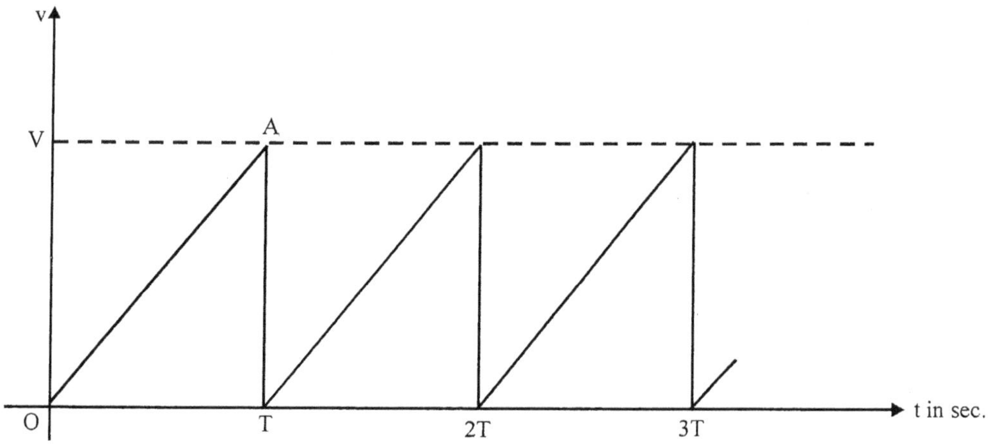

Consider the saw-tooth wave shown in the figure.

Slope of $OA = \frac{V}{T}$; But for general expression, slope $= \frac{v}{T}$

$$\therefore \frac{v}{t} = \frac{V}{T} \text{ or } v = \left(\frac{V}{T}\right)t$$

Time period = T sec.

(i) We have : Average value $= \frac{I}{T}\int_0^T v\,dt$

$$= \frac{I}{T}\int_0^T \left(\frac{V}{T}\right) t\,dt$$

$$= \frac{V}{T^2}\left|\frac{t^2}{2}\right|_0^T$$

$$= \frac{V}{2}$$

This is obvious, Since the curve Oa is linear, the average value is given as

$$\left(\frac{\text{Initial value+ Final Value}}{2}\right)$$

$$\therefore V_{av} = \frac{0+V}{2} = \frac{V}{2}, \text{ as above]}$$

(ii) We have : RMS value $= \sqrt{\frac{I}{T}\int_0^T v^2 dt}$

$$\sqrt{\frac{I}{T}\int_0^T \left(\frac{V}{T}\right)^2 t^2 dt}$$

$$= \sqrt{\left|\frac{V^2}{T^3}\right|\left|\frac{t^3}{3}\right|_0^T}$$

$$= \frac{V}{\sqrt{3}}$$

(iii) We have : Form factor $= \dfrac{V_{RMS}}{V_{av}} = \dfrac{V}{\sqrt{3}}\left(\dfrac{2}{V}\right) = 1.1547$

Note: The examples solved above are for some typical wave forms as given above. But whatever be the nature of the wave form by properly identifying the amplitude and period of the wave form, or the values of the alternating quantity at different values of time if there is any change in the pattern of the wave, we have to right the appropriate expressions for the given AC quantities for the different segments of time and then apply the general formula for RMS, Average Value and the form factor.

Alternatively the average value can be obtained by finding the areas of the curve. For RMS values the area of the square of the curve is considered.

However it is advantageous to make use of the mathematical expression for finding out the RMS, Average Values and Form factors.

3.5 DIFFERENT FORMS OF REPRESENTING ALTERNATINGQUANTITIES

Apart from wave forms the A.C. quantities can also be represented by instantaneous or average or RMS values. Alternating Quantities can be represented in different forms such as,

1. *Phasors*
2. *in rectangular coordinates using j notation or in complex quantity notation using real and imaginary parts.*
3. *in Trigonometrical forms*
4. *in Exponential form*
5. *in Polar form*

3.5.1 Phasor Representation of an Alternating Quantity

A phasor (or loosely called vector) of an alternating voltage or alternating current is the projection of the sinsoidal waveform on y axis i.e., the current or voltage axis which is assumed to rotate at an angular velocity of ω radians per second in the anti-clock wise direction which is conventionally taken to be the positive direction of the rotation. All the relationships between different alternating quantities can be represented by means of a phasor diagram which includes all the phasors corresponding to the different alternating quantities.

Suppose OA in Fig 3.13 (a) to represent to scale the maximum value of an alternating quantity say current; i.e., OA = I_m. Also, suppose OA to rotate anticlockwise about O at a uniform angular velocity. This is purely a conventional direction which has been universally adopted. An arrowhead is drawn at the outer end of the phasor, partly to indicate which end is assumed to move and partly to indicate the precise length of the phasor when two or more phasors happen to coincide.

Fig. 3.13 (a) shows OA when it has rotated through an angle θ from the position occupied when the current was passing through its zero value. If AB and AC are drawn perpendicular to the horizontal and vertical axes respectively :

$$OC = AB = OA \cos(90^o - \theta)$$
$$= OA \sin\theta$$
$$= I_m \sin\theta$$
$$= i = I_m \sin\theta \qquad \text{..........} (3.5.1)$$

namely the value of the current at that instant.

Hence, the projection of OA on the vertical axis represents to scale the instantaneous value of the current. Thus when θ = 30°, the projection is OA **Sin 30°** is equal to $I_m/2$, when, θ = 90°, the projection is OA and is equal to zero and corresponds to the current passing through zero from a positive to a negative value; when

$\theta = 210°$, the phasor is in position OA_1, and the projection = OD = $½OA_1 = -½I_m$; and when $\theta = 360°$, the projection is again zero and corresponds to the current passing through zero from a negative to a positive value. It follows that OA rotates through one revolution or 2π radians a sinusoidal waveform of one cycle of the current wave is traced as shown in Fig.3.13 (b).

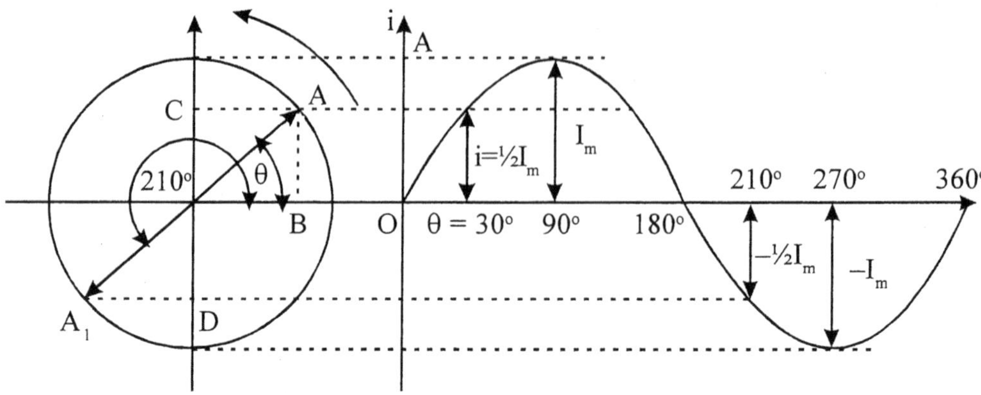

Fig. 3.13 Tracing of an Alternative Quantity

If f is the frequency in Hertz, then OA rotates through f revolutions of $2\pi f$ radians in 1s. Hence, the angular velocity of OA is $2\pi f$ radians per second and is denoted by the symbol ω (omega) i.e.,

$$\omega = 2\pi f \text{ radians per second} \qquad \text{.......... (3.5.2)}$$

If the time taken by OA in Fig. 3.13 to rotate an angle θ radians is t seconds, then,

$$\theta = \omega t = 2\pi f t \text{ radians}$$

We can therefore, express the instantaneous value of the current thus,

$$i = I_m \textbf{ Sin } \theta = I_m \textbf{ Sin } \omega t = I_m \textbf{ Sin } 2\pi f t \qquad \text{.......... (3.5.3)}$$

In Fig. 3.14 (a), OA represents the maximum value of the voltage and OB that of the current. Fig. 3.14 (b) shows the current leading the voltage waveform by an angle ϕ. The two quantities voltage and current can also be represented by a phasor diagram as given in Fig 3.14 (c) and Fig. 3.14 (d).

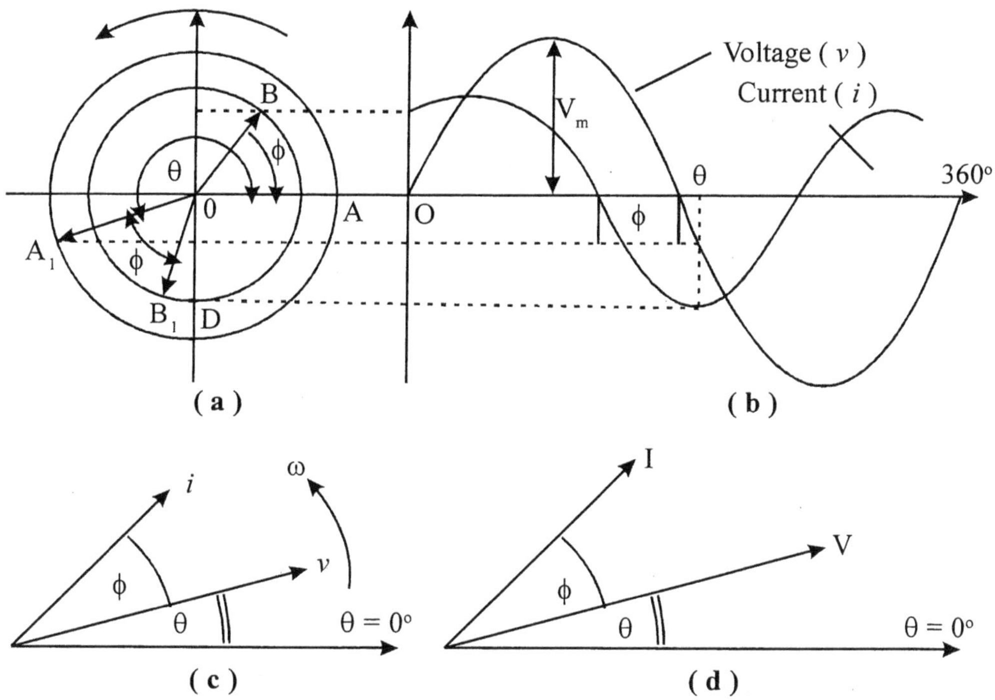

***Fig. 3.14** Tracing & Phasor representation of an Alternative Quantity*

When OA is along the horizontal axis, the voltage at that instant is zero and the value of the current is represented by the projection of OB on the vertical axis. These values correspond to instant t = 0, in Fig. 3.14 (b).

After the phasors have rotated through an angle θ, they occupy positions OA_1 and OB_1 respectively, with OB_1 still leading OA_1 by the same angle φ; and the instantaneous values of the current and voltage are again given by the projections of OA_1 and OB_1 on the vertical axis, as shown by the horizontal dotted lines.

If the instantaneous value of the voltage is represented by

$$v = V_m \, \textbf{Sin} \, \theta \qquad \text{.......... (3.5.4)}$$

then the instantaneous value of the current is represented by

$$i = I_m \, \textbf{Sin} \, (\theta + \phi) \qquad \text{.......... (3.5.5)}$$

where $V_m = OA$ and $I_m = OB$

in Fig. 3.14 (c).

The current in Fig. 3.14 is said to ***lead*** the voltage by an angle ϕ or the voltage is said to ***lag*** the current by an angle ϕ. The ***phase difference*** ϕ between the two phasors remains constant, irrespective of their positions or at any instance of time because the phasors rotate with the same angular velocity ω.

A phasor is a function of time where as a vector is a function spacial coordinates. Hence, it is proper that the term phasor is used instead of vectors.

$$v = V_m \textbf{Sin } \theta$$

$$i = I_m \textbf{Sin } (\theta + \phi)$$

These phasors are assumed to rotate in a anti-clockwise direction, considered to be positive direction at an angular velocity of ω radians/sec. The phasors can be drawn for instantaneous values or for R.M.S. or effective values. Generally, the phasors represent only R.M.S. values and are denoted by V or I phasors unless otherwise stated.

3.5.2 Lagging / Leading Waveforms With Phase & Phase Difference

The phase of an alternating quantity (current or voltage) at an instant is defined as the fractional part of a cycle through which the quantity has advanced from a selected origin. In actual practice, we are more concerned with the phase difference between the two alternating quantities rather than their absolute phase.

The two sets of waveforms e_1 & e_2 and v_1 & v_2 can also be represented by means of phasor diagrams as shown by the side of the waveforms in Fig. 3.15. Since all the phasors rotate at the same angular velocity, there is no need to represent the direction and velocity 'ω' of rotation in the figure. As all the phasors rotate at the same speed the relative angle between them remains the same without changing at any instant of time. When we say a waveform or phasor is lagging or leading it always means lag or lead is with reference to the reference phasor whose phase angle is taken to be zero.

When two alternating quantities having same frequency, attain their zero value at different instants of time, the quantities are said to have a phase difference. *This angle between zero points of two alternating quantities (while becoming negative to positive) is called the angle of* ***Phase Difference***. For the angle of phase difference we should also state whether it is lagging or leading.

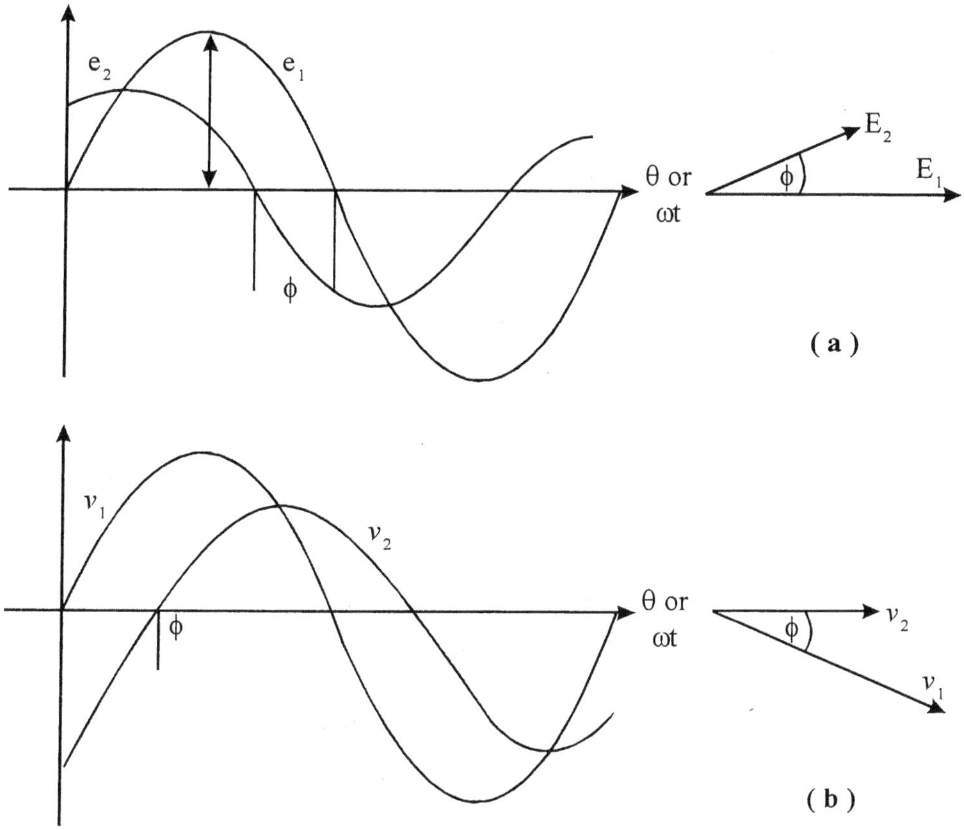

Fig. 3.15 Leading and Lagging Waveforms and their Phasors

3.5.3 Rectangular Coordinates Representation

3.5.3.1 j-Operator

The letter j is used to express operation of counterclockwise rotation of a phasor through 90°. If this operation is done twice on a vector, the vector gets rotated counterclockwise through 180° and hence reverses its sign i.e., gets multiplied by –1. Thus,

$$j\,j = j^2 = -1$$

or $$j = \sqrt{-1} \qquad \text{.......... (3.5.6)}$$

Thus when j operation is done on a phasor V, it results in the new vector jV as shown in Fig. 3.15. On a second j operation, the resulting phasor is

$$j^2V = -V. \qquad \text{.......... (3.5.6a)}$$

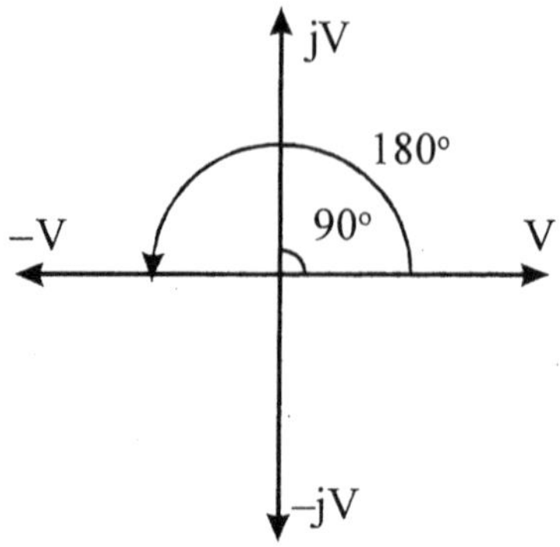

Fig. 3.16 *Application of j Operator*

Again if j operation is done on –V, the resulting phasor is $j^3 = -jV$ which is along –y direction. Finally, if j operation is done on –jV (or j^3V) the resulting phasor is

$$j^3V = -jV \qquad \text{.......... (3.5.7)}$$

$$j^4V = V \qquad \text{.......... (3.5.8)}$$

3.5.3.2 Phasor Represented in Rectangular Coordinate Components

Any vector may be resolved into its x-component and y-component. In Fig. 3.17 phasor V_1 has x_1 and y_1 as its X and Y components respectively. Hence, we may write,

$$V_1 = x_1 + jy_1 \qquad \text{.......... (3.5.9)}$$

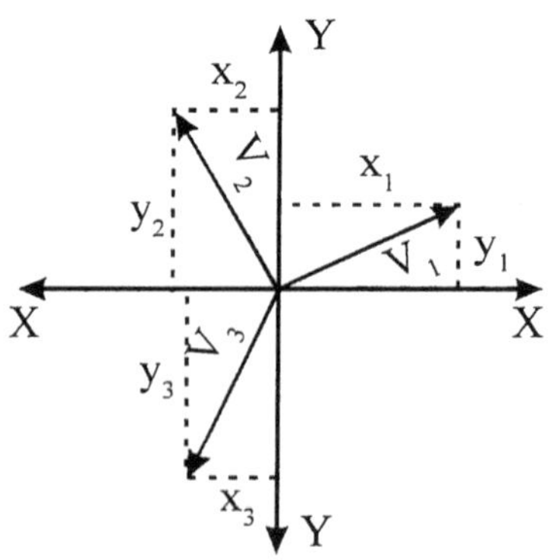

Fig. 3.17 x and y components of Phasors

The above equation expresses V_1 in the so called ***complex form***. In the language of mathematics, x_1 is said to constitute the ***real component*** and y_1 is said to constitute the ***imaginary component***. In the language of electrical engineering, x_1 is called the ***inphase component*** while y_1 is said to form ***quadrature component***. Similarly phasors V_2 and V_3 may be expressed as,

$$V_2 = -x_2 + jy_2 \qquad \text{.......... (3.5.10)}$$

$$V_3 = -x_3 - jy_3 \qquad \text{.......... (3.5.11)}$$

Obviously x_2 and x_3 both are negative and y_3 is negative.

Two complex numbers are said to be conjugate if they differ only in the algebraic sign of their imaginary or quadrature components. The complex number ($x + jy$) and ($x - jy$) are conjugates.

Obviously, then if we add two conjugate numbers we are left with only the real components or inphase components. On the other hand, if we subtract one conjugate number say ($x - jy$) from the other namely ($x + jy$) we are left with $j2y$ i.e., only the imaginary or quadrature components.

In general for addition considering appropriate signs for real and imaginary parts,

$$(x_1 + jy_1) + (x_2 + jy_2) = (x_1 + x_2) + j(y_1 + y_2) \qquad \text{.......... (3.5.12)}$$

3.5.3.3 Trigonometrical Form of Phasor Representation

Fig.3.18 shows a phasor V and its x and y components. Obviously the x-component is V **Cos θ** while the y-components is V **Sin θ**. Hence, we may write,

$$V = V(\mathbf{Cos}\ \theta + j\mathbf{Sin}\ \theta) \qquad \text{.......... (3.5.13)}$$

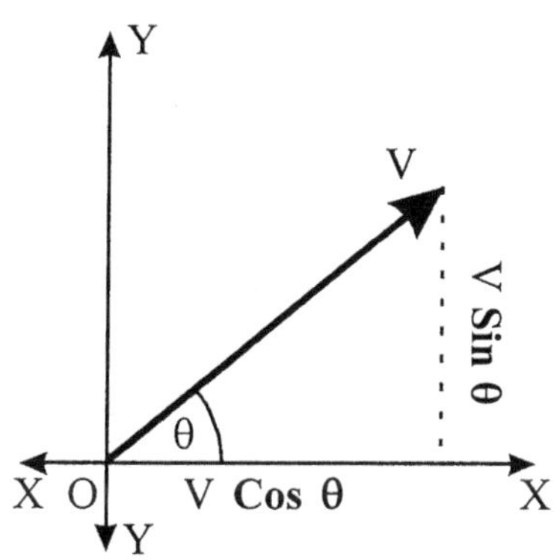

Fig. 3.18 *Phasor represention of Trigonometric Function*

Alternatively,

$$\frac{\mathbf{A}^3}{\sqrt{\mathbf{B}}} = \frac{1397.4 \angle 79.68^o}{4.64 \angle -10.9^o}$$

$$\frac{\mathbf{A}^3}{\sqrt{\mathbf{B}}} = \frac{1397.4\left[\mathbf{Cos\ 79.68^o} + j\ \mathbf{Sin\ 79.68^o}\right]}{4.64\ \left[\mathbf{Cos}\left(-\mathbf{10.9^o}\right) + j\ \mathbf{Sin}\left(-\mathbf{10.9^o}\right)\right]}$$

$$= \frac{1397.4(0.18 + j0.98)}{4.64(0.98 - j0.19)}$$

$$= \frac{251.532 + j1369.452}{4.547 - j0.882}$$

Rationalizing the fraction i.e., multiplying both the numerator and denominator by the complex conjugate of the denominator we have,

$$\frac{\mathbf{A}^3}{\sqrt{\mathbf{B}}} = \frac{(251.532 + j1369.452)(4.547 + j0.882)}{(4.547 - j0.882)(4.547 + j0.882)}$$

$$= \frac{\left[251.532 \times 4.547 + j^2(1369.452 \times 0.882)\right] + j\left[251.532 \times 0.882 + 1369.452 \times 4.547\right]}{(4.547)^2 + (0.882)^2}$$

$$= \frac{-64.14 + j\ 6448.74}{21.44} = -2.99 + j300.78$$

... in complex form

$$\frac{\mathbf{A}^3}{\sqrt{\mathbf{B}}} = \sqrt{(2.99)^2 + (300.78)^2} \angle \mathbf{Tan}^{-1}\left(\frac{\mathbf{300.78}}{\mathbf{-2.99}}\right)$$

$$= 300.75 \angle \mathbf{Tan^{-1}}\ (-100.59)$$

$$= 300.75 \angle 90.58^o$$

... in polar form

3.5.3.4 Exponential Form of Phasor Representation

Euler's Equation states that,

$$e^{\pm j\theta} = \mathbf{Cos}\ \theta \pm j\ \mathbf{Sin}\ \theta \qquad \text{.......... (3.5.14)}$$

Put seperately,

$$e^{+j\theta} = \mathbf{Cos}\ \theta + j\ \mathbf{Sin}\ \theta \qquad \text{.......... (3.5.15)}$$

and

$$e^{-j\theta} = \mathbf{Cos}\ \theta - j\ \mathbf{Sin}\ \theta \qquad \text{.......... (3.5.16)}$$

$$\mathbf{Cos}\ \theta = \frac{e^{j\theta} + e^{-j\theta}}{2} \qquad \text{.......... (3.5.17)}$$

$$\mathbf{Sin}\ \theta = \frac{e^{j\theta} - e^{-j\theta}}{2j} \qquad \text{.......... (3.5.18)}$$

Hence, phasor V may be expressed in the following exponential form :

$$\mathbf{V} = Ve^{\pm j\theta} \qquad \text{.......... (3.5.19)}$$

The above equation states that the phasor **V** has magnitude V and phase angle $\pm\theta$ with the reference axis.

3.6.3.5 Polar Form of Phasor Representation

Consider phasor V written in trigonometric form V (**Cos** θ + j **Sin** θ). This vector makes an angle θ with positive x-axis and has magnitude V. Hence it may also be written as $V\angle\theta$. Here angle θ is measured counterclockwise. However, if angle θ with positive x-axis is in the clockwise direction, the expression for vector V becomes $V\angle{-\theta}$. Hence, the general expression for vector V in polar form is $V\angle\pm\theta$. Evidently expression $V\angle\pm\theta$ forms a shorter or symbolic method of writing $Ve^{\pm j\theta}$.

Multiplication of any phasor by $e^{j\theta_1}$ *causes the phasor to rotate by an angle* θ_1 *in the positive direction i.e., in the anti-clockwise direction.*

For example,

if $\mathbf{V} = \mathbf{V} \times e^{j\theta}$ is multiplied by $e^{j\theta_1}$

then,

$$\mathbf{V} = \mathbf{V} \times e^{j(\theta+\theta_1)} \qquad \text{.......... (3.5.20)}$$

$$= \mathbf{V} \times \{ \mathbf{Cos}\ (\theta + \theta_1) + j\mathbf{Sin}\ (\theta + \theta_1) \}$$

$$\mathbf{V} = \mathbf{V}\angle(\theta + \theta_1) \qquad \text{.......... (3.5.21)}$$

Similarly, *multiplication of any phasor by* $e^{-j\theta_1}$ *causes the phasor to rotate by an angle* θ_1 *in the negative direction i.e., in the clockwise direction.*

For example,

if $\quad \mathbf{V} = \mathbf{V} \times e^{-j\theta}$ is multiplied by $e^{-j\theta_1}$

$$= \mathbf{V} \times e^{-j\theta_1} \times e^{-j\theta_2}$$

then,

$$\mathbf{V} = \mathbf{V} \times e^{-j(\theta_1 + \theta_2)} \qquad \text{.......... (3.5.22)}$$

$$= \mathbf{V} \times \{ \mathbf{Cos}\,[-(\theta_1 + \theta_2)] - j\mathbf{Sin}\,[-(\theta_1 + \theta_2)] \}$$

$$\mathbf{V} = \mathbf{V}\angle -(\theta_1 + \theta_2) \qquad \text{.......... (3.5.23)}$$

3.6 ARITHMETIC OPERATIONS OF PHASORS

3.6.1 Addition and Subtraction of Phasors

Addition and subtraction of phasors are easily done by using the rectangular coordinate components i.e., real and imaginary parts

$$\mathbf{A} = a_1 + jb_1 \qquad \text{.......... (3.6.1)}$$

$$\mathbf{B} = a_2 - jb_2 \qquad \text{.......... (3.6.1a)}$$

then

$$\mathbf{C} = \mathbf{A} + \mathbf{B}$$

$$= (a_1 + a_2) + j(b_1 - b_2) \qquad \text{.......... (3.6.2)}$$

$$= \sqrt{(a_1 + a_2)^2 + (b_1 - b_2)^2}$$

$$= C\angle\theta_1$$

$$\mathbf{C} = C\angle\, \mathbf{Tan}^{-1}\left(\frac{\mathbf{b_1} + \mathbf{b_2}}{\mathbf{a_1} + \mathbf{a_2}}\right) \qquad \text{.......... (3.6.2a)}$$

where

$$C = |\mathbf{C}| = \sqrt{(a_1 + a_2)^2 + (b_1 + b_2)^2}$$

$$\mathbf{D} = \mathbf{A} - \mathbf{B}$$

$$= (a_1 - a_2) + j(b_1 - (-b_2)) \qquad \text{.......... (3.6.3)}$$

$$= (a_1 - a_2) + j(b_1 + b_2)$$

$$= D\angle\theta_2$$

$$\mathbf{D} = D \angle \mathbf{Tan}^{-1}\left(\frac{\mathbf{b_1 + b_2}}{\mathbf{a_1 - a_2}}\right) \quad \text{.......... (3.6.3a)}$$

where $\quad D = |\mathbf{D}| = \sqrt{(a_1 - a_2)^2 + (b_1 + b_2)^2}$

3.6.2 Multiplication and Division of Phasors

Multiplication and division of phasors are easily done by using the polar coordinate components, i.e., using magnitude and angles for the phasors.

$$\mathbf{A} = a_1 + jb_1$$

$$= A \angle\alpha = A\,e^{j\alpha} \quad \text{.......... (3.6.4)}$$

where $\quad A = \sqrt{\left(a_1^2 + b_1^2\right)}$ and $\alpha = \mathbf{Tan}^{-1}\left(\frac{\mathbf{b_1}}{\mathbf{a_1}}\right)$

$$\mathbf{B} = a_2 - jb_2$$

$$= A \angle -\beta = Ae^{-j\beta} \quad \text{.......... (3.6.4a)}$$

where $\quad B = \sqrt{\left(a_2^2 + b_2^2\right)}$

and $\quad \beta = \mathbf{Tan}^{-1}\left(\frac{\mathbf{b_2}}{\mathbf{a_2}}\right)$

then, $\quad \mathbf{AB} = A \angle \alpha \times B \angle -\beta = AB \angle (\alpha - \beta)$

$$= AB\, e^{j(\alpha - \beta)}$$

$$= AB\, [\,\mathbf{Cos}\,(\alpha - \beta) + j\mathbf{Sin}\,(\alpha - \beta)\,) \quad \text{........ (3.6.5)}$$

Similarly, $\quad \frac{\mathbf{A}}{\mathbf{B}} = \frac{A\angle\alpha}{B\angle-\beta}$

$$= \frac{A}{B} \angle(\alpha - (-\beta))$$

$$= \frac{A}{B} \angle(\alpha + \beta)$$

$$= \frac{A}{B} e^{j(\alpha+\beta)}$$

$$\frac{\mathbf{A}}{\mathbf{B}} = \frac{A}{B}(\mathbf{Cos}\,(\alpha+\beta)+\mathrm{j}\;\mathbf{Sin}\,(\alpha+\beta)) \qquad \text{.......... (3.6.6)}$$

where A, B, α and β are as given above.

Alternatively,

$$\mathbf{AB} = (\ a_1 + jb_1\ )(\ a_2 - jb_2\ )$$
$$= a_1a_2 - ja_1b_2 + ja_2b_1 - j^2b_1b_2$$
$$= (\ a_1a_2 + b_1b_2\ ) + j(\ a_2b_1 - a_1b_2\ )$$

$$\mathbf{AB} = \sqrt{(a_1a_2 + b_1b_2)^2 + (a_2b_1 - a_1b_2)^2} \qquad \text{.......... (3.6.7)}$$

$$\mathbf{AB} = \angle\mathbf{Tan}^{-1}\left(\frac{(\mathbf{a_2b_1 - a_1b_2})}{(\mathbf{a_1a_2 + b_1b_2})}\right) \qquad \text{.......... (3.6.8)}$$

which is not so easy. Hence, *for multiplication or division of phasors use polar form.*

3.6.3 Power and Roots of Vectors

3.6.3.1 Powers

Consider the phasor

$$\mathbf{A} = A \angle \alpha.$$

Then n^{th} power of phasor **A** is obtained by multiplying the phasor **A** by itself n times. Hence,

$$\mathbf{A}^n = A\angle\alpha \times A\angle\alpha \times \ldots \times A\angle\alpha \ldots (\text{ n terms })$$
$$= A \times A \times \ldots \times A \angle \alpha + \alpha + \ldots + \alpha$$
$$= A^n \angle n\alpha \qquad \text{.......... (3.6.9)}$$

Similarly if

$$\mathbf{A} = A\angle\alpha$$

and $\mathbf{B} = B \angle \beta.$

then $\mathbf{A}^n\mathbf{B}^n = A^nB^n \angle n(\ \alpha + \beta\ )$ .... (3.6.10)

Similarly $\mathbf{A}^n\mathbf{B}^m = A^nB^m \angle n\alpha + m\beta$ .... (3.6.11)

3.6.3.2 Roots

Consider the phasor

$$\mathbf{A} = A \angle \alpha.$$

Then n^{th} root of phasor **A** is obtained by finding the nth root of magnitude **A** and dividing the phase angle α by n. Hence,

$$\sqrt[n]{\mathbf{A}} = \sqrt[n]{A} \angle (\alpha / n) \qquad \text{.......... (3.6.12)}$$

Similarly if

$$\mathbf{A} = A \angle \alpha$$

and

$$\mathbf{B} = B \angle \beta.$$

$$\sqrt[n]{\mathbf{A}}\,\sqrt[n]{\mathbf{B}} = \sqrt[n]{AB} \angle [(\alpha + \beta)/n] \qquad \text{.......... (3.6.13)}$$

3.6.4 Conjugate of a Phasor and Rationalization

Since a phasor is represented by a complex number the conjugate of a phasor will be the complex conjugate of the representation of the phasor in complex numbers or its magnitude will be the same at an angle equal to the negative of the original angle of the phasor.

Let

$$\mathbf{A} = A\angle\alpha \qquad \text{.......... (3.6.14)}$$

$$= A\,(\mathbf{Cos}\ \alpha + j\,\mathbf{Sin}\ \alpha)$$

$$= a + jb$$

then the conjugate of the phasor **A** represented by

$$\mathbf{A}^* = a - jb$$

$$= A\,(\mathbf{Cos}\ \alpha - j\,\mathbf{Sin}\ \alpha)$$

$$= A \angle -\alpha \qquad \text{.......... (3.6.15)}$$

The principle of rationalization of a complex number division is used for finding the division of two phasors in complex form.

Let

$$\mathbf{A} = A \angle \alpha$$

$$= A\,(\mathbf{Cos}\ \alpha + j\,\mathbf{Sin}\ \alpha)$$

$$= a + jb$$

$$\mathbf{B} = B \angle \beta$$

$$= B\,(\mathbf{Cos}\ \beta + j\,\mathbf{Sin}\ \beta)$$

$$= c + jd$$

then

$$\frac{\mathbf{A}}{\mathbf{B}} = \frac{a + jb}{c + jd}$$

which is rationalized by multiplying both the numerator and denominator with the complex conjugate of the denominator.

Hence,

$$\frac{\mathbf{A}}{\mathbf{B}} = \frac{(a+jb)(c-jd)}{(c+jd)(c-jd)}$$

$$= \frac{(a+jb)(c-jd)}{(c^2+d^2)}$$

$$\frac{\mathbf{A}}{\mathbf{B}} = \frac{(ac+bd)+j(bc-ad)}{(c^2+d^2)} \quad \text{.......... (3.6.16)}$$

which is difficult to evaluate. On the other hand, we use polar form of representation of phasors.

We have,

$$\frac{\mathbf{A}}{\mathbf{B}} = \frac{A\angle\alpha}{B\angle\beta}$$

$$= \frac{\mathbf{A}}{\mathbf{B}}\angle(\alpha-\beta)$$

$$\frac{\mathbf{A}}{\mathbf{B}} = \frac{\mathbf{A}}{\mathbf{B}}\{\, \mathbf{Cos}(\alpha-\beta) + j\mathbf{Sin}(\alpha-\beta) \,\}$$

.... (3.6.17)

which is easy. Hence, *for division of phasors it is better to use only polar form.*

Example 3.9:

Two phasors **A** and **B** are given by

$$\mathbf{A} = 10 + j5$$

$$\mathbf{B} = 20 - j8.$$

Find the phasors

(i) $\mathbf{A}^3\sqrt{\mathbf{B}}$ *and*

(ii) $\dfrac{\mathbf{A}^3}{\sqrt{\mathbf{B}}}$

Solution :

$$\mathbf{A} = 10 + j5$$

$$\mathbf{A} = \sqrt{10^2 + 5^2} \angle \text{Tan}^{-1}\left(\frac{5}{10}\right)$$

$$= 11.18 \angle 26.56^o$$

$$\mathbf{B} = 20 - j8$$

$$\mathbf{B} = \sqrt{20^2 + 8^2} \angle \text{Tan}^{-1}\left(\frac{-8}{20}\right)$$

$$= 21.54 \angle -21.8^o$$

$$\mathbf{A}^3 = (11.18)^3 \angle (3 \times 26.56^o)$$

$$= 1397.4 \angle 79.68^o$$

$$\sqrt{\mathbf{B}} = \sqrt{21.54} \angle \left(\frac{-21.8}{2}\right)$$

$$= 4.64 \angle -10.9^o$$

(i)

$$\mathbf{A}^3 \sqrt{\mathbf{B}} = 1397.4 \angle [79.68^o \times 4.64 \angle -10.9^o]$$

$$= 6483.94 \angle 79.68^o + (-10.9^o)$$

$$= 6483.94 \angle 68.78^o \quad \text{. . . in polar form}$$

$$= 6483.94 (\mathbf{Cos\ 68.78^o} + j\,\mathbf{Sin\ 68.78^o})$$

$$= 6483.94 (0.362 + j0.932)$$

$$= 2347.19 + j6043.03 \quad \text{. . . in complex form}$$

(ii)

$$\frac{\mathbf{A}^3}{\sqrt{\mathbf{B}}} = \frac{1397.4 \angle 79.68^o}{4.64 \angle -10.9^o}$$

$$= 301.12 \angle [79.68^o - (-10.9^o)]$$

$$= 301.12 \angle 90.58^o \quad \text{. . . in polar form}$$

$$= 301.12 (\mathbf{Cos\ 90.58^o} + j\,\mathbf{Sin\ 90.58^o})$$

$$= 301.12 (-0.01 + j0.999)$$

$$= -3.0112 + j300.82 \quad \text{. . . in complex form}$$

3.7 INSTANTANEOUS & AVERAGE POWER IN A.C. CIRCUITS

Power is the rate of doing work and is expressed in Joules per second. Power is measured in Watts. Power is denoted by symbol P. *When one coulomb of electrical charge moves through a potential difference of one volt in one second the the work done is one Joule/sec and in electrical engineering it is expressed as one Watt.*

In the case of D.C. circuits the power supplied by the source is given as,

$$P = E \times I \text{ Watts} \qquad \text{.......... (3.7.1)}$$

$$P = \frac{E^2}{R} \text{ Watts} \qquad \text{.......... (3.7.1a)}$$

$$P = I^2R \text{ Watts} \qquad \text{.......... (3.7.1b)}$$

where E is the source voltage

I is the current supplied

R is the resistance of the circuit.

Power expended in a resistance R,

$$P = V \times I \text{ Watts} \qquad \text{.......... (3.7.2)}$$

$$= \frac{V^2}{R} \text{ Watts} \qquad \text{.......... (3.7.2a)}$$

$$= I^2R \text{ Watts} \qquad \text{.......... (3.7.2b)}$$

where V is the voltage drop.

I is the current flowing

R is the resistance of the circuit.

Similarly in the case of A.C. circuits, we have the instantaneous power,

$$p = v \times i \text{ Watts} \qquad \text{.......... (3.7.3)}$$

$$= \frac{v^2}{R} \text{ Watts} \qquad \text{.......... (3.7.3a)}$$

$$= i^2R \text{ Watts} \qquad \text{.......... (3.7.3b)}$$

3.7.1 A.C. Circuit with Pure Resistance As Load

In the case of an A.C. circuit with only pure resistance R as an element the current *i*(t) will be in phase with the voltage *v*(t) across the resistance R.

The circuit diagram, the voltage and current waveforms and its phasor diagram are given in Fig. 3.19.

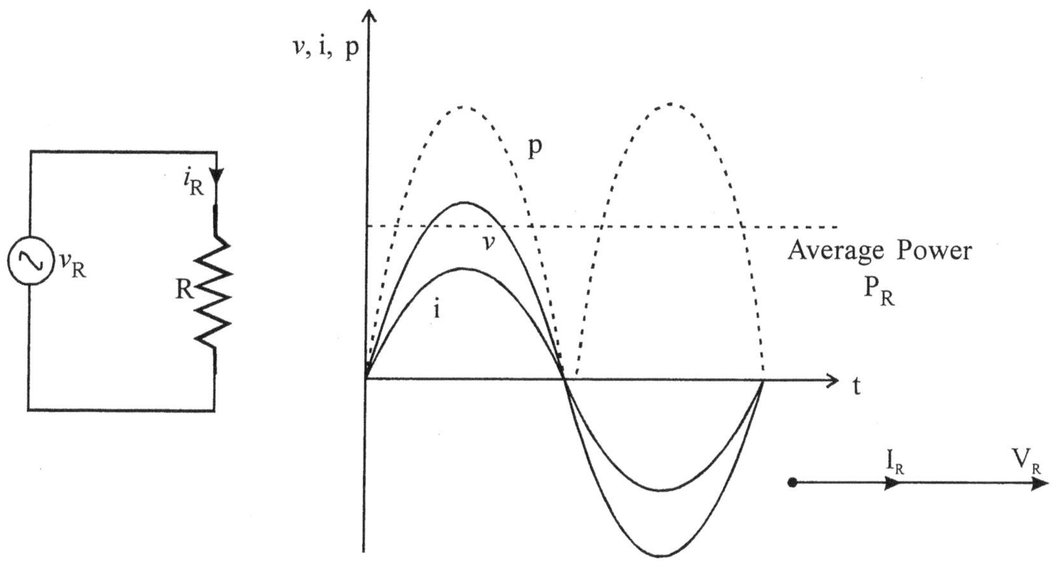

***Fig. 3.19** Waveforms of Voltage, Current and Power for Resistive Load*

For a pure resistance, voltage across the resistance,

$$v_R = V_m \textbf{Sin } \omega t \qquad \text{.......... (3.7.4)}$$

$$v_R = V_m \textbf{Sin } \theta \qquad \text{.......... (3.7.4a)}$$

and the current through the resistance,

$$i_R = \frac{v_R}{R}$$

$$= \frac{V_m \textbf{Sin } \omega t}{R}$$

$$= I_m \textbf{Sin } \omega t$$

$$i_R = I_m \textbf{Sin}\ \theta \qquad \text{.......... (3.7.5)}$$

where
$$I_m = \frac{V_m}{R} \qquad \text{.......... (3.7.5a)}$$

When the voltage across the resistance is zero, the current through the resistance is also zero. This shows that the current through the resistances is in phase with the voltage across the resistance and changes according to the voltage for a resistance. Hence I_R is the current in phase with the voltage,

$$I_R = \frac{V_R}{R} \qquad \text{.......... (3.7.6)}$$

$$= \frac{V_R\angle 0^\circ}{R\angle 0^\circ} = \frac{V_R}{R}\angle 0^\circ \qquad \text{.......... (3.7.7)}$$

The instantaneous power,

$$p = v \times i \text{ Watts} \qquad \text{.......... (3.7.8)}$$

The average power for one full cycle in Watts is given as,

$$P_R = \frac{1}{T}\int_0^T p_R\ dt$$

$$= \frac{1}{T}\int_0^T v(t)i(t)\ dt$$

$$= \frac{1}{2\pi}\int_0^{2\pi} (V_m \textbf{ Sin}\,\theta \times I_m \textbf{ Sin}\,\theta)\ d\theta \quad \text{Watts}$$

$$= \frac{V_m I_m}{2\pi}\int_0^{2\pi} \textbf{Sin}^2\theta\ d\theta$$

$$= \frac{V_m I_m}{2\pi}\int_0^{2\pi}\left(\frac{1-\textbf{Cos } 2\theta}{2}\right) d\theta$$

$$= \frac{V_m I_m}{2\pi}\left(\frac{\theta}{2} - \frac{\textbf{Sin } 2\theta}{4}\right)_0^{2\pi}$$

$$= \frac{V_m I_m}{2}$$

$$= \frac{V_m}{\sqrt{2}} \times \frac{I_m}{\sqrt{2}} \text{ Watts}$$

$$= V_{rms} I_{rms} \text{ Watts}$$

$$= \frac{V_{rms}^2}{R} \text{ Watts}$$

$$P = I_{rms}^2 R \text{ Watts} \qquad \text{.......... (3.7.9)}$$

The power associated with energy transfer from the electrical system to another system such as heat, light or mechanical drives is termed *Active Power* or *Real Power* or *Average Power* 'P' in Watts and is given by

$$P = I^2 R \qquad \text{.......... (3.7.10)}$$

where I is the R.M.S. value of the current in Amperes

and R is the resistance in Ohms.

Example 3.10:

A coil of wire carries simultaneously the instantaneous currents,

$$i_1 = 12 \textbf{ Sin } \boldsymbol{\omega t} \text{ Amps}$$

$$i_2 = 10 \textbf{ Sin } (\boldsymbol{\omega t + \pi/3}) \text{ Amps}$$

Find

1. Total Current
2. The Total Power expended in the coil if the resistance of the coil is 10Ω.
3. The energy consumed by the coil when the two currents flow through it for 2 Hrs.

Solution:

1. The RMS currents are represented by the phasor

$$I_1 = \frac{12}{\sqrt{2}} = 8.487\angle 0^\circ$$

and

$$I_2 = \frac{10}{\sqrt{2}} = 7.072\angle 60^\circ$$

The sum can be found by adding,

$$
\begin{aligned}
I &= I_1 + I_2 \\
&= 8.487 + 7.072\,(0.5 + j0.866) \\
&= 8.487 + 3.536 + j6.124 \\
&= 12.023 + j6.124 \\
&= 13.5\angle 27^\circ
\end{aligned}
$$

2. The total power expended in the coil,

$$
\begin{aligned}
P &= I^2R \\
&= (13.5)^2 \times 10 \\
&= 1822.5 \text{ W} \\
&= 1.8225 \text{ kW}
\end{aligned}
$$

3. The energy consumed by the coil is given by the product of the power and the time.

$$
\begin{aligned}
\text{Energy} &= 1.8225 \times 2 \\
&= 3.645 \text{ KWH}
\end{aligned}
$$

3.7.2 A.C. Circuit with Pure Inductance as Load

A coil will exhibit a resistance in series with an inductance. In the case of pure inductance the resistance of the coil will be equal to zero. The pure inductance is represented by the symbol L.

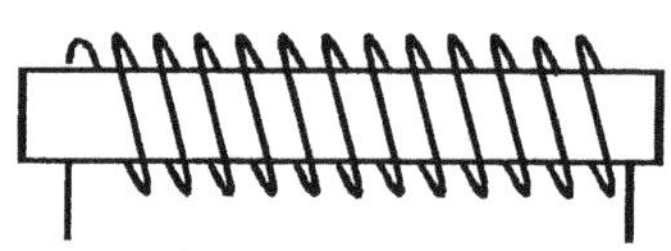

Fig. 3.20 Magnetic Coil

In a purely inductive circuit, if i_L is the current through the inductance then the voltage v_L across the inductance is given as

$$v_L = L\frac{di_L}{dt} \quad \text{.......... (3.7.11)}$$

$$i_L(t) = \frac{1}{L}\int v_L(t)\,dt \quad \text{.......... (3.7.12)}$$

For a sinusoidal supply,

$$v_L = V_m \textbf{ Sin } \boldsymbol{\omega t}$$

$$i_L = \frac{1}{L}\int V_m \textbf{ Sin } \boldsymbol{\omega t}\, dt$$

$$= -\frac{V_m}{\omega L} \textbf{ Cos } \boldsymbol{\omega t}$$

$$= \frac{V_m}{\omega L} \textbf{ Sin } (\boldsymbol{\omega t - \pi/2})$$

$$i_L = I_m \textbf{ Sin } (\boldsymbol{\omega t - \pi/2}) \qquad \text{.......... (3.7.13)}$$

where $$I_m = \frac{V_m}{\omega L} = \frac{V_m}{X_L}$$

$$X_L = \omega L$$

$$= 2\pi f\, L \quad \text{Ohms}$$

and is known as ***Inductive Reactance***.

Since current in an inductor lags behind the voltage across the inductor by 90° in phase,

$$\mathbf{I_L} = I_L \angle -90^\circ$$

$$= \frac{V_L \angle 0^\circ}{X_L \angle 90^\circ}$$

$$= \frac{V_L}{X_L} \angle -90^\circ$$

$$\mathbf{I_L} = -j\frac{V_L}{X_L} = \frac{V_l}{jX_L} \qquad \text{.......... (3.7.14)}$$

It may be noted that if the resistance is taken as reference phasor then the inductive reactance will be leading the resistance phasor by $\pi/2$ radians or 90°.

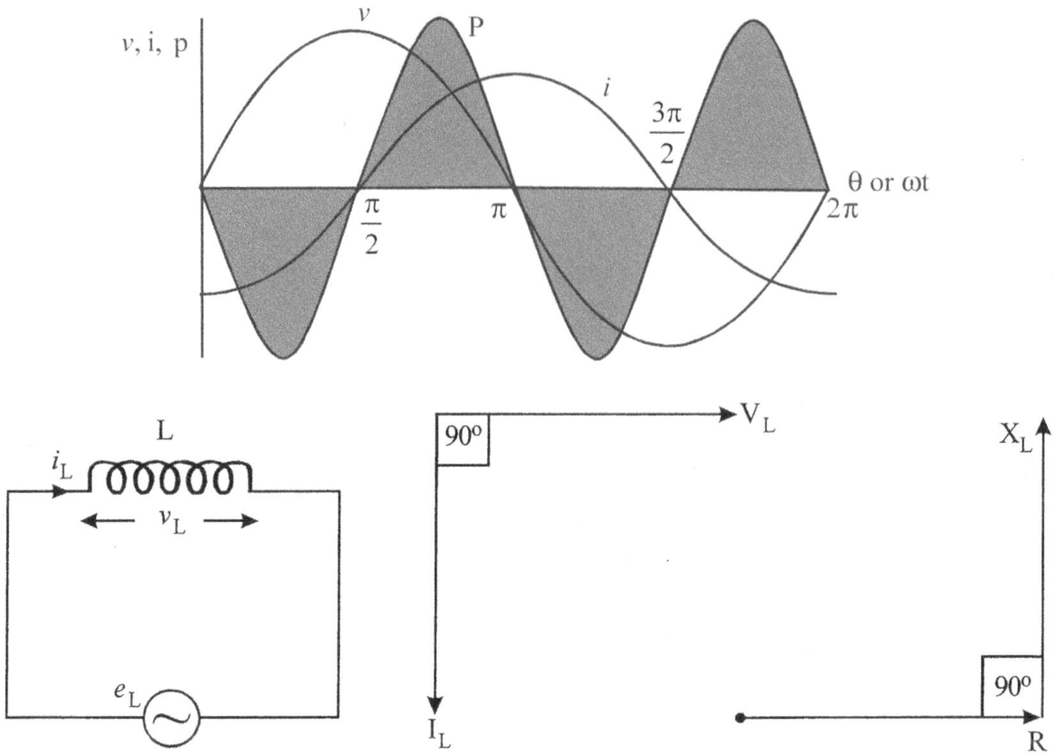

Fig. 3.21 Waveforms of Voltage, Current Power and Phasors for Inductive Load

Since $\quad I_L = \dfrac{V_L}{X_L} \angle -90^\circ$

From Eq. (3.7.15), it can be observed that the current waveform lags behind the voltage waveform by an angle $\pi/2$ radians. The same will be the case for the voltage and current phasors. That means the current in a pure inductance will lag behind the voltage by an angle $\pi/2$ radians or 90°. It can also be stated that in a pure inductance the voltage will lead the current by an angle $\pi/2$ radians or 90°.

$$p = v \times i \text{ Watts} \qquad \text{.......... (3.7.15)}$$

The circuit diagram, the voltage and current waveforms and its phasor diagram are given in Fig. 3.21.

The average power for one full cycle ,

$$P_L = \frac{1}{2\pi}\int_0^{2\pi}\left(V_m \,\mathbf{Sin}\,\boldsymbol{\theta} \times I_m \,\mathbf{Sin}\left(\theta - \frac{\pi}{2}\right)\right) d\theta$$

$$= \frac{1}{2\pi}\int_0^{2\pi}\left(V_m \mathbf{Sin}\,\theta \times I_m \mathbf{Cos}\theta\right) d\theta$$

$$= \frac{V_m I_m}{2\pi}\int_0^{2\pi}\left(\mathbf{Sin}\,\theta \times \mathbf{Cos}\theta\right) d\theta$$

$$= \frac{V_m I_m}{2\pi}\int_0^{2\pi}\left(\frac{\mathbf{Sin}\,2\theta}{2}\right) d\theta$$

$$= \frac{V_m I_m}{2\pi}\left(-\frac{\mathbf{Cos}\,2\theta}{4}\right)_0^{2\pi}$$

$$P_L = 0 \quad \text{for an inductor} \qquad \text{.......... (3.7.16)}$$

The average power over one full cycle for a pure inductance is zero.

3.7.3 A.C. Circuit with Pure Capacitance as Load

In a purely capacitive circuit, if v_C is the voltage across the capacitance then the current through the capacitance is given as

$$i_C(t) = C\frac{dv_C}{dt} \qquad \text{.......... (3.7.17)}$$

For a sinusoidal circuit,

$$v_C = V_m \mathbf{Sin}\ \omega t$$

then

$$i_C = C\frac{d}{dt}\left(V_m \mathbf{Sin}\,\omega t\right)$$

$$i_C = C\frac{d}{d\theta}\left(V_m \mathbf{Sin}\,\theta\right)$$

$$= \omega C V_m \mathbf{Cos}\ \omega t$$

$$i_C = \frac{1}{1/\omega C} V_m \mathbf{Cos}\ \omega t$$

$$= I_m \mathbf{Sin}\left(\omega t + \frac{\pi}{2}\right) \qquad \text{.......... (3.7.18)}$$

where $$I_m = \frac{1}{1/\omega C} V_m = \frac{V_m}{X_C}$$

and $$X_C = \frac{1}{\omega C}$$

where X_C is known as the capacitive reactance of the capacitance.

Since current in a pure capacitor leads the voltage across the capacitance by 90° in phase,

$$\mathbf{I}_C = I_C \angle 90^o$$

$$= \frac{V_C \angle 0^o}{X_C \angle -90^o}$$

$$= \frac{V_C}{X_C} \angle 90^o$$

$$\mathbf{I}_C = j\frac{V_C}{X_C} \qquad \text{.......... (3.7.19)}$$

It may be noted that if the resistance is taken as reference phasor then the capacitive reactance will be lagging the resistance phasor by $\pi/2$ radians or 90°.

From Eq. (3.7.19), it can be observed that the current waveform leads the voltage waveform by an angle $\pi/2$ radians or 90°. The same will be the case for the voltage and current phasors. That means the current in a pure capacitance will lead the voltage by an angle $\pi/2$ radians or 90°. It can also be stated that in a pure capacitance the voltage will lag behind the current by an angle $\pi/2$ radians or 90°.

$$p = v \times i \text{ Watts}$$

The average power for one full cycle,

$$P_C = \frac{1}{2\pi} \int_0^{2\pi} \left(V_m \, \mathbf{Sin}\, \boldsymbol{\theta} \times I_m \, \mathbf{Sin}\left(\boldsymbol{\theta} + \frac{\pi}{2} \right) \right) d\theta$$

$$= \frac{1}{2\pi} \int_0^{2\pi} \left(V_m \, \mathbf{Sin}\, \boldsymbol{\theta} \times I_m \, \mathbf{Cos}\boldsymbol{\theta} \right) d\theta$$

$$= \frac{V_m I_m}{2\pi} \int_0^{2\pi} (\mathbf{Sin}\,\theta \times \mathbf{Cos}\,\theta)\, d\theta$$

$$= \frac{V_m I_m}{2\pi} \int_0^{2\pi} \left(\frac{\mathbf{Sin}\,2\theta}{2} \right) d\theta$$

$$= \frac{V_m I_m}{2\pi} \left(-\frac{\mathbf{Cos}\,2\theta}{2} \right)_0^{2\pi}$$

$$P_C = 0 \qquad \text{for a capacitor} \qquad \text{..........} \ (3.7.20)$$

The average power over one full cycle for a pure capacitance is zero.

The circuit diagram, the voltage and current waveforms and its phasor diagram are given in Fig. 3.22.

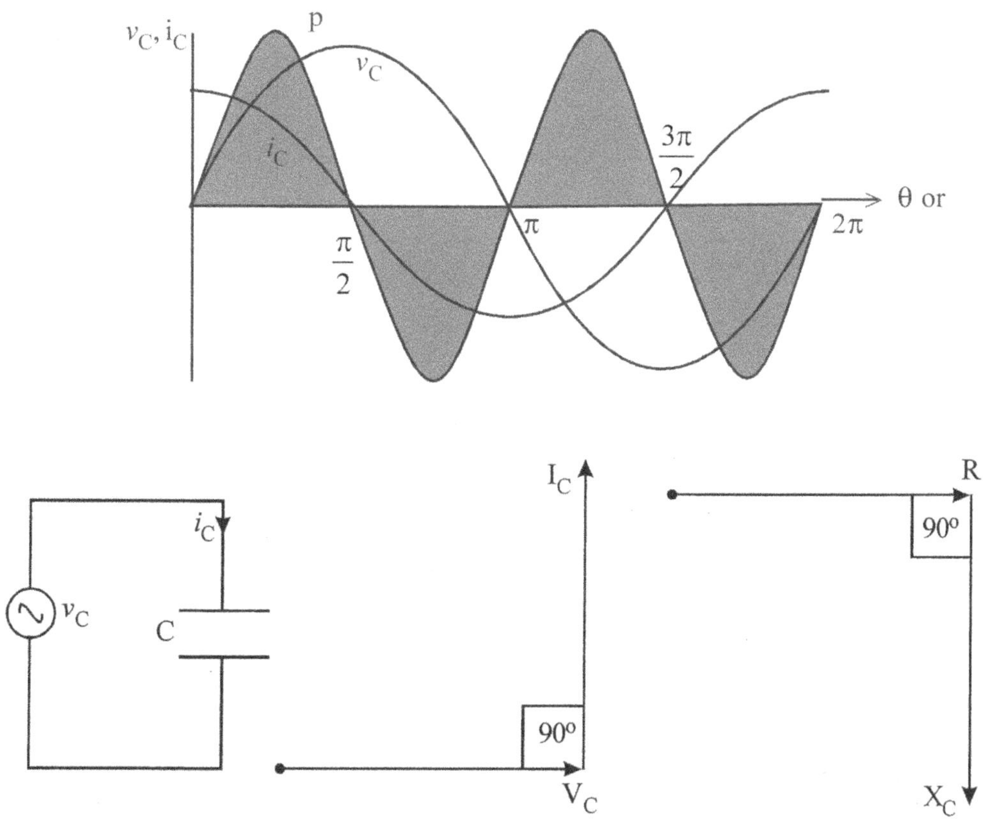

Fig. 3.22 Waveforms of Voltage, Current Power and Phasors for Capacitive Load

Example 3.11:

A 50 Hz sinusoidal voltage given by e = 200 **Sin ωt** is applied to a pure resistance of value 25Ω. Write the equation for the current and power. Find the RMS current and average power. Draw a phasor diagram.

Solution:

$$e_R = 200 \textbf{ Sin } \omega t$$

$$= E_m \textbf{ Sin } \omega t$$

$$\omega = 2\pi \times 50 = 314 \text{ radians}$$

$$E_R = \frac{E_m}{\sqrt{2}}$$

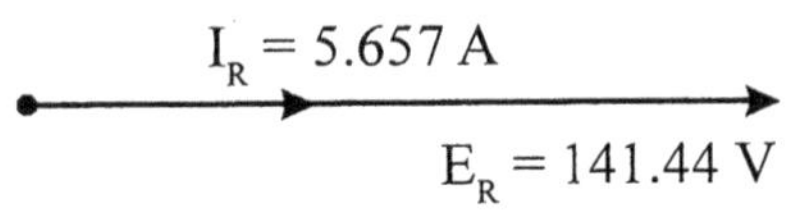

$$= \frac{200}{\sqrt{2}} = 141.44 \text{ V}$$

$$i_R = \frac{e}{R}$$

$$= 8 \textbf{ Sin 314t}$$

$$= I_m \textbf{ Sin } \omega t$$

RMS Value, $I_R = \frac{I_m}{\sqrt{2}}$

$$= \frac{8}{\sqrt{2}} = 5.657 \text{ A}$$

The given figure shows the voltage and current phasors.

Instantaneous Power,

$$p_R = e_R \times i_R \text{ Watts}$$

$$= (\ 200 \textbf{ Sin 314 })(\ 8 \textbf{ Sin 314t })$$

$$= 1600 \textbf{ Sin}^2 \textbf{ 314t} \text{ Watts}$$

Average Power,

$$P_R = E_R \times I_R = 141.44 \times 5.657 = 800 \text{ W}$$

or $$P_R = I_R^2 R = (5.657)^2 \times 25 = 800 \text{ W}$$

or $$P_R = \frac{E_R^2}{R} = \frac{(141.44)^2}{25} = 800 \text{ W}$$

Example 3.12:

A pure Inductance of 100 mH has an applied voltage e = 240 **Sin 314t**. Find the equation for the

(a) *Current*
(b) *Instantaneous Power*

Find also the

1. *Inductive Reactance*
2. *RMS Current*
3. *Average Power.*

Draw Phasor Diagram.

Solution:

Since, $L = 100 \text{ mH}$

Inductive Reactance,

$$X_L = \omega L$$
$$= 314 \times 0.1$$
$$= 31.4\Omega$$

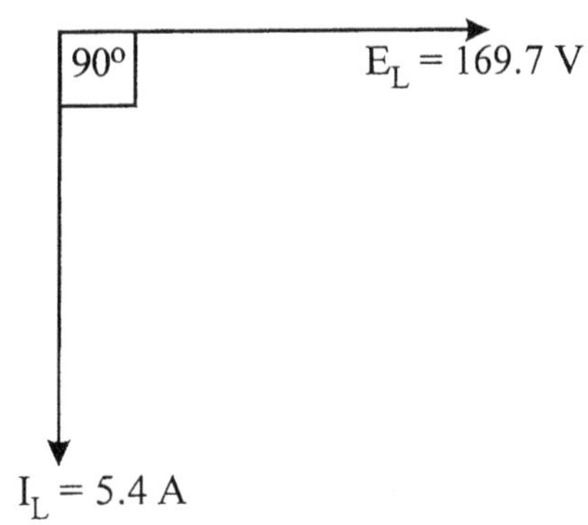

Since, $$E_L = \frac{240}{\sqrt{2}}$$
$$= 169.706 \text{ V}$$

RMS Current,

$$I_L = \frac{E_L}{X_L} = 5.4 \text{ A}$$

The phasor diagram is shown in the figure.

Since the current in an inductance lags the applied voltage by 90° radians.

$$i_L = I_m \textbf{ Sin}\left(\omega t - \frac{\pi}{2}\right)$$

$$= 5.4 \times \sqrt{2}\ \mathbf{Sin}\left(\mathbf{314t} - \frac{\pi}{2}\right)$$

$$i_L = 7.64\ \mathbf{Sin}\left(\mathbf{314t} - \frac{\pi}{2}\right)$$

Instantaneous Power,

$$p_L = (\ 240\ \mathbf{Sin\ 314t}\ )(\ 7.64\ \mathbf{Cos\ 314t}\ )$$

$$= \ 1833.6\ \mathbf{Sin\ 314t\ Cos\ 314t}$$

Average Power $P_L = 0$

Example 3.13:

A sinusoidal current, 10A, 60Hz is applied to a pure capacitor of value 50 mF. Write down the equation for the voltage across the capacitor. Find the RMS value of the voltage, capacitive reactance and the average power. Draw a phasor diagram.

Solution:

$$X_C = \frac{1}{\omega C} = \frac{1}{2 \times \pi \times f \times C}$$

$$= \frac{1}{2 \times \pi \times 60 \times 50 \times 10^{-6}} = 53.05$$

$$I_C = 10\ \text{A}$$

$$i_C = 10\sqrt{2}\ \mathbf{Sin}\ (\ \mathbf{2\pi\, f\, t}\ )$$

$$= 14.14\ \mathbf{Sin\ 377t}$$

$$E_C = I_C \times X_C = 10 \times 53.05$$

$$= 530.5\ \text{V}$$

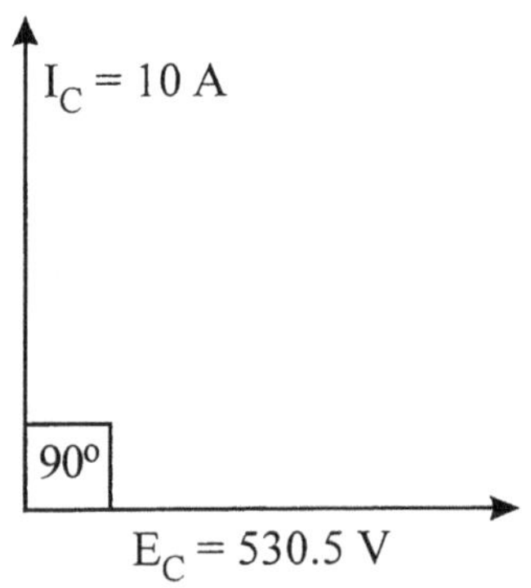

The phasor diagram is shown in the figure. Here I_C is the reference phasor.

The current in a capacitor leads the voltage by 90° or the voltage across the capacitor lags the current by 90°.

Therefore, $e_C = 530.5\sqrt{2}\ \mathbf{Sin}\ (\ \mathbf{377t - 90^o}\ )$

$$= 750.3\ \mathbf{Sin}\ (\ \mathbf{377t - 90^o}\ )\ \text{V}$$

Average Power, $P_C = 0$

* * * * *

COMPREHENSION - 3

Note : By answering all the questions given in this comprehension one can answer any objective type of questions easily whatever be the form of the question. Answers for these questions can be obtained by going through the topics in this chapter carefully.

1. What are the sources of DC voltages?
2. What will be the nature of the EMF induced in the coil of the DC generator?
3. What is an Alternator?
4. Draw DC and some typical voltage waveforms.
5. What is meant by periodic and aperiodic waveforms? Explain with figures.
6. Why is sinusoidal waveform popular?
7. Explain the generation of a sinusoidal waveform with figures.
8. What is Motional EMF or Dynamically Induced EMF?z
9. Draw a typical sinusoidal waveform and explain the following terms:

   (i) Instantaneous Value (ii) Amplitude

   (iii) Time Period (iv) Frequency
10. Obtain the relationship between frequency and time period.
11. Derive an expression for the EMF generated by a two pole alternator.
12. What is the difference between electrical and mechanical degrees?
13. What is the relationship between frequency and angular velocity?
14. Obtain the relationship between frequency, speed and number of pairs of poles.
15. Define the average value of an alternating current and obtain its value for a sinusoidal current.
16. Define the RMS or Effective value of an alternating current and obtain its value for a sinusoidal current.
17. Define the terms Form Factor, Peak Factor and Crest Factor.
18. What will be the RMS and average values for sinusoidal voltage in the case of

    (i) Half Wave Rectifier (ii) Full Wave Rectifier
19. What are the different forms of representing an alternating quantity.

20. Explain the phasor representation of two alternating quantities.
21. What is the angle of a reference phasor?
22. Why does the relative angle between two phasors remains the same at any times?
23. Define the phase of an alternating quantity. Explain lagging and leading between two alternating quantities in terms of waveforms and phasors.
24. What is a j-Operator? What are its properties? What is its use?
25. Explain the exponential form of phasor representation.
26. When a phasor is multiplied by $e^{j\theta_1}$, what is its effect on the phasor?
27. Explain the polar form of representation of alternating quantities. When is it useful?
28. Explain the addition, subtraction, multiplication, division, raising to a power, obtaining the nth root of alternating quantities. Use the best method that is suited in each case.
29. Give expressions for power in the case of
    (i) DC (ii) AC pure resistive circuit.
30. Trace the waveforms for voltage, current and power for an AC pure resistive circuit.
31. Trace the waveforms for voltage, current and power for an AC pure inductive circuit.
32. Trace the waveforms for voltage, current and power for an AC pure capacitive circuit.
33. Derive the expressions for average power in the case of AC pure resistive circuit.
34. Derive the expressions for average power in the case of AC pure inductive circuit.
35. Derive the expressions for average power in the case of AC pure capacitive circuit.
36. Draw the phasor diagrams between the voltage and current in the case of pure resistive circuit.
37. Draw the phasor diagrams between the voltage and current in the case of pure inductive circuit.
38. Draw the phasor diagrams between the voltage and current in the case of pure capacitive circuit.
39. Derive the expressions for inductive and capacitive reactance. Draw their phasors with resistance as the reference phasor.

* * * * *

EXERCISE - 3

Note: By answering all the questions given in this exercise one can answer subjective type of questions easily including problems. The worked examples in the chapter may help in solving these questions.

1. (a) How do you represent the instantaneous value of an alternating current? Explain the symbols used.

   (b) A supply voltage of 230V, 50Hz is fed to a residential building. Write down the equation for instantaneous value.

   [Ans: 325.27Sin (314.16t)]

   (c) A coil of 200 turns is rotated at 1500 RPM in a magnetic field having a uniform density of 0.05T, the axis of rotation being at right angles to the direction of the flux. The mean area per turn is $40cm^2$. Calculate

   (i) the frequency
   (ii) the period
   (iii) the maximum value of the generated E.M.F.
   (iv) the value of the generated E.M.F. when the coil has rotated through 45° from the position of zero E.M.F.

   [Ans: (i) 25Hz; (ii) 0.04sec; (iii) 12.56V; (iv) 8.87V]

2. (a) Define the terms Cycle, Time Period, Frequency, Amplitude, RMS Value and Average Value of an Alternating Quantity.

   (b) An alternating current is represented by i = 70.7 **Sin (520t)**. Determine (i) frequency, (ii) the current 0.0015sec after passing through zero, increasing positively, (iii) RMS and Average Value.

   [Ans: (i) 83Hz; (ii) 49.72A; (iii) 50A and 45A]

3. (a) Derive the RMS value and Average value of an alternating current.

   (b) Calculate the RMS Value, Average Value and Form Factor for the given voltage waveform given that for $0 < t < 0.2$,

   $$v = 20(1 - e^{-200t})$$

   [Ans: 19.62V; 19.5V 1.006]

4. A moving coil ammeter, a thermal ammeter, moving iron ammeter and a half wave rectifier are connected in series with a resistor across a 110V sinusoidal AC supply. The circuit has a resistance of 50Ω to current in

one direction and, due to the rectifier, an infinite resistance to current in the reverse direction. Calculate:

(i) the readings on the ammeters

(ii) the form and peak factors of the current wave.

Note: A moving coil ammeter is generally used for DC circuits. In the case of AC circuits, it reads the average value over one full cycle.

A thermal ammeter depends upon the heating effect of the current. So it reads the RMS value for full cycle.

A moving iron ammeter reads the RMS value of the Alternating Current.

The current during the negative half cycle of a half wave rectifier will be zero.

[Ans: (i) 0.99A, 1.56A, 1.56A; (ii) 1.57, 2.0, 2.0]

5. (a) Derive an expression for the form factor of a sinusoidal waveform.
   (b) Find the Average value of the periodic function shown in the figure.

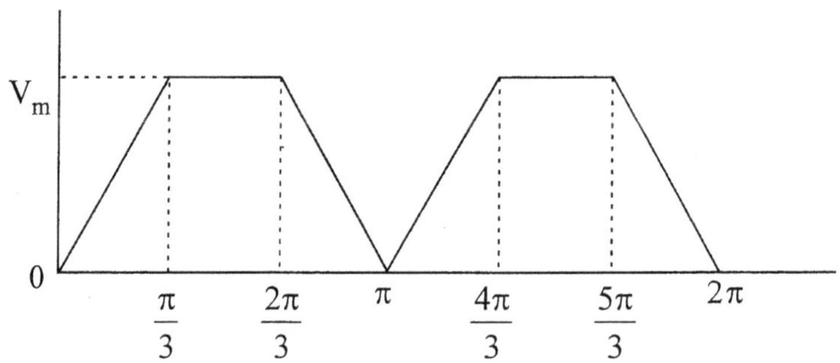

[Ans: $0.667V_m$]

6. Calculate the RMS and Average Value of the current *i* represented by the given figure.

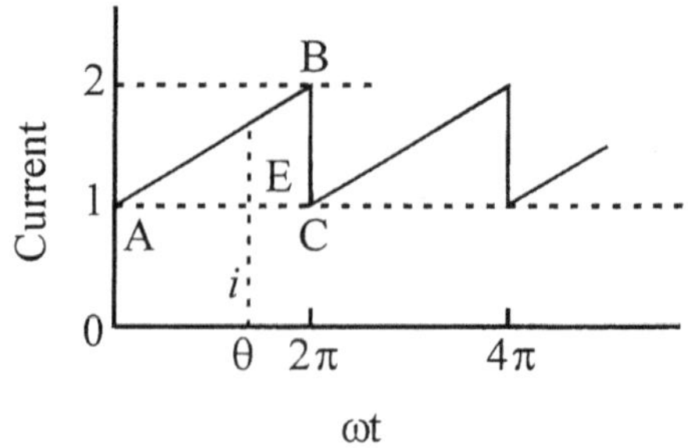

(i) When the triangular waveform has the time axis at $i = 0$ with maximum value of 1A

(ii) When the triangular waveform has a DC component of 1A i.e., the maximum value of the current will be (1 + 1) = 2A.

[Ans: (i) 0.5, 1.527; (ii) 1.5, 10.608]

7. (a) Explain the different forms of representing an Alternating Quantity.

(b) Perform the following mathematical operations and given the final answer in polar form.

(i) $\dfrac{(8+j12)}{j2}$ (ii) $10\angle 60^\circ + 20\angle 30^\circ$

(iii) $(3-j2)(1-j4)$ (iv) $3\angle 20^\circ - 8\angle 60^\circ$

[Ans: (i) $7.211\angle -33.7^\circ$; (ii) $29.093\angle 39.9^\circ$; (iii) $14.866\angle -109.65^\circ$; (iv) $6.02\angle -101.3^\circ$]

8. (a) Draw the waveforms and the phasors for the voltage and current in a pure resistive, inductive and capacitive circuits.

(b) Obtain the expressions for inductive and capacitive reactances.

(c) A voltage given by e = 200 **Sin ωt** when applied to a resistor is found to give a power of 100W. Find the value of the resistance and the equation for the current.

[Ans: (b) 200; (c) i = Sin ωt]

9. (a) A pure inductance of 254.65mH is connected to a 120V, 50Hz source. Calculate

(i) X_L (ii) Current

(iii) Maximum Power delivered (iv) Average Power

(v) Write the equation for the Voltage and Current.

[Ans: (i) j80Ω; (ii) –j1.5A; (iii) 180W; (iv) 0Watts; (v) e = $120\sqrt{2}$ Sin 314t; $i = -1.5\sqrt{2}$ Cos 314t]

(b) A 240V, 50Hz supply is given to a capacitance of value 26.53μF. Find the Capacitive reactance, current and maximum energy stored in the capacitance.

[Ans: –j120Ω; j2A; 1.528Joules]

* * * * *

4

A.C. CIRCUITS - I
SINGLE PHASE CIRCUITS

4.1 INTRODUCTION

So far we have studied the fundamentals of D.C. voltage sources, current sources and resistance components (or resistance parameters) which will generally be the load that will be connected in D.C. circuits.

We have also studied the solution of D.C. circuits like series, parallel and series parallel circuits and also the different theorems like Supeposition Theorem, Thevenin's Theorem, Norton's Theorem and Maximum Power Transfer Theorem as applicable to D.C. circuits. Analysis of D.C. circuits using loop and nodal analysis have been discussed in greater detail.

Then we have studied the fundamentals of A.C. circuits and the different parameters like Resistance, Inductance and Capacitance and their combinations which are used in A.C. circuits.

We have also studied the representation of A.C. quantities like voltage current etc., in different forms including wave forms. The phase difference betwen different wave forms and phasors have also been explained in terms of lagging and leading waveforms or lagging and leading phasors.

4.2 POWER AND POWER FACTOR IN SINGLE PHASE A.C. CIRCUITS

So far we have studied A.C. circuits with pure R, L and C components. In order to study the different types of power and power factor in A.C. circuits we have to study A.C. series, parallel and series parallel circuits with different combinations of R, L and C.

4.3 A.C. SERIES CIRCUITS

As in the case of D.C. circuits, we can have A.C. series circuits, A.C. parallel circuits and A.C. series-parallel circuits also with the elements R, L, C connected in different ways in a circuit.

4.3.1 R-L Series Circuit with A.C. Supply

For an A.C. circuit with a pure resistance R and pure inductance L in series the current wave form will lag behind the voltage waveform by an angle Φ.

where $$\Phi = \text{Tan}^{-1}\left(\frac{X_L}{R}\right) \text{ radians} \qquad \text{.......... (4.3.1)}$$

$X_L = \omega L = 2\pi f$ Ohms, f being frequency in hertzs.

R is resistance in the circuit in ohms.

Φ can also be expressed in degrees (electrical degrees).

Since the inductive impedance phasor leads the resistance phasor by 90°, the resistance and inductive reactance X_L will together be called ***Impedance Z*** and is denoted by the symbol

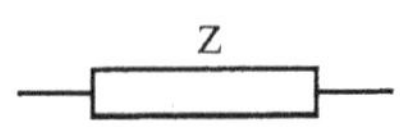

where $$\mathbf{Z} = R + jX_L$$

$$\mathbf{Z} = |Z| \angle \theta^{o} \qquad \text{.......... (4.3.2)}$$

where $$|Z| = \sqrt{R^2 + X_L^2}$$

and $$\theta = \text{Tan}^{-1}\left(\frac{X_L}{R}\right)$$

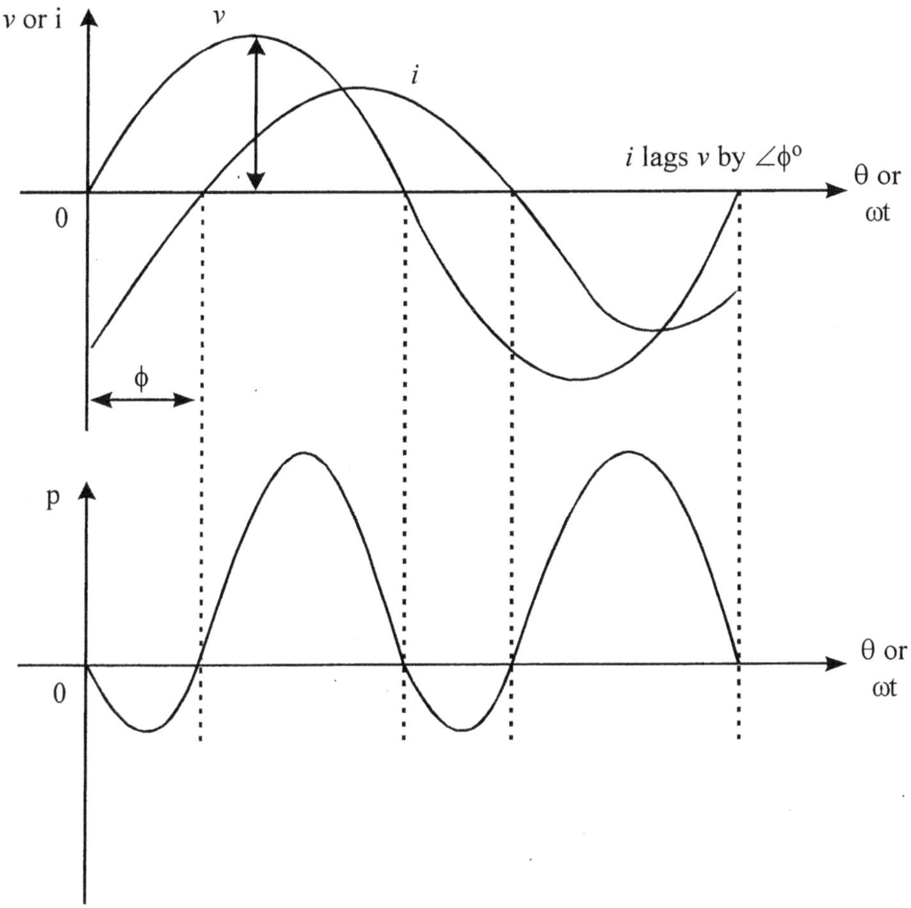

Fig. 4.1 *Voltage, Current & Power Curves for an Inductive Circuit (R-L Series Circuit)*

since X_L leads R by $\frac{\pi}{2}$ radians or 90°.

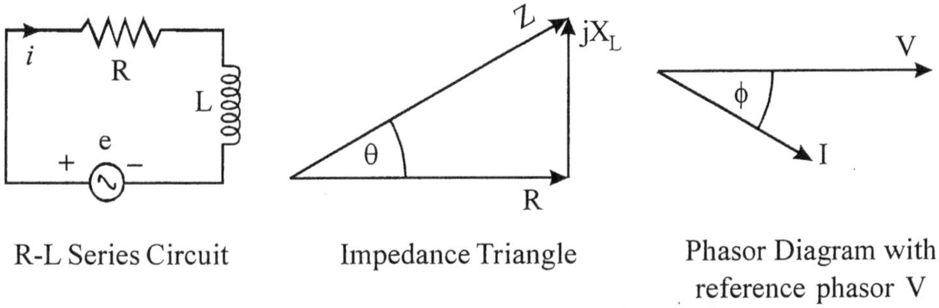

Fig. 4.2 *A.C. RL series circuit and phasor diagrams*

Voltage, Current and Instantaneous Power Waveforms for R-L Series circuit are given in Fig. 4.1. The impedance triangle, voltage drop triangle and phasor diagrams for R-L Series Circuit are drawn as given in Fig. 4.2.

Applying KVL to the R-L series circuit we have,

$$E = IR + jIX_L$$

i.e., Supply Voltage = Voltage Drop across the Impedance

= Voltage Drop across the Resistance + Voltage Drop across the reactance which will be leading the Voltage Drop across the resistance by an angle $\frac{\pi}{2}$ radians or 90°.

For an inductive circuit, the current phasor lags behind the voltage phasor by an angle ϕ where the phase angle ϕ is equal to the impedance angle θ.

Let the supply voltage for the series R-L circuit be,

$$v = V_m \textbf{Sin } \omega t$$

then

$$i = I_m \textbf{Sin } (\omega t - \phi)$$

$$= i\angle{-\phi^{o}} \quad \text{..........} (4.3.3)$$

where the angle between the instantaneous voltage and the instantaneous current phasors $\phi = \theta$

The instantaneous power in the series R-L circuit is,

$$vi = V_m \textbf{Sin } \omega t \times I_m \textbf{Sin } (\omega t - \phi)$$

$$\text{Since Sin A Sin B} = \frac{1}{2}\left[Cos(A-B) - Cos(A+B)\right]$$

$$vi = ½V_m I_m \{ \textbf{Cos } \phi - \textbf{Cos } (2\omega t - \phi) \}$$

Therefore, $p = ½V_m I_m \textbf{Cos } \phi - ½V_m I_m \textbf{Cos } (2\omega t - \phi)$ ....(4.3.4)

for RL series circuit

From the above expression, it is seen that the instantaneous value of the power consists of two components :

1. $½V_m I_m$ **Cos** ϕ, *which contains no reference to ωt and therefore remains constant in value.*
2. $½V_m I_m$ **Cos** ($2\omega t - \phi$), *the term $2\omega t$ indicating that it varies at twice the supply frequency; thus in Fig. 5.1 it is seen that the power undergoes two cycles of variation for one cycle of the voltage wave. Further more, since the average value of a cosine curve over a*

***complete cycle** is zero, it follows that this component does not contribute anything towards the average value of the power taken from the generator.*

Hence, average power over one cycle is,

$$p_{av} = \frac{1}{T}\int_0^T p\, dt$$

$$\tfrac{1}{2}V_m I_m \mathbf{Cos}\ \phi = \frac{V_m}{\sqrt{2}} \times \frac{I_m}{\sqrt{2}} \times \mathbf{Cos}\ \phi$$

$$\therefore \quad P = VI\ \mathbf{Cos}\ \phi \qquad \text{.......... (4.3.5)}$$

where V and I are the R.M.S. values of the voltage and current respectively.

Note that in the case of DC circuit the power

$$P = VI$$

Whereas, in the case of single phase AC circuits, the power,

$$P = VI\ \mathbf{Cos}\ \phi$$

Using R.M.S. values of voltage and current in RL series circuit,

$$I\angle -\phi = \frac{V}{R + jX_L} = \frac{V\angle 0^\circ}{Z\angle\theta} = \frac{V}{Z}\angle -\theta$$

Hence, the angle between the voltage and current phasors ϕ is equal to the angle between the resistance and the inductive reactance θ.

4.3.2 Power Factor

In the expression for average power,

$$P = VI\ \mathbf{Cos}\ \phi$$

the term **Cos** ϕ is known as power factor since, the real power P is obtained by multiplying V, I with the term **Cos** ϕ. The power factor is said to be lagging (Lagging Power Factor) when the current phasor lags behind the voltage phasor by an angle ϕ radians or degrees. In an inductive circuit (circuit with inductance), the power factor will be lagging. The angle ϕ by which the current lags behind the voltage is known as the power factor angle (ϕ^o lag).

$$\text{Power Factor} = \mathbf{Cos(\ \phi^o\ )}\ \text{lag} = \frac{R}{Z}\ \text{lag} \qquad \text{...... (4.3.6)}$$

When mentioning power factor and power factor angle, it should always be mentioned by the side whether it is lagging or leading (with reference to the reference phasor).

4.3.3 Voltage Drop Triangle

We know that the impedance consists of resistance and reactance. The reactance can be inductive or capacitive. The current flowing through an impedance will cause a voltage drop across the resistance component and a voltage drop across the reactance component of the impedance. They are also known as ***Resistance Drop*** and ***Reactance Drop***.

The total voltage drop across the impedance is known as ***Impedance Drop***. The impedance drop is the phasor sum of resistance drop and reactance drop. Together all these three phasors form a triangle which is known as ***Voltage Drop Triangle*** as shown in figure 4.3..

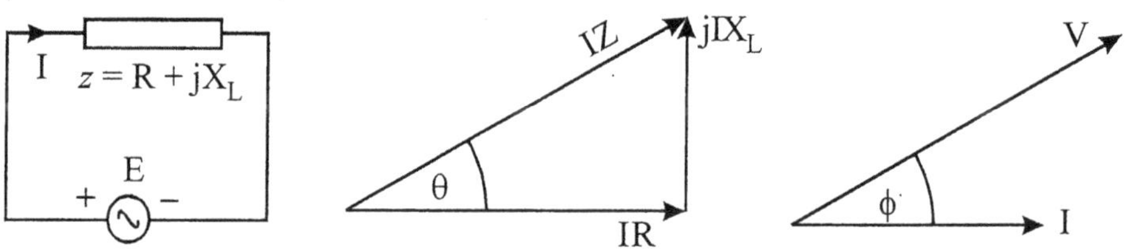

R-L Series Circuit　　Voltage Drop Triangle　　Phasor Diagram with reference phasor I

Fig. 4.3 *Impedance and Voltage Drop Triangles and Phasor Digarams for R-L Series Circuit*

The nature of the reactance decides the phase relationship between the resistance drop and reactance drop. The inductive reactance drop will be leading the resistance drop by angle 90°. The capacitive reactance drop will be lagging the resistance drop by angle 90°.

It may also be noted that the resistance drop will always be in phase with the current phasor.

This is the principle used in drawing the phasor diagrams of A.C. machines like Transformers, Alternators, etc.

4.3.4 Complex Power

In the case of D.C. circuits the power,

$$P = VI$$

gives the real power. Whereas, in A.C. circuits, the real power is given as

$$P = VI \mathbf{Cos}\ \phi$$

where ϕ is the angle between the voltage and current phasors.

In the case of inductive circuits, the angle ϕ which is equal to θ will be positive.

In A.C. circuits, there will be an additional term called ***Reactive Power*** which constitutes the imaginary part of the complex power given as

The Complex Power,

$$\mathbf{S} = \mathbf{V} \times \mathbf{I}^* \text{ VA (Volt-Amperes)}$$

$$\mathbf{S} = P + jQ$$

where $P = VI\ \mathbf{Cos}\ \phi$ Watts

$$Q = VI\ \mathbf{Sin}\ \phi \text{ VAR (Reactive Volt-Amperes)}$$

Sign Convention for the Reactive Part of the Complex Power:

There are two ways of writing the complex power.

1. $\mathbf{S} = \mathbf{V} \times \mathbf{I}^* = P + jQ$ .... (4.3.7)

where $P = \text{Re}(\mathbf{VI}^*)$ i.e., real part of $(\mathbf{VI}^*)$

and $Q = \text{Im}(\mathbf{VI}^*)$ i.e., imaginary part of $(\mathbf{VI}^*)$

Let $\mathbf{V} = a + jb$

$\mathbf{I} = c - jd$ for inductive circuit.

since the current in an inductive circuit is lagging behind the voltage.

$$\mathbf{S} = \mathbf{VI}^*$$

$$= (a + jb)(c + jd) = (ac + bd) + j(bc + ad)$$

$$\mathbf{S} = P + jQ$$

The inductive VAR (Q) is positive in $\mathbf{S} = \mathbf{VI}^*$ if we write $\mathbf{S} = P + jQ$

$$\mathbf{S} = \mathbf{VI}^*$$

2. $$\mathbf{S} = \mathbf{V}^* \times \mathbf{I} = P - jQ \qquad \text{.......... (4.3.8)}$$

where $P = \text{Re}\,(\,\mathbf{V}^*\mathbf{I}\,)$ i.e., real part of $(\,\mathbf{V}^*\mathbf{I}\,)$

and $Q = \text{Im}\,(\,\mathbf{V}^*\mathbf{I}\,)$ i.e., imaginary part of $(\,\mathbf{V}^*\mathbf{I}\,)$

Let $\mathbf{V} = a + jb$

$\mathbf{I} = c - jd$ for inductive circuit.

since the current in an inductive circuit is lagging behind the voltage.

$$\mathbf{S} = \mathbf{V}^*\mathbf{I}$$

$$= (\,a - jb\,)(\,c - jd\,) = (\,ac - bd\,) - j(\,bc + ad\,)$$

$$\mathbf{S} = P - jQ$$

The inductive VAR (Q) is negative in $\mathbf{S} = \mathbf{V}^*\mathbf{I}$ if we write $\mathbf{S} = P + jQ$

$$\mathbf{S} = \mathbf{V}^*\mathbf{I}$$

Generally, in Power Systems, the real power P and reactive power Q will be more and hence the real power P will be measured in Kilo-Watts (KW) or Mega-Watts (MW) and the reactive power Q will be measured in Reactive Kilo-Volt-Ampere (KVAR) or in Reactive Mega-Volt-Ampere (MVAR) and hence, the complex power S will be measured in Kilo-Volt-Ampere (KVA) or Mega-Volt-Ampere (MVA).

4.3.5 VA or KVA Triangle

Similar to the impedance triangle the VA or KVA triangle for inductive circuit or inductive load is also drawn as given in Fig. 4.4 for the two conventions.

$$S = VI^*$$

and $$S = V^*I$$

(**a**) For the Convention $S = VI^*$ (**b**) For the Convention $S = V^*I$

Fig. 4.4 *VA or KVA Triangle for an inductive circuit*

The real power (P) is also known as ***True Power***, ***Active Power*** or ***Inphase Power***.

The reactive power (Q) is also known as ***Imaginary Power*** or ***Quadrature Power***

The complex power (S) is known as ***Apparent Power***.

4.3.6 Wattfull and Wattless Component of Current

From the expression for complex power, we observe that the complex power consists of two components.

1. Real Component

and **2.** Imaginary Component or Reactive Component.

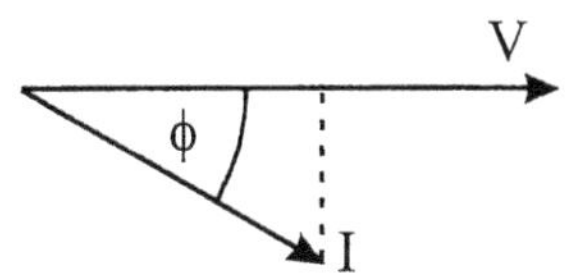

Phasor Diagram for Inductive Cicuit with reference Phasor V

$$S = P + jQ$$

in the convention $S = VI^*$

and $S = P - jQ$

in the convention $S = V^*I$

and we also know that the real power or useful power or average power,

$$P = VI \textbf{ Cos } \phi \qquad \text{..........(4.3.9)}$$

and the reactive power or the imaginary power,

$$Q = VI \textbf{ Sin } \phi \qquad \text{......... (4.3.10)}$$

For an inductive circuit (R-L series circuit), we have the phasor diagram given below.

Comparing the Eq. (4.3.9) and Eq. (4.3.10), we observe that, if V is multiplied by the projection of current phasor I on the voltage phasor V, we obtain, the real power P. Hence this projection of I on V is known as the real component of current or the *Wattfull Component* of current or inphase component of current (in phase with V).

$$I_P = I \textbf{ Cos } \phi$$

Similarly if V is multiplied by the quadrature component of current which is perpendicular to V, we obtain the reactive power Q. Hence this component of current which is perpendicular to V is known as the imaginary component of current or the *Wattless Component* of current or quadrature component of current.

$$I_Q = I \textbf{ Sin } \phi$$

This Quadrature component of current for inductive circuit is positive.

Example 4.1:

A sinusoidal voltage $V = 230\angle 15^\circ$ of frequency 50Hz is applied to a R-L series circuit consisting of $R = 4\Omega$ and $L = 0.05H$. Calculate

1. The RMS current and its phase angle
2. Power Factor
3. Average Power drawn by the circuit
4. Reactive & Apparent Power.

Solution:

$$X_L = 2\pi \times f \times L$$

$$= 2\pi \times 50 \times 0.05 = 15.71\Omega$$

$$Z = \sqrt{R^2 + X^2} = \sqrt{4^2 + 15.71^2}$$

$$= 16.21\Omega$$

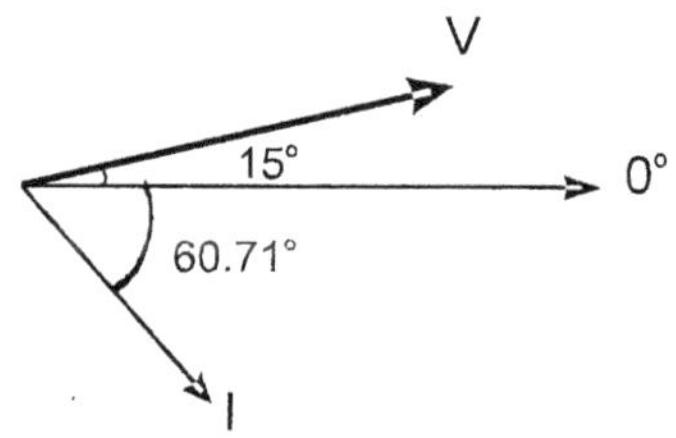

$$Z = 16.21\angle \text{ArcTan}\,\frac{X_L}{R}$$

$$= 16.21\angle \text{ArcTan}\,\frac{15.71}{4}$$

$$= 16.2\angle 75.71^\circ$$

1. $$I = \frac{V}{Z} = \frac{230\angle 15^\circ}{16.21\angle 75.71^\circ}$$

$= 14.19\angle -60.71^\circ$ with respect to the reference phasor (0^0)

$= 14.19\angle -75.71^\circ$ with respect to the voltage phasor

2. Phase angle between voltage and current is the same as the angle of Z namely 75.71° with current lagging,

Power Factor = **Cos (75.71⁰)** $= 0.2468$ lag

3. Average Power $= I^2R = (14.19)^2 \times 15.71$

$$= 805W$$

4. Reactive Power $= I^2X_L = (14.19)^2 \times 15.71$

$$= 3163VAR$$

Example 4.2:

An inductive circuit draws 11A and 0.8KW from a 220V, 50Hz supply. Determine

1. The impedance of the circuit in Cartesian Form
2. Impedance in Polar Form
3. The Power Factor
4. Reactive & Apparent Power.

Solution:

$$Z = \frac{220}{11} = 20\Omega$$

$$P = I^2R$$

or $$800 = (11)^2R$$

Hence, $$R = \frac{800}{121} = 6.61\Omega$$

$$X_L = \sqrt{Z^2 - R^2} = \sqrt{20^2 - 6.61^2}$$

$$= 18.8\Omega$$

1. $$Z = 6.61 + j18.8\Omega$$

2. $$|Z| = 20\Omega$$

$$\textbf{Tan}\ \phi = \frac{18.8}{6.61} = 2.844$$

Hence, $$\phi = 70.6^{\circ}\ \text{lag}$$

Hence, $$Z = 20\angle 70.6^{\circ}$$

3. Power Factor $=$ **Cos** ϕ $=$ **Cos 70.6°** $= 0.332$ lag

4. Reactive Power $=$ VI **Sin** ϕ $= 220 \times 11 \times$ **Sin 70.6°**

$$= 2282VAR = 2.282\ KVAR$$

Apparent Power $= VI = 220 \times 11 = 2420VA. = 2.42\ KVA$

4.3.7. R-C Series Circuit with A.C. Supply

For an A.C. circuit with a pure resistance R and pure Capacitance C in series the current wave form will lead the voltage waveform by an angle θ.

where $$\phi = \mathrm{Tan}^{-1}\left(\frac{X_C}{R}\right) \text{ radians lead} \qquad \text{.......... (4.3.11)}$$

ϕ can also be expressed in degrees (electrical degrees).

Since the capacitive impedance phasor lags the resistance phasor by 90°, the resistance and capacitive reactance X_C will together be called ***Impedance Z*** and is denoted by the symbol

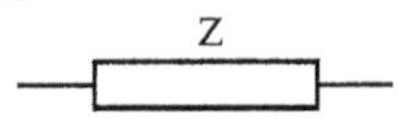

where $$\mathbf{Z} = R - jX_C$$

$$\mathbf{Z} = |\,Z\,|\angle{-\theta^\circ} \qquad \text{.......... (4.3.12)}$$

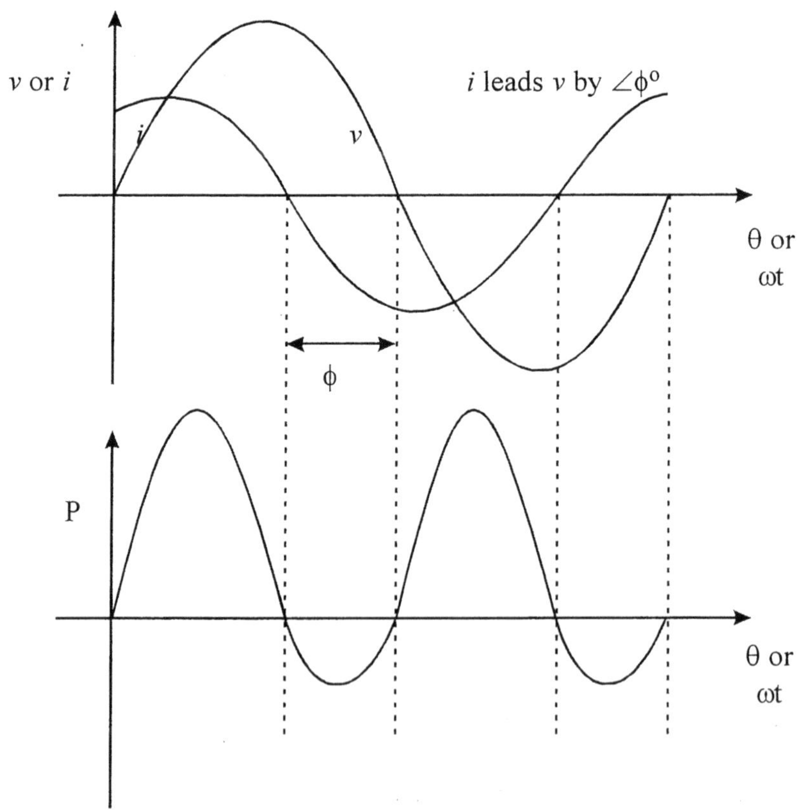

Fig. 4.5 *Voltage, Current & Power Curve for a Capacitve Circuit (R-C Series Circuit)*

where $|Z| = \sqrt{R^2 + X_C^2}$

and $\theta = \text{Tan}^{-1}\left(\frac{X_C}{R}\right)$

since X_C lags R by $\frac{\pi}{2}$ radians or 90°.

Voltage, Current and Instantaneous Power Waveforms for R-C Series circuit are given in Fig. 4.5.

Applying KVL to the R-C series circuit we have,

$$E = IR - jIX_C$$

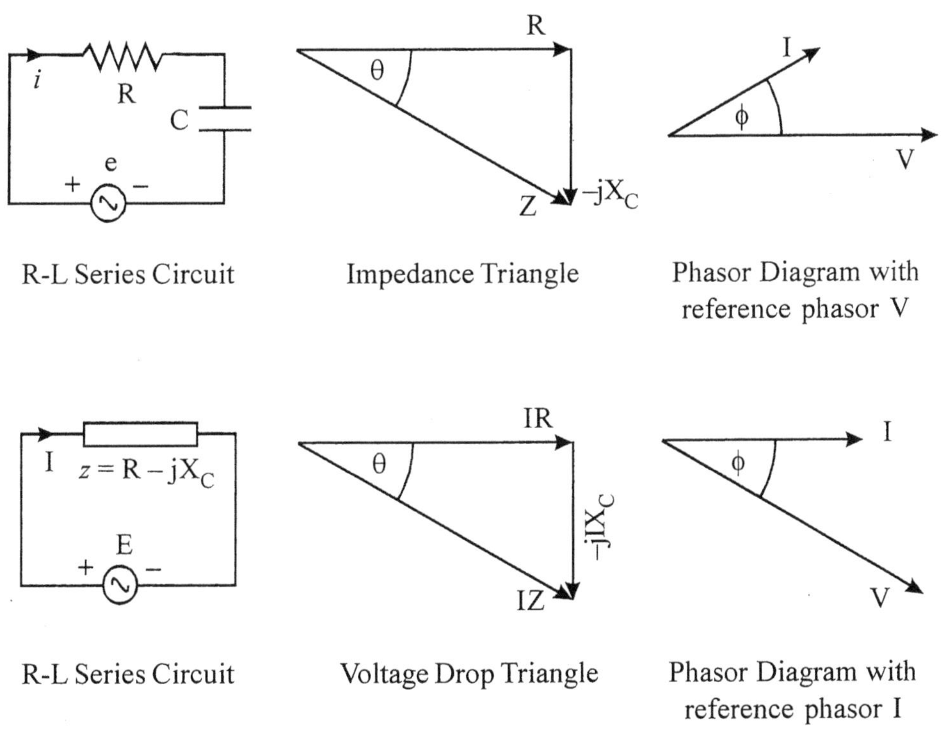

Fig. 4.6 Impedance and Voltage Drop Triangles and Phasor Digarams for R-C Series Circuit

i.e., Supply Voltage = Voltage Drop across the Impedance

= Voltage Drop across the Resistance + Voltage Drop across the reactance which will be lagging the Voltage Drop across the resistance by an angle $\frac{\pi}{2}$ radians or 90°.

The impedance triangle, voltage drop triangle and phasor diagrams for R-C Series Circuit are drawn as given in Fig. 4.6

For a capacitive circuit, the current phasor leads the voltage phasor by an angle ϕ where the phase angle ϕ is equal to the impedance angle θ.

Let the supply voltage for the series R-C circuit be,

$$v = V_m \mathbf{Sin}\ \omega t$$

then

$$i = I_m \mathbf{Sin}\ (\omega t + \phi)$$

$$= i\angle\phi^{o} \qquad \text{..........} \ (4.3.13)$$

where the angle between the instantaneous voltage and the instantaneous current phasors $\phi = \theta$

The instantaneous power in the series R-C circuit is,

$$vi = V_m \mathbf{Sin}\ \omega t \times I_m \mathbf{Sin}\ (\omega t + \phi)$$

$$= V_m \mathbf{Sin}\ \omega t \times I_m [\ \mathbf{Sin}\ \omega t\ \mathbf{Cos}\ \phi + \mathbf{Cos}\ \omega t\ \mathbf{Sin}\ \phi\ ]$$

$$= V_m I_m [\ \mathbf{Sin}^2\ \omega t\ \mathbf{Cos}\ \phi + \mathbf{Sin}\ \omega t\ \mathbf{Cos}\ \omega t\ \mathbf{Sin}\ \phi\ ]$$

$$= \tfrac{1}{2} V_m I_m [\ \mathbf{Cos}\ \phi - \mathbf{Cos}\ (2\omega t - \phi)\ ]$$

$$= \tfrac{1}{2} V_m I_m \mathbf{Cos}\ \phi - \tfrac{1}{2} V_m I_m \mathbf{Cos}\ (2\omega t - \phi)$$

$$\text{..........} \ (4.3.14)$$

for R-C series circuit

From the above expression, it is seen that the instantaneous value of the power consists of two components :

1. $\frac{1}{2}V_m I_m$ **Cos** ϕ, *which contains no reference to* ωt *and therefore remains constant in value.*
2. $\frac{1}{2}V_m I_m$ **Cos** $(2\omega t - \phi)$, *the term* $2\omega t$ *indicating that it varies at twice the supply frequency; thus in Fig. 5.4 it is seen that the power undergoes two cycles of variation for one cycle of the voltage wave.*

Further more, since the average value of a cosine curve over a ***complete cycle*** *is zero, it follows that this component does not contribute anything towards the average value of the power taken from the generator.*

Hence, average power over one cycle is,

$$p_{av} = \frac{1}{T}\int_0^T p\,dt$$

$$\tfrac{1}{2}V_m I_m \mathbf{Cos}\ \phi = \frac{V_m}{\sqrt{2}} \times \frac{I_m}{\sqrt{2}} \times \mathbf{Cos}\ \phi$$

$$\therefore \quad P = VI\ \mathbf{Cos}\ \phi \qquad \text{.......... (4.3.15)}$$

where V and I are the R.M.S. values of the voltage and current respectively.

Note that in the case of DC circuit the power

$$P = VI$$

Whereas, in the case of single phase AC circuits, the power,

$$P = VI\ \mathbf{Cos}\ \phi$$

Using R.M.S. values of voltage and current in RC series circuit,

$$I\angle\phi = \frac{V}{R - jX_C} = \frac{V\angle 0^\circ}{Z\angle -\theta} = \frac{V}{Z}\angle\theta$$

Hence, the angle between the voltage and current phasors ϕ is equal to the angle between the resistance and the capacitive reactance θ.

4.3.8 Power Factor

The power factor is said to be leading (Leading Power Factor) when the current phasor leads the voltage phasor by an angle ϕ radians or degrees. In a capacitive circuit (circuit with capacitance), the power factor will be leading. The angle ϕ by which the current leads the voltage is known as the power factor angle and is written as,

$$\text{Power Factor Angle} = \phi^o \text{ lead.}$$

When mentioning power factor and power factor angle, it should always be mentioned by the side whether it is lagging or leading (with reference to the reference phasor).

4.3.9 Voltage Drop Triangle

The nature of the reactance decides the phase relationship between the resistance drop and reactance drop in the Voltage Drop Triangle. The capacitive reactance drop will be lagging the resistance drop by angle 90°.

4.3.10 Complex Power

The Complex Power,

$$\mathbf{S} = \mathbf{V} \times \mathbf{I}^* \text{ VA (Volt-Amperes)}$$

$$\mathbf{S} = \text{P} + \text{jQ}$$

where $\text{P} = \text{VI } \mathbf{Cos}\ \phi \text{ Watts}$

$$\text{Q} = \text{VI } \mathbf{Sin}\ \phi \text{ VAR (Reactive Volt-Amperes)}$$

Sign Convention for the Reactive Part of the Complex Power:

There are two ways of writing the complex power.

1. $\mathbf{S} = \mathbf{V} \times \mathbf{I}^* = \text{P} + \text{jQ}$

where $\text{P} = \text{Re (} \mathbf{VI}^* \text{)}$ i.e., real part of ($\mathbf{VI}^*$)

and $\text{Q} = \text{Im (} \mathbf{VI}^* \text{)}$ i.e., imaginary part of ($\mathbf{VI}^*$)

Let $\mathbf{V} = \text{a} + \text{jb}$

$\mathbf{I} = \text{c} + \text{jd}$ for capacitive circuit.

since the current in an capacitive circuit is leading the voltage.

$$\mathbf{S} = \mathbf{VI}^*$$

$$= (\text{ a} + \text{jb })(\text{ c} - \text{jd }) = (\text{ ac} + \text{bd }) - \text{j}(\text{ ad - bc })$$

$$= \text{P} - \text{jQ}$$

The capacitive VAR (Q) is negative in the convention

$$\mathbf{S} = \mathbf{VI}^*$$

2. $\mathbf{S} = \mathbf{V}^* \times \mathbf{I} = \text{P} + \text{jQ}$

where $\text{P} = \text{Re (} \mathbf{V}^*\mathbf{I} \text{)}$ i.e., real part of ($\mathbf{V}^*\mathbf{I}$)

and $\text{Q} = \text{Im (} \mathbf{V}^*\mathbf{I} \text{)}$ i.e., imaginary part of ($\mathbf{V}^*\mathbf{I}$)

Let $\mathbf{V} = \text{a} + \text{jb}$

$\mathbf{I} = \text{c} + \text{jd}$ for capacitive circuit.

since the current in an capacitive circuit is leads the voltage.

$$\mathbf{S = V^*I}$$

$$= (a - jb)(c - jd) = (ac + bd) + j(ad - bc)$$

$$= P + jQ$$

The capacitive VAR (Q) is positive in the convention

$$\mathbf{S = V^*I}$$

4.3.11 VA or KVA Triangle

Similar to the impedance triangle the VA or KVA triangle for capacitive circuit or capacitive load is also drawn as given in Fig. 4.7 for the two conventions of S.

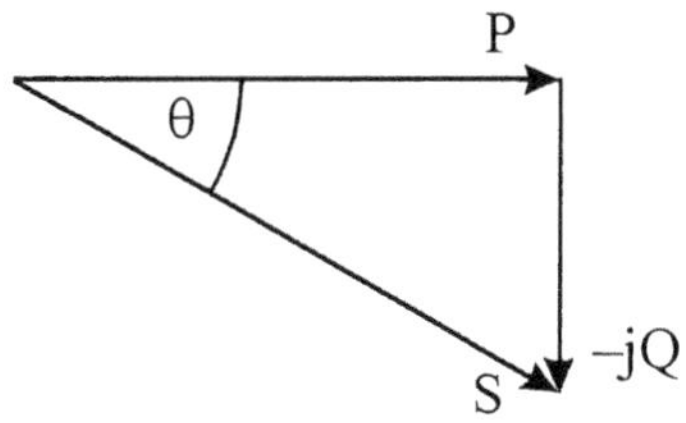

(**a**) For the Convention $S = VI^*$

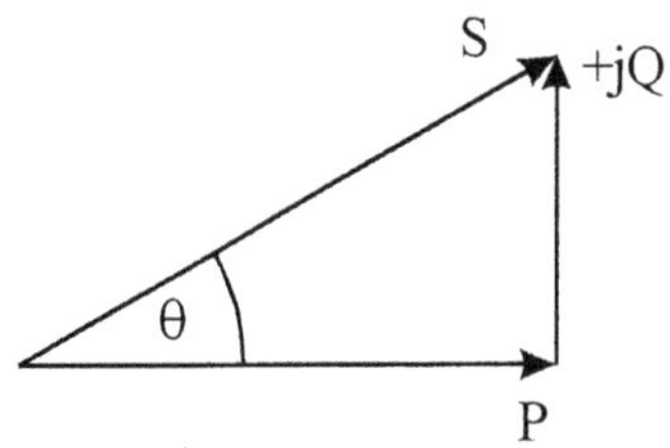

(**b**) For the Convention $S = V^*I$

Fig. 4.7 *VA or KVA Triangle for a capacitive circuit*

4.3.12 Wattfull and Wattless Component of Current

For an capacitive circuit (R-C series circuit), we have the phasor diagram given below.

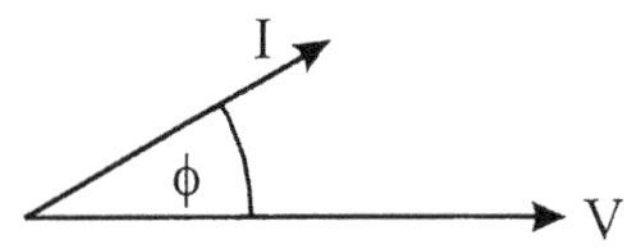

Phasor Diagram for Inductive Cicuit with reference Phasor V

The projection of I on V is known as the real component of current or the *Wattfull Component* of current or inphase component of current (in phase with V).

$$I_P = I \textbf{ Cos } \phi$$

The component of current which is perpendicular to to V is known as the imaginary component of current or the *Wattless Component* of current or Quadrature component of current.

$$I_Q = I \textbf{ Sin } \phi$$

This Quadrature component of current for capacitive circuit is positive.

Example 4.3:

A voltage of 125V at 50Hz is applied across a non inductive resistor (pure resistance) connected in series with a condenser. The current in the circuit is 2.2A. The power loss in the resistor is 96.8W and that in the condenser is negligible. Calculate the resistance and the capacitance.

Solution:

Applied Voltage,

$$V = 125V;$$

Supply Frequency,

$$f = 50Hz$$

Current in the circuit,

$$I = 2.2A$$

Power Loss in the resistor,

$$P = 96.8W$$

Power Loss in Resistor,

$$P = I^2R$$

Therefore, Circuit Resistance,

$$R = \frac{P}{I^2} = \frac{96.8}{2.2^2}$$

$$= 20\Omega$$

Circuit Impedance,

$$Z = \frac{V}{I} = \frac{125}{2.2}$$

$$= 56.82\Omega$$

Capacitive Reactance,

$$X_C = \sqrt{Z^2 - R^2} = \sqrt{56.82^2 - 20^2}$$

$$= 53.18\Omega$$

Capacitance of the Capacitor,

$$C = \frac{1}{2\times\pi\times f\times X_C} = \frac{1}{2\times\pi\times 50\times 53.18}$$

$$= 59.85\mu F$$

Example 4.4:

In a circuit, the applied voltage is 100V and is found to lag the current of 10A by 30°. Find the power factor, the active and reactive power. Is the circuit inductive or capacitive?

Solution:

The voltage lags the current, i.e., the current leads the voltage.

Therefore, Power Factor is leading and the circuit is capacitive.

$$PF = \mathbf{Cos\ 30^o} = 0.866 \text{ lead.}$$

$$\text{Active Power} = |E||I|\mathbf{Cos\ \phi} = 100\times 10\times 0.866$$

$$= 866W$$

$$\text{Reactive Power} = |E||I|\mathbf{Sin\ \phi} = 100\times 10\times 0.5$$

$$= 500W$$

4.3.13 Admittance & Admittance Triangle

The impedance **Z** used in A.C. circuits is similar to the resistance in R used in D.C. circuits. But **Z** is a complex number consisting of both resistance and reactance parts. The reactance can be negative for capacitive reactance and positive for inductive reactance. Similar to the conductance G used in DC circuits, we have in A.C. circuits *admittance* **Y**, which will have both conductance and susceptance parts.

$$\mathbf{Y} = \frac{1}{\mathbf{Z}} \text{ mhos} \qquad \text{.......... (4.3.16)}$$

The inductive admittance,

$$\mathbf{Y_L} = \frac{1}{R + jX_L}$$

$$= \frac{(R - jX_L)}{(R + jX_L)(R - jX_L)}$$

$$= \frac{R}{(R^2 + X_L^2)} - j\frac{X_L}{(R^2 + X_L^2)}$$

$$\mathbf{Y_L} = (G_L - jB_L) \qquad \text{.......... (4.3.17)}$$

where G_L is the conductance and is not equal to $\frac{1}{R}$.

$$G_L = \frac{R}{\left(R^2 + X_L^2\right)}$$

and B_L is the susceptance for the inductive reactance.

$$B_L = \frac{X_L}{\left(R^2 + X_L^2\right)}$$

The capacitive admittance,

$$\mathbf{Y}_C = \frac{1}{R - jX_C}$$

$$= \frac{(R + jX_C)}{(R - jX_C)(R + jX_C)}$$

$$= \frac{R}{\left(R^2 + X_C^2\right)} + j\frac{X_C}{\left(R^2 + X_C^2\right)}$$

$$\mathbf{Y}_C = (G_C - jB_C) \qquad \text{.......... (4.3.18)}$$

where G_C is the conductance and is not equal to $\frac{1}{R}$.

$$G_C = \frac{R}{\left(R^2 + X_C^2\right)}$$

B_C is the capacitive susceptance

where $$B_C = \frac{X_C}{\left(R^2 + X_C^2\right)}$$

Note: *For inductive circuit the reactance will be positive and is given by jX_L. The inductive susceptance will be negative and is given by $-jB_L$.*

$$\mathbf{Z}_L = R + jX_L$$

and $$\mathbf{Y}_L = G - jB_L$$

For capacitive circuit the reactance will be negative and is given by $-jX_C$. The capacitive susceptance will be positive and is given by jB_C.

$$\mathbf{Z}_C = R - jX_C$$

and $$\mathbf{Y}_C = G + jB_C$$

Similar to the impedance triangles, the admittance triangles for Inductive and Capacitive Reactances are shown in Fig. 4.8.

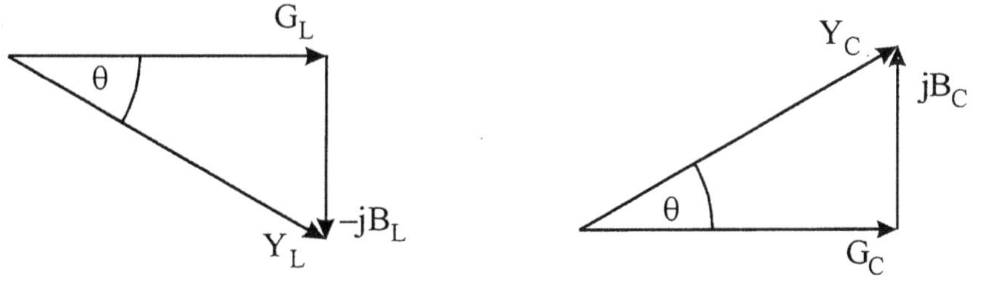

(a) For the Inductive Reactance **(b)** For the Capacitve Reactance

Fig. 4.8 *Admittance Triangles for Inductive and Capacitive Reactances*

Admittance can be represented by the symbol,

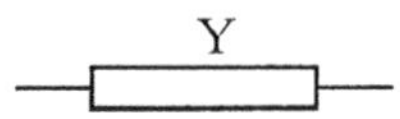

4.3.14 R-L-C Series Circuit

The R-L-C series circuit can be treated as a circuit having a resistance and also a reactance which will be equal to net reactance of inductive and capacitive reactances. The R-L-C series circuit and its impedance triangle are given in Fig. 4.9.

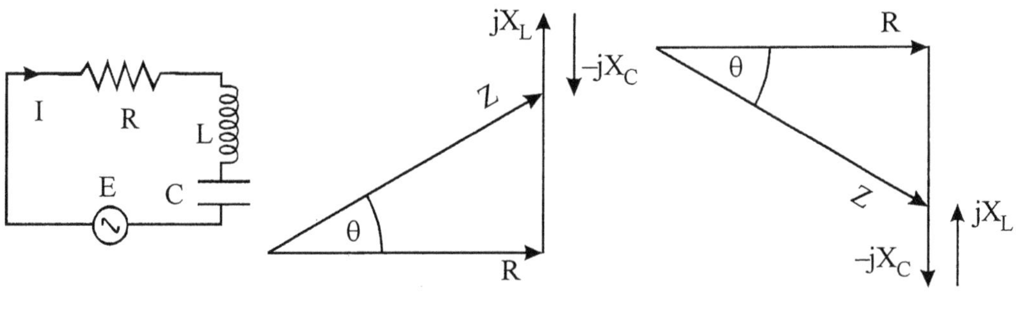

(a) RLC Circuit **(b)** for $X_L > X_C$ **(c)** for $X_C > X_L$

Fig. 4.9 *R-L-C Series Circuits and its impedance triangles*

Since the inductive reactance is taken to be positive and capacitive reactance is taken to be negative, the net reactance of the both the components put together will be positive if the inductive reactance X_L is greater than the capacitive reactance X_C, then, the circuit becomes an effective R-L series circuit with

$$X = (X_L - X_C) \qquad \text{.......... (4.3.19)}$$

$$Z = R + jX \qquad \text{.......... (4.3.20)}$$

Since the inductive reactance is taken to be positive and capacitive reactance is taken to be negative, the net reactance of the both the components put together will be negative if the capacitive reactance X_C is greater than the inductive reactance X_L, then, the circuit becomes an effective R-C series circuit with

$$X = (X_C - X_L) \qquad \text{.......... (4.3.21)}$$

$$Z = R + jX$$

For the R-L-C series circuit, in general,

$$Z = R + jX_L - jX_C$$

$$Z = R + j(X_L - X_C) \qquad \text{.......... (4.3.22)}$$

In the case of series circuits the resulting impedance for impedances which are in series is given as

$$\mathbf{Z_{Se}} = \mathbf{Z_1} + \mathbf{Z_2} + \ldots$$

where $\mathbf{Z_{Se}}$ stands for resultant series impedance.

On the other hand, if admittances are used then for series circuit

$$\mathbf{Z_{Se}} = \frac{1}{\mathbf{Y_{Se}}} = \frac{1}{\mathbf{Y_1}} + \frac{1}{\mathbf{Y_2}} + \ldots$$

which will be difficult to calculate. $\mathbf{Y_{Se}}$ stands for the admittance of the resultant series impedance.

Hence in series circuits, use of impedances is advantageous.

Sometimes, the term *immittance* is used to denote the fact that what is stated can be either impedance or admittance.

The impedance can also be represented by using a small rectangular box symbol for either impedance or admittance with the legend ($R \pm jX$) for impedance ($G \mp jB$). +jX stands for inductive reactance and the corresponding –jB stands for inductive susceptance. –jX stands for capacitive reactance and the corresponding +jB stands for capacitive susceptance.

4.3.15 A.C. Series Circuit with Two or More Impedances

When two or more impedances are connected in series, then the total series impedance is given as,

$$Z_{Se} = \sum_{i=1}^{n} Z_i \qquad \text{.......... (4.3.23)}$$

$$Z_{Se} = \sum_{i=1}^{n} (R_i + jX_i) \qquad \text{.......... (4.3.23a)}$$

where Z_i is the impedance of the ith element

R_i is the resistance of the ith element

X_i is the inductive or capacitive reactance of the ith element as the case may be. For inductive reactance X_i is positive and for capacitive reactance X_i is negative. X_i is zero, if the ith element is purely resistive.

Depending upon the net reactance the current phasor will be lagging or leading the voltage phasor if the net reactance is inductive or capacitive respectively.

Example 4.5:

In the given circuit, Z_1, Z_2 and Z_3 with impedances $(4 + j3)\Omega$, $(3 + j5)\Omega$ and $(2 - j6)\Omega$ respectively, are connected in series to a voltage source V. The voltage drop across Z_1 is $(20 + j0)$ Volts.

Calculate

1. Current in the circuit
2. Voltage drops across Z_2 and Z_3
3. Power Factor of the entire circuit

Solution:

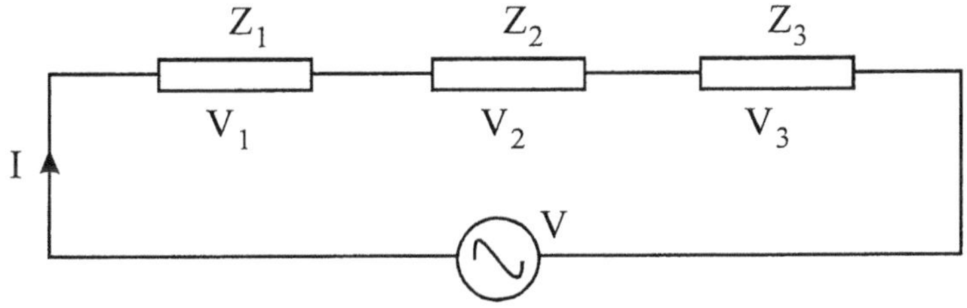

$$\mathbf{I} = \frac{20 + j0}{4 + j3} = \frac{20\angle 0^\circ}{5\angle 36.87^\circ}$$

$$= 4\angle -36.87^\circ \text{ A}$$

$$= 4(\ \mathbf{Cos\ 36.87^\circ} - j\mathbf{Sin\ 36.87^\circ}\ )$$

$$= (\ 3.2 - j2.4\ ) \text{ A}$$

$$V_2 = IZ_2 = (\ 3.2 - j2.4\ )(\ 3 + j5\ )$$

$$= (\ 21.6 + j8.8\ )\text{V}$$

$$V_3 = IZ_3 = (\ 3.2 - j2.4\ )(\ 2 - j6\ )$$

$$= (\ -8.0 - j24\ )\text{V}$$

$$V = V_1 + V_2 + V_3$$

$$= (\ 20 + j0\ ) + (\ 21.6 + j8.8\ ) + (\ -8.0 - j24\ )$$

$$= (\ 33.6 - j15.2\ )\text{V}$$

$$= 36.88\angle -24.34^\circ \text{ V}$$

$$Z = \frac{V}{I} = \frac{36.88\angle -24.34^\circ}{4\angle -36.87}$$

$$= 9.22\angle 12.53^\circ\ \Omega \text{ .... (Note:} Z = Z_1 + Z_2 + Z_3)$$

Hence, Power Factor,

$$\text{PF} = \mathbf{Cos}\ \phi = \mathbf{Cos}\ \theta = \mathbf{Cos}\ (\ \mathbf{12.53^\circ}\ )$$

$$= 0.976 \text{ lag}$$

4.4 A.C. PARALLEL CIRCUITS

A circuit with more than one element connected in parallel and with A.C. supply (voltage or current) is known as ***A.C. Parallel Circuit.***

Solution of A.C. parallel circuits is similar to the solution of D.C. parallel circuits except that in the A.C. parallel circuit we have impedances instead of pure resistances and admittances instance of pure conductances, which are all complex quantities. Also the voltages and currents in the A.C. parallel circuits will be the R.M.S. values of the voltages and currents instead of only real quantities. The power in A.C. parallel circuits will also be the complex powers, the real part of which will be the real power and imaginary part of which will be the reactive power.

The angle between the voltage and corresponding current will be the power factor angle and the cosine of that angle will be the power factor.

4.4.1 A.C. Parallel Circuit with Two or More Impedances

Instead of R-L-C Parallel Circuits, we may have three impedances Z_1, Z_2 and Z_3 in parallel wherein each parallel element may be an inductive reactance or a capacitive reactance or a pure resistance. Calculations are done on similar lines of R-L-C parallel circuit except that we use Z_1, Z_2 and Z_3 instead of R_1, L_2 and C_3 respectively.

$$\mathbf{Y_{Eq} = Y_1 + Y_2 + Y_3} \quad \text{..........} \ (4.4.1)$$

and

$$\mathbf{I = VY_{Eq}} \quad \text{..........} \ (4.4.2)$$

$$\mathbf{I_1 = VY_1} \quad \text{..........} \ (4.4.2a)$$

$$\mathbf{I_2 = VY_2} \quad \text{..........} \ (4.4.2b)$$

$$\mathbf{I_3 = VY_3} \quad \text{..........} \ (4.4.2c)$$

The net current **I** will be lagging the voltage phasor if the total susceptance of $\mathbf{Y_{Eq}}$ is negative. The net current **I** will be leading the voltage phasor if the total susceptance of $\mathbf{Y_{Eq}}$ is positive and will be in phase if the total susceptance is zero. Individual currents will be lagging or leading the voltage V depending upon the nature of the susceptance of that branch.

Example 4.6:

A resistance of 20Ω, an inductance of j5Ω and a capacitance of –j10Ω, an impedance of $(6 + j8)\Omega$ and an impedance of $(4 - j3)\Omega$ are all connected in parallel. To a voltage source of 200 $\angle 45^0$ Volts. Calculate

1. The over all impedance and
2. The individual and over all currents.

Solution:

$$Z_1 = 20 + j0$$

$$Y_1 = \frac{1}{20} = 0.05\angle 0^o = 0.05 + j0$$

$$I_1 = EY_1$$

$$= (200\angle 45^o)0.05 = 10\angle 45^o = 7.07 + j7.07A$$

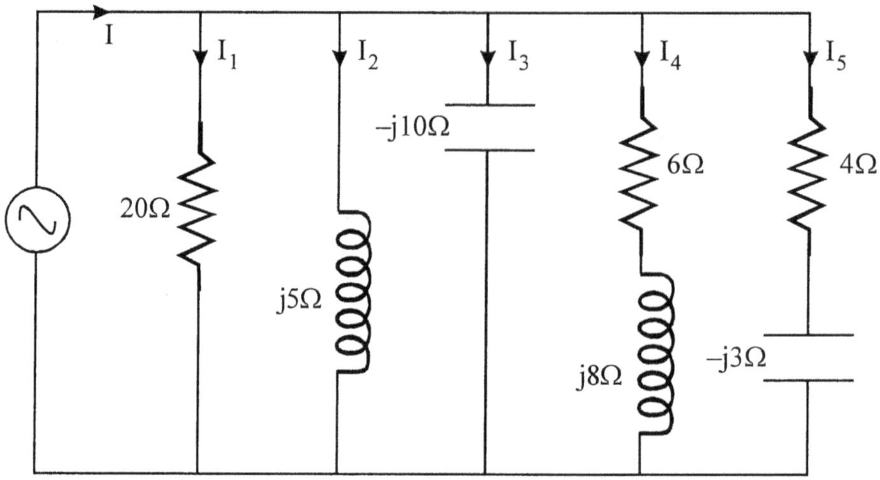

$$Z_2 = 0 + j5$$

$$Y_2 = \frac{1}{j5} = 0 - j0.2 = 0.2\angle{-90^o}$$

$$I_2 = EY_2$$
$$= (200\angle 45^o)(0.2\angle{-90^o})$$
$$= 40\angle{-45^o} = 28.2843 - j28.2843\,A$$

$$Z_3 = 0 - j10$$

$$Y_3 = \frac{1}{-j10} = 0 + j0.1 = 0.1\angle 90^o$$

$$I_3 = EY_3$$
$$= (200\angle 45^o)(0.1\angle 90^o)$$
$$= 20\angle 135^o = -14.1421 + j14.1421\,A$$

$$Z_4 = 6 + j8$$

$$Y_4 = \frac{1}{6 + j8} = 0.06 - j0.08 = 0.1\angle{-53.13^o}$$

$$I_4 = EY_4$$

$$= (200\angle 45^\circ)(0.1\angle -53.13^\circ)$$

$$= 20\angle -8.13^\circ = 19.8 - j2.83\text{A}$$

$$Z_5 = 4 - j3$$

$$Y_5 = \frac{1}{4-j3} = 0.16 + j0.12 = 0.2\angle 36.87^\circ$$

$$I_5 = EY_5$$

$$= (200\angle 45^\circ)(0.2\angle 36.87^\circ)$$

$$= 40\angle 81.87^\circ = 5.66 + j39.6\text{A}$$

Total Admittance,

$$Y = Y_1 + Y_2 + Y_3 + Y_4 + Y_5$$

$$= 0.05 - j0.2 + j0.1 + 0.06 - j0.08 + 0.16 + j0.12$$

$$= 0.27 - j0.06$$

$$= 0.2766\angle -12.529^\circ$$

Overall Impedance,

$$Z = \frac{1}{Y}$$

$$= \frac{1}{0.2766\angle -12.529^\circ}$$

$$= 3.5294 + j0.7843$$

$$\text{Total Current} = I_1 + I_2 + I_3 + I_4 + I_5$$

$$= 46.67 + j29.7$$

$$= 55.32\angle 32.471^\circ$$

4.5 A.C. SERIES - PARALLEL CIRCUIT

The parallel equivalent impedance of the parallel combination is obtained as follows.

$$\mathbf{Y_{Eq}} = \text{Sum of the admittances which are in parallel}$$

$$\mathbf{Z_{Eq}} = \frac{1}{\mathbf{Y_{Eq}}} \quad \text{.......... (4.5.1)}$$

This $\mathbf{Z_{Eq}}$ will be in series with other series impedances. The total impedance of the circuit $\mathbf{Z_T}$ is calculated as follows

$\mathbf{Z_{Se}}$ = the phasor sum of the impedances which are in series.

$$\mathbf{Z_T = Z_{Eq} + Z_{Se} = R + jX} \qquad \text{.......... (4.5.2)}$$

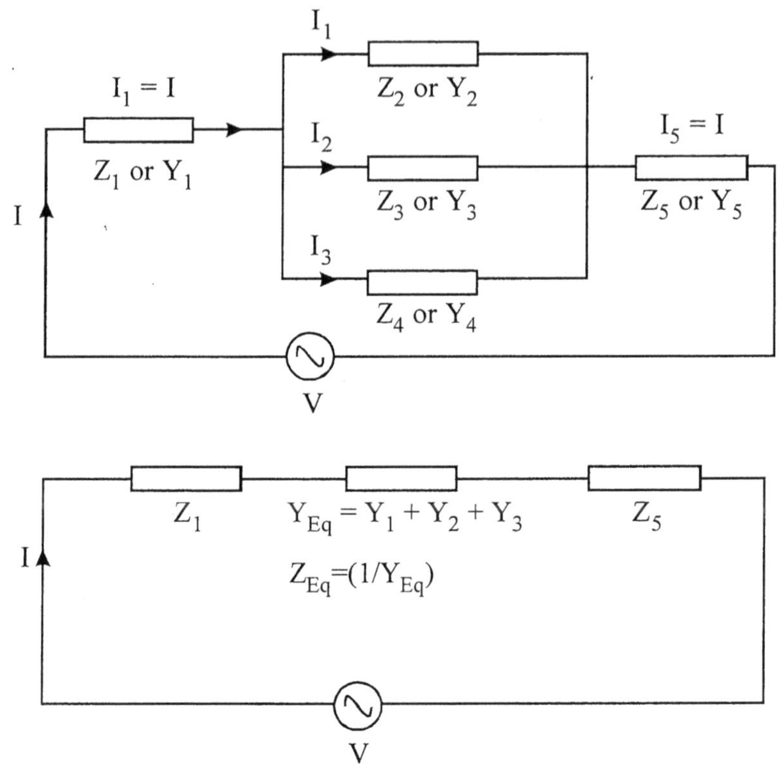

Fig. 4.10 *A.C. Series-Parallel Circuit*

The current,

$$\mathbf{I} = \frac{\mathbf{V}}{\mathbf{Z_T}} \qquad \text{.......... (4.5.3)}$$

where, **V** is the total voltage across the series-parallel combination.

The individual branch currents in the parallel paths are calculated as,

$$\mathbf{I_{Br}} = \frac{\text{Voltage drop across the parallel combination } \mathbf{Z_{Eq}}}{\text{Impedance of that particular branch } \mathbf{Z_{Br}}}$$

The real, imaginary and complex powers and also the power factor etc., are calculated as

$$\mathbf{S} = \mathbf{VI^*} \text{ or } \mathbf{V^*I}$$

Real Power,

$$P = \text{Re}(\mathbf{VI^*}) \text{ or } \text{Re}(\mathbf{V^*I}) \qquad \text{.......... (4.5.4)}$$

Reactive Power,

$$Q = \text{Im}(\mathbf{VI^*}) \text{ or } \text{Im}(\mathbf{V^*I}) \qquad \text{.......... (4.5.5)}$$

$$\text{KVA} = \sqrt{P^2 + Q^2} \qquad \text{.......... (4.5.6)}$$

Power Factor,

$$\text{PF} = \mathbf{Cos}\left(\frac{\mathbf{Q}}{\mathbf{S}}\right) \quad \text{or } \text{Cos}\left(\frac{R}{Z}\right) \qquad \text{.......... (4.5.7)}$$

or from the impedance triangle and admittance triangle, we have,

$$\text{PF} = \mathbf{Cos}\ \boldsymbol{\Phi} = \mathbf{Cos}\ \boldsymbol{\theta} = \frac{R}{Z} \qquad \text{..... (4.5.7a)}$$

or

$$\text{PF} = \mathbf{Cos}\ \boldsymbol{\Phi} = \mathbf{Cos}\ \boldsymbol{\theta} = \frac{G}{Y} \qquad \text{...... (4.5.7b)}$$

Power Factor Angle,

$$\phi = \mathbf{Tan}^{-1}\left(\frac{\mathbf{Q}}{\mathbf{P}}\right) \text{ or } \mathbf{Tan}^{-1}\left(\frac{\mathbf{X}}{\mathbf{R}}\right) \qquad \text{....... (4.5.7c)}$$

or

$$\phi = \mathbf{Tan}^{-1}\left(\frac{\mathbf{B}}{\mathbf{G}}\right) \qquad \text{.......... (4.5.7d)}$$

Note: *If V is the voltage across a particular impedance, I is the current through the impedance, and θ is the impedance angle then, the Power Factor applies to that particular impedance. The power factor of the total circuit is given by cosine of the angle between the total voltage and total current phasors of the circuit. The power factor angle of the total circuit is given by the angle between the total voltage and total current phasors of the circuit.*

Example 4.7:

In the circuit shown in the figure, determine what 50Hz voltage must be applied across AB in order that a current of 10A may flow in the capacitor.

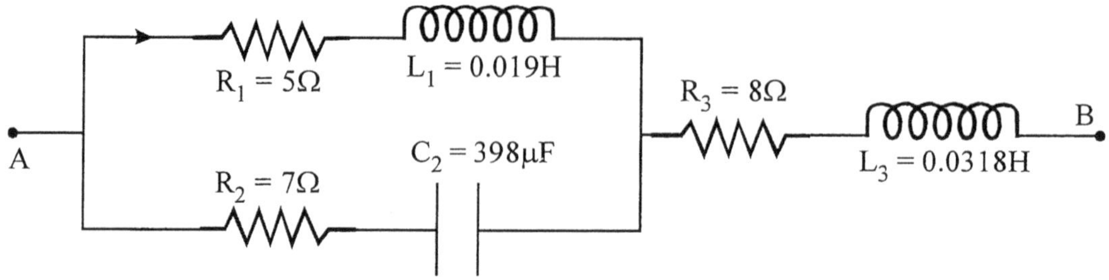

Solution:

$$f = 50Hz$$

$$X_{L1} = 2 \times \pi \times f \times L_1 = 6\Omega$$

$$X_{C1} = \frac{1}{2 \times \pi \times f \times C} = 8\Omega$$

$$X_{L3} = 2 \times \pi \times f \times L_3 = 10\Omega$$

$$\mathbf{Z_1} = (R_1 + jX_{L1}) = (5 + j6)\Omega = 7.81\angle 50.19^o$$

$$\mathbf{Z_2} = (R_2 - jX_{C2}) = (7 - j8)\Omega = 10.63\angle -48.81^o$$

$$\mathbf{Z_3} = (R_3 + jX_{L3}) = (8 + j10)\Omega = 12.8\angle 51.34^o$$

It is not necessary that the supply voltage should be the reference phasor. Any other voltage or current phasor can also be taken as reference phasor for convenience. Then all other phasor angles will be with reference to this reference phasor. Here, we take I_2 as the reference phasor.

$$\mathbf{I_2} = (10 + j0) = 10\angle 0^o$$

The voltage across each parallel branch is given as,

$$\mathbf{V_1} = \mathbf{V_2} = \mathbf{I_2 Z_2} = 106.3\angle -48.81^o = (70 - j80)V$$

$$\mathbf{I_1} = \frac{\mathbf{V_1}}{\mathbf{Z_1}} = \frac{106.3\angle -48.81^o}{7.81\angle 50.19^o} = 13.61\angle -90^o$$

$$= (-2.13 - j13.44)A$$

The total current **I** flowing through $\mathbf{Z_3}$ is,

$$\mathbf{I} = \mathbf{I_1} + \mathbf{I_2} = (-2.13 - j13.44) + (10 + j0)$$

$$= (7.87 - j13.44) = 15.57\angle{-59.65^{\circ}}$$

The voltage $\mathbf{V_3}$ across the impedance $\mathbf{Z_3}$ is,

$$\mathbf{V_3} = \mathbf{IZ_3} = (15.57\angle{-59.65^{\circ}})(12.8\angle{51.34^{\circ}})$$

$$= 199.3\angle{-8.31^{\circ}} = (197.2 - j28.8)\text{V}$$

The supply voltage V across AB is,

$$\mathbf{V_{AB}} = \mathbf{V_2} + \mathbf{V_3} = (70 - j80) + (197.2 - j28.8)$$

$$= 267.2 - j108.8 = 288.5\angle{-22.15^{\circ}}\text{V with respect to phasor } I_2$$

RMS Value of $V_{AB} = 288.5$V

4.6 A.C. STAR - DELTA CONNECTED CIRCUITS

Just as we have star - delta connected DC circuits, we can have star-delta connected single phase AC circuits also. The connection diagrams, the expression for equivalent star for given delta and equivalent delta for given star will all be the same as in the case of DC circuits except that in the place of resistance R in DC circuits, we have the impedance Z in AC circuits.

Accordingly, we have the equivalent star impedances for the given delta impedances,

$$Z_1 = \frac{Z_{12}Z_{13}}{\sum Z_{12}} \qquad \text{.......... (4.6.1}$$

$$Z_2 = \frac{Z_{21}Z_{23}}{\sum Z_{12}} \qquad \text{.......... (4.6.2)}$$

$$Z_3 = \frac{Z_{31}Z_{32}}{\sum Z_{12}} \qquad \text{.......... (4.6.3)}$$

Note that we observe that, *the equivalent star impedances connected to a given terminal is given by the product of the two Delta impedances that are connected to that terminal, the product being divided by the sum of the three Delta connected impedance.*

The same statement can be extended to more than three impedances connected in Mesh and its equivalent Star !! (check up)

Similarly, we have the equivalent delta impedances for the given star impedances

$$Z_{12} = Z_1 + Z_2 + \frac{Z_1 Z_2}{Z_3} \qquad \text{.......... (4.6.4)}$$

$$Z_{23} = Z_2 + Z_3 + \frac{Z_2 Z_3}{Z_1} \qquad \text{.......... (4.6.5)}$$

$$Z_{31} = Z_3 + Z_1 + \frac{Z_3 Z_1}{Z_2} \qquad \text{.......... (4.6.6)}$$

Note that we observe that, *the equivalent Delta impedance connected between any two given terminals is given by the sum of the two Star impedance that are connected between those two terminals plus the product of those two impedances connected between those two terminals divided by the third Star impedances.*

In terms of admittances the equivalent mesh element can be given in terms of the star elements for star or delta having three or more impedances as

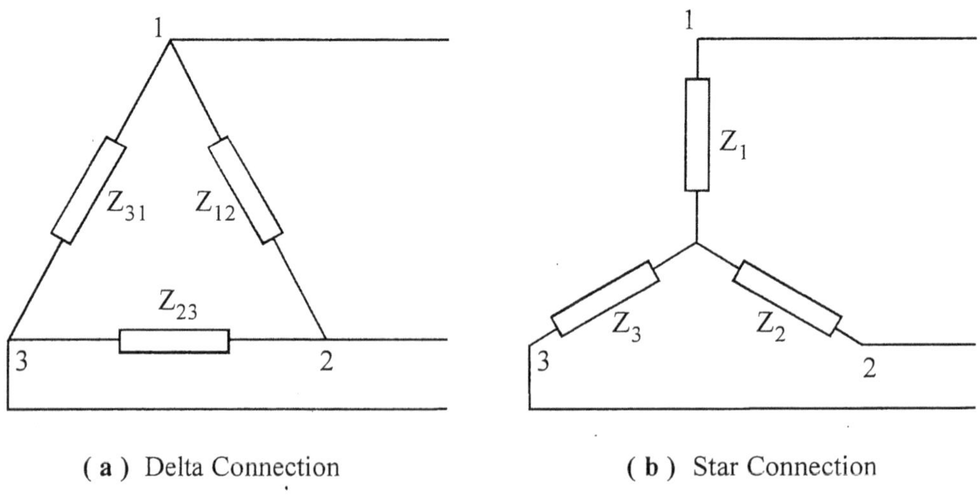

Fig. 4.11 A.C. Star or Delta Connected Circuit

$$Y_{ij} = \frac{Y_i Y_j}{\sum Y_i} \quad \text{!! (check up)} \qquad \text{.......... (4.6.7)}$$

If $Z_{12} = Z_{23} = Z_{31} = Z$ (say), then, the equivalent star impedances will be,

$$Z_1 = Z_2 = Z_3 = \frac{Z}{3} \qquad \text{.......... (4.6.8)}$$

If $Z_1 = Z_2 = Z_3 = Z$ (say), then, the equivalent delta impedances will be,

$$Z_{12} = Z_{23} = Z_{31} = 3Z \qquad \text{.......... (4.6.9)}$$

Example 4.8:

Determine the current drawn from the supply shown in the network given.

Solution:

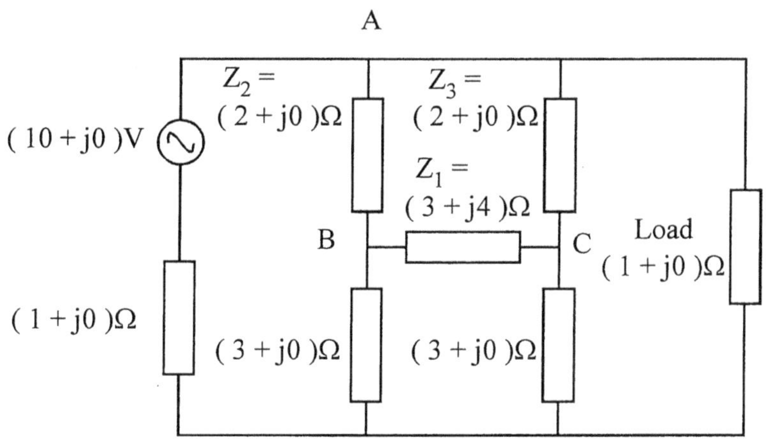

In this example, the solution would be simple if star-delta or delta-star conversion is used. Here we need to observe that the two (2 + j0)Ω impedances and the (3 + j4)Ω impedance; denoted as Z_2, Z_3 and $Z_1$are connected in delta. Let us transform them into the equivalent star.

$$\mathbf{Z_A} = \frac{\mathbf{Z_2 Z_3}}{\mathbf{Z_1 + Z_2 + Z_3}}$$

$$\mathbf{Z_A} = \frac{(2+j0)+(2+j0)}{(2+j0)+(2+j0)+(3+j4)}$$

$$= \frac{(4+j0)}{(7+j4)} = \frac{28-j16}{65}$$

$$= (\,0.43 - j0.25\,)\Omega$$

$$\mathbf{Z_B} = \mathbf{Z_C} = \frac{(2+j0)+(3+j4)}{(2+j0)+(2+j0)+(3+j4)}$$

$$= \frac{(5+j4)}{(7+j4)} = \frac{(51-j8)}{65}$$

$$= (\,0.78 - j0.12\,)\Omega$$

The network therefore can be redrawn as shown in the figure given below.

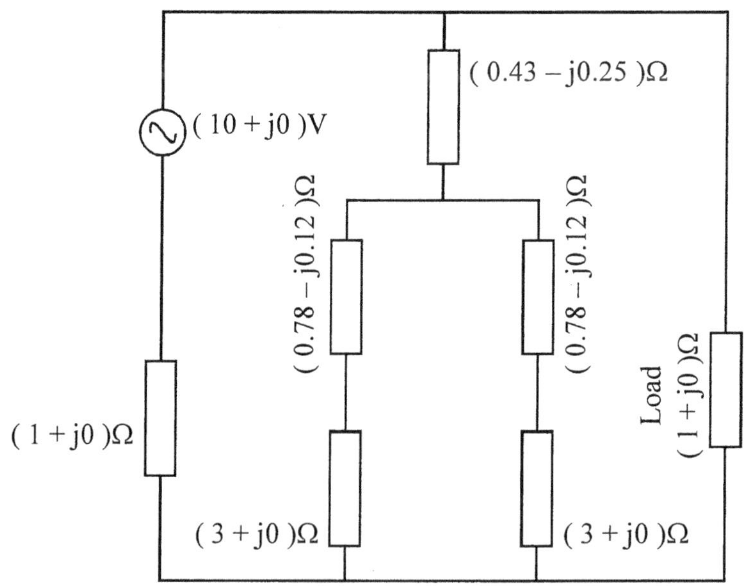

The series impedances can be added to give (3.78 – j0.12)Ω in each of the parallel branches. There being two identical branches in parallel, they can be replaced by an equivalent impedance of (1.89 – j0.06)Ω as shown in the figure.

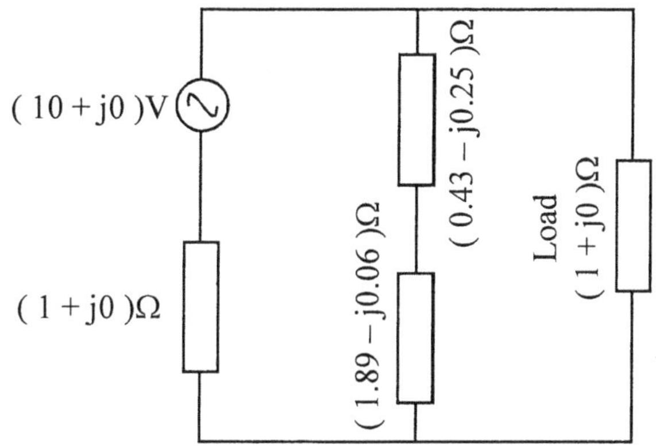

The series impedances can be summed to give,

$$\mathbf{Z_{Se}} = (\ 2.32 - j0.31\ )\Omega$$

This $\mathbf{Z_{Se}}$ is in parallel with load impedance (1 + j0). The equivalent impedance of this parallel combination is,

$$\mathbf{Z_{Eq}} = \frac{(2.32 - j0.31)(1 + j0)}{(2.32 - j0.31) + (1 + j0)}$$

$$= \frac{(2.32 - j0.31)}{(3.32 - j0.31)}$$

$$= (\ 0.70 - j0.03\ )\Omega$$

This $\mathbf{Z_{Eq}}$ is in series with the impedance (1 + j0), which is in series with the voltage source.

The total impedance of the circuit is,

$$\mathbf{Z_T} = (\ 3.32 - j0.31\ ) + (\ 1 + j0\ ) = (\ 1.70 - j0.03\ )\Omega$$

$$= 1.7\angle 1^o$$

The current drawn from the supply,

$$\mathbf{I} = \frac{\mathbf{E}}{\mathbf{Z_T}} = \frac{(10 + j0)}{(1.70 - j0.03)}$$

$$= 5.88\angle 1^o \text{A}$$

$$\text{I} = (5.88\text{A} + j0) \text{ Amp approx.}$$

4.7 IMPORTANCE OF POWER FACTOR

The power factor of an A.C. circuit plays an important role in power system. Since power of an A.C. circuit is given by the relation:

$$P = VI \cos \phi$$

$$I = \frac{P}{V\cos \varphi}$$

From the above relation, it is clear that for a fixed power at constant voltage, the current drawn by the circuit increases with decrease in Power Factor.

Thus at low Power Factor, A.C. circuits draw more current from their mains and results in the following disadvantages:

1. Greater Conductor Size:

At low power factor conductors are to carry more current for same power, therefore, they require larger area of cross section.

2. Poor Efficiency:

At low power factor conductors have to carry larger current which increases copper losses (I^2R) and results in poor efficiency.

3. Larger Voltage Drop:

At low power factor conductors have to carry larger current which increases voltage drop (IR) in the system and results in poor regulation.

4. Large KVA Rating of Equipment:

The KVA rating of electrical machines and equipment connected in the power system such as alternators, transformers, switch gears etc., will be more at lower power factors since it is inversely proportional to power factor ($KVA = KW/\cos \phi$).

To improve the power factor of an A.C. circuit a capacitor is connected across the circuit i.e., parallel to the circuit.

4.8 FEATURES AND USES OF A.C. SINGLE PHASE CIRCUITS

1. Solution of Single Phase AC circuit is just similar to the solution of DC circuit except that, instead of resistance in DC circuit, we use impedance in AC circuits. The voltage, current are in RMS values unless otherwise stated. The voltage, current and impedances are also a complex quantities.

2. Unlike in DC circuit, in which the current will be in phase with the voltage, in the case of AC circuits, the current phasor will be in phase with the voltage phasor if the circuit is resistive, or the current will be lagging by

an angle ϕ equalling to the impedance angle θ if the circuit is inductive or the current phasor will be leading the voltage phasor by an angle ϕ equalling to the impedance angle θ if the circuit is capacitive.

3. The power, in Volt-Amperes, in the case of AC circuit is a complex quantity consisting of real power or average power, P = VI **Cos** ϕWatts and a quadrature component Q = VI **Sin** ϕVAR where **Cos** ϕ is the power factor and ϕ is the angle between the voltage and current phasors.
4. The power factor is lagging in the case of inductive circuits and leading in the case of capacitive circuits.
5. If the complex power is expressed as $\mathbf{S} = P + jQ$ then Q will be positive if the convention, $\mathbf{S} = \mathbf{VI}^*$ is used for the complex power and Q will be negative if the convention, $\mathbf{S} = \mathbf{V}^*\mathbf{I}$ is used for the complex power.
6. The impedance, admittance and complex power can be represented by an impedance triangle, admittance triangle and KVA triangle respectively.
7. The current in AC circuits will have two components called the Wattfull component which will be in phase with the voltage and Wattless component which will be in quadrature with the voltage phasor.
8. The study of analysis of single phase AC circuit helps us in analyzing the performance of single phase AC machines by drawing the equivalent single phase AC circuit for the given machine.
9. Balanced three phase circuit can also be represented in terms of equivalent single phase circuit and can be analyzed as a single phase AC circuit

4.9 SINGLE PHASE CIRCUITS WITH VARIABLE ELEMENTS - RESONANCE

So far we have studied the circuit analysis using lumped parameters which have fixed values. In this chapter we will be studying the circuits in which we have variable parameters. By varying the resistance parameter, the nature of the circuit like whether it is inductive or capacitive does not alter except the value the of impedance changes. On the other hand, if we vary the energy consuming elements like resistance or capacitance, the nature of the impedance may change.

By making the inductive reactance equal to the capacitive reactance, the net reactance becomes zero and the circuit becomes purely resistive. Such a condition is known as *Resonance Condition.* Also by changing the frequency of the supply, the value of the inductive reactance and capacitive reactance can be changed and resonance condition can be obtained.

4.10 RESONANCE IN A.C. CIRCUITS

In an A.C. circuit with R, L and C elements either series or parallel, if the inductive reactance and the capacitive reactance are equal in the case of series circuit or inductive susceptance and capacitive susceptance are equal in the case of parallel circuit, then the net impedance in the circuit will be purely resistive and the power factor of the circuit becomes unity (**Cos** $\phi = 1$). Such a situation is known as ***Resonance*** and is of importance in the case of tuning a circuit in radio, televisions, etc.

4.10.1 Frequency Variation

When introducing A.C. Circuits, we defined the supply by its voltage and frequency. If we are thinking about the supply which we have at home, then it is reasonable to think of both of these as constant. This would even be reasonable in many other applications ranging from heavy power utilization to data and control communications circuit.

However, most communications systems involve circuits in which either the supply voltage (in this application it is usually called a ***signal***) operates with a varying frequency, or a number of signals operate together, each with its own frequency.

In recent times, power applications have also taken to varying the frequency of operation, usually with a view to the speed control of motors.

If we wish to learn about communications systems including radio, television and telephones, as well as machine control systems, we first need to investigate the effects on a circuit of varying the frequency. In particular, we need to investigate the condition known as ***Resonance***. This we can introduce by analyzing the effect of frequency variation in an RLC circuit.

4.10.2 Series Resonance

In the case of a series R-L circuit, the series impedance,

$$Z_{Se} = R + j(X_L - X_C)\Omega$$

1. *Z_{Se} will be inductive, if $X_L > X_C$ and the current will be lagging behind the voltage.*

2. *Z_{Se} will be capacitive, if $X_C > X_L$ and the current will be leading the voltage.*

3. *Z_{Se} will be purely resistive, if $X_L = X_C$ and the current will be in phase with the voltage. The power factor of the circuit will be unity. This condition is called* ***Series Resonance****. Also it will be observed that the current will be maximum under resonance since the total impedance comes down.*

For series resonance, the condition is $X_L = X_C$ which can be obtained by any one of the following methods, since

$$X_L = \omega L = 2\pi f L$$

and
$$X_C = \frac{1}{\omega C} = \frac{1}{2\pi f C}.$$

Methods

1. *by varying the value of the inductance.*
2. *by varying the value of the capacitance.*
3. *by varying the frequency f of the supply.*

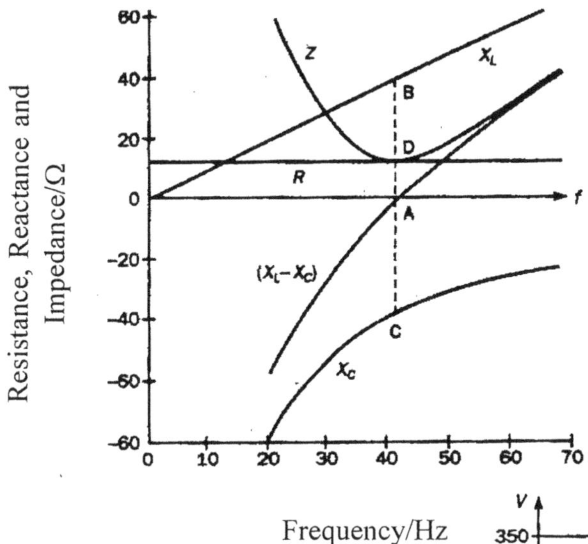

(a) Variation of Reactances and Impedance with Frequency

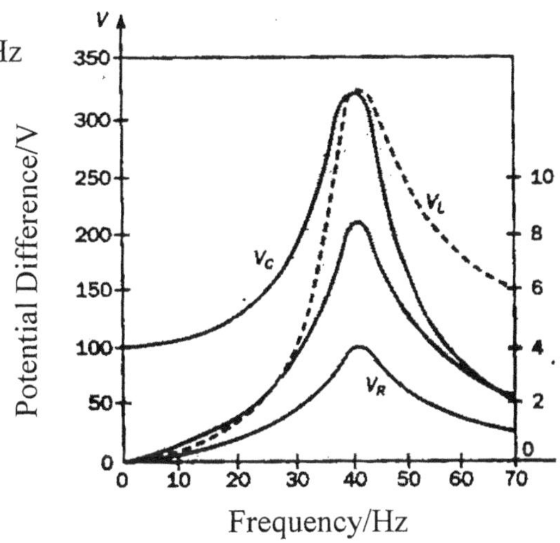

(b) Effect of Frequency Variation upon Voltages across R, L & C in Series

Fig. 4.12 *Series Resonance*

In the case of a fixed coil or a fixed capacitance, it is not possible to change the value of L or C. Hence, the only alternative to obtain the resonance is to vary the frequency of the supply.

The variation of the inductive reactance and capacitive reactance and the impedance of the circuit with respect to frequency and the resonant conditions are shown in Fig. 4.12.

4.10.3 Series Resonance Frequency

When the frequency is zero, i.e., corresponding to a D.C. supply,

$$X_L = 2\pi f L = 0$$

and

$$X_C = \frac{1}{2\pi f C} = \infty$$

As the frequency increases, X_L increases in direct proportion, while X_C decreases inversely. For resonance:

$$V_L = V_C$$

$$IX_L = IX_C$$

$$X_L = X_C$$

i.e., *for series resonance the inductive reactance is equal to capacitive reactance.*

$$2\pi f L = \frac{1}{2\pi f C}$$

then

$$f = \frac{1}{2\pi\sqrt{LC}}$$

At this frequency, Z = R and I = V/R. If the resistance is small compared with the inductive and capacitive reactances, the potential differences across the latter namely $X_L I$ and $X_C I$, are many times the supply voltage, and represented by the phasors in Fig. 4.13

Again from the general graph of reactance against frequency shown in Fig. 4.14 it is seen that the condition $X_L = X_C$ is satisfied when the frequency has a value of f_r which is termed as the ***Resonant Frequency***. It follows that,

$$f_r = \frac{1}{2\pi\sqrt{LC}} \qquad \text{.......... (4.10.1)}$$

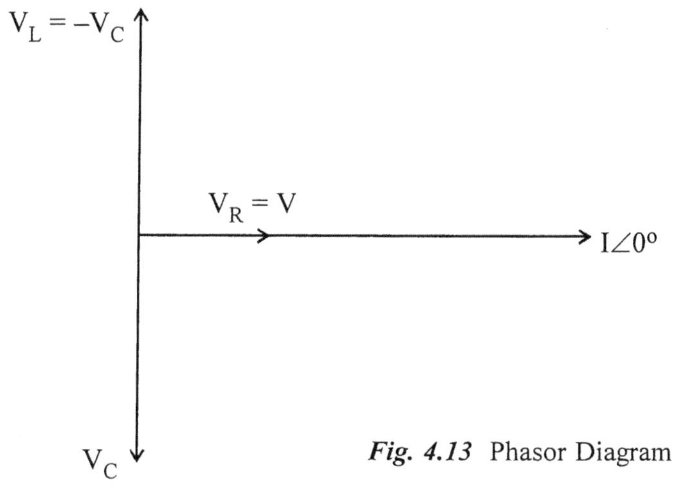

Fig. 4.13 Phasor Diagram for Series Circuit at Resonance

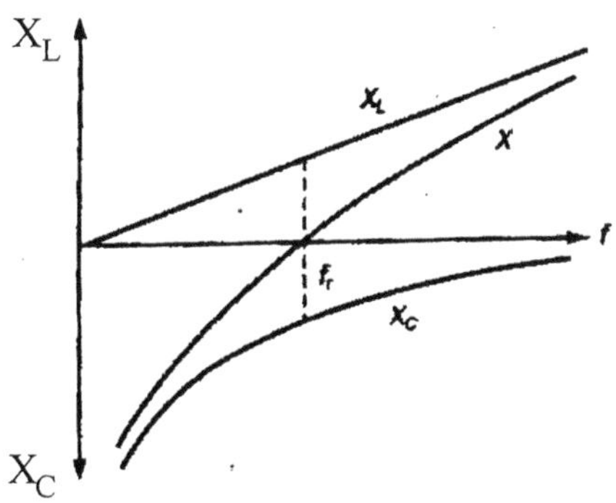

Fig. 4.14 *Variation of Reactances with Frequency*

For the condition of the reactances of the inductor and the capacitor being equal, it has therefore been noted that V and I are in phase and that the circuit impedance is a minimum being equal to the resistance R. The resonant current is dependent on the resistance and the value of current, given by,

$$I_r = \frac{V}{R} \qquad \text{.......... (14.10.2)}$$

may be considerably higher than that at other frequencies. A consequence of this is that the voltage drops across the reactive components may reach very high values - many times that of the supply voltage. This was illustrated in Fig. 4.35. It may be added that the power dissipation in the circuit is consequently a maximum while the peak rates of energy storage for the reactors become equal and maximum.

Owing to the peak rates of energy storage in the reactors, it is necessary to inquire into the relation of this statement to the operation of the circuit at unity power factor, which infers that the network has no reactive power. The energy stored by the reactors is a constant and it oscillates between the electric and magnetic modes of storage. It therefore remains with the reactors and it is this oscillation of energy that makes the circuit appear to be purely resistive.

A circuit is said to resonate whenever this effect of energy oscillation predominates; this is usually interpreted to be when the peak rate of energy storage in each of the reactors - given by I^2X - is at least 10 times the power associated with the circuit resistance. If this condition is not met, then the circuit is a power circuit (usually) that has had its power factor adjusted to unity.

To the electronics and communications engineers, the degree of predomination of the energy oscillation is important since the better the ratio of predominance then the better the circuit is able to accept current (and power) at the resonant frequency to the exclusion of other frequencies. This selection of frequency is used in radio and television receivers and is the process of tuning, e.g., accepting radio1 to the exclusion of other stations, each of which has its particular transmitting frequency.

The degree of predomination is given a factor of goodness termed the ***Q Factor*** or ***Quality Factor***. Thus,

$$\text{Q Factor} = \frac{I^2X_L}{I^2R}$$

or

$$\text{Q Factor} = \frac{I^2X_C}{I^2R}$$

Therefore,

$$\text{Q Factor} = \frac{\omega_r L}{R} \qquad \text{.......... (4.10.3)}$$

which is the voltage magnification at resonance. It follows that voltage magnification at resonance is,

$$V_M = \frac{\text{Voltage Across L (or C)}}{\text{Supply Voltage}}$$

$$V_M = \frac{2\pi f\, LI}{RI} = \frac{2\pi f\, L}{R} \qquad \text{.......... (4.10.4)}$$

$$V_M = \text{Q Factor of the Circuit.} \qquad \text{.:........ (4.10.4a)}$$

Since, $2\pi f_r = \dfrac{1}{\sqrt{LC}}$ at resonance

Therefore,

$$\text{Q Factor} = \frac{2\pi f\,L}{R}$$

Substituting the expression of resonance frequency for f we have,

$$\text{Q Factor} = \frac{1}{R}\sqrt{\frac{L}{C}} \qquad \text{.......... (4.10.5)}$$

A series resonant circuit is often referred to as an ***Acceptor***, since the current is a maximum at resonance.

It can be observed that in a series resonance circuit, the voltage (voltage drop) across the inductance or voltage (voltage drop) across capacitance can even be more than the supply voltage because the two drops get cancelled as they will be in phase opposition i.e., the phasor voltage drop across the capacitance is exactly 180° out of phase with reference to the phasor voltage drop across the inductance.

The series resonance circuit is also known as ***Voltage Magnification Circuit*** because voltage across the inductance or capacitance will be maximum. The ***Voltage Magnification of the circuit*** is equal to the ratio of the voltage across the inductance or capacitance to the supply voltage.

Voltage Magnification,

$$V_M = \frac{\text{Voltage across the Inductance}}{\text{Supply Voltage}}$$

$$= \frac{I_M X_L}{V}$$

$$V_M = \frac{I_M\,\omega L}{I_M R} = \frac{\omega L}{R} \qquad \text{.......... (4.10.6)}$$

where I_M is the maximum current achieved at resonance

4.10.4 Effects of Series Resonance

The following are the main effects of series resonance:

1. At resonance $X_L = X_C$, therefore, the impedance of the circuit is minimum and is reduced to the resistance of the circuit, i.e.,

$$Z_r = R$$

2. Since, impedance is minimum, the circuit current is maximum at resonance,

i.e.,
$$I_r = \frac{V}{Z_r}$$

$$I_r = \frac{V}{R}$$

The resistance R includes the resistance of the inductance coil also.

3. Power Taken by the circuit is maximum, as I_r is maximum i.e.,

$$P_r = I_r^2 R$$

4. As the current drawn by the circuit, at resonance, is very large (maximum), the voltage drop across L, i.e.,

$$V_L = IX_L = I \times 2\pi f_r L$$

and
$$V_C = IX_C = I \times \frac{1}{2\pi f_r C}$$

are also very large.

In power system, at resonance, the excessive voltage built up across the inductive and capacitive components (such as circuit breakers, reactors, etc.) may cause damage. Therefore, series resonance should be avoided in power system. However, in some of the electronic devices (such as antenna circuit of radio and TV receiver, tuning circuits etc.,), the principle of series resonance is used to increase the signal voltage and current at a desired frequency (f_r).

Since a series resonant circuit has the capability to draw heavy current and power from the mains, it is often regarded as ***Acceptor Circuit***.

The series resonance circuit is also known as ***Acceptor Circuit*** because it causes maximum at the resonance frequency and the signals at the resonant frequency are accepted. It is also known as ***Selector Circuit*** because it selects the signals at the resonance frequency.

Example 4.9:

A series RLC circuit with R = 10Ω, L = 1mH and C = 100 pico-Farads connected across a sinusoidal source of 20V with variable frequency. Compute

(a) the Resonant Frequency of the circuit

(b) the Q factor of the circuit at the resonant frequency.

(c) Determine the half power frequencies.

Solution:

$$R = 10\Omega;$$

$$L = 1mH;$$

$$C = 1000pF;$$

$$E = 20V$$

(a) $$f_0 = \frac{1}{2\pi\sqrt{LC}}$$

$$= \frac{1}{2\pi\sqrt{1\times10^{-3}\times1000\times10^{-12}}}$$

$$= 159155 cps$$

(b) $$Q\text{ factor} = \frac{1}{R}\sqrt{\frac{L}{C}}$$

$$= \frac{1}{10}\sqrt{1\times\frac{10^{-3}}{1000}\times10^{-12}} = 100$$

Assume that $\left(\frac{R}{2L}\right)^2 << \frac{1}{LC}$, then,

(c) $$f_1 = f_0 - \frac{R}{4\pi L}$$

$$= 159155 - \frac{10}{4\pi\times10^{-3}}$$

$$= 159155 - 796$$

$$= 158359\text{cps}$$

$$f_2 = f_0 + \frac{R}{4\pi L}$$

$$= 159155 + \frac{10}{4\pi \times 10^{-3}}$$

$$= 159155 + 796$$

$$= 159951\text{cps}$$

$$= 160 \text{ KHz (approx.)}$$

4.10.5 Selectivity & Bandwidth

The current versus frequency curve of an RLC series circuit is symmetrical about the resonance frequency. At frequency f_0, the current is maximum given by E/R. The will be two frequencies f_1, f_2 to either side of resonant frequency at which the power is half the power at resonance. So they are called half power frequencies. f_1 is called *Lower Half Power Frequency* and f_2 is called *Upper Half Power Frequency* as shown in Fig.4.15.

Power at points 1 and 2,

$$P_1 = I_1^2 R$$

and

$$P_2 = I_2^2 R$$

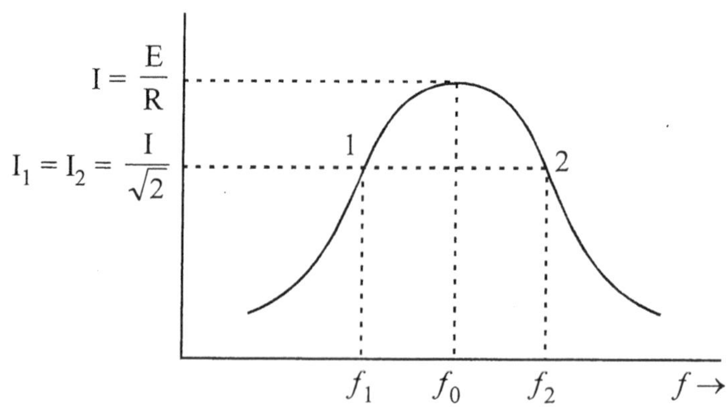

Fig 4.15 Half Power Points and Band Width

At the points where the power becomes half, the current becomes $\frac{I}{\sqrt{2}}$, since P is proportional to the square of the current.

Hence at half power points 1 and 2, we have,

$$P = \frac{P_M}{2}$$

$$I = \frac{I_M}{\sqrt{2}} = 0.707 I_M \qquad \text{.......... (4.10.7)}$$

where I_M is the maximum current at the resonance frequency.

The lower half power point, f_1 is also known as *Lower Cut-off Frequency* and may denoted by f_l. The upper half power point, f_2 is also known as *Upper Cutoff Frequency* and may denoted by f_u.

The band of frequencies between the lower and upper half power points or cutoff frequencies is known as the ***Bandwidth***.

At the lower power point f_1, the power factor is leading since X_C will be greater than X_L as $X_C = 1/\omega C$ and at the upper power point f_2, the power factor is lagging since X_L will be greater than X_C as $X_L = \omega L$.

At resonance frequency,

$$X_L = X_C$$

and

$$I_M = \frac{E}{R}$$

At half power points the power factor will be 0.707 since,

$$P = VI \cos\phi = \frac{P_M}{2}$$

$$I = \frac{I_M}{\sqrt{2}}$$

At half power points,

$$|I_1| = |I_2| = \frac{|E|}{|Z|} = \frac{I}{\sqrt{2}} \times \frac{E}{R}$$

Hence, $|Z| = \sqrt{2} \times R$

or $Z^2 = 2R^2$

$$R^2 + (X_L - X_C)^2 = 2R^2$$

At f_1, we have,

$$(X_L - X_C) = -R$$

or $$\omega_1 L - \frac{1}{\omega_1 C} = -R$$

$$\omega_1^2 LC + \omega_1 RC - 1 = 0$$

or $$\omega_1 = \frac{-RC \pm \sqrt{(RC)^2 + 4LC}}{2LC}$$

$$= \frac{-R}{2L} \pm \sqrt{\left(\frac{R}{2L}\right)^2 + \frac{1}{LC}}$$

$$= \frac{-R}{2L} + \sqrt{\left(\frac{R}{2L}\right)^2 + \frac{1}{LC}}$$

Neglecting $\left(\frac{R}{2L}\right)^2$ compared to $\frac{1}{LC}$,

we get,

$$\omega_1 = \frac{-R}{2L} + \sqrt{\frac{1}{LC}}$$

$$= \frac{-R}{2L} + \omega_0$$

where ω_0 is the resonant frequency in radians

Therefore,

$$f_1 = \frac{\omega_1}{2\pi}$$

$$= \frac{-R}{4\pi L} + \frac{1}{2\pi\sqrt{LC}}$$

$$f_1 = \frac{-R}{4\pi L} + f_0 \qquad \text{.......... (4.10.8)}$$

At f_2,

$$(X_L - X_C) = R$$

or
$$\omega_2 L - \frac{1}{\omega_2 C} = R$$

or
$$\omega_2^2 LC - \omega_2 RC - 1 = 0$$

$$\omega_2 = \frac{RC \pm \sqrt{(RC)^2 + 4LC}}{2LC}$$

$$= \frac{R}{2L} + \sqrt{\left(\frac{R}{2L}\right)^2 + \frac{1}{LC}}$$

Again, neglecting $\left(\frac{R}{2L}\right)^2$,

we get,

$$\omega_2 = \frac{R}{2L} + \sqrt{\frac{1}{LC}}$$

$$= \frac{R}{2L} + \omega_0$$

Therefore, $f_2 = \frac{\omega_2}{2\pi}$

$$= \frac{-R}{4\pi L} + \frac{\omega_0}{2\pi}$$

$$f_2 = \frac{R}{4\pi L} + f_0 \qquad \text{.......... (4.10.9)}$$

The bandwidth is defined as the band of frequency between f_2 and f_1.

$$\text{The bandwidth} = f_2 - f_1 = \frac{R}{2\pi L} \qquad \text{.......... (4.10.10)}$$

The cutoff frequencies are also known as 3-dB points because the current at these frequencies falls to $\frac{1}{\sqrt{2}}$ of the current I_M at resonance.

The frequencies that fall within the bandwidth are selected by the series resonance circuit without any distortion. Hence, the bandwidth is a measure of selectivity. *The smaller the bandwidth, the greater the selectivity* because the required frequency signal to which the circuit is tuned will give the maximum current. Or in other words, if the tuning is sharper, i.e., if the bandwidth is less, the circuit will be sensitive to the tuned frequency.

The sharpness of the tuning depends on ratio $\frac{R}{L}$. A small value of $\frac{R}{L}$ indicates a high degree of sensitivity.

Ratio f_0 to $(f_2 - f_1)$ or ratio ω_0 to $(\omega_2 - \omega_1)$ is called the Quality Factor of the coil represented by Q.

$$Q = \frac{f_0}{f_2 - f_1}$$

$$= \frac{f_0 2\pi L}{R}$$

$$Q = \frac{\omega_0 L}{R} \qquad \text{.......... (4.10.11)}$$

$$Q = \frac{1}{\omega_0 CR} \qquad \text{.......... (4.10.12)}$$

It can be proved from the expressions for ω_1 and ω_2 that,

$$\omega_1 \omega_2 = \omega_0^2 \qquad \text{.......... (4.10.13)}$$

4.10.6 Effect of Resistance in Series Resonance

At series resonance the current is limited only by the resistance. As shown in Fig.4.16, if the resistance is large, the maximum value of the current is less and also the curve representing the variation of the current with frequency will not be sharp. As the resistance decreases the curve becomes sharper and sharper and when R = 0, the maximum value of the curve tends to infinity. Hence to have good selectivity or sharp tuning, the value of the resistance should as small as possible, but not zero. Generally the coils used in the electronic circuits will definitely have some resistance. Hence the possibility of current becoming infinity is ruled out.

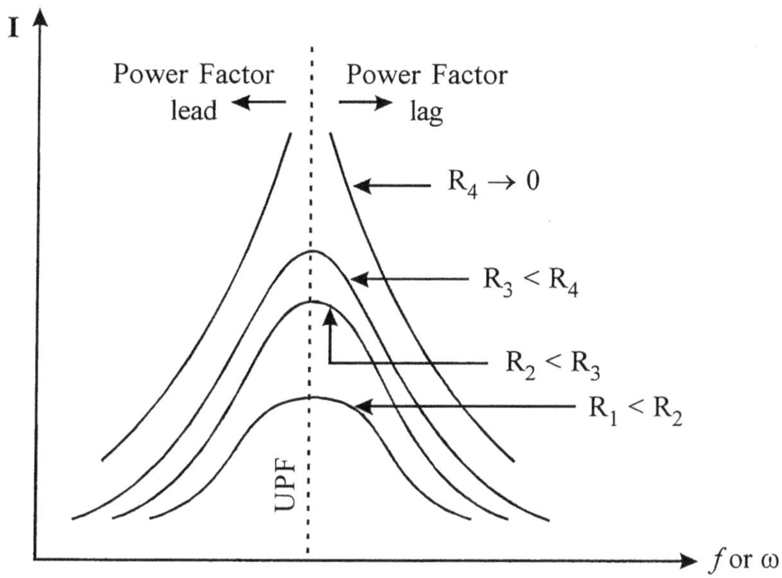

Fig. 4.16 *Resonance Curves for different values of resistances*

Example 4.10:

A series RLC circuit with Q = 250 is resonant at 1.5MHz. Find the frequencies at half power points and also bandwidth.

Solution:

$$Q = \frac{f_0}{BW} = \frac{f_0}{f_2 - f_1} = 250$$

Given, $f_0 = 1.5 \times 10^6$ cps

$$f_2 - f_1 = \frac{f_0}{250} = \frac{1.5 \times 10^6}{250}$$

$$\text{BW} = f_2 - f_1 = 6000\text{cps} \qquad \text{.......... (1)}$$

we know $f_1 \times f_2 = f_0^2 = 2.25 \times 10^{12}$

$$f_1 + f_2 = \sqrt{(f_2 - f_1)^2 + 4 f_1 f_2}$$

$$= \sqrt{6000^2 + 4 \times 2.25 \times 10^{12}}$$

$$= 3000006 \qquad \text{.......... (2)}$$

From Eq. (1) and Eq. (2),

$$2f_2 = 3006006$$

$$f_2 = 1503003\text{cps}$$

$$f_1 = \frac{f_0^2}{f_2} = \frac{2.25 \times 10^{12}}{1503003}$$

$$= 1497003\text{cps} = 1.497\text{MHz.}$$

Example 4.11:

A coil of resistance 40Ω and inductance 0.75H form part of a series circuit for which the resonant frequency is 55Hz. If the supply is 250V, 40Hz. Find the line current, power factor and voltage across the coil.

Solution:

Given, $\text{R} = 40\Omega;\ \text{L} = 0.75\text{H};\ f_0 = 55\text{Hz};$

$\text{E} = 250\text{V};\ f = 50\text{Hz}$

$$f_0 = \frac{1}{2\pi\sqrt{\text{LC}}}$$

$$\text{C} = \frac{1}{(2\pi f_0)^2 \text{L}} = \frac{1}{(2\pi \times 55)^2 \times 0.75}$$

$$= 11.16 \times 10^{-6}\text{F}$$

$$\text{X}_\text{L} = 2\pi f \text{L} = 235.6\Omega.$$

$$X_C = \frac{1}{2\pi f\, C} = 285.2\Omega.$$

$$|Z| = \sqrt{R^2 + (X_L - X_C)^2} = 63.72\Omega \text{ (capacitive)}$$

Therefore, Line current,

$$|I| = \frac{|E|}{|Z|} = \frac{250}{63.72}$$

$$= 3.92\text{A (leading)}$$

$$\text{Power Factor} = \mathbf{Cos}\ \phi = \frac{R}{Z}$$

$$= 0.628 \text{ lead}$$

Impedance of the coil,

$$|Z_{Coil}| = \sqrt{R^2 + X_L^2}$$

$$= 239\Omega \text{ (inductive)}$$

Voltage across the coil,

$$|E_{Coil}| = |I| \times |Z_{Coil}|$$

$$= 937\text{V}$$

4.10.7 Parallel Resonance

An A.C. circuit containing an inductor and capacitor in parallel is said to be in parallel resonance when the circuit current is in phase with the applied voltage.

Consider an inductor of L henry having some resistance R Ohms connected in parallel with a capacitor of capacitance C Farads across a supply voltage of V volts as shown in Fig. 4.17 (a). The phasor diagram of the circuit is shown in Fig. 4.17 (b).

The circuit current I_r will only be in phase with the supply voltage then,

$$I_C = I_L \mathbf{Sin}\ \phi_L \qquad \text{.......... (4.10.14)}$$

Since, at resonance, the reactive component of current is suppressed, the circuit draws ***minimum current*** under this condition.

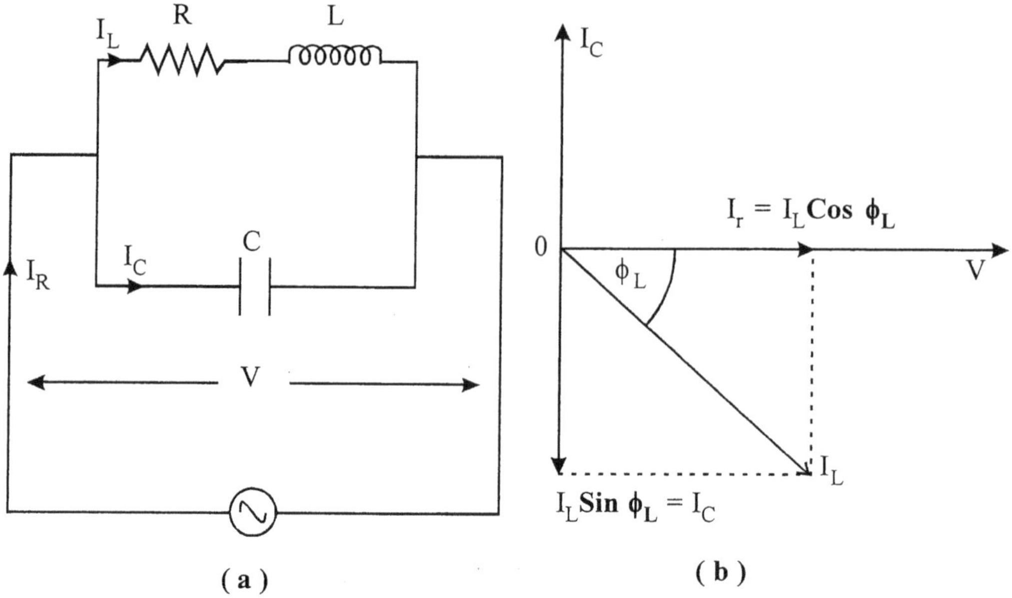

Fig. 4.17 Parallel Resonance and Phasor Diagram for a practical circuit

4.10.8 Parallel Resonant Frequency

Case I: Practical Circuit (Fig. 4.40)

The value of

$$X_L = 2\pi f L$$

$$X_L = \frac{1}{2\pi f C}$$

can be changed by changing the supply frequency.

When frequency increases, the value of X_L and consequently the value of Z_L increases. This decreases the magnitude of current I_2 which also lags behind the voltage V by a progressively greater angle. On the other hand, the value of X_C decreases and consequently the value of I_C increases. At some frequency f_r (called ***resonant frequency***),

$$I_C = I_L \textbf{Sin } \phi_L$$

and resonance occurs.

Therefore, at parallel resonance,

$$I_C = I_L \textbf{Sin } \phi_L$$

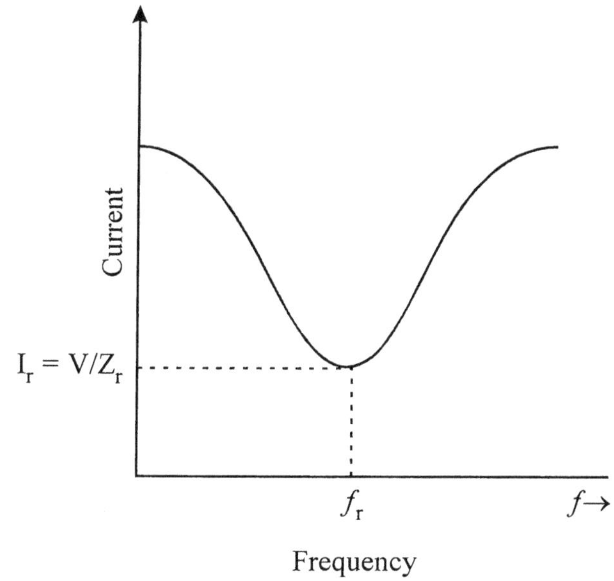

Fig. 4.18 *Resonance Curve*

where $$I_L = \frac{V}{Z_L}$$

$$\textbf{Sin } \phi_L = \frac{X_L}{Z_L}$$

$$I_C = \frac{V}{X_C}$$

$$\frac{V}{X_C} = \frac{V}{Z_L} \times \frac{X_L}{Z_L}$$

or $$X_L X_C = Z_L^2$$

Alternatively, inductive susceptance is equal to capacitive susceptance

$$\frac{X_L}{R^2 + X_L^2} = \frac{X_C}{0 + X_C^2}$$

or $$\frac{\omega L}{\omega C} = Z_L^2 = (R^2 + X_L^2)$$

or $$\frac{L}{C} = R^2 + (2\pi f_r L)^2$$

or $$2\pi f_r L = \sqrt{\frac{L}{C} - R^2}$$

or $$f_r = \frac{1}{2\pi L}\sqrt{\frac{L}{C} - R^2}$$

$$f_r = \frac{1}{2\pi}\sqrt{\frac{1}{LC} - \frac{R^2}{L^2}} \qquad \text{.......... (4.10.15)}$$

If R is very small as compared to L, then, Resonant Frequency,

$$f_r = \frac{1}{2\pi}\sqrt{\frac{1}{LC} - \frac{R^2}{L^2}}$$

$$f_r = \frac{1}{2\pi\sqrt{LC}} \qquad \text{.......... (4.10.15a)}$$

A current frequency curve for a typical parallel resonant circuit is shown in Fig. 4.18. The value of line current, I_r is minimum at resonance.

4.10.9 Effects of Parallel Resonance

At parallel resonance, Line current,

$$I_r = I_L \text{Cos } \phi$$

or $$\frac{V}{Z_r} = \frac{V}{Z_L} \times \frac{R}{Z_L}$$

or $$\frac{1}{Z_r} = \frac{R}{Z_L^2}$$

or $$\frac{1}{Z_r} = \frac{R}{\frac{L}{C}}$$

$$= \frac{CR}{L}$$

Therefore, Circuit Impedance,

$$Z_r = \frac{L}{CR} \qquad \text{.......... (4.10.16)}$$

It shows that,

1. *Circuit impedance, Z_r is purely resistive because there is no frequency term present. If the value of L, R and C is in Henry, Ohm and Farads then the value of Z_r is in Ohms.*
2. *The value of Z_r is very high because the ratio of L and C is very large at parallel resonance.*
3. *The value of circuit current I_r is very small because the value of Z_r is very high.*
4. *The current flowing through the capacitor and coil is much greater than the line current because the impedance of each branch is quite low than circuit impedance Z_r.*

The series resonance circuit is also known as ***Acceptor Circuit*** because it causes maximum at the resonance frequency and the signals at the resonant frequency are accepted. It is also known as ***Selector Circuit*** because it selects the signals at the resonance frequency.

Since a parallel resonant circuit can draw a very small current and power from the mains, it is often regarded as ***Rejector Circuit****.*

4.10.10 Quality Factor of a Parallel Resonant Circuit

We have seen that at parallel resonance, the current circulating between the two branches is many times greater than the line current drawn from the mains. This current amplification produced by the resonance is called ***Q Factor*** of the parallel resonant circuit.

$$\text{Q Factor} = \frac{\text{Current Circulating between L and C}}{\text{Line Current}}$$

$$\text{Q Factor} = \frac{I_C}{I_r}$$

Now, $$I_C = \frac{V}{X_C} = 2\pi f_r CV$$

$$I_r = \frac{V}{\frac{L}{CR}}$$

$$\text{Q Factor} = \frac{2\pi f_r CV}{V} \times \frac{L}{CR}$$

$$= \frac{2\pi f_r L}{R}$$

or
$$\text{Q Factor} = \frac{2\pi L}{R} \times \frac{1}{2\pi\sqrt{LC}}$$

or
$$\text{Q Factor} = \frac{1}{R}\sqrt{\frac{L}{C}} \qquad \text{.......... (4.10.17)}$$

Case II: RLC in Parallel

At parallel resonance,

inductive susceptance = capacitive susceptance

or total susceptance = 0

$$\frac{1}{X_C} - \frac{1}{X_L} = 0$$

or
$$\omega_0 C - \frac{1}{\omega_0 L} = 0$$

or
$$\omega_0^2 = \frac{1}{LC}$$

or
$$\omega_0 = \frac{1}{\sqrt{LC}}$$

or
$$f_0 = \frac{1}{2\pi\sqrt{LC}}$$

Case III: R_1C in Parallel with R_2L

At resonance,

total susceptance = 0

$$\frac{X_C}{R_2^2+X_C^2}-\frac{X_L}{R_1^2+X_L^2}=0$$

$$X_C(R_1^2+X_L^2)=X_L(R_2^2+X_C^2)$$

$$\frac{1}{\omega_0 C}\left(R_1^2+\omega_0^2L^2\right)=\omega_0 L\left(R_2^2+\frac{1}{\omega_0^2C^2}\right)$$

Simplifying above equation, we get,

$$\omega_0^2=\frac{1}{LC}\left(\frac{\left(R_1^2-\frac{L}{C}\right)}{\left(R_2^2-\frac{L}{C}\right)}\right)$$

$$\omega_0^2=\frac{1}{\sqrt{LC}}\sqrt{\left(\frac{\left(R_1^2-\frac{L}{C}\right)}{\left(R_2^2-\frac{L}{C}\right)}\right)} \qquad \text{.......... (4.10.18)}$$

Case IV: L and C in Parallel

At parallel resonance,

inductive susceptance = capacitive susceptance

$$\frac{1}{X_L}=\frac{1}{X_C}$$

i.e.,

$$\frac{1}{\omega_0 L}=\omega_0 C$$

$$\omega_0^2=\frac{1}{LC}$$

$$\omega_0=\frac{1}{\sqrt{LC}}$$

or

$$f_0=\frac{1}{2\pi\sqrt{LC}}$$

4.10.11 Comparison of Series And Parallel Resonant Circuits

The comparison of series and parallel resonant circuits is given in Table 4.1

Table 4.1

SNo.	Particulars	Series Circuit	Parallel Circuit
1.	Impedance	Minimum ($Z_r = R$)	Maximum ($Z_r = \frac{L}{CR}$)
2.	Current	Maximum ($I_r = \frac{V}{R}$)	Minimum ($I_r = \frac{V}{Z_R}$)
3.	Resonant Frequency	$f_r = \frac{1}{2\pi\sqrt{LC}}$	$f_r = \frac{1}{2\pi}\sqrt{\frac{1}{LC} - \frac{R^2}{L^2}}$
4.	Power Factor	Unity	Unity
5.	Q Factor	$\frac{X_L}{R}$	$\frac{X_L}{R}$
6.	Amplification	Amplifies Voltage	Amplifies Current

It can be observed that in a parallel resonance circuit, the current through the inductive coil or current through capacitive coil can even be more than the supply current because the two currents get cancelled as they will be in phase opposition i.e., the phasor current through the capacitive coil is exactly 180° out of phase with reference to the phasor current through the inductive coil.

The parallel resonance circuit is also known as ***Current Magnification Circuit*** because current through the inductive coil or capacitive coil will be maximum. The ***Current Magnification of the circuit*** is equal to the ratio of the current through the inductive coil or capacitive coil to the current drawn from the supply.

$$\textit{Current Magnification} = \frac{\text{Current through the Inductive coil}}{\text{Supply Current}}$$

$$= \frac{I\omega L}{IR} = \frac{\omega L}{R}$$

Example 4.12:

A coil of 200Ω resistance has an inductance of 0.2H and is connected in parallel with a condenser of 100μF capacitance. Calculate the frequency at which the circuit will act as a noninductive resistance of RΩ. Find also the value of R.

Solution:

This is a practical case of parallel resonance. The circuit will be acting as a noninductive resistance at a resonant frequency,

$$f_0 = \frac{1}{2\pi}\sqrt{\frac{1}{LC} - \frac{R_1^2}{L^2}} \qquad \text{.......... (1)}$$

At resonance,

$$Y = \frac{R_1}{R_1^2 + X_L^2} \qquad \text{.......... (2)}$$

Substituting the values in Eq. (1), we get,

$$f_0 = \frac{1}{2\pi}\sqrt{\frac{1}{0.2 \times 100 \times 10^{-6}} - \frac{20^2}{0.2^2}}$$

$$= 31.83 \text{Hz}$$

From Eq. (2),

$$Z = \frac{1}{Y} = \frac{R_1^2 + X_L^2}{R_1}$$

$$= \frac{20^2 \times (2\pi \times 31.8 \times 0.2)^2}{20} = 100\Omega.$$

Example 4.13:

An impedance of (100 + j10)Ω is connected in parallel with an impedance consisting of a variable resistance R and a capacitive reactance of -j2Ω. Find the value of R for resonance. Draw the circuit diagram.

Solution:

The total admittance,

$$Y = Y_1 + Y_2$$

$$= \frac{1}{10 + j10} + \frac{1}{R - j2}$$

$$= \frac{10 - j10}{200} + \frac{R + j2}{R^2 + 4}$$

$$= \left(0.05 + \frac{R}{R^2 + 4}\right) + j\left(\frac{2}{R^2 + 4} - \frac{1}{20}\right)$$

At parallel resonance,

total susceptance = 0

$$\frac{2}{R^2 + 4} - \frac{1}{20} = 0$$

$$\frac{2}{R^2 + 4} = \frac{1}{20}$$

$$R^2 + 4 = 40$$

$$R^2 = 36$$

$$R = 6\Omega.$$

Example 4.14:

A current source is applied to a parallel combination of R, L and C, where $R = 10\Omega$, $L = 1H$ and $C = 1\mu F$.

(1) Compute the resonant frequency

(2) Find the Quality Factor.

(3) Calculate the value of the bandwidth.

(4) Compute the lower and upper frequency points of the bandwidth.

Solution:

(1) Resonant Frequency,

$$\omega_0 = \frac{1}{\sqrt{LC}} = 1000 \text{ rad/sec}$$

(2) Quality Factor,

$$Q = \omega_0 RC$$

$$= 0.01$$

(3) Bandwidth,

$$\Delta\omega = \frac{\omega_0}{Q}$$

$$= 10^5 \text{ rad/sec.}$$

(4) Lower Frequency Point,

$$\omega_1 = \omega_0 \left(\sqrt{1 + \left(\frac{1}{2Q}\right)^2} - \frac{1}{2Q} \right)$$

$$= 9.999 \text{ rad/sec}$$

Upper Frequency Point,

$$\omega_2 = \omega_0 \left(\sqrt{1 + \left(\frac{1}{2Q}\right)^2} - \frac{1}{2Q} \right)$$

$$= 100010 \text{ rad/sec}$$

4.11 SINGLE PHASE CIRCUITS WITH VARIABLE ELEMENTS - CURRENT LOCUS DIAGRAMS

The path traced by the tip of the current phasor of an A.C. circuit for variation in R, L or C with constant frequency of supply is known as the ***Locus Diagram.***

It is known that as the frequency changes the reactance of an inductance,

$$X_L = \omega L = 2\pi f L$$

will change in direct proportion causing the change in impedance of the circuit So the current drawn by the circuit also changes since

$$\mathbf{I} = \frac{\mathbf{V}}{\mathbf{Z}}$$

where $\mathbf{Z} = R + jX_L$

Similarly, as the frequency changes the reactance of a capacitance,

$$X_C = \frac{1}{\omega C}$$

$$X_C = \frac{1}{2\pi f C}$$

will change in inverse proportion causing the change in impedance of the circuit.

So the current drawn by the circuit also changes since

$$\mathbf{I} = \frac{\mathbf{V}}{\mathbf{Z}}$$

where $\mathbf{Z} = R - jX_C$

Also, if the resistance of an A.C. circuit is varied, the current also will vary in inverse proportion since,

$$\mathbf{I} = \frac{\mathbf{V}}{\mathbf{Z}}$$

where $\mathbf{Z} = R \pm jX$

Hence, we can draw locus diagrams

1. *Variable R and constant L in R-L series circuit*
2. *Variable R and constant C in R-C series circuit*
3. *Variable L and constant R in R-L series circuit*
4. *Variable C and constant R in R-C series circuit*

4.11.1 Series RL And RC Circuits with Variable R & Constant Reactance

Consider a series circuit having a constant reactance X but variable resistance R which may vary from zero to infinity. Let such a circuit be driven by a voltage source of constant voltage V volts and constant frequency *f* Hz. Fig. 4.19 shows two such circuits, an R-L circuit and an R-C circuit.

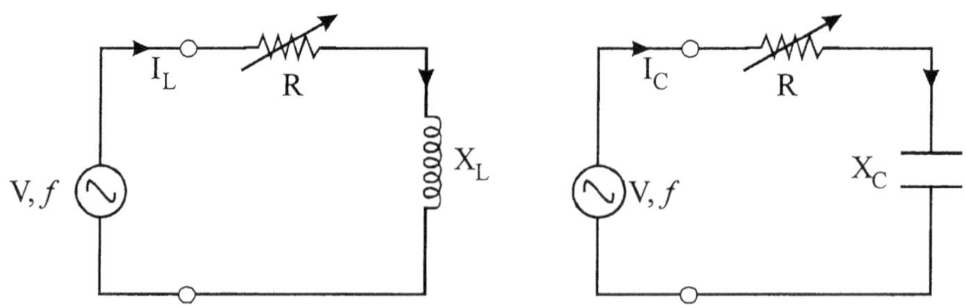

(a) Series RL Circuit (b) Series RC Circuit

***Fig. 4.19** Series RL and RC circuits driven by Constant X, Variable R*

4.11.1.1 RL Circuit with variable resistance

From Fig. 4.19 (a), current I_L is given by,

$$I_L = \frac{V}{R + jX_L}$$

$$= \frac{V}{\sqrt{R^2 + X_L^2}} \angle -\theta_L$$

$$= \frac{V}{Z_L} \angle -\theta_L$$

$$= I_L \angle -\theta_L \quad \text{.......... (4.11.1)}$$

where $$Z_L = \sqrt{R^2 + X_L^2} \quad \text{.......... (4.11.2)}$$

$$I_L = \frac{V}{Z_L} = \frac{V}{\sqrt{R^2 + X_L^2}} \quad \text{.......... (4.11.3)}$$

and $$\theta_L = \mathbf{Tan}^{-1}\left(\frac{\mathbf{X_L}}{\mathbf{R}}\right) \text{ lagging} \quad \text{.......... (4.11.4)}$$

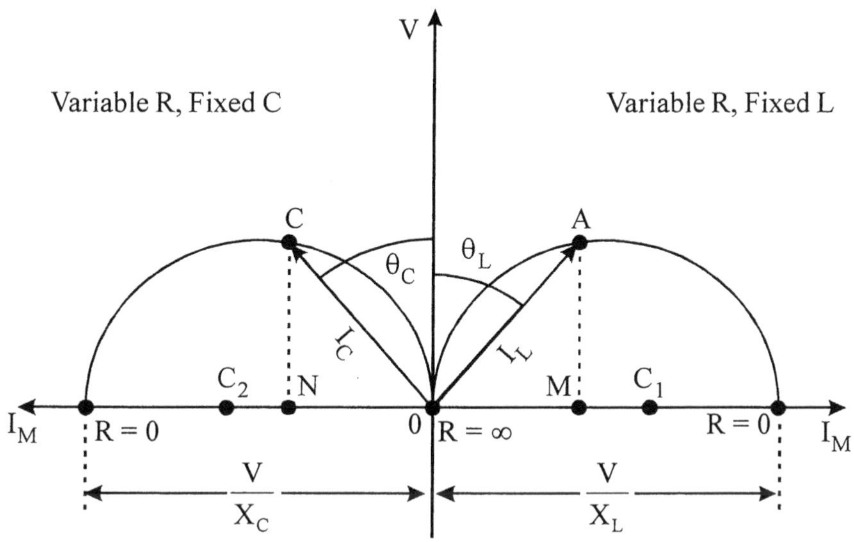

Fig. 4.20 Locus of Current Vector in Series RL and RC circuits with X fixed and R Variable

In the Fig. 4.20, the reference phasor V is taken at an angle 90° with the X-axis. Some authors take reference phasor V along X-axis, in which case, the entire figure will be 90° clockwise.

It is evident that as R increases, current I_L and phase angle θ_L decrease. It may be readily established that the locus of the end point A of current vector I_L represented by OA, is a semicircle. We chose the coordinate of point A with respect to the origin 0, Y-axis along voltage vector V and X-axis at right angle as shown in Fig. 4.20.

Then, $$y = I_L \textbf{Cos } \theta_L$$

$$= \frac{V}{Z_L} \times \frac{R}{Z_L}$$

$$= \frac{VR}{Z_L^2} = V \times \frac{V}{R^2 + X_L^2} \qquad \text{.......... (4.11.5)}$$

Similarly $$x = I_L \textbf{Sin } \theta_L$$

$$= \frac{V}{Z_L} \times \frac{X_L}{Z_L}$$

$$= V \times \frac{X_L}{Z_L^2} = V \times \frac{X_L}{R^2 + X_L^2} \qquad \text{.......... (4.11.6)}$$

Squaring and adding Eq. (4.36.5) and Eq. (4.36.6) we get,

$$x^2 + y^2 = \frac{V^2}{\left(R^2 + X_L^2\right)^2} \times \left(R^2 + X_L^2\right)$$

$$= \frac{V^2}{\left(R^2 + X_L^2\right)} \qquad \text{.......... (4.11.7)}$$

X_L is constant while R is variable. We eliminate R from Eq. (4.11.7). Thus Eq. (4.11.6) becomes,

$$\frac{VX_L}{x} = (R^2 + X_L^2) \qquad \text{.......... (4.11.8)}$$

Substituting this value of ($R^2 + X_L^2$) in Eq. (4.11.7) we get,

$$x^2 + y^2 = \frac{V^2}{\frac{VX_L}{x}} \qquad \text{.......... (4.11.9)}$$

Eq. (4.11.9) may be put as,

$$y^2 + \left(x - \frac{V}{2X_L}\right)^2 = \frac{V^2}{4X_L^2} \qquad \text{.......... (4.11.10)}$$

Eq. (4.11.10) is the equation of a circle of radius $\frac{V}{2X_L}$ with centre C_1 at,

$$y = 0,$$

and $$x = \frac{V}{2X_L}$$

Fig. 4.20 shows the locus of the end point of current vector I_L represented by a semicircle OAB. For series RL Circuit, from Eq. (4.11.4), we find that angle θ_L is lagging and current I_L lags behind the voltage **V** by angle θ_L. The locus of the apex of the current vector I_L is a semicircle on the RHS of vector OV.

Consider the extreme case when R = 0. Then from Eq. (4.11.1), current is at maximum value,

$$I_m = \frac{V}{X_L}$$

at point B and $\theta_L = 90°$, i.e., current I_L lags behind voltage **V** by 90°. Consider the other extreme case when $R = \infty$. Then from Eq. (4.11.1), current $I_L = 0$ at origin 0 and $\theta_L = 0$.

For a given angle θ_L,

$$y = AM = I_L \mathbf{Cos}\ \boldsymbol{\theta_L}.$$

But power,

$$P = VI\mathbf{Cos}\ \boldsymbol{\theta_L}.$$

Hence for constant V, length AM or ordinate y, on a suitable scale, represents the power consumed by the RL circuit.

4.11.1.2 RC Circuit with Variable Resistance

Fig. 4.19 (b), shows a series RC circuit driven by a constant voltage source V. Reactance X_C is constant while R is variable.

From Fig. 4.19 (b), current I_C is given by,

$$I_C = \frac{V}{R - jX_C}$$

$$= \frac{V}{\sqrt{R^2 + X_C^2}} \angle\theta_C$$

$$= \frac{V}{Z_C} \angle\theta_C$$

$$= I_C \angle\theta_C \qquad \text{.......... (4.11.11)}$$

where $$Z_C = \sqrt{R^2 + X_C^2} \qquad \text{.......... (4.11.12)}$$

$$I_C = \frac{V}{Z_C} = \frac{V}{\sqrt{R^2 + X_C^2}} \qquad \text{.......... (4.11.13)}$$

and $$\theta_C = \mathbf{Tan}^{-1}\left(\frac{\mathbf{X_C}}{\mathbf{R}}\right) \text{ leading} \qquad \text{.......... (4.11.14)}$$

As R increases, current I_C and phase angle θ_C decreases. Proceeding as in the case of RL circuit, it may be established that locus of the end point C of current vector I_C represented by OC is a semicircle. We chose coordinates of point C with respect to the origin O, Y-axis along the voltage vector V and X-axis at right angles as shown in Fig. 4.20

Then, $$y = I_C \mathbf{Cos}\ \theta_C$$

$$= \frac{V}{Z_C} \times \frac{R}{Z_C}$$

$$= \frac{VR}{Z_C^2} = V \times \frac{V}{R^2 + X_C^2} \qquad \text{.......... (4.11.15)}$$

Similarly $$x = -I_C \mathbf{Sin}\ \theta_C$$

$$= \frac{V}{Z_C} \times \frac{-X_C}{Z_C}$$

$$x = V \times \frac{X_C}{Z_C^2} = V \times \frac{-X_L}{R^2 + X_C^2} \qquad \text{.......... (4.11.16)}$$

Squaring and adding Eq. (4.11.15) and Eq. (4.11.16) we get,

$$x^2 + y^2 = \frac{V^2}{\left(R^2 + X_C^2\right)^2} \times \left(R^2 + X_C^2\right)$$

$$= \frac{V^2}{\left(R^2 + X_C^2\right)} \qquad \text{.......... (4.11.17)}$$

X_C is constant while R is variable. We eliminate R from Eq. (4.11.17). Thus Eq. (4.11.16) becomes,

$$\frac{-VX_C}{x} = (R^2 + X_C^2) \qquad \text{.......... (4.11.18)}$$

Substituting this value of ($R^2 + X_C^2$) in Eq. (4.11.17) we get,

$$x^2 + y^2 = \frac{V^2}{\frac{-VX_C}{x}} = \frac{-xV}{X_C} \qquad \text{.......... (4.11.19)}$$

Eq. (4.11.19) may be put as,

$$y^2 + \left(x + \frac{V}{2X_C}\right)^2 = \frac{V^2}{4X_C^2} \qquad \text{.......... (4.11.20)}$$

Eq. (4.11.20) is the equation of a circle of radius $\frac{V}{2X_C}$ with centre C_2 at,

$$y = 0,$$

and $$x = \frac{-V}{2X_C}$$

Fig. 4.21 shows the locus of the end point of current vector I_C represented by a semicircle OCD.

For series RC Circuit, from Eq. (4.11.14), we find that angle θ_C is leading and current I_C leads the voltage **V** by angle θ_C. The locus of the apex of the current vector I_C is a semicircle on the LHS of vector OV.

Consider the extreme case when R = 0. Then from Eq. (4.11.11), current is at maximum value,

$$I_m = \frac{V}{X_C}$$

at point D and $\theta_C = 90°$, i.e., current I_C leads voltage **V** by 90°.

Consider the other extreme case when $R = \infty$. Then from Eq. (4.11.13), current $I_C = 0$ at origin 0 and $\theta_C = 0$.

For a given angle θ_C,

$$y = AM = I_C \mathbf{Cos}\ \theta_C.$$

But power,

$$P = VI\mathbf{Cos}\ \theta_C.$$

Hence for constant V, length CN or ordinate y, on a suitable scale, represents the power consumed by the RC circuit.

4.11.2 Locus of Voltage

4.11.2.1 RL Circuit with variable resistance

In an series RL Circuit, voltage drop $I.jX_L$ across X_L leads the voltage drop I. R across the resistor by 90°. Hence for a given value of resistance R, the phasor diagram for voltage IR and j. IX_L is a right angled triangle with applied voltage V as the hypotenuse as shown in Fig. 4.21.

For a constant source voltage V and constant reactance X_L, as the resistance R varies, the other move along a semicircle OEV with voltage V as the hypotenuse as shown in Fig. 4.21. Hence the locus of the apex of the right angled triangle circuit, this semicircular locus lies on the right hand side of the hypotenuse OV.

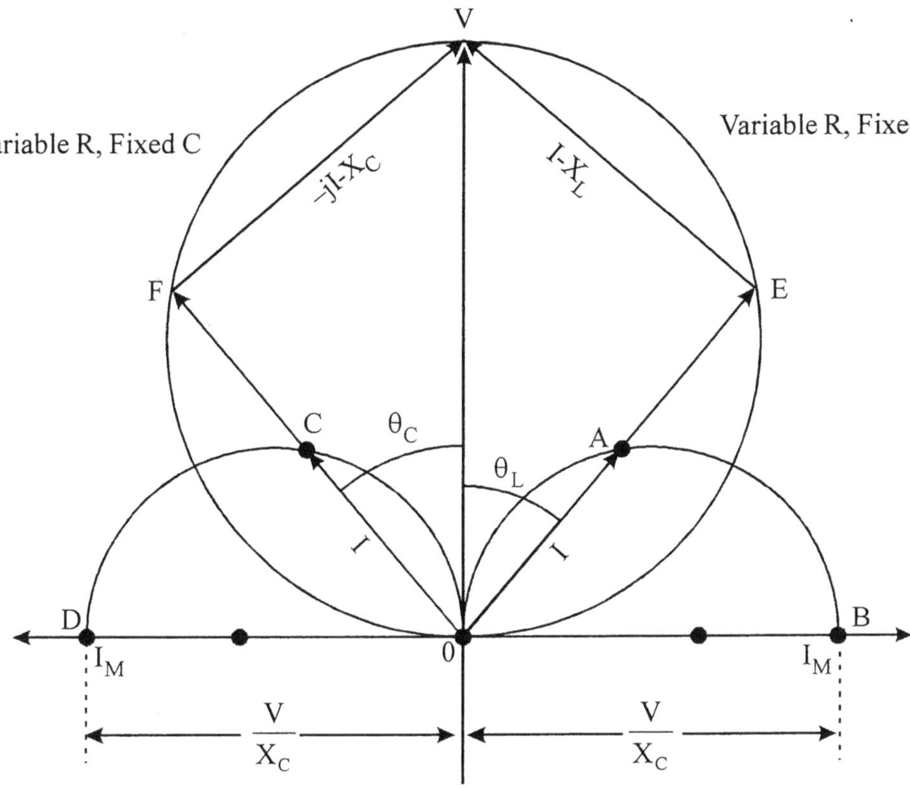

Fig. 4.21 Locus of Voltage Vectors IR and IX for constant V and Reactance

4.11.2.2 RC Circuit with variable resistance

Similarly for a series RC Circuit, voltage drop I. ($-jX_C$) across X_C lags the voltage drop IR across the resistor R by 90°. Hence for a given value of R, the phasor diagram for voltages IR and $-jIX_C$ is a right angled triangle with applied voltage V as the hypotenuse as shown in Fig. 4.21.

For a constant source voltage V and reactance X_C, as the resistance R varies, the V as the hypotenuse as shown in Fig. 4.21. Hence the locus of the apex of the right angled triangle OFV is a semicircle on the hypotenuse. For the RC circuit, this semicircular locus lies to the left of the hypotenuse OV.

4.11.2.3 Series RL and RC Circuits with Variable Reactance & Constant R

Fig. 4.22 (a) shows a series RL circuit with constant R and X_L variable. Fig. 4.22 (b) shows a series RC circuit with R fixed and X_C variable. The reactance X (X_L or X_C) may vary from zero to infinity. Each of these circuits is driven by a constant voltage constant frequency source.

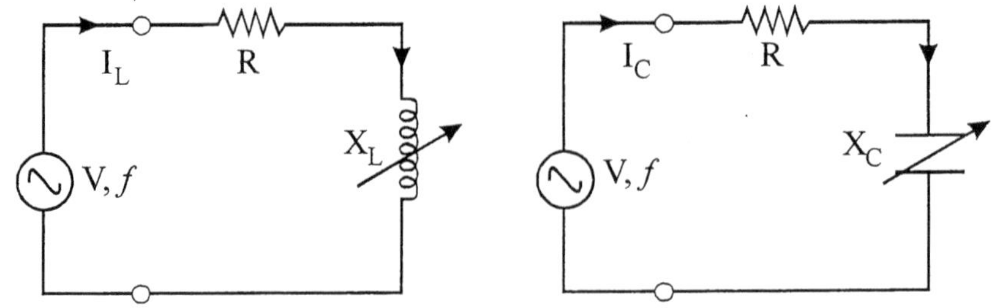

(a) Series RL Circuit (b) Series RC Circuit

Fig. 4.22 Series RL and RC circuits driven by Constant R, Variable X

4.11.2.4 RL Circuit with variable reactance

As shown in Section 6.3.1

$$I_L = \frac{V}{R + jX_L}$$

$$= \frac{V}{\sqrt{R^2 + X_L^2}} \angle -\theta_L$$

$$= \frac{V}{Z_L} \angle -\theta_L$$

$$= I_L \angle -\theta_L \qquad \text{.......... (4.11.21)}$$

where

$$Z_L = \sqrt{R^2 + X_L^2} \qquad \text{.......... (4.11.22)}$$

$$I_L = \frac{V}{Z_L} = \frac{V}{\sqrt{R^2 + X_L^2}} \qquad \text{.......... (4.11.23)}$$

and

$$\theta_L = \mathbf{Tan}^{-1}\left(\frac{\mathbf{X_L}}{\mathbf{R}}\right) \text{ lagging} \qquad \text{.......... (4.11.24)}$$

It is evident that as X_L increases, current I_L and phase angle θ_L decrease. It may be readily established that the locus of the end point A of current vector I_L represented by OA in Fig. 4.21, is a semicircle. We chose the coordinate of point A with respect to the origin 0, X-axis along voltage vector V and X-axis at right angle as shown in Fig. 4.21.

Then, $y = I_L \mathbf{Cos}\ \theta_L$

$$= \frac{V}{Z_L} \times \frac{R}{Z_L}$$

$$= \frac{VR}{Z_L^2} = V \times \frac{V}{R^2 + X_L^2} \qquad \text{......... (4.11.25)}$$

Similarly $x = I_L \mathbf{Sin}\ \theta_L$

$$= \frac{V}{Z_L} \times \frac{X_L}{Z_L}$$

$$= V \times \frac{X_L}{Z_L^2} = V \times \frac{X_L}{R^2 + X_L^2} \qquad \text{....... (4.11.26)}$$

Squaring and adding Eq. (4.11.25) and Eq. (4.11.26) we get,

$$x^2 + y^2 = \frac{V^2}{\left(R^2 + X_L^2\right)^2} \times \left(R^2 + X_L^2\right)$$

$$= \frac{V^2}{\left(R^2 + X_L^2\right)} \qquad \text{.......... (4.11.27)}$$

R is constant while X_L is variable. We eliminate X_L from Eq. (4.11.27). Thus Eq. (4.11.26) becomes,

$$\frac{VR}{y} = (R^2 + X_L^2) \qquad \text{.......... (4.11.28)}$$

Substituting this value of ($R^2 + X_L^2$) in Eq. (4.11.27) we get,

$$x^2 + y^2 = \frac{V^2}{\frac{VR}{x}} \qquad \text{.......... (4.11.29)}$$

Eq. (4.11.29) may be put as,

$$x^2 + \left(y - \frac{V}{2X_L}\right)^2 = \frac{V^2}{4R^2} \qquad \text{.......... (4.11.30)}$$

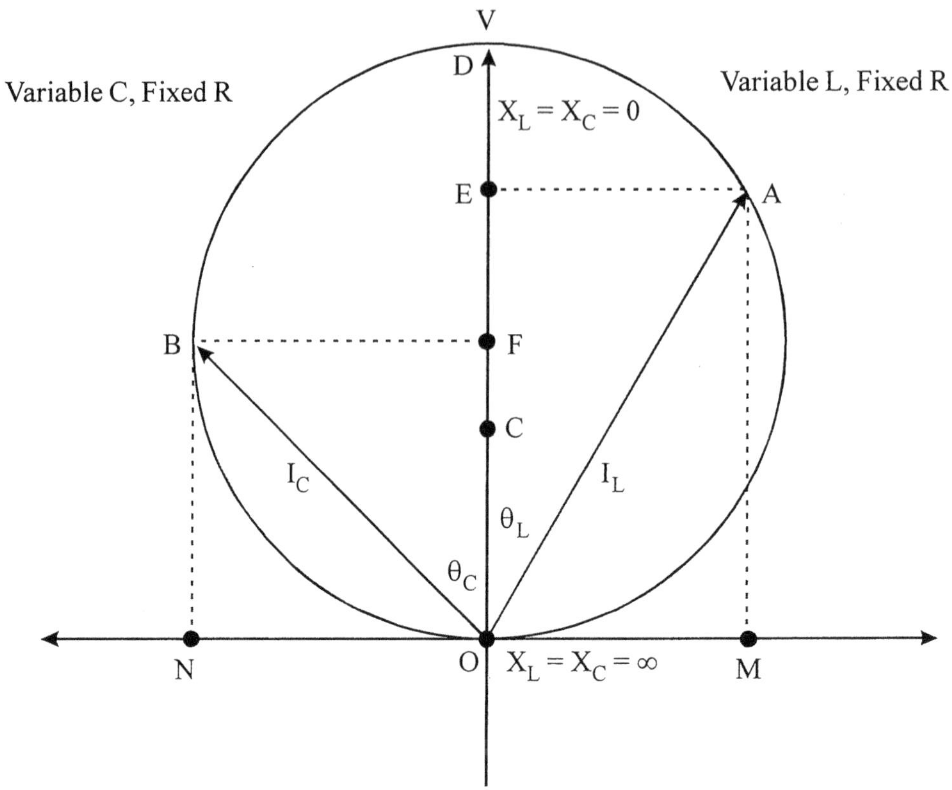

Fig. 4.23 Locus of Current Vectors I_L and I_C for constant V and R

Eq. (4.11.30) is the equation of a circle of radius $\frac{V}{2R}$ with centre C at,

$$x = 0,$$

and
$$x = \frac{V}{2R}$$

For series RL Circuit, from Eq. (4.11.24), we find that angle θ_L is lagging and current I_L lags behind the voltage **V** by angle θ_L. The locus of the apex of the current vector I_L is a semicircle on the RHS of vector OD in Fig. 4.23.

Consider the extreme case when $X_L = 0$. Then from Eq. (4.11.21), current is at maximum value,

$$I_m = \frac{V}{R}$$

at point D and $\theta_L = 0°$, i.e., current I_L is in phase with Voltage V.

Consider the other extreme case when $X_L = \infty$. Then from Eq. (4.11.21), current $I_L = 0$ at origin 0 and $\theta_L = 90^o$.

For a given angle θ_L,

$$y = AM = I_L \mathbf{Cos}\ \boldsymbol{\theta}_L.$$

But power,

$$P = V I \mathbf{Cos}\ \boldsymbol{\theta}_L.$$

Hence for constant V, length AM or ordinate y, on a suitable scale, represents the power consumed by the RL circuit.

4.11.2.5 RC Circuit with variable reactance :

Fig. 4.22 (b), shows a series RC circuit driven by a constant voltage source V. Here, resistance R is constant while reactance X_C is variable.

From Fig. 4.22 (b), current I_C is given by,

$$I_C = \frac{V}{R - jX_C}$$

$$= \frac{V}{\sqrt{R^2 + X_C^2}} \angle\theta_C$$

$$= \frac{V}{Z_C} \angle\theta_C$$

$$= I_C \angle\theta_C \qquad \text{.......... (4.11.31)}$$

where $$Z_C = \sqrt{R^2 + X_C^2} \qquad \text{.......... (4.11.32)}$$

$$I_C = \frac{V}{Z_C} = \frac{V}{\sqrt{R^2 + X_C^2}} \qquad \text{.......... (4.11.33)}$$

and $$\theta_C = \mathbf{Tan}^{-1}\left(\frac{\mathbf{X_C}}{\mathbf{R}}\right) \text{ leading} \qquad \text{.......... (4.11.34)}$$

As X_C increases, current I_C and phase angle θ_C decreases.

Proceeding as in the case of RL circuit, it may be established that locus of the apex B of current vector I_C represented by OC is a semicircle. We chose coordinates of point C with respect to the origin O, Y-axis along the voltage vector y and X-axis at right angles as shown in Fig. 4.23.

Then, $y = I_C \mathbf{Cos}\ \theta_C$

$$= \frac{V}{Z_C} \times \frac{R}{Z_C}$$

$$= \frac{VR}{Z_C^2} = V \times \frac{V}{R^2 + X_C^2} \qquad \text{.......... (4.11.35)}$$

Similarly $x = -I_C \mathbf{Sin}\ \theta_C$

$$= \frac{V}{Z_C} \times \frac{-X_C}{Z_C}$$

$$x = V \times \frac{X_C}{Z_C^2} = V \times \frac{-X_L}{R^2 + X_C^2} \qquad \text{.......... (4.11.36)}$$

Squaring and adding Eq. (4.11.35) and Eq. (4.11.36) we get,

$$x^2 + y^2 = \frac{V^2}{\left(R^2 + X_C^2\right)^2} \times \left(R^2 + X_C^2\right)$$

$$= \frac{V^2}{\left(R^2 + X_C^2\right)} \qquad \text{.......... (4.11.37)}$$

R is constant while X_C is variable. We eliminate X_C from Eq. (4.36.37). Thus Eq. (4.11.36) becomes,

$$\frac{VR}{y} = (R^2 + X_C^2) \qquad \text{......... (4.11.38)}$$

Substituting this value of ($R^2 + X_C^2$) in Eq. (4.11.37) we get,

$$x^2 + y^2 = \frac{V^2}{\frac{VR}{y}} = \frac{yV}{R} \qquad \text{.......... (4.11.39)}$$

Eq. (4.11.39) may be put as,

$$x^2 + \left(y - \frac{V}{2R}\right)^2 = \frac{V^2}{4R^2} \qquad \text{.......... (4.11.40)}$$

Eq. (4.11.40) is the equation of a circle of radius $\frac{V}{2R}$ with centre C_2 at,

$$x = 0,$$

and $$y = \frac{V}{2R}$$

This centre is the same as for the semicircle of RL circuit.

For series RC Circuit, from Eq. (4.11.34), we find that angle θ_C is leading and current I_C leads the voltage **V** by angle θ_C. The locus of the apex of the current vector I_C is a semicircle on the LHS of vector OD in Fig. 4.23

Consider the extreme case when $X_C = 0$. Then from Eq. (4.11.31), current is at maximum value,

$$I_m = \frac{V}{R}$$

at point D and $\theta_C = 0°$, i.e., current I_C is in phase with applied voltage V.

Consider the other extreme case when $X_C = \infty$. Then from Eq. (4.11.31), current $I_C = 0$ at origin 0 and $\theta_C = 90°$.

For a given angle θ_C,

$$y = BN = I_C \mathbf{Cos}\ \boldsymbol{\theta}_C.$$

But power,

$$P = V I \mathbf{Cos}\ \boldsymbol{\theta}_C.$$

Hence for constant V, length CN or ordinate y, on a suitable scale, represents the power consumed by the RC circuit.

4.11.2.6 Properties of Constant Reactance, Variable Resistance

From Fig. 4.23, the following properties of series RL or RC circuits with R variable become obvious.

4.11.2.7 Maximum Current

Current is maximum when R = 0 and this maximum value is given by,

$$I_m = \frac{V}{X_L} = \frac{V}{\omega L}$$

for series RL Circuit.

$$I_m = \frac{V}{-X_C} = -\omega CV$$

for series RC Circuit.

4.11.2.8 Maximum Power Supplied to the Circuit

Power, $P = VI\cos\theta$

If V is constant, then $P \propto I\cos\theta$, P is proportional to the ordinate of the current semicircle. Hence the maximum power is V times the maximum ordinate. Evidently the maximum ordinate equals the radius of the current semicircle, i.e., equal to $V/(2X_L)$ or $V/(2X_C)$ or $\frac{1}{2}I_m$.

Thus Maximum Power,

$$P_m = V \times \frac{V}{2X} = \tfrac{1}{2}I_m \qquad \text{.......... (4.11.41)}$$

For RL Circuit,

$$P_m = \frac{V}{2} \times \frac{V}{X_L}$$

$$= \frac{V^2}{2\omega L} \qquad \text{.......... (4.11.42)}$$

For RC Circuit,

$$P_m = \frac{V}{2} \times \frac{V}{X_C}$$

$$= \frac{V^2\omega C}{2} \qquad \text{.......... (4.11.43)}$$

4.11.2.9 Power Factor at Maximum Power Condition

Thus refer to a current semicircle shown in Fig. 4.24. It is evident that the ordinate passes through the centre of the current semicircle. In this condition, phase angle $\theta = 45°$. Hence angle ϕ in triangle OA C_1 is also 45°.

Hence under maximum power condition, power factor being equal to **Cos θ** is

Cos 45° = 0.707.

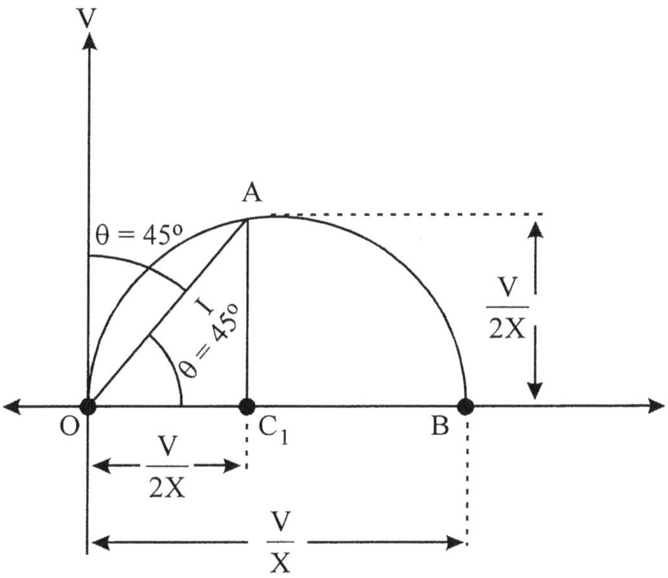

Fig. 4.24 Maximum Power Condition

It may be seen that the triangle OAC_1 is an isosceles triangle.

This implies that the voltage drop IR across R is equal to the voltage drop IX across the reactance X and each is equal to $\frac{V}{\sqrt{2}}$ or 0.707V where V is the supply voltage. Also,

$$R = X_L \text{ (or } X_C \text{).}$$

The above properties apply equally well to the circuit in which R in constant and X (X_C or X_L) is variable.

4.11.3 Properties of Constant Resistance, Variable Reactance

4.11.3.1 Maximum Current

Current is maximum when $X_L = 0$ in the case of RL Circuit and $X_C = 0$ in the case of RC Circuit. This maximum current is given by,

$$I_m = \frac{V}{R} \qquad \text{.......... (4.11.44)}$$

4.11.3.2 Maximum Power Supplied to the Circuit

Power,

$$P = VICos\ \theta$$

If V is constant, then $P \propto$ **ICos** θ, P is proportional to the ordinate of the current semicircle. Hence the maximum power is V times the maximum ordinate. Evidently the maximum ordinate equals the radius of the current semicircle, i.e., equal to V/R.

Thus Maximum Power in both RL Circuit and RC Circuit is given by,

$$P_m = V \times \frac{V}{R} = \frac{V^2}{R} \qquad \text{.......... (4.11.45)}$$

4.11.3.3 Power Factor at Maximum Power Condition

In the case of both RL Circuit and RC Circuits, under maximum power condition, $\theta_L = \theta_C = 0$. Hence,

Power Factor = **Cos 0°** = 1.

4.11.3.4 Points to be Considered in Locus Diagrams

1. To Obtain the Resonance

The locus diagram of any existing circuit can be obtained as detailed above. We know that for resonance the reactive component of the total current should be equal to zero. Hence, the capacitive or the inductive component of the current that has to flow through the element connected in parallel to obtain resonance can be calculated. Knowing the scale of the current locus, we can calculate the component required to obtain resonance.

2. To Obtain the Locus Diagram for Parallel Circuits

If one limb of the parallel circuit is RL or RC Series circuit with variable resistance R or variable reactance, the current locus of that limb is drawn with the tip of the current phasor of the other limb as the origin, then the sum of this two current phasors will be measured from the original origin i.e., from the starting of the current phasor of the other limb.

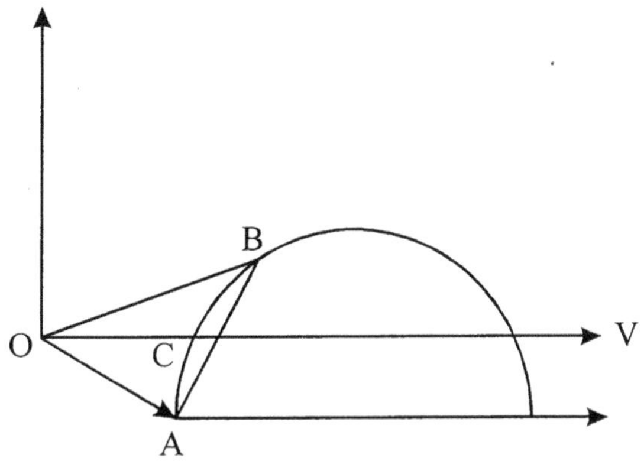

Fig. 4.25 *Composite Locus Diagram*

Fig. 4.25 gives the current locus diagram of a parallel circuit with one limb consisting of a resistance and inductive reactance whose current phasor OA is lagging behind the voltage phasor V, taken as reference phasor along the X-axis. With A as the new origin the current locus is drawn for the second limb which has variable R and fixed C. The total current of the parallel circuit is

$$\mathbf{OA + AB = OB}$$

measured from the original origin O.

The resonant condition can be obtained from the point C and the current at resonance will be OC. Then AC is the current phasor for the second limb (i.e., variable R, fixed C), because the total current phasor will be in phase with voltage phasor V at resonance.

3. On similar lines, the current locus can be drawn for any number of parallel limbs, but all measurements of the currents should be made from the original origin. For the individual limbs, the corresponding origins will be considered.

Example 4.15:

A coil of unknown reactance $X_L \Omega$ and resistance $R_L \Omega$ with a series variable resistor R is connected across a 200V, 50Hz supply. The current locus diagram has diameter of 10A. When the value of series resistor R is 12Ω, the power dissipated is maximum.

Determine the

(1) reactance X_L and resistance R_L of the coil

(2) value of maximum power

(3) maximum current

and (4) current under maximum power condition

Solution:

$$\text{Diameter of current semicircle} = \frac{V}{X_L}$$

Hence, $$10 = \frac{200}{X_L}$$

$$X_L = 20\Omega$$

Power is maximum when total resistance ($R + R_L$) is equal to X_L. i.e.,

$$12 + R_L = 20$$

$$R_L = 8\Omega.$$

$$\text{Maximum Power} = \frac{V^2}{2X_L} = 1000\text{W}$$

$$\text{Maximum Current} = \frac{200^2}{\sqrt{200^2 + 8^2}} = 9.28\text{A}$$

Current under maximum power condition,

$$= \frac{V}{\sqrt{20^2 + 20^2}} = 7.07\text{A}$$

Example 4.16:

A 200V, 50Hz source feeds a series circuit consisting of a constant reactance 25Ω and variable resistor R. For the given circuit determine maximum power condition, current, power and value of R when the power factor is 0.3

Solution:

Maximum Power Condition,

$$I_M = \frac{V}{X} = 8A$$

The current locus for the RL series circuit with a variable R will be a semicircle with center at $\left(\frac{V}{2X}, 0\right)$ and radius $\frac{V}{2X}$ when the reference phasor V is taken along the Y-axis. With a current scale of 1cm = 2Amp, the current, the current locus diagram is drawn with a center at (2, 0) and radius 2cm. The current phasor is drawn at an angle of 72.5° with respect to phasor V. I is measured as 7.63A.

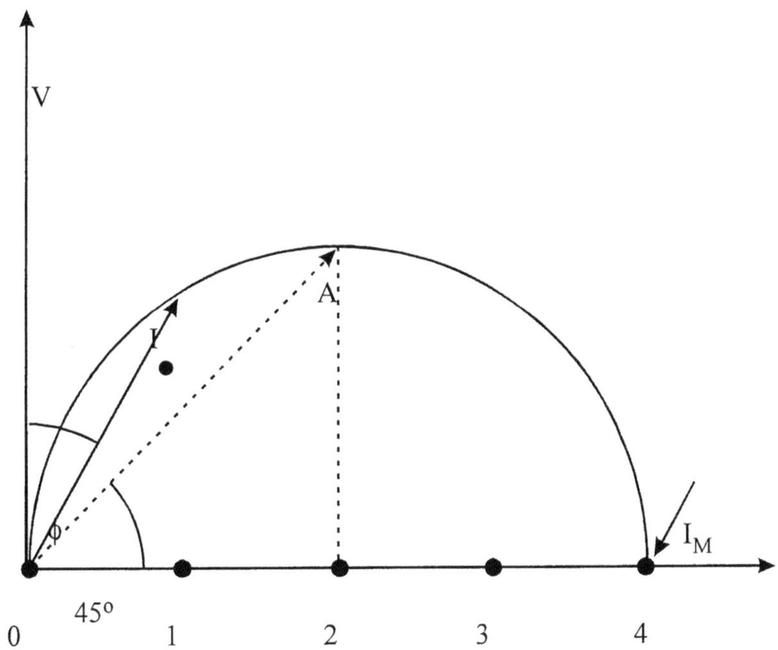

$$P_M = \tfrac{1}{2}VI_M = 800W$$

Current, $I = \dfrac{I_M}{\sqrt{2}} = 5.657A$

Power Factor,

$$\mathbf{Cos}\ \phi = \mathbf{Cos}\ 45^\circ = 0.707$$

Under maximum power condition,

$$R = X = 25\Omega.$$

When $\mathbf{Cos}\ \theta = 0.3$

$$\theta = 72.5^\circ$$

Current, $I = I_M\mathbf{Cos}(\ 90^\circ - 72.5^\circ\ ) = 7.63A$

Power $= VI\mathbf{Cos}\ \theta = 458W$

But Power $= I^2R$

Hence, $R = \dfrac{Power}{I^2} = 7.87\Omega.$

* * * * *

COMPREHENSION - 4

Note: **By answering all the questions given in this comprehension one can answer any objective type of questions easily whatever be the form of the question. Answers for these questions can be obtained by going through the topics in this chapter carefully.**

1. a) Explain the different forms of representing A.C quantities.
   b) What are the differences between a DC circuit and a Single Phase purely resistive AC circuit?
2. What are the types of AC circuits?
3. Give an expression for the impedance of a single phase R-L series circuit.
4. Draw the voltage, current waveforms and also the phasor diagrams for the single phase R-L series circuit.
5. Derive an expression for the instantaneous power for a single phase R-L series circuit. Draw its waveform.
6. What is the expression for real power in the case of single phase AC circuit.
7. Give an expression for the impedance of a single phase R-C series circuit.
8. Draw the voltage, current waveforms and also the phasor diagrams for the single phase R-C series circuit.
9. Derive an expression for the instantaneous power for a single phase R-C series circuit. Draw its waveform.
10. What is meant by the terms impedance, admittance and immittance?
11. Draw the impedance triangle and the admittance triangle for an inductive circuit and for a capacitive circuit.
12. What will be the nature of power in the case of AC circuit? Give different forms for the complex power in the AC circuit.
13. What are the two conventions that are followed in representing the complex power of an AC circuit?
14. Draw the KVA triangle for an inductive circuit and for a capacitive AC circuit.
15. What are the expressions for conductance G and susceptance B of an admittance? What will be their nature of signs for inductive and capacitive admittances?

16. What is meant by power factor? What is its importance?
17. What is the condition to be followed while expressing a power factor?
18. What is meant by Wattfull and Wattless components of current? What are its significances? Draw a phasor diagram and explain.
19. How do you obtain the power factor and power factor angle from
    - (a) impedance triangle
    - (b) admittance triangle
    - (c) KVA triangle
    - (d) Wattfull and Wattless components of current.
20. Give an expression for the impedance and admittance of a single phase R-L-C series circuit.
21. How do you solve an R-L-C series circuit?
22. Obtain an expression for the series impedance of a single phase AC circuit having more than one impedance in series.
23. How do you solve an AC parallel circuit?
24. Obtain an expression for the equivalent impedance and equivalent admittance for an AC parallel circuit having more than one impedances in parallel.
25. Obtain the expressions for the equivalent admittance for
    - (a) R-L Parallel Circuit
    - (b) R-C Parallel Circuit
    - (c) R-L-C Parallel Circuit
26. How do you obtain the current distribution in different parallel limbs of a parallel circuit from the total current drawn by the parallel combination?
27. Draw the phasor diagrams for a parallel circuit with
    - (a) two inductive impedances in parallel
    - (b) two capacitive impedances in parallel
28. How do you solve the AC Series-Parallel circuit?
29. How do you obtain the equivalent star impedances for a given delta or mesh connected impedances and vice-versa with
    - (a) three impedances
    - (b) more than three impedances
30. What are the features and uses of a Single Phase AC Circuit?

31. What is meant by Resonance?
32. What are the different methods of obtaining Resonance in a circuit?
33. Derive the condition for series resonance.
34. Explain the variation of the parameters R, L and C with change in frequency with a diagram.
35. Draw the variation of voltages across R, L and C in series with variation in frequency.
36. Can the voltage across inductance or capacitance in a RLC series circuit be greater than the supply voltage? If so, when?
37. What is meant by Q-factor?
38. Obtain an expression for the Q-Factor of a RLC Series circuit?
39. What are the other names for series resonance circuit and Why?
40. Give an expression for voltage magnification.
41. What will be the nature of the power in a series resonance circuit?
42. Give an expression for the resonant frequency of a series resonant circuit.
43. What is meant by upper and lower cutoff frequencies? Derive expressions for them in the case of a series resonance circuit.
44. What is meant by bandwidth? What is its effect on the selectivity of a circuit for a signal frequency?
45. What is the effect of resistance on the selectivity or tuning of a series resonance circuit?
46. When is a tuned circuit is highly sensitive?
47. What are the different cases of parallel resonance?
48. Draw a phasor diagram and derive the expression for resonant frequency in the case of

    1. Practical Parallel Circuit
    2. Parallel RLC Circuit
    2. Inductive and Capacitive Impedances in parallel.
    3. Pure LC parallel Circuit

49. What are the features of a parallel resonant circuit?
50. Draw the current response with respect to frequency for a parallel resonance circuit?

51. What is current magnification? When do you obtain it?
52. Compare the series and parallel resonance circuits.
53. What is meant by locus diagrams?
54. What are the different loci that are generally drawn?
55. Prove that the current locus of a RL series circuit with fixed R and variable L is a semicircle. What will be its center and radius?
56. Prove that the current locus of a RC series circuit with fixed R and variable C is a semicircle. What will be its center and radius?
57. Prove that the current locus of a RL series circuit with fixed L and variable R is a semicircle. What will be its center and radius?
58. Prove that the current locus of a RC series circuit with fixed C and variable R is a semicircle. What will be its center and radius?
59. How do you draw the voltage locus for RL, RC series circuits?
60. What are the expressions for maximum current in the case of RL and RC Circuits?
61. Obtain an expression for the maximum power supply to the RL and RC series circuits?
62. Obtain the power factor for maximum power condition from the current locus of variable R and fixed L series circuit.
63. Obtain the power factor for maximum power condition from the current locus of variable R and fixed C series circuit.
64. How do you obtain the locus diagram of a parallel circuit? How do you obtain the resonant condition from the locus diagrams?

* * * * *

EXERCISE - 4

Note: **By answering all the questions given in this exercise one can answer subjective type of questions easily including problems. The worked examples in the chapter may help in solving these questions.**

1. (a) Draw the waveforms and phasor diagrams for a single phase AC circuit having two resistances in series.

   (b) Two resistances $R_1 = 5\Omega$ and $R_2 = 15\Omega$ are connected in series to a 230V, 50Hz single phase AC supply. What will be the current and the voltage drop across each resistance? Draw the phasor diagram for the circuit.

   [Ans: I = 11.5A; V_1 = 57.5V, V_2 = 172.5V]

2. (a) Draw the impedance triangle, admittance triangle and the voltage drop triangle for a single phase R-L series circuit.

   (b) Draw the waveforms and phasor diagram for the circuit.

   (c) What will be the Wattfull and Wattless components of the current drawn by the circuit?

   (d) Obtain expressions for the complex power of the circuit using both the conventions.

3. (a) A 50Hz AC voltage has the value (3 + j4)V. When applied to a circuit the current is (1 + j1)A. Write down the equation for the voltage and current. Find also (i) the impedance in the circuit, (ii) Power Factor, (iii) real and reactive powers

   [Ans: e = $5\sqrt{2}$Sin(314t + 53.1°)V; (i) 3.536W; (ii) 0.99 lag; (iii) 7W, 1VAR]

   (b) A series R-L circuit has R = 20W and L = 0.05H and is connected to 250V, 50Hz source. Calculate (i) the impedance, (ii) current, (iii) power and (iv) power factor. Draw the phasor diagram.

   [Ans: (i) $25.43\angle 38.15°\Omega$; (ii) $9.831\angle -38.15°$A; (iii) 1932.7W; (iv) 0.786 lag]

4. (a) A coil A draws current of 10A and power of 200W when driven by a voltage of 110V, 60Hz. The coil B draws current of 11A and power of 400W when driven by the same source. What current

and power will be drawn from the same voltage source by coil A and B connected in series?

[Ans: 5.27A; 147.2W]

(b) A sinusoidal voltage $v(t) = 57.85\mathbf{Sin}(\mathbf{314t} + \mathbf{30^o})$ is applied to a coil having resistance of 3Ω and inductive reactance of 0.6Ω at 30Hz. Obtain equations for the instantaneous values of current I, voltage V_R across resistor, and voltage V_l across inductive reactance?

[Ans: I = 26.83Sin(314t + 11.6);

V_R = 80.5Sin(314t + 11.6°);

V_L = 26.83Sin(314t + 101.6)]

5. (a) A coil when connected to 200V DC supply dissipates 2000W of power. When connected across 200V AC supply of frequency 50cps., it dissipates 800W of pwer. Calculate the value of resistance and inductance of the coil.

[Ans: R = 20Ω; L = 0.078H]

(b) A coil A having resistance of 10Ω and inductance of 0.2H is connected in series with coil B having resistance of 30W, and inductance of 0.1H. The two coil in series are fed from 220V, 50Hz supply. Determine the voltage across each coil, power dissipated in each coil and power factor of the whole circuit.

[Ans: 136.66V, 93.3V; 46.14W, 138.4W; 0.39 lag]

6. (a) Draw the impedance triangle, admittance triangle and the voltage drop triangle for a single phase R-C series circuit.

(b) Draw the waveforms and phasor diagram for the circuit.

(c) What will be the Wattfull and Wattless components of the current drawn by the circuit?

(d) Obtain expressions for the complex power of the circuit using both the conventions.

7. (a) A series R-C circuit has R = 20W and C = 127μF has a 160V, 50Hz supply connected to it. Calculate the impedance, current, power and power factor. Draw the phasor Diagram.

[Ans: $32\angle{-51.3^o}$; $5\angle{51.3}$A; 500W; 0625 lead]

(b) A 400Hz generator has an induced EMF of 100V and an internal impedance $(5 + j0)\Omega$. If it supplies an impedance consisting of a 40Ω capacitive reactance in series with 10W resistance, what is the magnitude of the current passing through it? Determine also the voltage at the terminals of the generator.

[Ans: 2.341A; 96.52V]

8. (a) A capacitor of $8\mu F$ takes a current of 1.0A when the alternating voltage applied across it is 250V. Calculate

   (i) the frequency of the applied voltage;
   (ii) the resistance to be connected in series with the capacitor to reduce the current in the circuit to 0.5A at the same frequency.
   (iii) the phase angle of the resulting circuit

[Ans: 79.5Hz; 433Ω; 30° lead;]

(b) A metal-filament lamp, rated at 750W, 100V, is to be connected in series with a capacitor across a 230V, 60Hz supply. Draw the phasor diagram and calculate

   (i) the capacitance required
   (ii) the phase angle between the current and the supply voltage.

[Ans: 96μF; 64.12°]

9. (a) A 120V, 60W lamp is to be operated on 220V, 50Hz supply mains. Calculate what value of

   (i) noninductive resistance
   (ii) pure inductance and
   (iii) pure capacitance

would be required in order that lamp is run on correct voltage. Which method is preferred and Why? Draw relevant phasor diagram.

[Ans: 200Ω; 1.174H; 8.63μF]

(b) A given load consisting of a resistor R and a capacitor C, takes a power of 4800W from 200V, 60Hz supply mains. Given that the voltage drop across the resistor is 120V. Calculate the current, impedance, power factor, resistance and capacitance. Write the equations for the voltage and current.

[Ans: 40A; 5Ω; 0.6 lead; 3Ω; 663μF; 282.84Sin(377t); 56.57Sin(377t + 53.1)]

10. (a) How do you solve R-L-C Series circuit?

    (b) Draw the phasor diagram, impedance and KVA triangles for R-L-C series circuit if $X_L > X_C$ and $X_C > X_L$.

    (c) What will happen if happen if $X_L = X_C$.

11. (a) A coil of resistance 15Ω and inductance 0.2H is connected in series with a capacitor of 100mF across a 220V, 50Hz supply. Calculate

    (i) inductive reactance
    (ii) capacitive reactance
    (iii) impedance of series circuit
    (iv) current
    (v) power factor
    (vi) voltage across the coil
    (vii) voltage across the capacitor

    [Ans: 62.8Ω; 31.83Ω; 34.41Ω; 6.39A; 0.436 lag; 64.57Ω; 412.6V]

    (b) A voltage (100 + j60)V drives a current (4 – j5)A through a series R-L-C circuit. Determine

    (i) the complex expression for the impedance

    (ii) the power consumed

    (iii) Power factor

    Determine the magnitude of capacitive reactance to be added in series to increase the power factor to 0.95.

    [Ans: 2.44 + j18.05; 100 + j740; 0.134; 17.7Ω]

12. (a) Obtain an expression for the equivalent impedance of single phase AC circuit with two impedances connected in parallel.

    (b) How is the total current distributed in each of the parallel branches? Derive the expressions.

    (c) How do you solve a single phase AC circuit having more than two impedances in parallel?

13. (a) Draw the phasor diagram for a single phase AC circuit with

    (i) two inductive impedances in parallel
    (ii) two capacitive impedances in parallel
    (iii) R, L and C in parallel

(b) Explain the two conventions of obtaining the complex power of a single phase AC circuit. What will be the nature of the reactive power in each convention when you express the complex power as **S** = (P + jQ)? Draw the corresponding KVA triangles.

14. (a) An AC circuit draws active and lagging reactive currents of 20A and 10A respectively from a 200V, 50Hz supply. Calculate

(i) the conductance, the susceptance, the admittance
(ii) power factor of the circuit

What would be the resistance and inductance of a coil if it draws the same current from the same supply at the same power factor?

[Ans: 0.1S, –0.05S, 0.1118S; 0.894 lag; 8Ω, 127mH]

(b) A circuit consists of a 120Ω resistor in parallel with a 40μF capacitor and is connected to a 240V, 50Hz supply. Calculate

(i) the branch currents and the supply currents
(ii) the circuit phase angle
(iii) the circuit impedance

[Ans: 2.0A; 80Ω; 3.0A; 3.6A; 56.3° lead; 66.7Ω]

15. (a) The circuits A and B are connected in parallel to a 230V, 50Hz supply. Circuit A consists of resistance 20Ω in series with an inductive reactance of 20Ω and circuit B consists of resistance 40Ω in series with a capacitive reactance of 20Ω. Determine

(i) the current drawn by each circuit
(ii) the total current drawn from the mains.

[Ans: 8.13A, 5.143A; 10.91A]

(b) Three branches possessing a resistance of 50Ω, and inductance of 0.15H and a capacitance of 100μF respectively are connected in parallel across a 100V, 50Hz supply. Calculate

(i) the current in each branch
(ii) the supply current
(iii) the phase angle between the supply current and the supply voltage.

[Ans: 2A; 2.125A; 3.14A]

16. (a) How do you solve the single phase AC Series-Parallel Circuit?

(b) An impedance of (10 + j8)Ω is connected in series with a parallel combination of two impedances (3 + j4)Ω and (10 – j12)Ω draws a total current of (10 + j0)Ω. Calculate the branch currents, voltages across each impedance and the total voltage, complex power for each of the three impedances.

[Ans: 10.24∠–18.58°A, 3.28∠84.73°A; 12.81∠38.66°A, 51.2∠34.54°V, 179.2∠37.47°V; (1422 – j1090)W, (1000 – j800)W, (315 – j419)W, (107 + j129)W]

17. For the RL series circuit with a variable R varying from 0 to 50Ω and and inductance of j25Ω, plot the locus of the current, mark the range of I for maximum and minimum values of R, and the maximum power consumed in the circuit, when supplied with a single phase AC voltage of 200V, 50Hz supply.

    [Ans: 8∠–90°; 3.777∠–27.76°; 800W]

18. A 10μF capacitor, an coil (L = 0.1H, R = 10Ω) and a variable resistor R_1 are placed in series and connected to a 220V, 50Hz supply. Calculate the value of variable resistor R_1 when the power is maximum; power and current under this maximum power condition and maximum possible current.

    [Ans: R_1 = 276.9Ω; 83.3W; 0.766A]

19. A 20W resistor is connected in series with a variable inductive reactance. Obtain the locus of apex of the current vector when the circuit is connected to 200V, 50Hz supply. Determine the value of the current, power factor when the reactance is 10Ω, 20Ω and 30Ω.

    [Ans: 10A; 8.94∠–26.6°, 0.89 lag, 1591W; 7.07∠–5.63°, 0.707 lag, 1000W; 5.55∠–56.3°, 0.55 lag. 610W]

20. (a) For fixed R and variable C series circuit, draw the current locus and mark the current range. X_C varies from 0 to 50Ω. R = 10Ω.

    [Ans: 40∠0°; 7.716∠77.8°]

    (b) A resistance of 25Ω is connected with a impedance consisting of variable resistance varying between 10 and 100Ω and a inductive reactance of 25Ω are connected in parallel to a 200V, 50Hz supply. Draw the current locus and obtain the currents in the parallel branches and the total current when R_1 = 100Ω.

    [Ans: 0.94A; 8A; 1.94∠–14°]

* * * * *

5

NETWORK THEOREMS

5.1 D.C. NETWORK THEOREMS

5.1.1 Introduction to Network Theorems

We have already studied the general methods of network analysis. However, in a large and complex network, these methods tend to become laborious and time-consuming. In many cases, it is advantageous to use special techniques to reduce the quantum of labour involved in circuit solution. We term such techniques as **Network Theorems.** The other important features of the theorems, besides saving labour and time, are :

(a) *they are applicable to a useful and fairly wide class of networks*

(b) *their conclusion are simple*

(c) *they sometimes provide good physical insight into the problems.*

The theorems are useful when the problem is limited like finding the current in a single element, the value of the load for maximum power etc. In this chapter we shall study some important theorems, viz., *Reciprocity Theorem, Superposition Theorem, Substitution Theorem, Thevenin's Theorem, Norton's Theorem. Maximum Power Transfer Theorem, Compensation Theorem, Millman's Theorem and Tellegen's Theorem.*

While discussing these theorems the applicability of these theorems to certain kinds of networks only, must be clearly understood. The network elements could be classified as *Linear, Limped, Finite, Passive and Bilateral* and *Bilateral* usually abbreviated as LLFPB elements.

5.1.2 Theorems in A.C. Circuits

Just as we applied Superposition Theorem, Thevenin's Theorem, Norton's Theorem and Maximum Power Transfer Theorem. We can apply these theorems to A.C. circuits also. It's light modification that the Voltage and Current Sources will be A.C. sources and in addition to resistances in the circuits we may have inductances and capacitances also in the circuit.

Hence in the statement of the theorems, instead of resistances we will have to use impedances. Instead of conductances we have to use admittances.

5.2 NETWORK CLASSIFICATION

Electric Networks can be classified mainly in two ways. The first one is based on the kind of elements in the network e.g. time-invariant or non-linear or time variant etc. The second one is based on excitation (supply source) and response (output) i.e. for a given excitation the response decides the type of network.

Here we shall study classification based on the second approach.

5.2.1 Linear Networks

Suppose the network to be classified is under relaxed condition i.e., with no initial conditions and an excitation $v_1(t)$ is applied for which response is $c_1(t)$ and for excitation $v_2(t)$ the response is $c_2(t)$. Then the network is classified as linear if for excitation $v_1(t) + v_2(t)$ the response is $c_1(t) + c_2(t)$. This shows that a linear network follows superpositions principle.

5.2.2 Passive Networks

Some networks have the property of absorbing / dissipating or storing energy. These are able to return the energy previously stored to the external network. However, this energy returned is never more than the energy stored. Such networks are known as **passive networks.**

Let network N be connected to a source S. Suppose the network is initially relaxed. If $v(t)$ is the voltage of the source and $i(t)$ is the resulting current into the network N_2 the average power delivered into the network is

$$p(t) = v(t)\, i(t)$$

and the energy delivered w(t) over time t is given by

$$w(t) = \int_{\infty}^{t} v(t)\ i(t)\ dr \qquad \text{...... (5.2.1)}$$

This energy will always be non-negative for a passive netowrk i.e., over time t there will be always be some energy delivered to the network or atleast there will not be any energy when the netowrk would have returned all the energy back to the source or it would have stored some energy. It will never be a situation for a passive network when it would supply more energy what it had drawn previously. Mathematically for a passive network.

$$w(t) = \int_{\infty}^{t} v(t)\ i(t)\ dt \geq 0 \qquad \text{........ (5.2.2)}$$

However, if the network has some initial energy stored equal to w(t_0) before it is switched on to the supply at $t = t_0$ then for a passive network energy delivered to the network from the source should satisfy the inequality at any time t.

$$w(t) = \int_{\infty}^{t} v(t)\ i(t)\ dt + w(t_0) \geq 0$$

A circuit which does not satisfy this condition is known as active circuit. That is for an active circuit.

$$\int_{\infty}^{t} v(t)\ i(t)\ dt < 0 \qquad \text{.........(5.2.3)}$$

5.2.3 Lumped Networks

Many devices in electric system are distributed in space e.g. transmission lines, the windings of transofrmers or that of the generators, are distributed in a way. Whenever these devices are energized, the effect is not experienced by the line length or winding lengths instantly because of finite velocity of electric signals.

However, if we are interested in steady state and terminal quantities, it is sufficient to assume the parameters to be lumped rather than distributed. Sometimes when we are interested i the intermediate values and point to point variation of electric signals, we talk true nature of these devices i.e., the distributed nature. Otherwise, we assume the parameters to be lumped.

5.2.4 Bilateral Networks

The elements to be considered for electric network are assumed to be bilateral which means the voltage and current relations are same irrespective of direction of flow of current e.g. resistance, inductance, etc, However, for unilateral elements the voltage and current are different for two possible direction of flow of currents e.g. diodes.

5.2.5 Time Invariant Networks

A network is said to be time0invariant when there is some response to a certain excilation irrespective of time of application of excitation. Suppose for certain excitation v_1 (t) the response is c_1 (t), if now the excitation is v(t + t_0) the response would be c(t + t_0). Here the values of the parameters are assumed to be constant at all times and do not change with time.

5.2.6 Reciprocity

A network is said toa be reciprocal when excitation and response teriminals could be interchanged, without any change in the performance of the network.

If excitation v_1 produces a response i_2 and if the input and output are interchanged then with new $v_2 = v_1$, the new response i_1 will be equal to i_2. E.g., a transmission line as shown in fig. 5.1

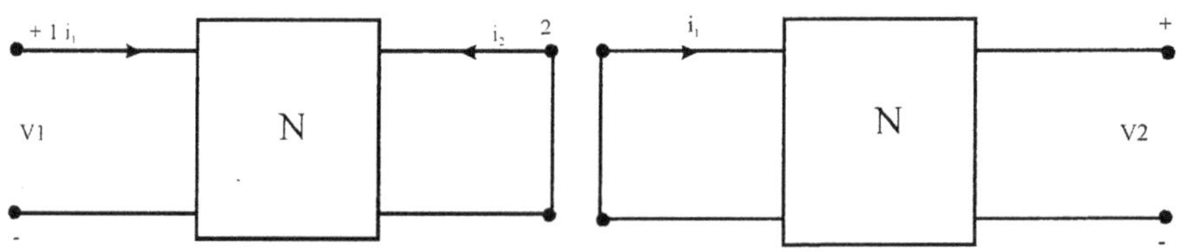

Fig.5.1 Networks to describe Reciprocity

5.3 SUPERPOSITION THEOREM FOR D.C. EXCITATION

This theorem is applicabe to linear networks where the response is proportional to the excitation.

Statement:

In a linear network if more than one number of excitations are acting, the response in a certain branch can be obtained by superimposing algebrecially the responses due to various sources taking one excitation at a time considering the remaining voltage sources, short circuited if they are ideal sources or replaced by their equivalent internal Resistances and current sources are open circuited if they are ideal sources or replaced by their equivalent resistances.

Proof:

In a linear network, if for an excitation E_1, the response I_1 of the given network is related by

$$I_1 = CE_3 \qquad \text{..... where 'C' is conductance}$$

with initial condition equals zero and if for an excitatio E2, the response I_2 of the given network is related by

$$I_2 = CE_2$$

with initial condition equals zero then for excitation ($E_1 + E_2$), the response will be ($I_1 + I_2$) for the given linear network. It is to be noted that if the excitation and response are related by the linear equation,

$$I_1 = CE_1 + d_1 \qquad \text{..... where 'd}_1 \text{ is conductance}$$

the network is not linear since for double the excitation the response will not be doubled. Thus a circuit with R,L,C elements with constant values with no initial conditions.

i.e.,

$$I_1(0^+) = 0$$

$$V_C(0^+) = 0$$

and

However, the initial conditions if any can be considered as sources or excitation and the overall response of the circuit can be obtained.

Example 5.1:

Using superposition principle determine the current through 2Ω resistor connected between A and B shown in the circuit here.

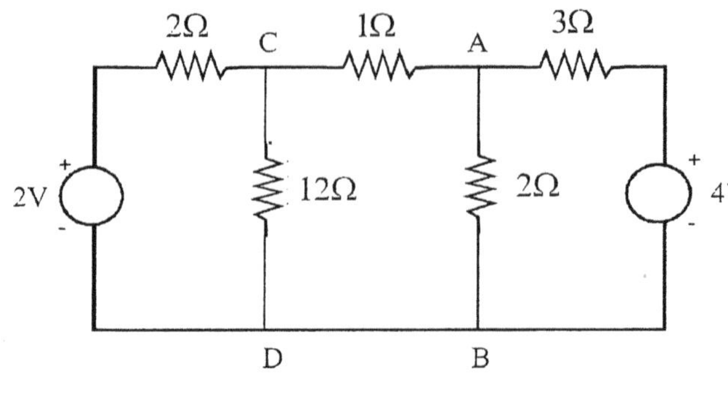

Solution:

Since there are two voltage sources in the circuit, there will be two circuits for analysis purposes where in one source will be present and the other short circuited being the voltage source.

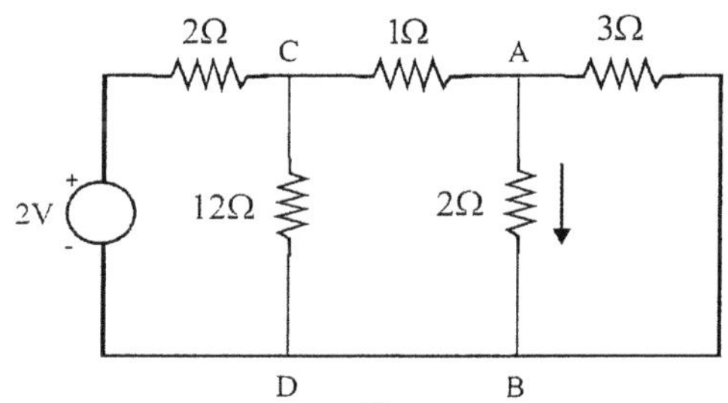

(i)

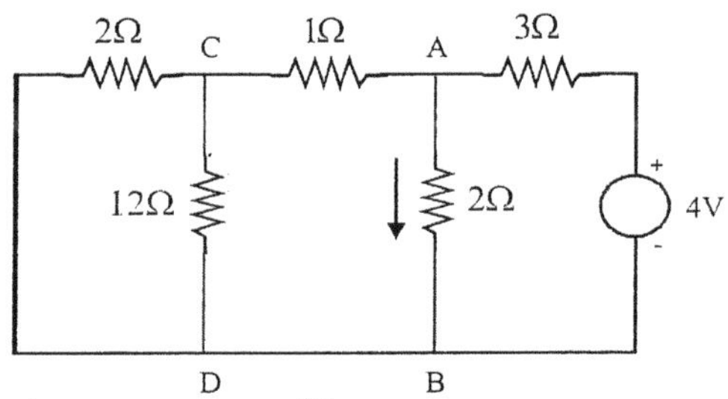

(ii)

in circuit (i) the resistance between CD exlcuding 12Ω resistance

$$1+\frac{6}{5}=1+1.2=2.2\Omega,$$ including 12Ω resistance

$$\frac{72}{274}+\frac{76}{137}=\frac{224}{274}=0.84\,℧$$

Total resistnace $2+\frac{132}{71}=\frac{274}{71}\Omega$

Therefore, current through the source is $\frac{2\times71}{274}$ A

The current through 1Ω resistor is

$$\frac{2\times 71}{274}\times\frac{12\times 10}{142}=\frac{120}{274}\text{ A}$$

Therefore current through 2Ω resistor

$$\frac{120}{274}\times\frac{3}{5}=\frac{72}{274}A$$

in circuit (ii)

$$1+\frac{24}{14}=\frac{28}{14}\Omega$$

$$\frac{1}{R}=\frac{1}{2}+\frac{14}{28}=\frac{33}{38}\ \mho$$

or $$R=\frac{38}{33}\Omega$$

Total resistance $3+\frac{38}{33}=\frac{137}{33}\Omega$

Current through the source $\frac{4\times 33}{137}A$

Current through 2Ω resistor

$$\frac{4\times 33}{137}\times\frac{28}{14}\times\frac{14}{66}=\frac{76}{137}A$$

Therefore, total current through 2Ω resistor is

$$\frac{72}{274}+\frac{76}{137}=\frac{224}{274}=0.8A$$

Example 5.2:

Determine the current through the voltage source and the voltage across the current source in the network shown here.

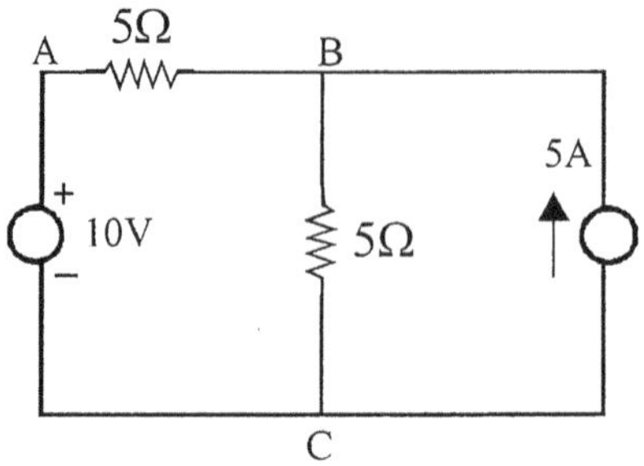

Solution:

Since we have two sources (excitations) one voltage and another current we have to look for two circuits where in one circuit, voltage source will be short circuited and in another, current source will be open circuited.

To find out current through the voltage source i.e., through branch AB we proceed as follows :

(a) (i) When current source alone is acting using parallel currents theorem.

Current through BA is,

$$5 \times \frac{5}{(5+5)} = 2.5A$$ in the direction from B to A

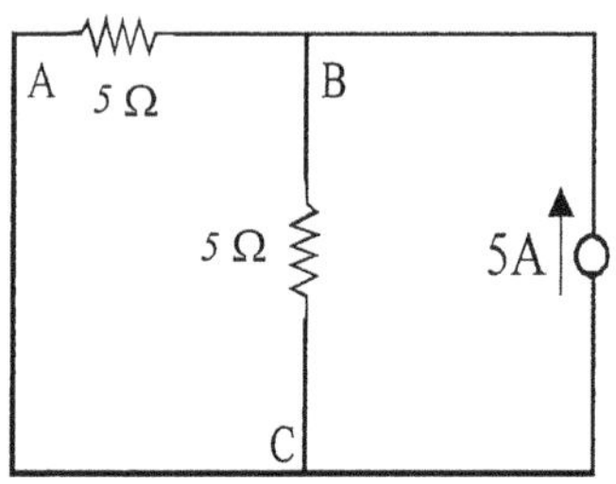

(i) Voltage Source Short Circuited
Current source operating

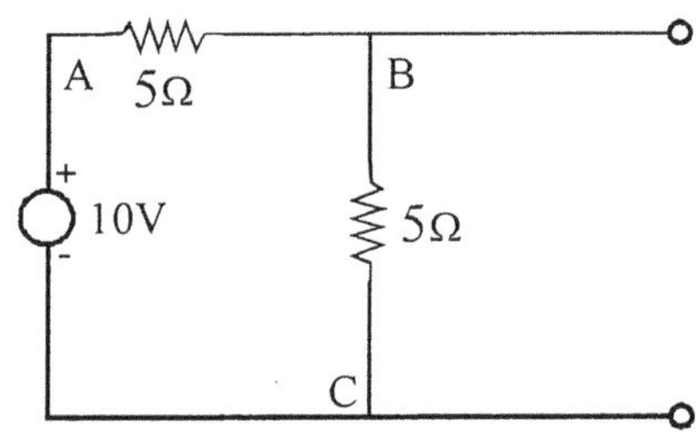

(ii) Current Source Open Circuited
Voltage source operating

(ii) When voltage source alone is acting in the circuit,

Current through AB is,

$$\frac{(5+5)}{10} = 1 \text{ A. in the direction of A to B.}$$

Therefore, the current through AB branch when both sources are simultaneously acting, is given by,

$$(-BA) \longrightarrow \quad (+) \quad AB \longrightarrow$$

$$-2.5 \quad + \quad 1.0 = -2.5 - 1.0$$

$$= -1.5A$$

hence the direction of the current is from B to A.

(b) To find out voltage across the current source we find current through the branch BC.

Current through branch BC due to current source alone is 2.5 A from B to C. Due to voltage source it is 1A. again from B to C. Therefore,

$$\text{Total Current (I)} = 2.5 + 1$$

$$= 3.5A$$

Therefore, voltage across the current source is,

$$I \times R_{BC} = 3.5 \times 5 = 17.5\ V$$

5.4 THEVENIN'S THEOREM FOR D.C. EXCITATION

Sometimes it is desired to determine the current through or voltage across any one branch in a network without calculating currents or voltages in other branches of the network. in such cases it is not necessary to write complete set of equations for the whole network and then solving these equations for this single current or voltage in a particular branch. In other words, it is always not necessary to analyse the complete circuit. Thevenin's Theorem may conveniently be used for such problems. Thevinin's Theorem is frequently used in electronic circuits.

Statement:

A two terminal linear network N consisting of independent and / or dependent voltage or current sources and impedances can be replaced with an equivalent circuit consisting of a voltage source V_{Th} in series with an impedance R_{Th} (Z_{Th} in A.C.).

*The value of V_{Th} is the open circuit voltage between the terminals of the network and R_{th} is the resistance measured between the terminals of the network with all enregy sources eliminated (but not their internal resistances). This is also called the **Voltage Source Equivalent Circuit or Thevinin's Equivalent Circuit.***

Proof :

In determining the R_{Th} of the circuit, all the ideal voltage sources are short circuited. The practical voltage sources are replaced by their internal *resistances*. Al the ideal current sources are open circuited and all the practical current sources are replaced by their internal *resistances*. In the case of A.C. networks, we use the notation V_{Th} and Z_{Th} instead of V_{Th} and R_{Th}.

The current through any branch in the circuit can be calculated using Thevinin's Theorem, denoting that branch as the load branch R_L in the case of D.C. networks and finding the Thevinin's Equivalent Network for the remaining part of the network.

Thevenin's Theorem can be explained with the help of two networks A and B connected as shown in Fig. 5.2

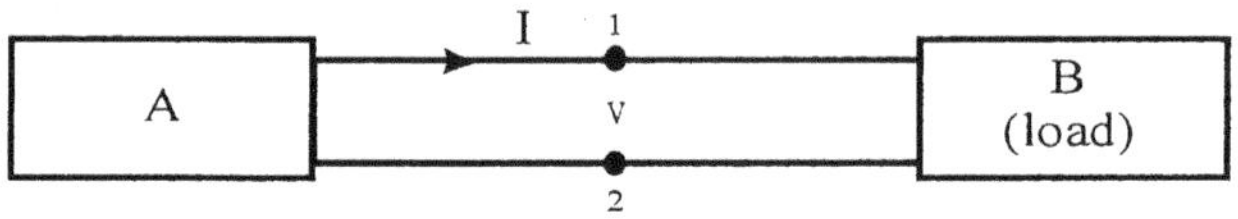

Fig. 5.2 *Two networks "A and B" connected*

The original network can be considered as a combination and connection of networks A and B. Network A contains linear elements which may have initial condition. It may also have an independent and dependent sources. However, there is no mutual magnetic source in A couple to B.

For network B, the elements need not necessarily be linear. These could be non-linear or time varying or both. However, in our analysis we will assume that the network elements are linear and many a times this will be a single branch (load) of the network

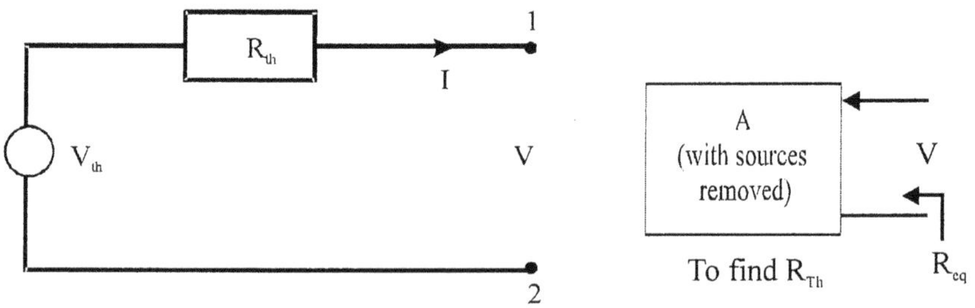

Fig. 5.3 *Thevinin Equivalent of the circuit in Fig. 5.2 (Portion A)*

Network A is to be replaced by an equivalent network under the condition that the current I and voltage V identified in the figure remain invariant when the replacement is made.This amounts to replacing network A by an equivalent Thevenin's Voltage source and an equivalent Thevenin's series impedance (resistance) as shown in Fig.5.3

The following procedure is followed to find Thevenin's equivalent circuit.

To determine V_{Th} we remove the network B (could be a load resistor or the element through which current is required) and solve the network A to find out voltage across the open circuited terminals. This voltage is the Thevinin's Equivalent Voltage Source V_{Th}. Also R_{Th} is the resistance as seen in the network between the two open circuited terminals replacing the voltage/current sources by their internal resistances. The resistance as seen is known as Thevenin's Equivalent series resistance. While finding Thevenin's Equivalent, dependent sources continue to operate in the network.

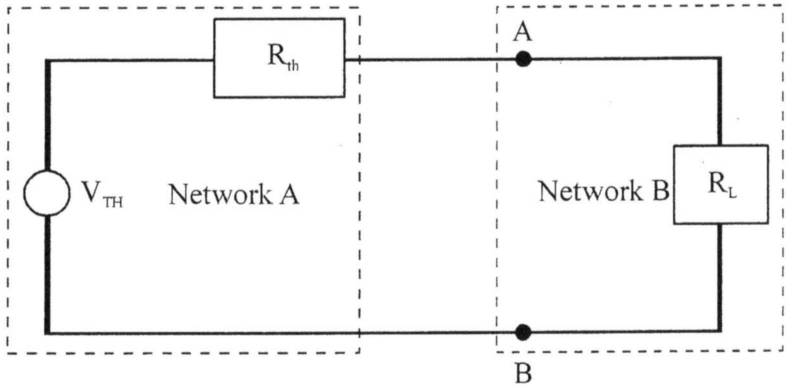

Fig. 5.4 *The current through the load impedance is given by*

If Z_L is the load impedance and Z_{Eq} the Thevenin's Equivalent series impedance

$$I = \frac{V_{TH}}{Z_{TH} + Z_1} \qquad \text{........... (5.4.1)}$$

Example 5.3:

Determine the current I in the network by using Thevenin's Theorem.

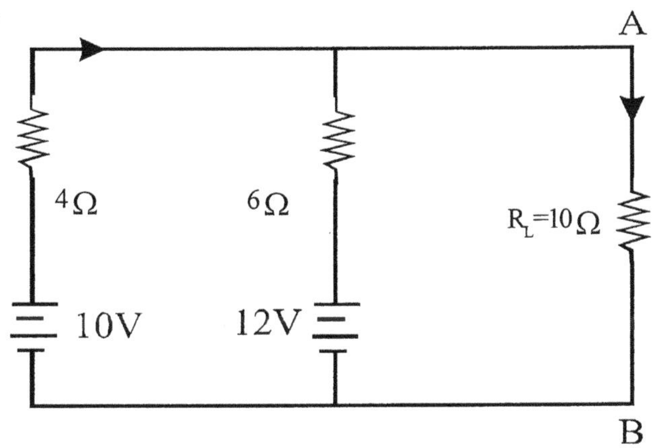

Solution:

Step 1:

The Thevinin's equivalent circuit is,

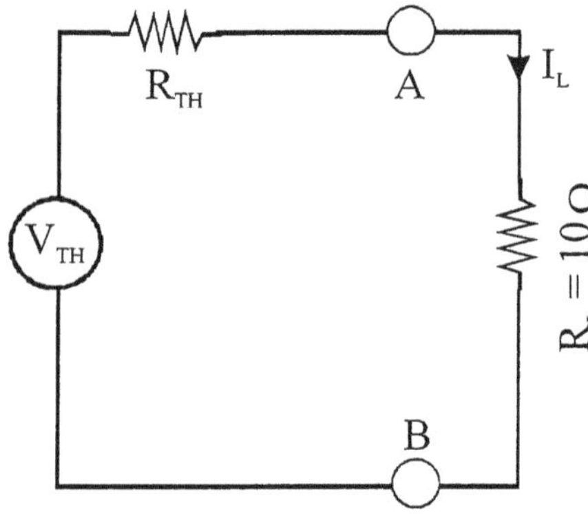

$$I = \frac{V_{TH}}{Z_{TH} + Z_l} = \frac{V_{TH}}{R_{TH} + R_L}$$

Step 2: **To find V_{TH} :**

From the given circuit, disconnect

$R_L = 10\Omega$

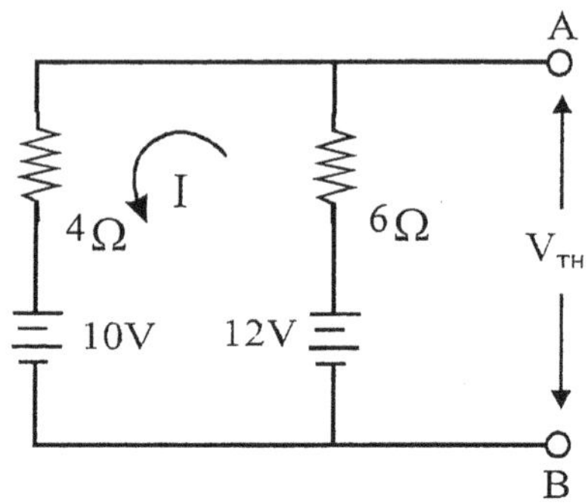

$$I = \frac{12-10}{4+6} = 0.2A$$

$$\begin{aligned} V_{TH} = V_{AB} &= 12 - 6 \times I \\ &= 6 \times 0.2 + 6 \times 0.2 \\ &= +10.8V \end{aligned}$$

B is negative polarity and A is positive polarity.

Step 3: **To Calculate R_{TH} :**

From the given circuit, kill the sources. The resultant circuit is,

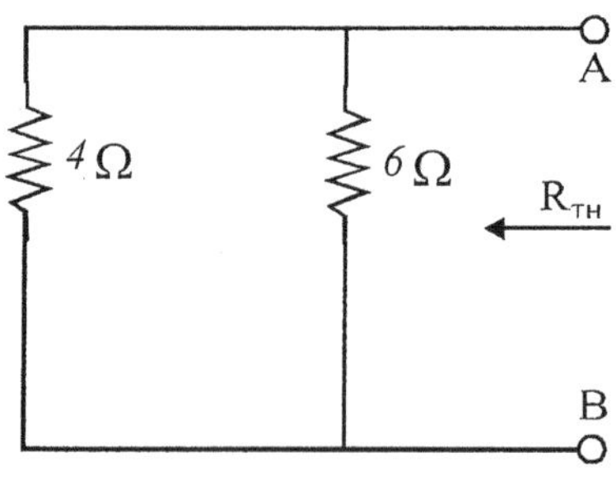

Circuit to find R_{TH}

$$R_{TH} = R_{AB} = 4 \parallel 6 = \frac{4 \times 6}{4+6} = 2.4\Omega$$

Step 4:

$$I_L = \frac{V_{TH}}{R_{TH} + R_L} = \frac{10.8}{2.4+10} = 0.871A$$

Example 5.4:

Find Thevenin's equivalent of the network shown in figure given at terminals A.B. Determine the current through the load resistor of 4 Ω connected across terminals A.B.

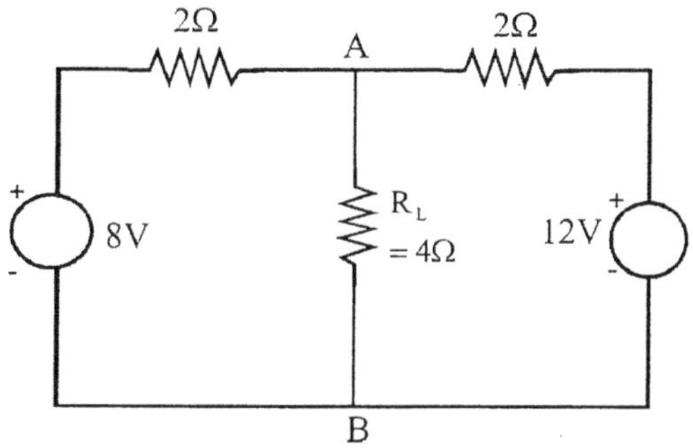

Solution:

Without the 4 Ω load resitor, the given circuit reduces to a single loop circuit. There is a net voltage of (12-8) = 4V acting anti-clock wise, and this drives a current of $\left(\dfrac{4}{2+2}\right) = 1A$

Hence, open circuit voltage across terminals A, B $V_{OC} = 12 - 2 \times I = 10\ V$

(or, O.C. voltage at A, B = 8 + 2 $\times$ 1 = 10V).

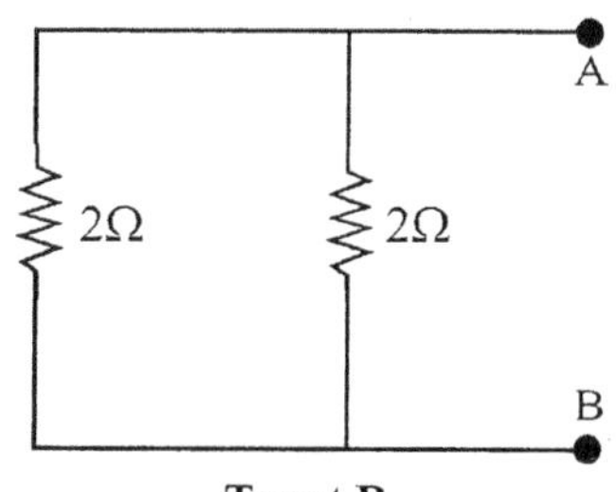

To get R_{th}

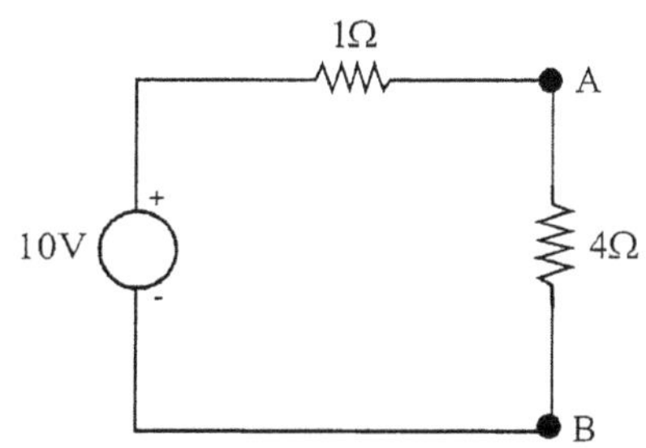

Thevenin's Equivalent Circuit

To determine the Thevenin equivalent resitance, the two voltage sources are replaced by short-circuits, and the circuit would then be as shown in figure.

We have, Resistance between terminals is given as : $R_{TH} = \dfrac{2 \times 2}{2+2} = 1\Omega$

Hence, from the Thevenin's equivalent circuit with a load resistance of 4 Ω acrosss A, B we have :

Total reistance = 1 + 4 = 5 Ω

$$\text{Load current } I_L = \frac{V_{Th}}{(R_{Th} + R_L)} = \frac{10}{1+4} = \frac{10}{5} = 2.00\text{ A}$$

Example 5.5:

Determine the current through the ammeter of 2 Ω resistance connected in a unbalanced Wheat Stone Bridge of the given figure.

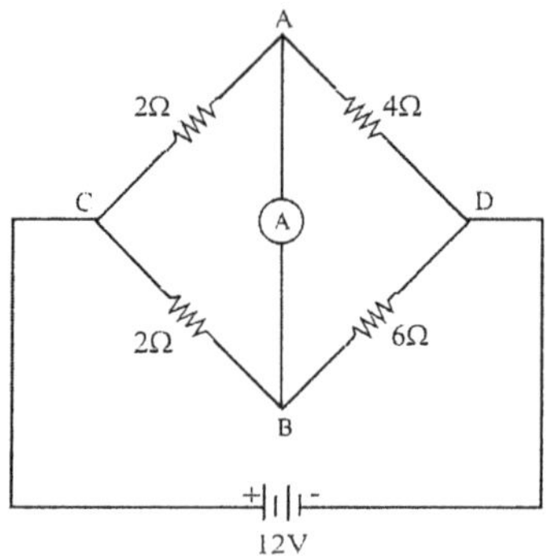

Solution:

To find the Thevenin's voltage V_{Th} across AB, with the ammeter disconnected, the given circuit reduces as shown in figure.

There are two resistances, one of 2 + 4 = 6 Ω and the other of 2 + 6 = 8 Ω value joined in parallel. We have :

$$\text{Total resistance} = \frac{6\times 8}{6+8} = \frac{48}{14}\Omega$$

$$\therefore \text{ Total current} = \frac{12}{(48/14)} = 3.5\text{A}$$

This divides between the two branches CAD and CBD in the ratio 8 : 6 i.e., 4 : 3.

$\therefore$ Current through branch CAD = 2A and

Current through branch CBD = 1.5A

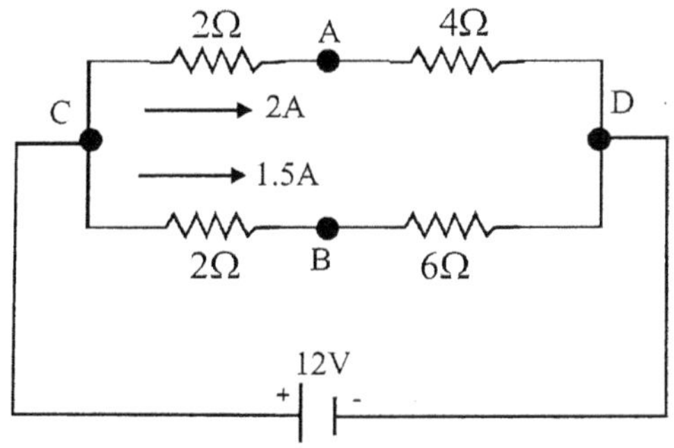

Voltage drop from A to D = 4 × 2 = 8V

Voltage rise from D to B = 6 × 1.5 = 9V

Voltage between B and A = 9 - 8 = 1V

(B is at higher potential)

i.e., open circuit voltage, V_{OC} between B,A = 1V

To find R_{TH} **:**

With the ammeter still disconnected, let the 12V battery to replaced by a short-circuit. The circuit reduces as shown in the figure.

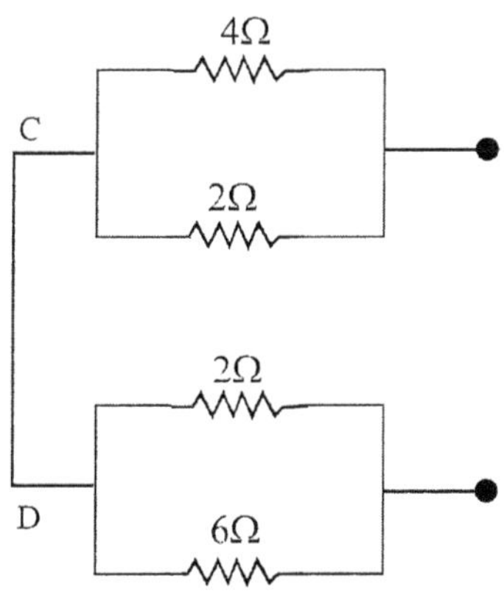

Resistance between terminals A,B

$$= \left(\frac{4\times 2}{4+2}\right) + \left(\frac{2\times 6}{2+6}\right) = \frac{17}{6}\Omega$$

The Thevenin's equivalent circuit is as shown in figure.

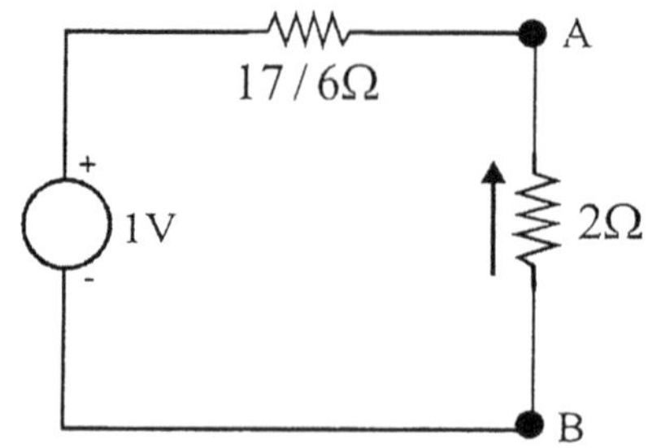

Thevenin's Equivalent Circuit

Let the ammeter of 2 Ω resitance be connected across AB.

We have :

$$\text{total resistance} = \frac{17}{6} + 2 = \frac{29}{6}\Omega$$

$$\therefore \text{ Current through the ammeter} = \frac{1}{(29/6)}$$

or $\quad I = 0.207\text{ A (from B to A)}$

Example 5.6:

Find the current through the 1 Ohm resistance across AB in the given circuit making use of Thevenin's Theorem.

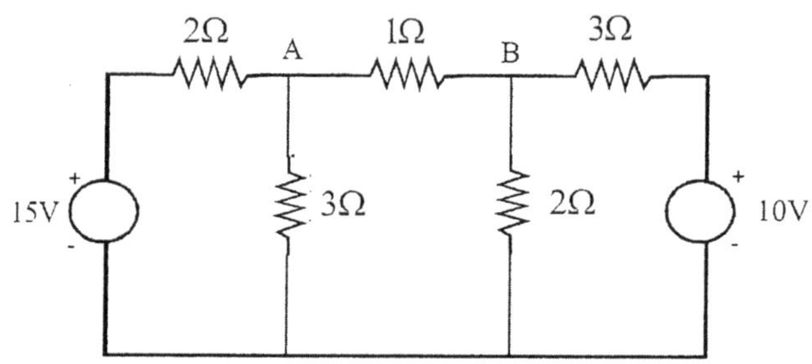

Solution:

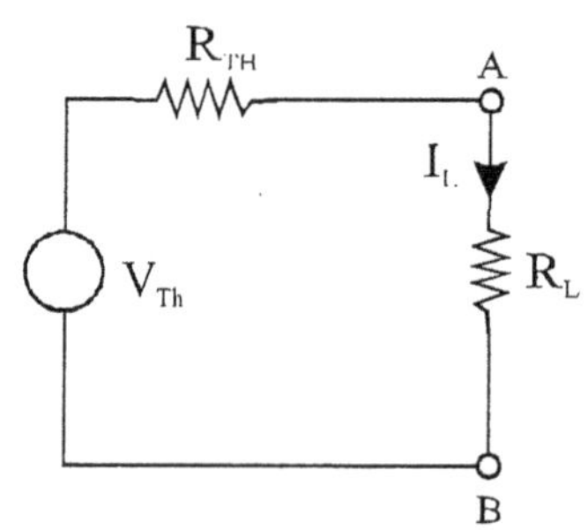

Thevenin's Equivalent Circuit

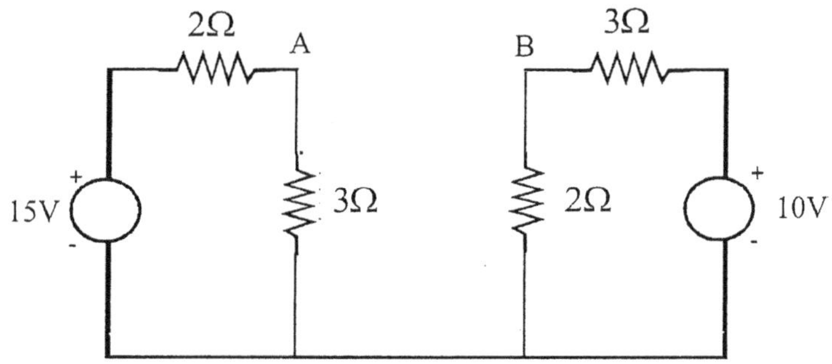

Circuit to get V_{Th}

$$V_{Th} = V_{OC} = V_{AB} = V_A - V_B = 9 - 4 = +\ 5V$$

$$V_A = \frac{15}{(2+3)} \times 3 = +9V \ ; \qquad V_B = \frac{10}{(3+2)} \times 2 = +4V$$

$$V_{Th} = V_{OC} = V_{AB} = V_A - V_B = 9 - 4 = +5V$$

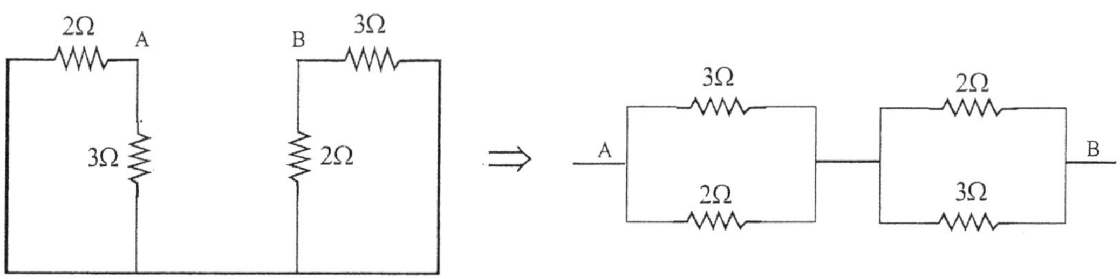

Circuit to get R_{Th}

$$R_{Th} = \frac{3 \times 2}{(3+2)} + \frac{2 \times 3}{2+3} = 1.2 + 1.2 = 2.4\Omega$$

$$I_L = \frac{V_{Th}}{(R_{Th} + R_L)} = \frac{5}{2.4+1} = \frac{5}{3.4} = 1.4706A$$

Example 5.7:

It is required to find current through the 0.1 Ω resistor in the figure, using Thevenin's Method.

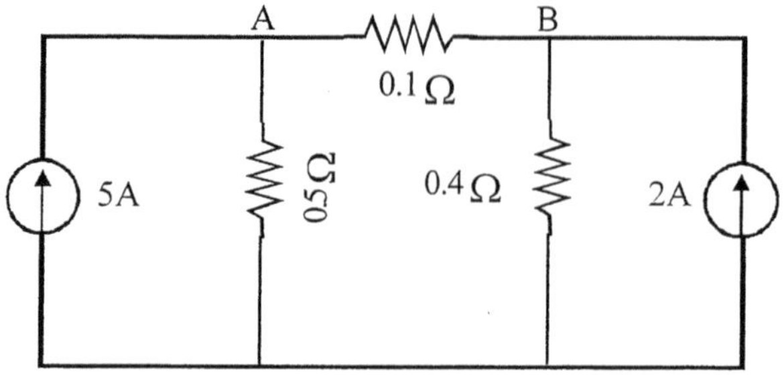

Solution:

Step 1:

The Thevenin's equivalent circuit is,

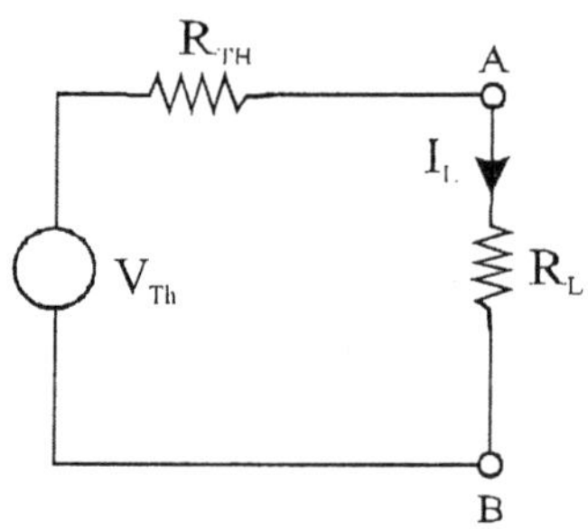

Step 2: **To find V_{TH} :**

From the given circuit, disconnect $R_L = 0.1\ \Omega$

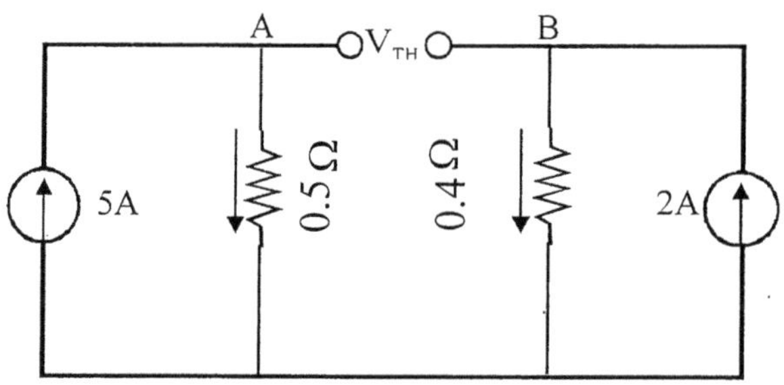

***Circuit to find* V_{TH} :**

$$V_{TH} = V_{AB} = V_A - V_B = 5 \times 0.5 - 2 \times 0.4$$

$$= +1.7\ V$$

B is negative polarity and A is positive polarity.

Step 3: To calculate R_{TH} :

From the given circuit, kill the current. The resultant circuit is,

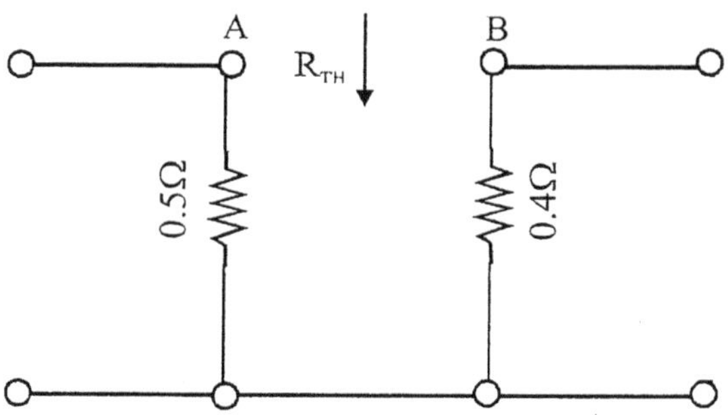

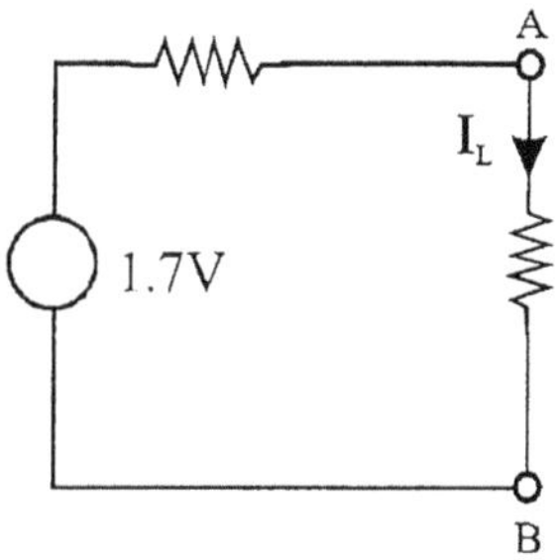

Circuit to find R_{TH}

$$R_{TH} = R_{AB}$$

$$= 0.5 + 0.4 = 0.9\,\Omega$$

Step 4:

$$I_L = \frac{V_{TH}}{R_{TH} + R_L} \qquad = \frac{1.7}{0.9 + 0.1} = 1.7A$$

5.5 NORTON'S THEOREM FOR D.C. EXCITATION

Norton's Theorem is the dual of Thevinin's Theorem and similar to Thevenin's Theorem, it is used in finding out the current through any particular branch of the network instead of solving the entire network. Norton's Theorem is also frequently used in electronic circuits.

Statement:

Any two terminal linear network can be replaced by an equivalent network consisting of a current source I_{SC} in parallel with a resistance R_{SC}. The short circuited resistance R_{SC} is the same as the Thevenin's Resistance as defined earlier and I_{SC} is the short circuit current which flows through the short circuit when these two terminals are short circuited.

Proof: In the Thevenin's equivalent circuit given in Fig. 5.4 the voltage supply $V_{AB} = V_{TH} - I\,R_{TH}$ can be replaced with a current source of $I = I_{SC}$ in parallel with a resistance equal to the Thevenin's resistance. This current I_{SC} is obtained by short circuiting the terminals A and B i.e., by making $V_{AB} = 0$. This leads to the development of Norton's Theorem.

The circuit of fig. 5.5 can be represented as circuit given in fig.5.4 replaced by using Norton's theorem which is a dual of Thevenin's theorem.

In determining the load current I_L, use is made of the formula for current division in any parallel branch knowing the total current for the parallel combination.

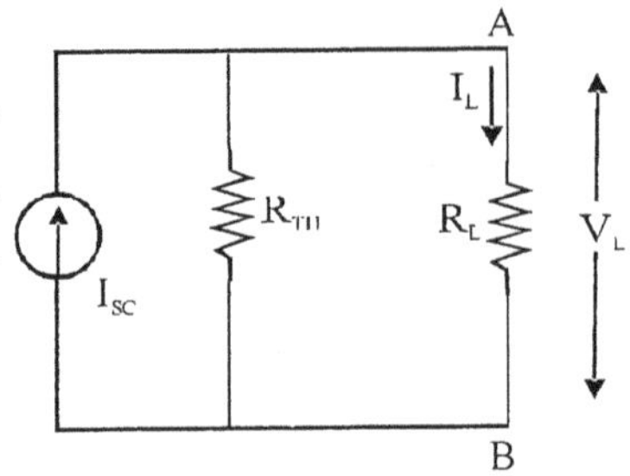

Fig : 5.5 *Norton's equivalent circuit*

$$I_L = \frac{I_{SC} R_{TH}}{R_{TH} + R_L}$$

Example 5.8:

Determine the voltage across 200 Ω resistor in circuit by Norton's Theorem.

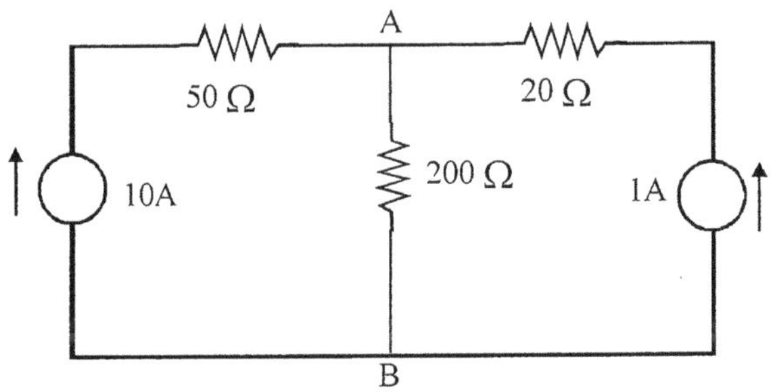

Solution:

Step 1:

To find the short circuit current I_{SC}, replace the 200 Ω by short-circuit. The current through short circuited A, B is I_{SC}.

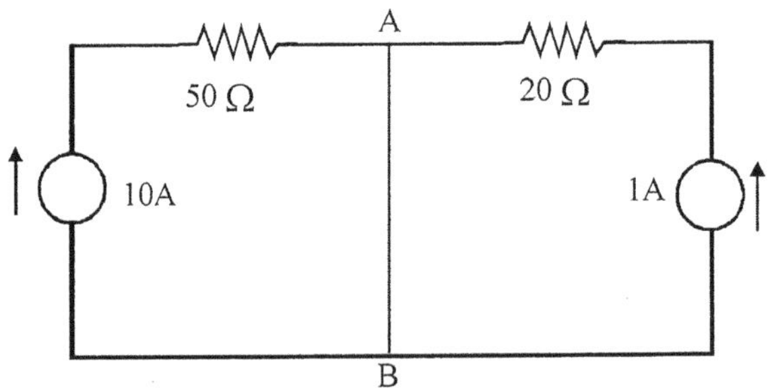

The modified circuit is shown below.

The voltage source will drive a current of 10/50 = 0.2A, whcih flows through short-circuited A, B only.

Similarly, the current of 1A flows through 20 and then through short-circuited A, B only. 0.2A will not flow through 20 Ω and IA will not flow through 50 Ω. It is because of short-circuit. So,

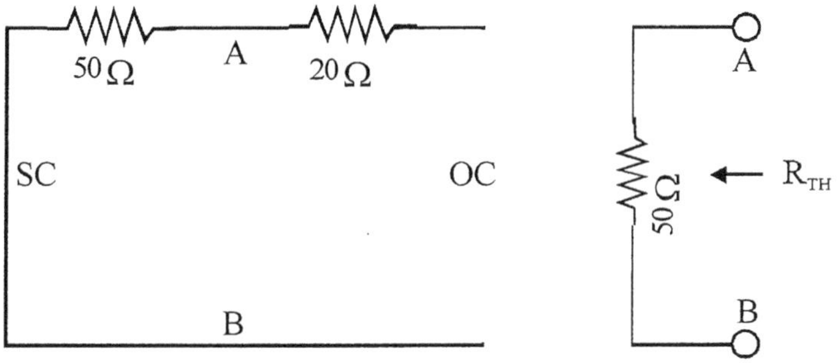

$$R_{TH} = 50\,\Omega$$

$$I_{SC} = 0.2 + 1 = 1.2A$$

Step 2: To find R_{TH}:

From the given circuit, disconnect $R_L = 200\,\Omega$ between A and B and also kill the sources. The resultant circuit becomes as follows.

Step 3:

Drawing Norton's equivalent circuit and showing the load resistance R_L we get the following circuit.

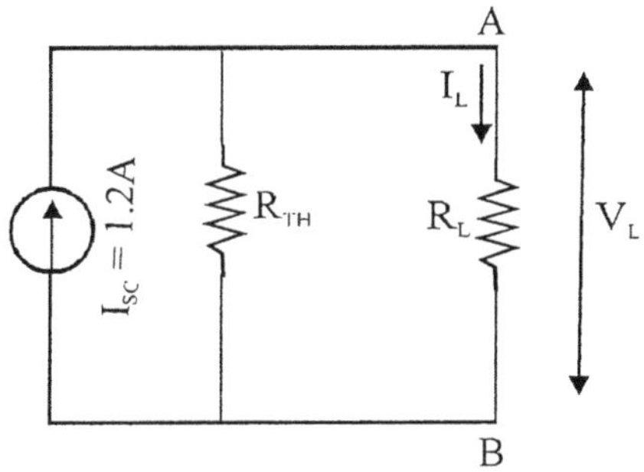

$$I_L = \frac{I_{SC} R_{TH}}{R_{TH} + R_L}$$

$$= \frac{1.2x50}{50+200} = 0.24A$$

$$V_1 = I_L R_L = 0.24 \text{ x } 200 = 48V$$

Example 5.9:

Find the current through the 2 Ohm resistance across AB in the given circuit using Norton's Theorem.

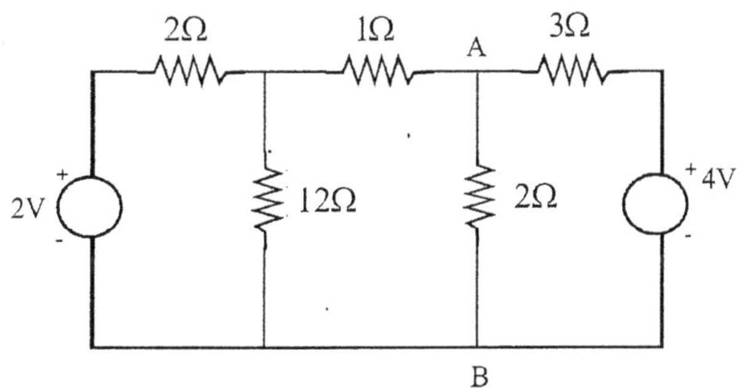

Norton's equivalent current source will be the current through the short circuited branch AB.

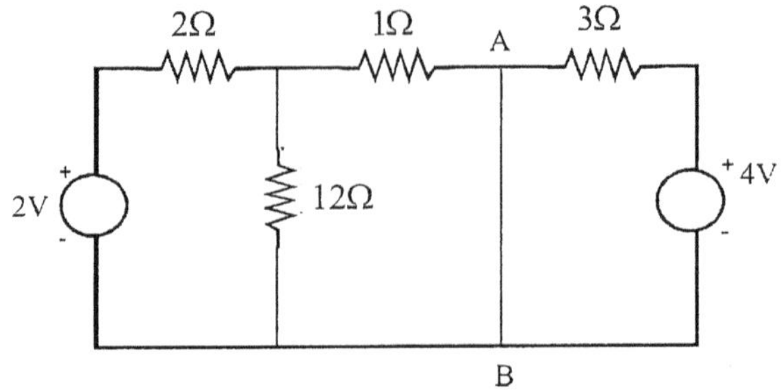

Using superposition principle to find current through branch AB we first short circuit 2V battery, the current through AB due to 4 volt will be $\frac{4}{3}A$. When 4 volt battery is shorted, the current supplied by 2V battery is $\frac{2}{2+\frac{12}{13}}=\frac{26}{38}$

Therefore current through AB will be $\frac{26}{38}\times\frac{12}{13}=\frac{12}{19}A$ so total current is

$$\frac{4}{3}+\frac{12}{19}=\frac{112}{57}A$$

So that Norton's equivalent current source is 112/57 A. The Norton's equivalent resistance is same as in case of Thevenins equivalent resistance except that it is to be connected across the current source. Along with this the resistance through which current is required should be connected across the current source. This forms the complete equivalent Norton's circuit and is drawn here.

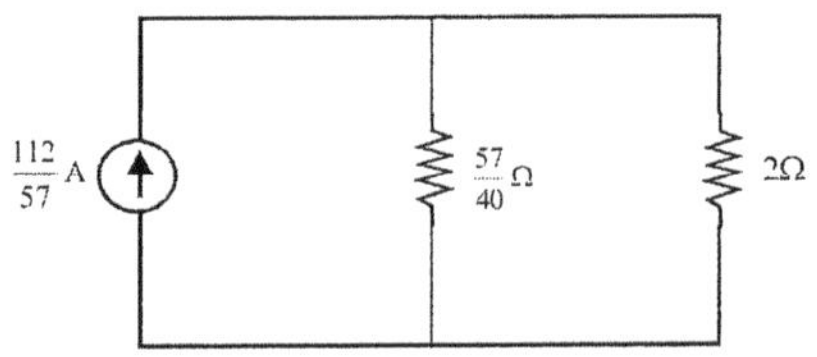

Current through 2 Ω resistor is

$$\frac{112}{57}\times\frac{57}{40}\times\frac{1}{\frac{57}{40}+2}=\frac{112}{137}=0.82A$$

5.6 MAXIMUM POWER TRANSFER THEOREM FOR D.C. EXCITATION

In electronic circuits, it is desirable to transfer maximum power through a source to a load or from one stage to another stage in the case of multi-stage amplifiers because the currents and also power in the circuits will be very small of the order of milli-Amperes (mA) and milli-watts (mW). This is what is done by matching the load impedance wih the source impedance. But in the case of Electrical Power circuits, maximum power transfer is not tried because the power is in the order of Mega Watts (MW).

Statement :

In an electric circui maximum power is transferred to the load when the Load Resistance is equal to the Thevenin's Resistance in the case of D.C. circuits.. In the case of A.C. circuits maximum power is transferred to the load when the load impedance is equal to the complex conjugate of the Thevenin's Impedance. The maximum power that can be transferred to the load will be 50% of the Power supplied to the circuit.

An ideal voltage source V_s will give the full voltage to a load connected to it. However, if the source has an internal resistance R_s then the voltage across the load R_L falls short of V_S. As it is desirable to get as much power from V_s as possible the problem that arises is one of optimising the load for maximum transfer of power. Depending on the parameters of the load the relation between R_s and R_L for maximising the power to the load will differ. There are three situations which are of significance.

When the internal resistance R_s is purely resistive then R_L should be $R_L = R_{TH}$ since any circuit upto the load can be represented by Thevenin's voltage V_{Th} in series with Thevenin's resistance R_{Th}. This V_{Th} is similar to V_s and R_{Th} is similar to R_s.

For the Thevenin's equivalent Circuit $I = \left(\frac{V_{Th}}{R_{Th} + R_L}\right)$

and power in the load

$$P_L = I^2 R_L = \left(\frac{V_{Th}}{R_{Th} + R_L}\right)^2 R_L$$

For maximising P_L obtain $\frac{dP_L}{dR_L}$ and equate it to zero.

$$\frac{dP_L}{dR_L} = \frac{d}{dR_L}\left(\frac{V_{Th}}{R_{Th} + R_L}\right)^2 R_L = 0$$

$$= V_{Th}^2 \left[\frac{1}{(R_{Th} + R_L)^2} - \frac{2R_L}{(R_{Th} + R_L)^3} \right] = 0$$

$$= V_{Th}^2 \left[\frac{R_{Th} + R_L - 2R_L}{(R_{Th} + R_L)^3} \right] = 0$$

Hence equating the numerator to zero we get

$$V_{Th}^2 [(R_{Th} + R_L - 2R_L] = 0$$

$$= V_{Th}^2 [(R_{Th} - R_L] = 0$$

Hence for maximum power to be transferred to the load R_L we have the condition that

$$R_L = R_{Th}$$

The load resistance should be equal to the Thevenin's Resistance for maximum power transfer to the load in the case of D.C. circuit.

In the case of A.C. circuits for maximum to be transferred to the load impedance Z_L we have the condition that the load impedance should be equal to the complex conjugate of the Thevenin's Impedance.

$$Z_L = Z_{Th}^*$$

Example 5.10:

The circuit shown in the figure R, absorbs maximum power. Compute the value of R and maximum power.

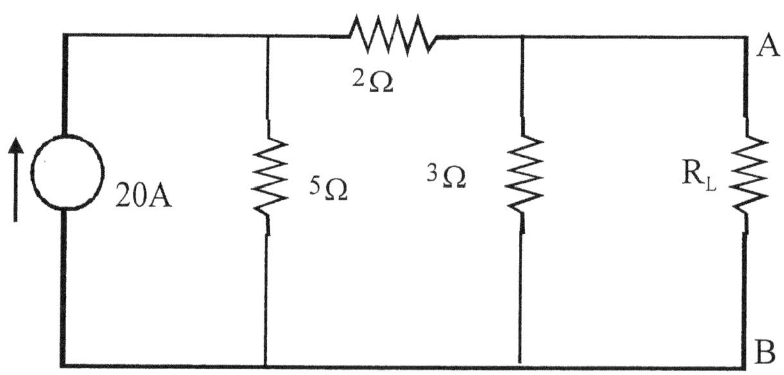

Solution:

For maximum power load to the load R_L the condition is R_L R_{Th} and the circuit will be as given below.

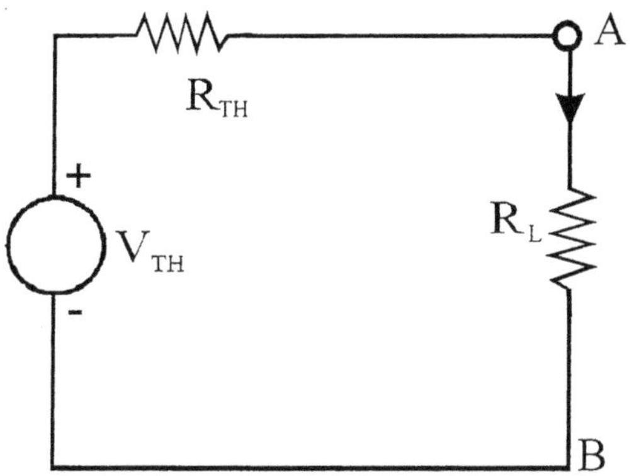

To find V_{TH} :

Disconnect R from the original circuit

By parallel current division theorem the current through 3 Ohms resistance is

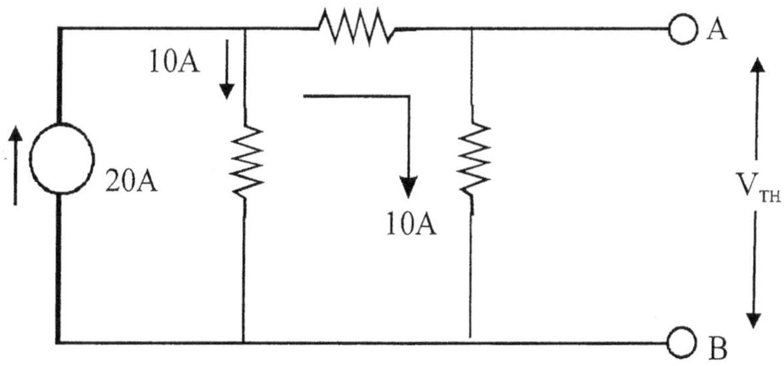

$$I_2 = I \times \frac{5/2}{(2+3)} = 20 \times \frac{1}{2} = 10A$$

$$V_{TH} = V_{AB}$$

$= 10 \times 3 = 30V$

$= 30V$

B is negative polarity and A is positive polarity.

To find R_{TH} :

From the above circuit, open circuit the current source. Thus,

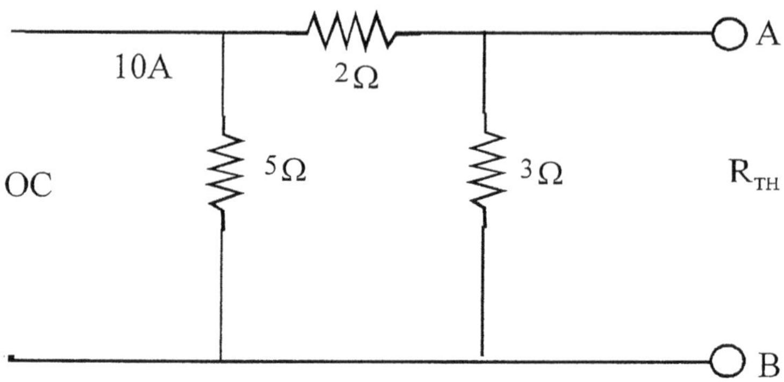

$$R_{TH} = \frac{7X3}{7+3} = 2.1\Omega$$

According to the statement of Maximum Power Transfer Theorem, the value of R for maximum power transfer to it.

$$R_L = R_{Th} = 2.1\Omega$$

and the maximum power transfered

$$P_{max} = I^2R_L = \left(\frac{V_{Th}}{2R_L}\right)^2 R_L$$

$$= \frac{V_{TH}^2}{4R_L} = \frac{V_{TH}^2}{4R_{TH}} = \frac{30^2}{4 \times 2.1}$$

$$= 107W$$

5.7 SUPERPOSITION THEOREM FOR A.C. EXCITATION

This theorem is applicabe to linear networks where the response is proportional to the excitation.

Statement :

In a linear network if more than one number of excitations are acting, the response in a certain branch can be obtained by superimposing algebrecially the responses due to various sources taking one excitation at a time considering the remaining voltage sources, short circuited if they are ideal sources or replaced by their equivalent internal impedances and current sources are open circuited if they are ideal sources or replaced by their equivalent admittances.

Example 5.11:

Using Superposition Theorem, find the current in branch AB of the given network.

Solution:

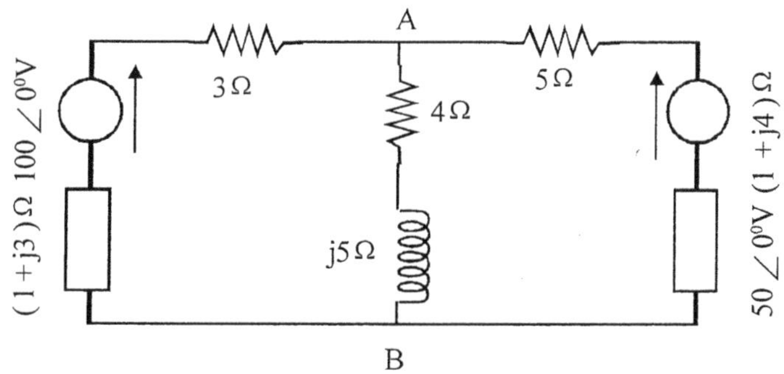

The given network can be modified by drawing series impedances and the modified network is shown below.

(i) Let the 100 $\angle 0^0$ source be set equal to ZERO.

The circuit reduces as shown in the figure below.

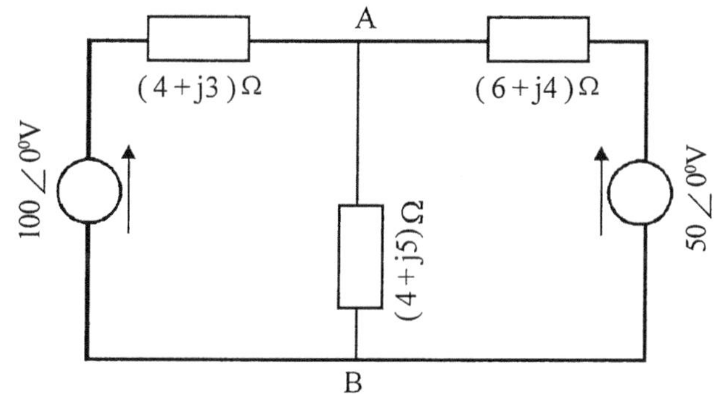

Let x be the current through the (4 + j5) impedance.

Equivalent impedance of the parallel circuit.

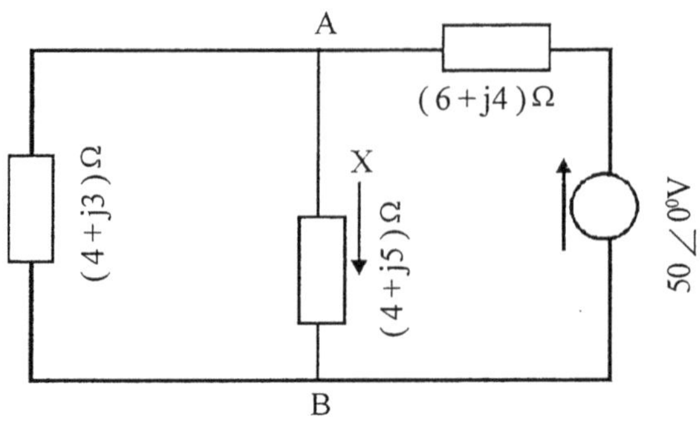

$$= \frac{(4+j3)(4+j5)}{(4+j3)+(4+j5)} = \frac{(5\angle 36.87^0(6.403\angle 51.34^0)}{8+j8\ (or 11.314\angle 45^0)}$$

$$2.83\angle 43.21^0\Omega\ = (2.063 + j1.938)\Omega$$

This comes in series with the (6 + j4) Ω impedance.

Total Impedance = (2.063 + j1.938) + (6+j4)

= (8.063 + j5.938) Ω

$$\text{Total Current} = \frac{50\angle 0^0}{8.063 + j5.938}$$

$$= \frac{50\angle 0^0}{10.0136\angle 36.37^0}$$

$$\text{Current (x)} = \frac{(4.9932\angle -36.37^0)(2.83\angle 43.21^0)}{(4+j5)(\,or\ 6.403\angle 51.34^0)}$$

x = 2.2069 ∠ -44^{0}A. from A to B

(ii) Let the 50 $\angle 0^0$ source be set equal to ZERO.

The circuit reduces as shown in the figure below.

Let y be the current in the branch AB.

The (6 + j4) Ω and (4 + j5) Ω impedances now come in parallel.

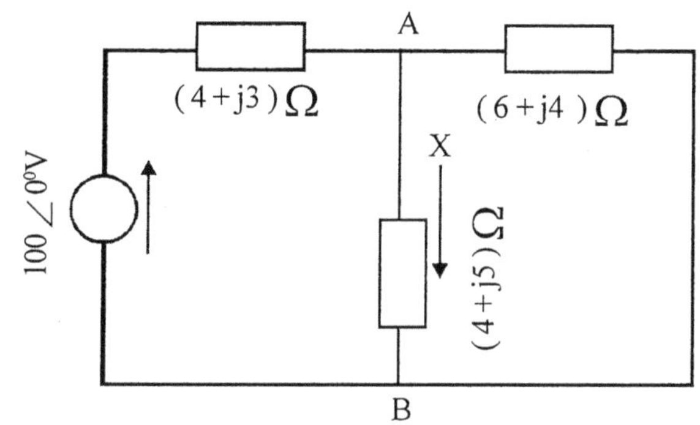

$$\text{Their equivalent impedance} = \frac{(6+j4)(4+j5)}{(6+j4)+(4+j5)}$$

$$= \frac{(7.21\angle 33.69^0)(6.403\angle 51.34^o)}{(10+j9)(or13.45\angle 41.99^0)}$$

$= 3.432 \angle 43.04^0\ \Omega$

$= (2.5084 + j2.3424)\ \Omega$

This is in series with the (4 + j3) impedance.

Total Impedance $= (2.5084 + j2.3424) + (4 + j3)$

$= (6.5084 + j5.3424)\ \Omega$

$$\text{Total Current} = \frac{110\angle 0^0}{8.42\angle 39.381^0\Omega} = 11.8765\angle -39.381^0 A$$

$$\text{Current (y)} = \frac{(11.8765\angle -39.381^0)(3.432\angle 43.04^0)}{(4+j5)(or6.403\angle 51.34^0)}$$

or y = 6.3658 $\angle -47.681^0$ A from A to B

It is observed both currents x and y flow in the same direction. Hence the total current through AB due to both voltage sources acting together = x + y.

$$I_{AB} = x + y = (2.2069 \angle -44.5^0) + (6.3658 \angle -47.681^0)$$

$$= 8.6164 \angle 47.15^0 \text{ A}$$

5.8 THEVENIN'S THEOREM FOR A.C. EXCITATION

Statement :

A two terminal linear network N consisting of independent and / or dependent voltage or current sources and impedances can be replaced with an equivalent circuit consisting of a voltage source V_{Th} in series with an impedance $Z_{Th.}$

*The value of V_{Th} is the open circuit voltage between the terminals of the network and Z_{th} is the impedance measured between the terminals of the network with all enregy sources eliminated (but not their impedances). This is also called the **Voltage Source Equivalent Circuit or Thevinin's Equivalent Circuit.***

Proof as in section 5.4 with R_{Th} and R_L replaced by Z_{Th} and $Z_{L.}$

Example 5.12:

Using Thevenin's theorem, find the current through impedance Z_L in the circuit shown in the figure given below, when $Z_L = (10 - j7.5)\,\Omega$.

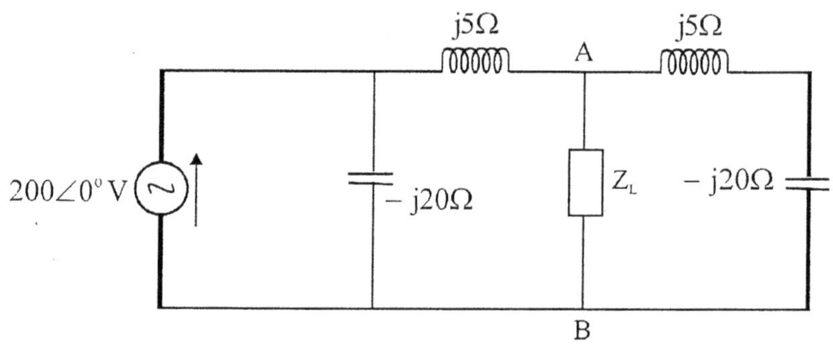

Solution:

Let Z_L be removed. The circuit reduces as shown in figure below Let i_1 and i_2 denote the mesh currents.

Applying KVL, we get :

$$-200 - j20\,(i_1 - i_2) = 0$$

or $$-j20 i_1 + j20 i_2 = 200 \qquad .....(1)$$

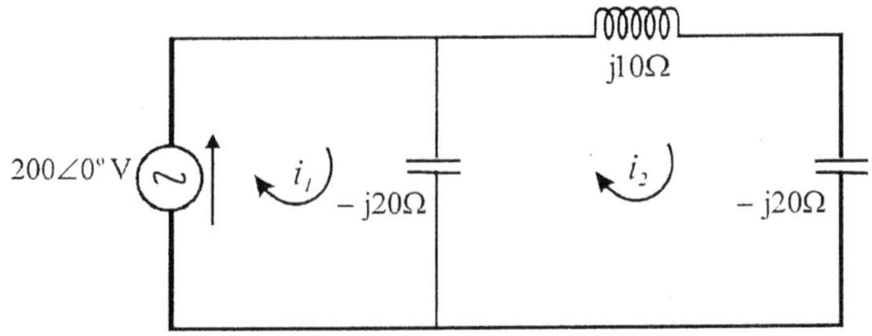

Also, $j20\,(i_1 - i_2) + j10i_2 - j20i_2 = 0$

or $j20i_1 - j30i_2 = 0$ .....(2)

Solving (1) and (2) we get,

$i_1 = j30A$ and $i_2 = j20A$

Referring to the original figure we have,

Potential of A w.r.t B, with Z_L removed $= 200\angle0° - j5i_2$

$= 200 - j5\,(j20)$

$= 300\ \angle0°\,V$

$\therefore V_{OC} = 300\angle0°\,V$

To find Z_{Th} :

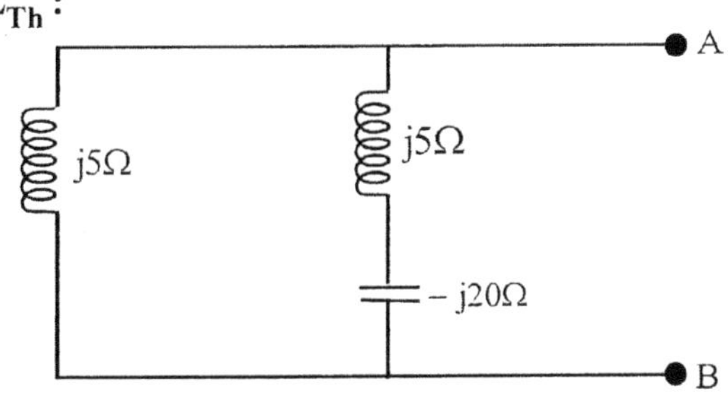

Replacing the $200\angle0°\,V$ source by a short-circuit, the circuit reduces as shown in the figure to find the Z_{Th}.

Equivalent impedance of the circuit

$$= \frac{j5(j5 - j20)}{j5 + j5 - j20}$$

$\therefore Z_{TH} = j7.5\Omega$

The Thevenin equivalent circuit is given below.

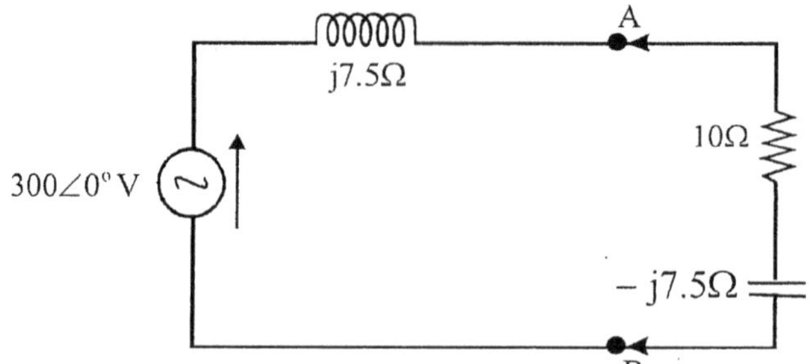

If the load impedance $Z_L = (10 - j7.5)\,\Omega$ is connected across AB, we get:
Total impedance of the circuit.

$$Z_t = j7.5 + (10 - j7.5) = 10\,\Omega$$

$$\therefore \quad \text{Current through } Z_L = \frac{300\angle 0^\circ}{10}$$

or $\quad I = 30\angle 0^\circ\ \text{A}$

5.9 NORTON'S THEOREM FOR A.C. EXCITATION

Statement :

Any two terminal linear network can be replaced by an equivalent network consisting of a current source I_{SC} in parallel with a network of impedance Z_{SC}. The impedance is the same as the Thevenin's Impedance as defined earlier and I_{SC} is the current between the two terminals when these therminals are short circuited.

Proof as in Norton's theorem for D.C. Excitation with R_{Th} and R_L replaced with Z_{Th} and Z_L.

Example 5.13:

Find the current I in the circuit given below by using Norton's Theorem.

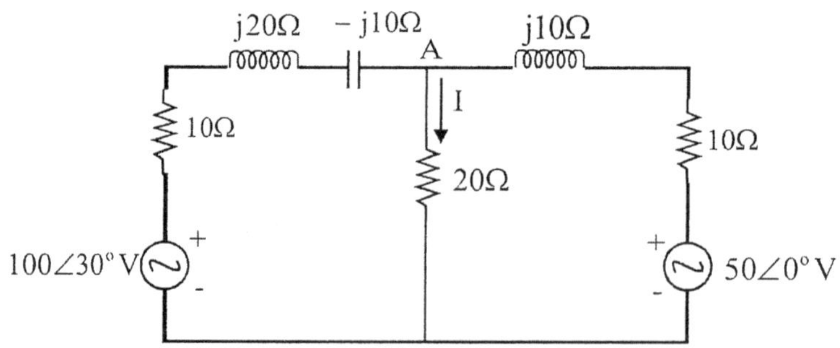

Solution:

Let the 20 Ω resistor be disconnected.. The circuit reduces as shown in the figure, if the two voltage sources are replaced by short-circuits.

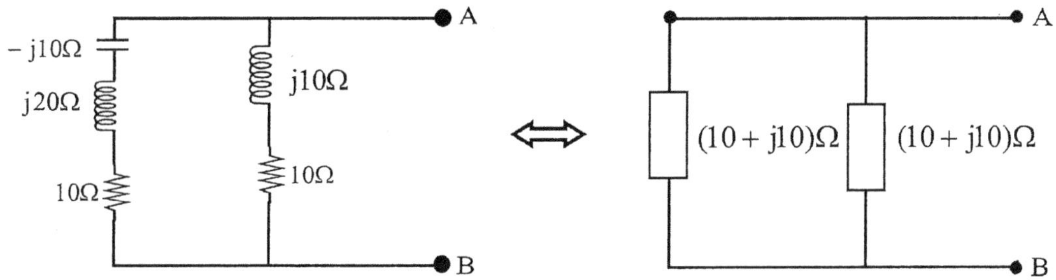

we have, equivalent impedance = (1/2) (10 + j10) Ω

$$\text{i.e.} Z_{Th} = (5 + j5)\,\Omega$$

$$\text{or } Z_{Th} = 7.071\angle 45°\,\Omega$$

To find I_{SC}

Let terminals A and B be shorted, as shown in the figure given below.

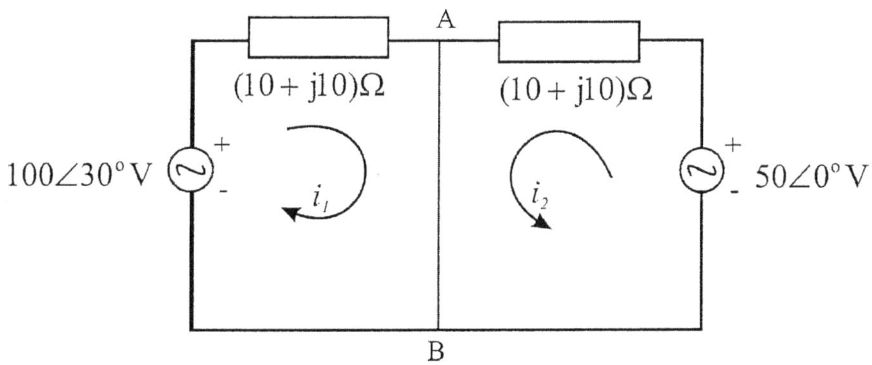

Let x and y be the loop currents, We have :

$$i_1 = \frac{100\angle 30°}{10 + j10}, \text{ and}$$

$$i_2 = \frac{50\angle 0°}{10 + j10}$$

$$\therefore I_{SC} = i_1 + i_2 = \frac{100\angle 30^\circ + 50\angle 0^\circ}{10 + j10}$$

$$= \frac{(86.6 + j50) + (50 + j0)}{10 + j10}$$

$$= \frac{136.6 + j50}{10 + j10} = \frac{145.563\angle 20.104^\circ}{14.142\angle 45^\circ}$$

or $I_{SC} = 10.286\angle -24.896^\circ$ A

The Norton equivalent cirucit is as shown.

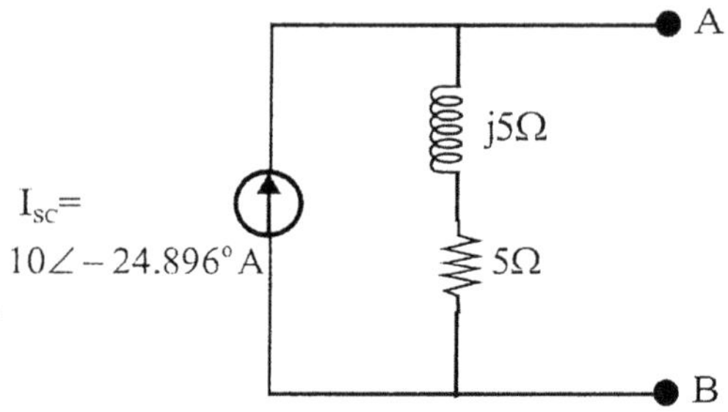

To find current I :

With the 20 Ω resistor connected across AB-

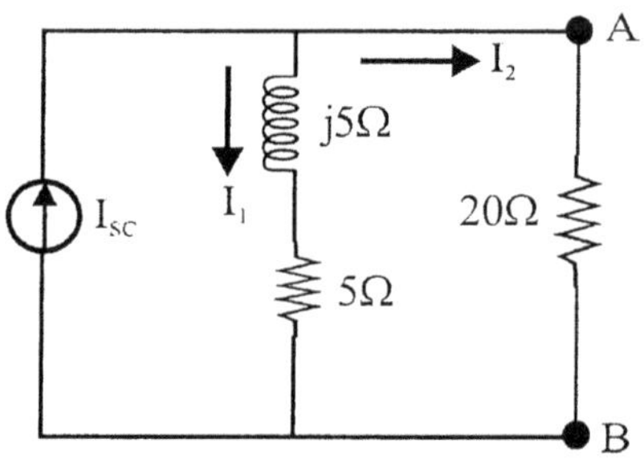

By applying parallel current distribution theorem, we have

$$I_2 = I_{SC} = \frac{Z_1}{(Z_1 + Z_2)}$$

$$I_2 = 10[\text{Cos}(-24.896^\circ) + j\text{Sin}(-24.896^\circ)] \times \frac{(5 + j15)}{(5 + j15 + 20)}$$

$$\therefore I = 2.8528\angle -5.391^\circ \text{A}$$

5.10 MAXIMUM POWER TRANSFER THEOREM FOR A.C. EXCITATION

Statement:

In an electric circui maximum power is transferred to the load when the Load Resistance is equal to the Thevenin's Resistance in the case of D.C. circuits and in the case of A.C. circuits maximum power is transferred to the load when the load impedance is equal to the complex conjugate of the Thevenin's Impedance. The maximum power that can be transferred to the load will be 50% of the Power supplied to the circuit. If the load is to be a pure resistance R_L *(or* $R_L + j0$*) then* $R_L = |Z_{Th}|$ *for maximum power to be transferred to the load.*

Example 5.14:

In the network shown figure, the load consists of a fixed capacitive reactance of 15 Ω and a variable resistance R_L connected across terminals A and B. Determine the value of R_L for which the power transformed is maximum. Also find the maximum power.

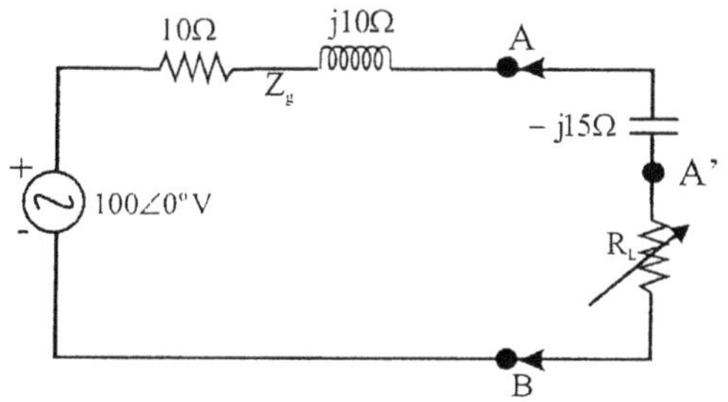

Solution:

For a load impedance of fixed reactance and variable resistance, the condition for maximum power transfer is -

$$R_L = |Z_g + jX_L|, \text{ where}$$

Z_g = active network impedance upto the load terminals and

X_L = the fixed load reactance.

Given $Z_g = (10+j10)\,\Omega$ and $X_L = -j15\,\Omega$

$$R_L = |(10+j10) - j15| = |10 - j5| = \sqrt{10^2 + 5^2} = 11.18\Omega$$

For this value of R_L, we have :

Total circuit impedance $= (10 + j10) + (11.18 - j15)$

$= (21.18 - j5)\,\Omega$

$= 21.762\ \angle -13.283°\,\Omega$

$$\text{Circuit current, } I = \frac{100\angle 0^\circ}{21.762\angle -13.283^\circ}$$

$$= 4.5952 \angle 13.283^\circ \text{ A}$$

$$\text{Maximum power delivered to the load} = RR_L$$

$$= 4.5952^2 \text{ x } 11.18.$$

$$\text{or } P_{max} = 236 \text{ Watts.}$$

Note :

(i) If the load is an impedance load it should be equal to the complex conjugate of the Thevenin's impedance for maximum power transfer to the load.

(ii) If the load is pure resistance then $R_L = |Z_{Th}|$ for maximum power transfer to the load.

(iii) In this problem it can also be solved alternatively for max. power transfer to the load as follows :

Since the load consists of a fixed reactance and variable resistance, the problem is treated as having only a variable resistance load and the fixed reactance of the load is taken into the remaining circuit as part of the Thevenin's impedance calculation circuit. And the open circuit voltage across the load terminals A and B when the load is removed (V_{OC}) is replaced with V_{OC} across A' and B (across the resistance load terminals only when the resistance load is removed)

Example 5.15:

In the network shown in the figure given below two voltage sources act on the load impedance connected to the terminals A, B. If this load is variable in both reactance and resistance, what load Z_L will receive maximum power? What is the value of the maximum power?

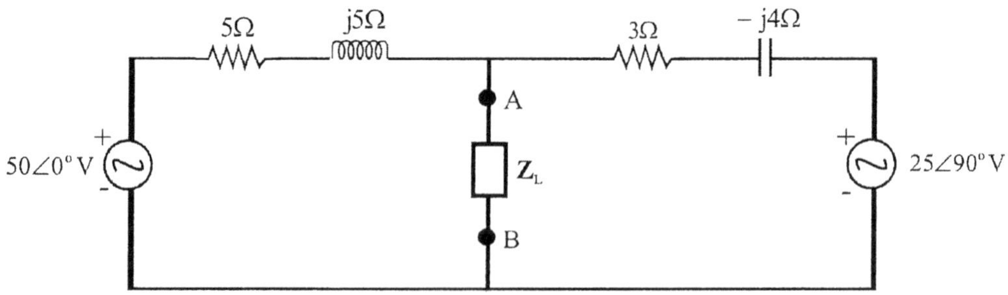

Solution:

Since the load is variable in both resistance and reactance, the condition to be satisfied for maximum power transfer is that the load impedance must be the complex conjugate of the active network impedance (upto the load terminals).

if $Z_L = R_L + jX_L$ then for maximum power transfer to Z_L

$R_L = R_{Th}$ and $jX_L = -jX_{Th}$ if $Z_{Th} = R_{Th} + JX_{Th}$

To find R_{Th} and X_{Th} :

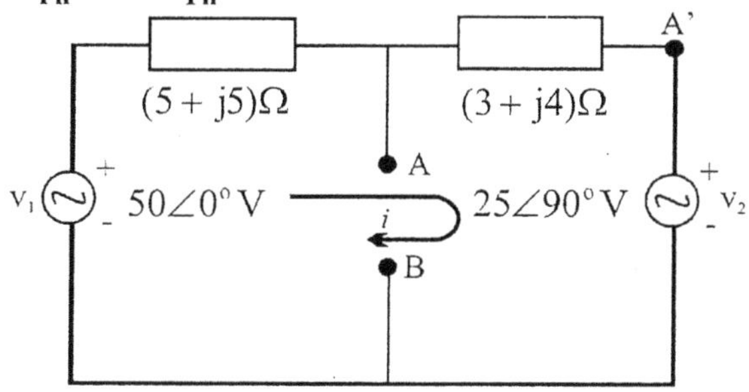

Let the load Z_L acrosss AB be removed

∴ Open circuit voltage of the network at the terminals A, B is

$V_{OC} = V_{AB} = V_{AA'} + V_{A'B} = V_{AA'} + V_2$

Using potential divider circuit formula for $V_{AA'}$

$$V_{AB} = \frac{50\angle 0^\circ - 25\angle 90^\circ}{(5+j5)+(3-j4)}(3-j4) + 25\angle 90^\circ$$

$$= 9.81\angle 78.71^\circ \text{ V}$$

To infd Z_{AB} ($= Z_{Th}$) :

Let the voltage sources be replaced by short-circuits. The circuit reduces as shown.

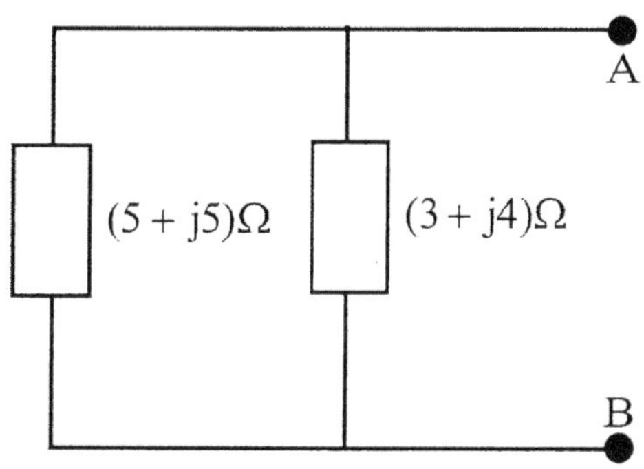

$$Z_{Th} = \frac{(5+j5)(3-j4)}{(5+j5)+(3-j4)}$$

$$= \frac{(7.071\angle 45^\circ)(5\angle -53.13^\circ)}{(8+j1)\,\text{or}\,8.062\angle 7.125^\circ}$$

$$= 4.385\angle -15.255^\circ\,\Omega$$

or $\quad Z_{Th} = (4.23 - j1.154)\ \Omega$

Hence for maximum power transfer $Z_L = Z_{Th}^*$

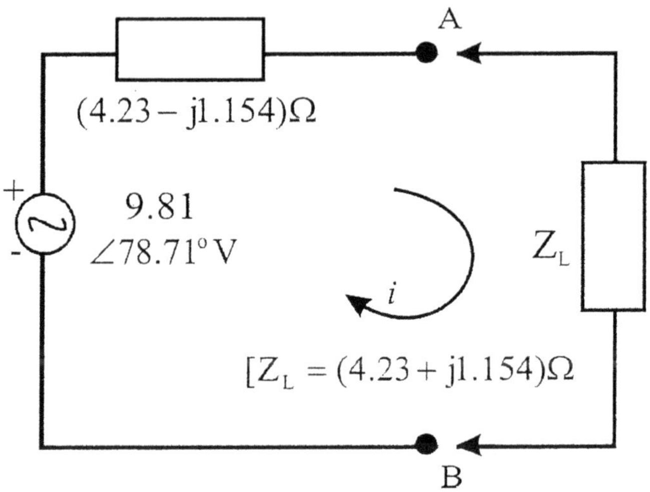

$\therefore R_L = 4.23\ \Omega$ and $X_L = 1.154\ \Omega$

The Thevenin equivalent circuit is as shown in the figure.

With Z_L connected across AB, we have, total impedance of the circuit.

$$= Z_{Th} + Z_L = (4.23 - j1.154) + (4.23 + j1.154)$$

$$= 8.46\ \Omega$$

$$\text{Current } I = \frac{9.81\angle 78.71^\circ}{8.46} A$$

$$= 1.1596\angle 78.71^\circ A$$

$\therefore$ Maximum power delivered to the load is obtained as :

$$P_{max} = I^2 R_L = 1.1596^2 \times 4.23 = 5.688 \text{ Watts.}$$

5.11 RECIPROCITY THEOREM FOR A.C AND D.C. EXCITATIONS

Reciprocity is an important property of linear, passive, time invariant, bilateral networks which implies that the input and output can be interchanged without altering the response of the network to a certain input. Then the network is also said to be bilateral. It is applicable to resistors and capacitors, inductors with and without coupling and transformers. However, both dependent and independent sources are not permissible. Also, we are considering only the zero state response by taking all initial conditions to be zero.

Statement:

If any source of E.M.F.E, located at one point in a network consisting of linear, passive, time invariant, bilateral elements, produces a current I at a second point in the network, the same source of E.M.F.E acting at the second point of the same network will produce the same current I at the first point.

Proof:

Let us consider a network N having only one driving voltage source $E = E_k$ in loop k and the current response in loop m. In the loop analysis we have,

$$[E_{loop}] = [Z_{loop}]\,[I_{loop}]$$

$$[I_{loop}] = [Z_{loop}]^{-1}\,[E_{loop}]$$

$$[I_{loop}] = [Y_{loop}]\,[E_{loop}]$$

then, for loop m,

$$I_m = Y_{km}\,E_k \qquad .....(5.11.1)$$

Next, interchanging the positions of cause and effect, i.e., placing the same voltage source $E = E_m$ in loop m, we get the current response in loop k as

$$I_k = Y_{km}\,E_m \qquad .....(5.11.2)$$

Then, from Eq. (5.11.1) and (5.11.2), I_k will be equal to I_m, provided

$$Y_{km} = Y_{mk} \qquad .....(5.11.3)$$

Then is the condition for reciprocity, $Y_{km} = Y_{mk}$ for all m and k signifies that the admittance matrix Y is symmetric.

Note: Reciprocity Theorem is applicable to D.C. circuits also except that the impedances are replaced by resistance and the A.C. quantities are replaced by corresponding D.C. quantities. Admittance Y is replaced with conductance G in D.C. excitations.

Example 5.16:

Verify the reciprocity theorem in a circuit shown in the figure. Also calculate the transfer resistance.

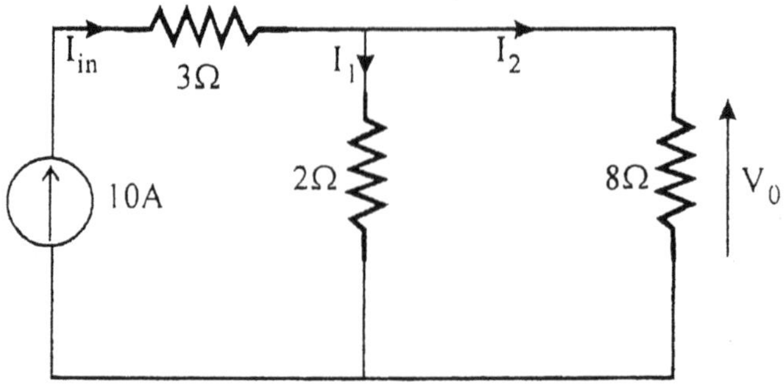

Solution:

For the given network, the output voltage,

$$V_0 = (I_2 \times 8)\text{ V}$$

By Current Division in Partial Circuits,

$$I_2 = \frac{10 \times 2}{2+8} = 2\text{ A}$$

$$V_0 = 16\text{ V}$$

According to Reciprocity Theorem, when the source and the response are interchanged,

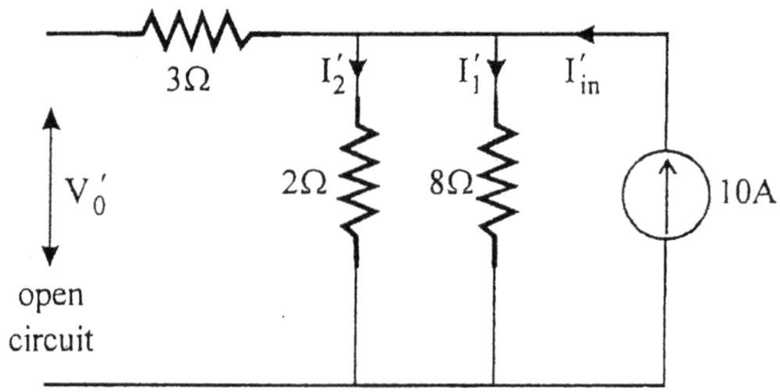

we have,

$$V_0' = I_2' \times 2 = \left(\frac{10 \times 8}{2+8}\right) \times 2 = 16\text{ V}$$

Hence, the Reciprocity Theorem is proved.

$$\text{Transfer Impedance} = \frac{\textit{Output voltage}}{\textit{Input current}} = \frac{16}{10} = 1.6\ \Omega$$

Example 5.17:

For the circuit shown in the figure, find I_3 and verify the Reciprocity Theorem.

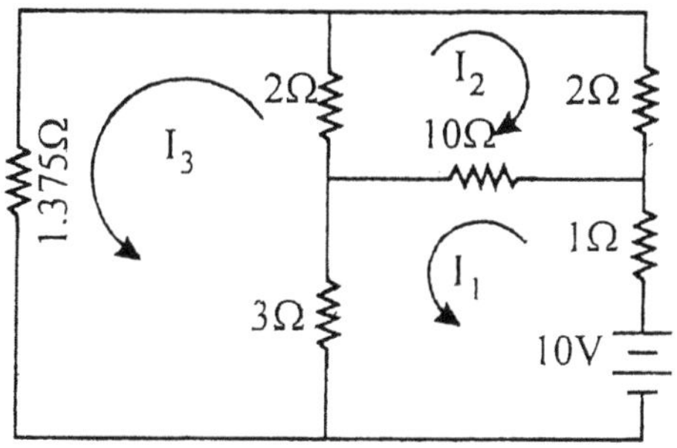

Solution:

For the given network, it can be found by loop analysis,

$$I_3 = 2A.$$

To prove the Reciprocity Theorem, interchanging the source and the response

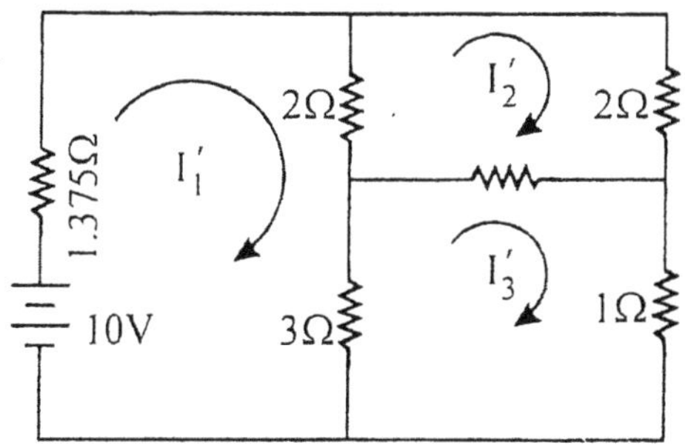

we have,

By Loop Analysis,

$$I_3' = 2A$$

Hence, the Reciprocity Theorem is proved.

Example 5.18:

The ladder network is shown in the figure. Verify the Reciprocity Theorem for the circuit.

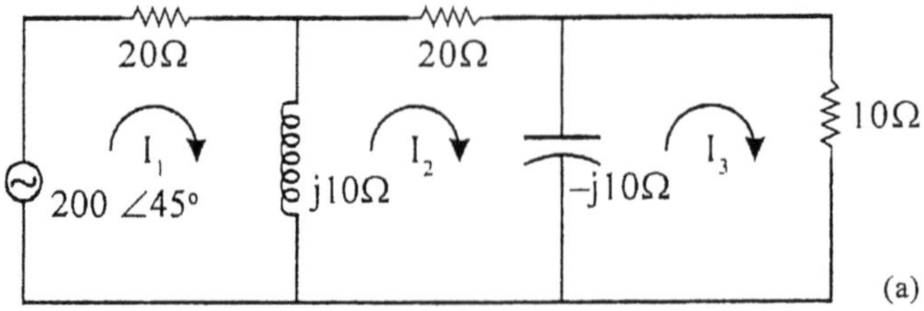

(a)

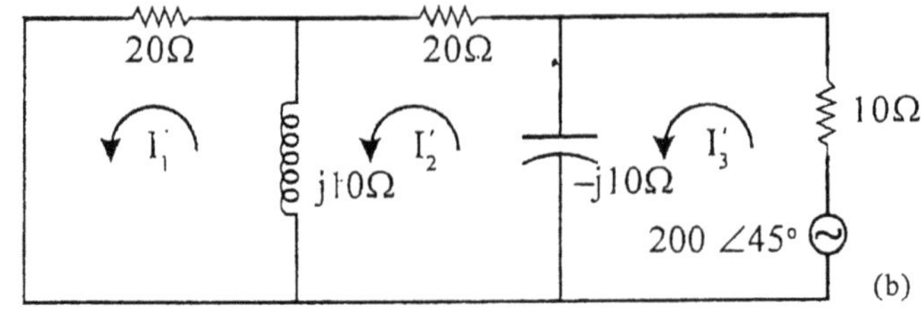

(b)

Solution:

Let us use mesh analysis with the currents shown in the figure (a). We have to first determine I_3.

$$I_3 = \frac{\begin{vmatrix} 20+j10 & -j10 & 200\angle 45^\circ \\ -j10 & 20 & 0 \\ 0 & j10 & 0 \end{vmatrix}}{\begin{vmatrix} 20+j10 & -j10 & 0 \\ -j10 & 20 & j10 \\ 0 & j10 & 10-j10 \end{vmatrix}}$$

$$= 2.16\ \angle 57.5^\circ \text{A}$$

For verification of the reciprocity theorem, we change the points of excitation and responses, as shown in figure (b).

Once again, using mesh analysis with mesh currents shown in the figure the response is,

$$I_1' = \frac{\begin{vmatrix} 0 & -j10 & 0 \\ 0 & 20 & j10 \\ 200\angle 45^\circ & j10 & 10-j10 \end{vmatrix}}{\begin{vmatrix} 20+j10 & -j10 & 0 \\ -j10 & 20 & j10 \\ 0 & j10 & 10-j10 \end{vmatrix}}$$

$$= 2.16\ \angle 57.5^\circ \text{A}$$

As $I_3 = I'$, the network is reciprocal.

5.12 TELLEGEN'S THEOREM FOR D.C AND A.C. EXCITATIONS

Tellegen's theorem is remarkable and one of the most general theorems of circuit theory. It is applicable for any lumped network having elements which are linear or non-linear, active or passive, time-varying or time-invariant. The theorem is based on the two Kirchoff's Laws. It is completely independent of the nature of elements and is only concerned with the graph of the network.

Consider an arbitrary lumped network whose graph G has b branches and n nodes. Suppose, to each branch of the graph, we assign arbitrarily a branch voltage v_k and a branch current i_k for $k = 1, 2, \ldots b$ and suppose that they are measured with respect to arbitrarily chosen associated reference directions. If the branch voltages $v_1, v_2, \ldots v_n$ satisfy all the constraints imposed by KVL and if the branch currents i_1, $i_2, \ldots, i_n$ satisfy all the constraints imposed by KCL, then,

$$\sum_{k=1}^{b} v_k i_k = 0$$

Note:

1. The theorem is not concerned with the type of circuit elements.
2. The theorem is only based on the two Kirchoff's Laws.
3. The reference directions of the branch voltages and currents are arbitrary, except that they have to satisfy kirchoff's Laws.
4. For any network. Tellegen's Theorem states

$$\sum_{k=1}^{b} v_k(t) i_k(t) = 0$$

For all values of t.

This implies that the sum of the power delivered to all branches of a network is zero.

In a linear time-invariant network composed of energy sources and passive elements under steady-state sine-wave excitation, the conservation of power is depicted by Tellegen's Theorem. Let us consider network N in Fig. 5.6 (a), having m energy sources and (b – m) remaining RLC branches.

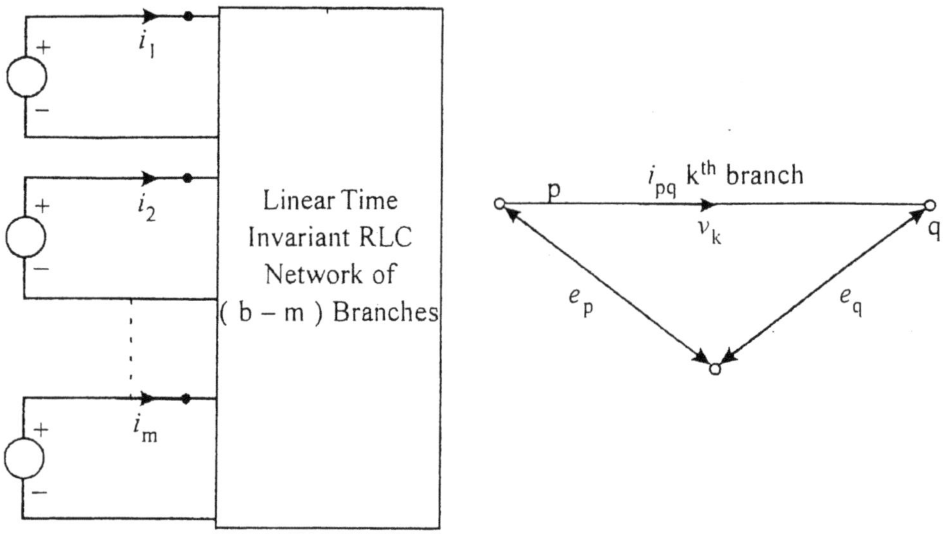

Fig. 5.6 Network illustration of Tellegen's Theorem

Let the branch voltage and current phasors be v_k and i_k, $k = 1, 2, ..., m, m+1...,$ b satisfy KVL and KCL. Then the conjugate currents i_k^*,, $k = 1, 2 ...., b$ will also satisfy KCL. From Tellegen's Theorem.

$$\sum_{k=1}^{b} \frac{1}{2} V_k i_k = 0$$

or

$$-\sum_{k=1}^{m} \frac{1}{2} V_k I_k^* = \sum_{k=m+1}^{b} \frac{1}{2} V_k I^*$$

The above equation signifies that the power delivered by the m sources is equal to the power absorbed by the remaining (b – m) passive branches of the network. This consideration is true whether the energy source is of voltage or current type.

Proof:

Let us assume that the graph of the network under consideration is connected. Moreover, between any two nodes, we assume only one branch is present. In case there is more than one branch in parallel, we combine them into a single branch. Let us also take node 1 as the datum node, i.e., $e_1 = 0$. We assign branch voltages v_1, v_2, ..., v_b and branch currents i_1, i_2 .....i_k, ... i_b where k = 1, 2, ...b, according the KVL and KCL, respectively.

This uniquely specifies the node voltages $e_1, e_2, ...e_p, ...e_q$..... In Fig. 5.9, let us assume that k^{th} branch connects nodes p and q, whose voltages are e_p and e_q respectively. The branch current is i_{pq} flowing from p to q. Then,

$$v_k i_k = (e_p - e_q) i_{pq}$$

This can also be written as,

$$v_k i_k = (e_q - e_p) i_{qp}$$

where i_{qp} is the current from node q to node p. Addition of both the above equations gives,

$$v_k i_k = \tfrac{1}{2} [(e_p - e_q) i_{pq} + (e_q - e_p) i_{qp}]$$

Now if the left and right hand sides of the above equation are summed for all the b branches,

$$\sum_{k=1}^{b} v_k i_k = \frac{1}{2} \sum_{p=1}^{b} \sum_{q=1}^{b} \left(e_p - e_q\right) i_{pq}$$

which covers all the nodes. In case a particular branch is not connected, then its current $i_{xy} = 0$ and $i_{yx} = 0$. We can rewrite above equation as,

$$\sum_{k=1}^{b} v_k i_k = \frac{1}{2} \sum_{p=1}^{b} e_p \sum_{q=1}^{b} i_{pq} - \frac{1}{2} \sum_{q=1}^{b} e_q \sum_{p=1}^{b} i_{pq}$$

Now at each node, according to KCL, the algebraic sum of currents is zero.

Hence,

$$\sum_{q=1}^{b} i_{pq} = \sum_{p=1}^{b} i_{pq} = 0$$

Therefore, $$\sum_{k=1}^{b} v_k i_k = 0$$

which is Tellegen's Theorem.

Example 5.19:

Verify Tellegen's Theorem in respect of the two networks given which have identical graphs.

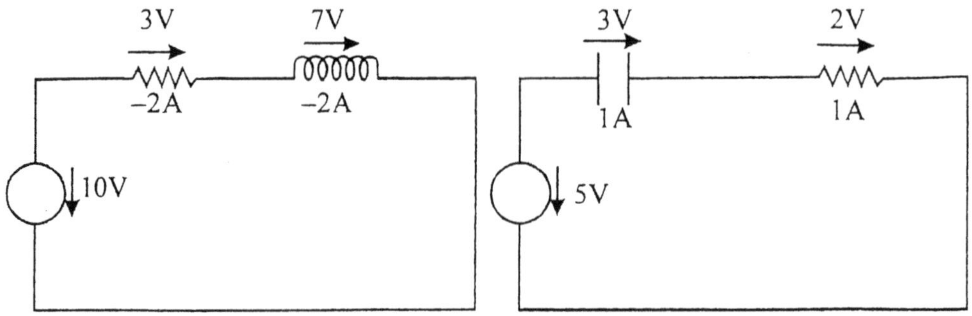

Solution:

For the individual networks

(i) $\sum_{i=1}^{3} V_k i_k = 10\times2-3\times2-7\times2=0$

(ii) $\sum_{k=1}^{3} V_k i_k = 5\times1-3\times1-2\times1=0$

Voltage of network (1) and current of network (2).

$$\sum_{k=1}^{3} V_{1k} i_{2k} = 10\times1-3\times1-7\times1=0$$

$$\sum V_{2k} i_{1k} = 5\times2-3\times2-2\times2=0$$

Hence verified.

5.13 MILLMAN'S THEOREM FOR A.C. AND D.C. EXCITATIONS

When a circuit contains number of loops, instead of writing a number of loop equations and solving the circuit, the current through any branch can be easily obtained by use of Millman's Theorem from which the potential of the common point to which all the impedances are connected. Knowing the potentials of the other end of the branches which are connected in common, with the knowledge of the potential of the common point, the current through any branch can easily be obtained by simple application of Ohm's Law for each branch.

Millman's theorem is very useful in solving unbalanced poly-phase A.C. networks, wherein the impedances are all connected in star. Millman's Theorem helps in finding the potential of the star point or the neutral point.

If there are large number of sources with E.M.F.s E_1, ..., E_n and internal impedance Z_1, ... Z_n and these are connected in parallel as shown, the potential of the common point is given by Millman's Theorem as,

$$V = \frac{E_1 Y_1 + E_2 Y_2 + ... + E_n Y_n}{Y_1 + Y_2 + ... + Y_n}$$

where $$Y_n = \frac{1}{Z_n}$$

For D.C excitation we use $G = \frac{1}{R}$ instead of Y.

Consider Fig. 5.7.

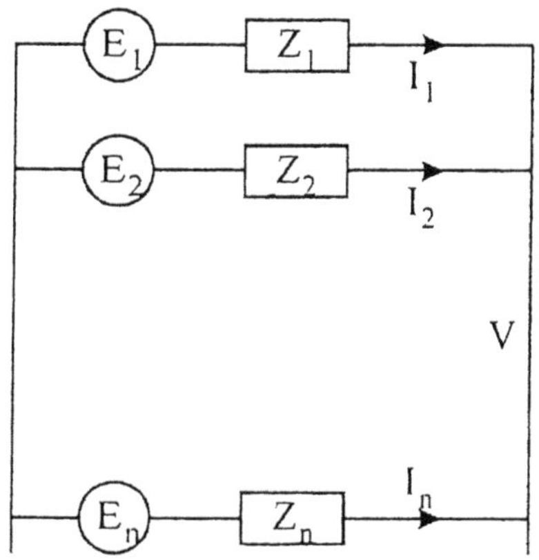

Fig. 5.7 *Sources connected in parallel*

From Fig. 5.7

$$I_1 = \frac{E_1 - V}{Z_1}$$

$$I_2 = \frac{E_2 - V}{Z_2}$$

and $$I_n = \frac{E_n - V}{Z_n}$$

According to Kirchoff's Current Law.

$$I_1 + I_2 + ... + I_n = 0$$

$$(E_1 - V)Y_1 + (E_2 - V)Y_2 + ... + (E_n - V)Y_n = 0$$

or $$E_1 Y_1 + E_2 Y_2 + ... + E_n Y_n = V(Y_1 + Y_2 + ... + Y_n)$$

or $$V = \frac{E_1Y_1 + E_2Y_2 + ... + E_nY_n}{Y_1 + Y_2 + ... + Y_n}$$

Example 5.20:

Determine the current through the branch AB using Millman's Theorem.

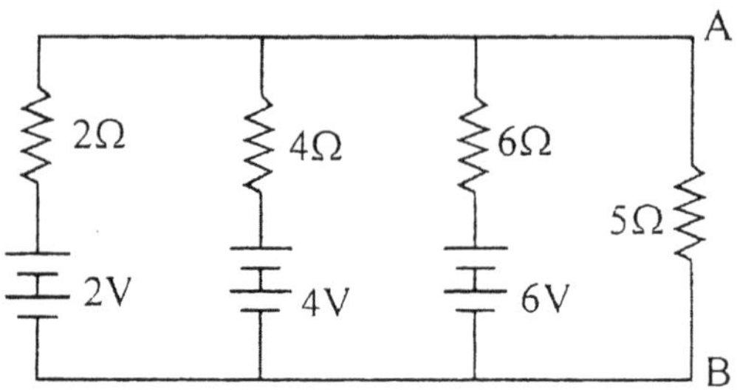

Solution:

Using Millman's Theorem,

$$V_{AB} = \frac{\frac{2}{2} + \frac{4}{4} + \frac{6}{6}}{\frac{1}{2} + \frac{1}{4} + \frac{1}{6}} = \frac{36}{11} V$$

and the equivalent impedance between AB.

$$Z = \frac{12}{11} \Omega$$

$\therefore$ current $$I_{AB} = \frac{(36/11)}{(5 + 12/11)} = \frac{36}{67}$$

5.14 COMPENSATION THEOREM FOR A.C. AND D.C. EXCITATIONS

In some problems, we are interested in finding the corresponding changes in various voltages and currents of a network subjected to a change in one of its branches. For example, we may be interested in finding the effect of using a resistor in a network which is not of the exact value due to component tolerance. The compensation theorem provides us a convenient method for determining such effects.

In a linear network N, if the current in a branch is 1 and the impedance Z of the branch is increased by δZ, then the increment of voltage and current in each branch of the network is that voltage or current that would be produced by an opposing voltage source of value V_c introduced into the altered branch after the modification. The compensation theorem is based on the superposition principle, and the network is required to be linear. For D.C excitations we use R instead of Z.

Consider the network N in Fig. 5.8 (a), having branch impedance Z. Let the current through Z be I and its voltage be V.

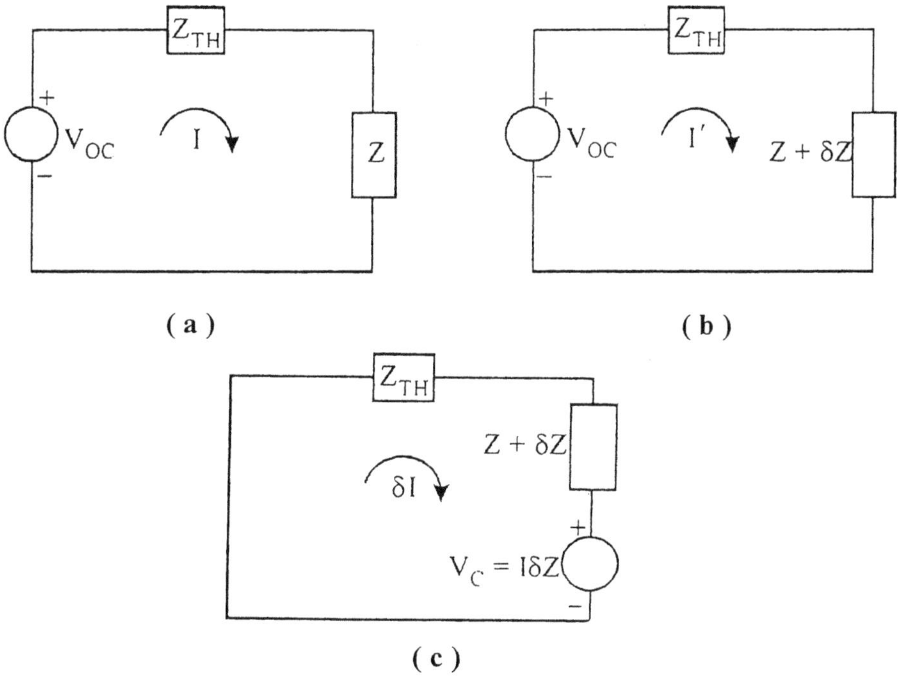

***Fig. 5.8** Network illustration of Compensation Theorem*

$$I = \frac{V_{OC}}{Z + Z_{TH}}$$

Let δZ be the change in Z. Then I, the new current can be written as

$$I' = \frac{V_{OC}}{Z + \delta Z + Z_{TH}}$$

$$\delta I = I' - I$$

$$= \frac{V_{OC}}{Z + \delta Z + Z_{TH}} - \frac{V_{OC}}{Z + Z_{TH}}$$

$$= \frac{-V_{OC}\ (Z + \delta Z + Z_{TH} - Z - Z_{TH})}{(Z + Z_{TH})\ (Z + \delta Z + Z_{TH})}$$

$$= -\left(\frac{V_{OC}}{Z + Z_{TH}}\right)\frac{\delta Z}{(Z + \delta Z + Z_{TH})}$$

$$= -\frac{I\delta Z}{Z + \delta Z + Z_{TH}} = -\frac{V_C}{Z + \delta Z + Z_{TH}}$$

Addition of $(-V_C)$ maintains same I as before, even after changing Z to $(Z + \delta Z)$

Example 5.21:

Calculate the change in the current in the circuit of the given figure, by using Compensation Theorem, when the reactance has changed to $j35\,\Omega$.

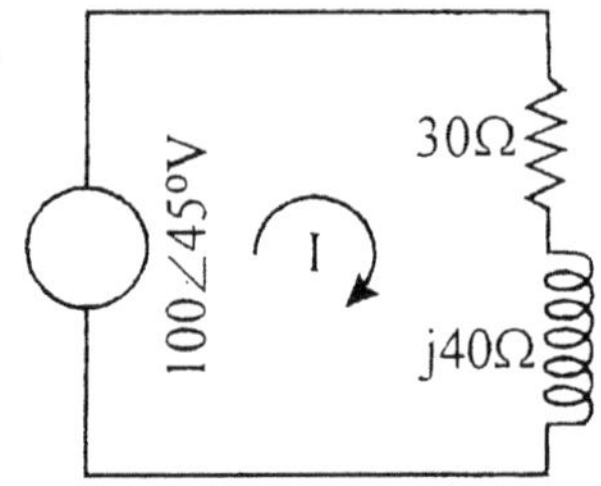

Solution:

According to Compensation Theorem, we have, change of current due to a change of impedance.

$$\delta I = \frac{I\delta Z}{Z_{Total}}$$

Given $\quad Z = (30 + j40)W$

and $\quad V_{OC} = 100\ \angle 45°\,V$

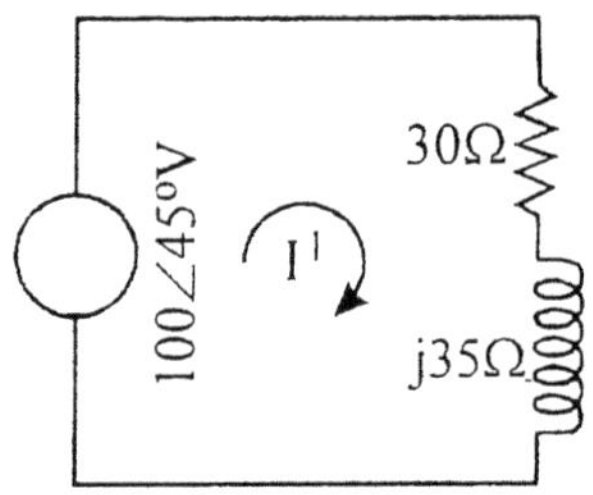

$\therefore$ Current, $\quad I = \frac{100\angle 45°}{30 + j40} = \frac{100\angle 45°}{50\angle 53.13°}$

$$= 2\angle -8.13°\,A$$

Reactance is changed from $j40\,\Omega$ to $j35\ \Omega$

$\therefore$ Change in reactance value $= j40 - j35 = j5\,\Omega = \delta Z$

The new network is as shown

$$Z_{Total} = (30 + j35)\,\Omega = 46.1\ \angle 49.4°\,\Omega$$

$$\therefore \quad \delta I = \frac{I\delta Z}{Z_{Total}} = \frac{(2\angle -8.13°)(j5)}{46.1\angle 49.4°} = 0.217\angle 32.47°\,A$$

COMPREHENSION - 5

Note: **By answering all the questions given in this comprehension one can answer any objective type of questions easily whatever be the form of the question. Answers for these questions can be obtained by going through the topics in this chapter carefully.**

1. What are the important features of Network Theorems?
2. What are the different classifications of Networks?
3. State and Prove Superpostion Theorem for D.C and A.C excitations.
4. When and How is the Superposition Theorem used?
5. State and Prove Thevenin's Theorem for D.C and A.C excitations.
6. Explain the terms Thevenin's Voltage and Thevenin's Impedance. How do you obtain them?
7. What is the importance of Thevenin's Theorem? Where it is used generally?
8. Is the Thevenin's Theorem applicable for AC and for Transients?
9. What is the Dual of Thevenin's Theorem?
10. State & Prove Norton's Theorem for D.C and A.C excitations.
11. Explain the terms Norton's Short-circuit Current. How do you obtain it?
12. What is the importance of Norton's Theorem? Where it is used generally?
13. Is the Norton's Theorem applicable for AC and for Transients?
14. How is the remaining part of the circuit other than the load represented in Norton's Theorem?
15. State & Prove Maximum Power Transfer Theorem for D.C and A.C excitations.
16. What is the condition for Maximum Power Transfer to take place in

    (i) DC Circuits (ii) AC Circuits?
17. What is the importance of Maximum Power Transfer? Where is it used?
18. State & Prove Reciprocity Theorem.
19. When is the Reciprocity Theorem Applicable?
20. State and Prove Tellegen's Theorem.

21. Is the Tellegen's Theorem applicable for AC circuits?
22. Explain the features made use of an Tellegen's Theorem?
23. State and Prove Millman's Theorem.
24. Is the Millman's Theorem applicable for AC circuits?
25. When do you use Millman's Theorem in 3-phase or Multi-phase circuits?
26. State and Prove Compensation Theorem.
27. Is the Compensation Theorem applicable for AC circuits?
28. How do you apply Compensation Theorem in

    (i) DC Circuits (ii) AC Circuits

* * * * *

EXERCISE - 5

Note: **By answering all the questions given in this exercise one can answer subjective type of questions easily including problems. The worked examples in the chapter may help in solving these questions.**

1. Using Superposition Theorem, determine the current through the branch AB in Fig. 1. Verify the result by Loop Analysis.

[Ans: 0.8A]

2. Using Superposition Theorem, determine the current through the $3\,\Omega$ resistor and the power loss in it in Fig. 2.

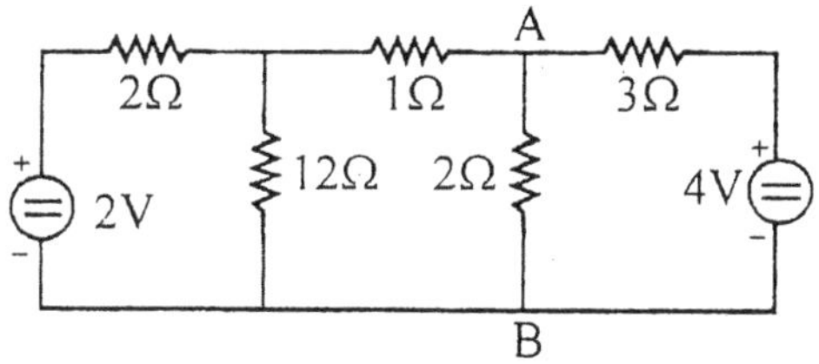

Fig. 1

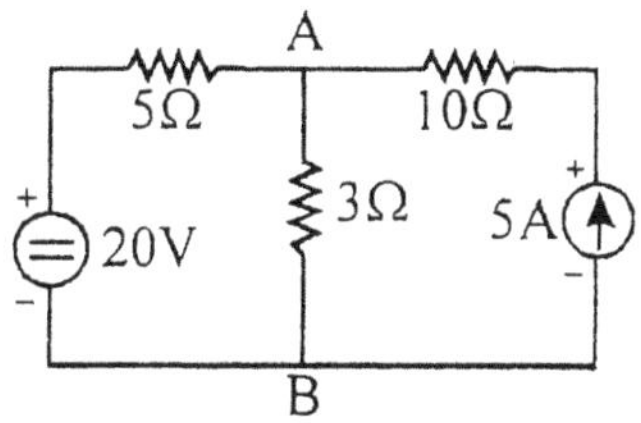

Fig. 2

[Ans: 5.62A; 94.92W]

3. Using Thevenin's Theorem, determine the current through $2\,\Omega$ resistor connected between A and B in the circuit of Fig. (3.3.).

[Ans: 0.82 A]

4. Determine the Thevenin's Equivalent across the Terminats A, B, in Fig. 4.

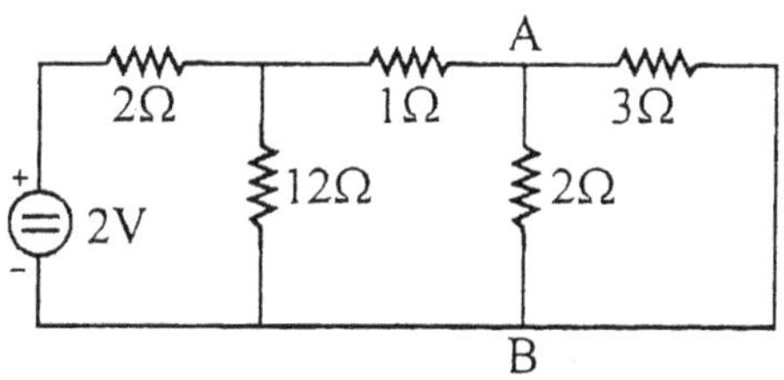

Fig. 3

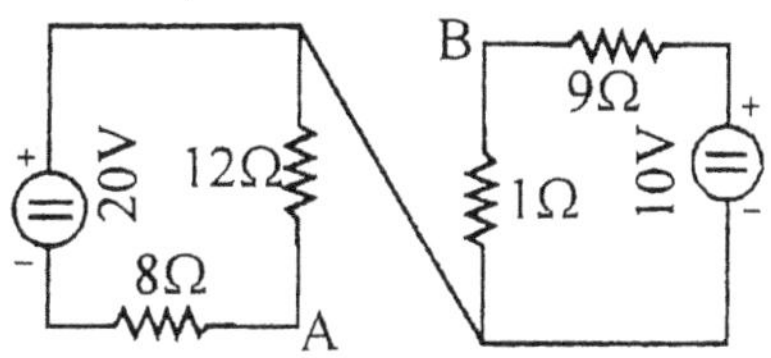

Fig. 4

[Ans: $V_{TH} = 11V$; $R_{TH} = 5.7\,\Omega$]

5. Using Thevenin's Theorem, determine the current through the ammeter of 2 Ω connected in unbalanced Weat Stone Bridge of Fig. 5.

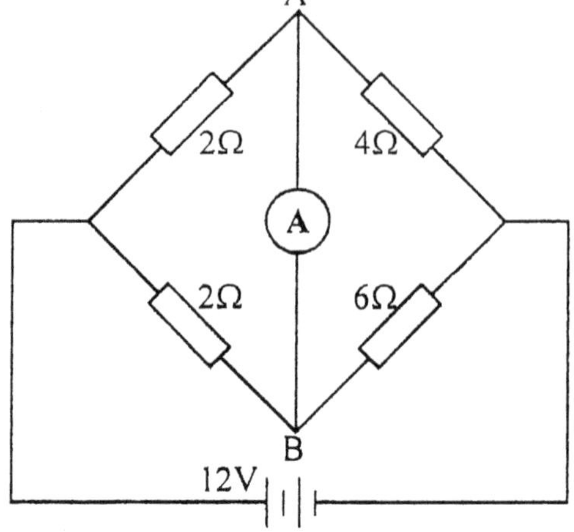

Fig. 5

[Ans: 0.207A]

6. Using Norton's Theorem, Solve Problem No.3.
7. Using Norton's Theorem, Solve Problem No.4.
8. Using Norton's Theorem, Solve Problem No. 5
9. Using Maximum Power Transfer Theorem, find the resistance R_L for the maximum power transfer in the circuit shown in Fig. 6. Calculate the Maximum Power also. (Hint: Resonance AB is a redundant element).

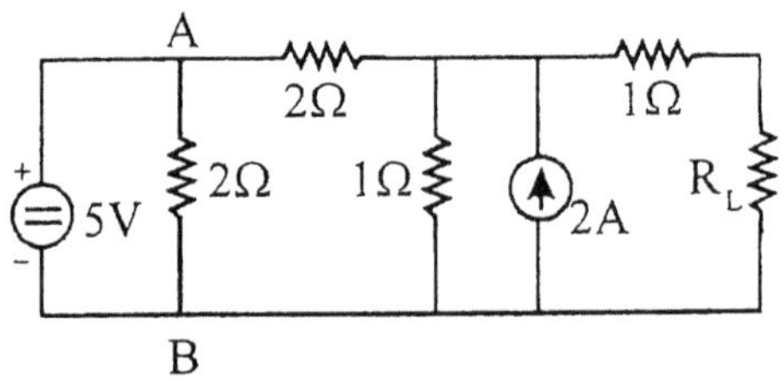

Fig. 6

[Ans: R_L = 1.75 Ω; P_L = 1.08V]

10. Using superposition theorem, find the current in branch AB of the network given in the figure.

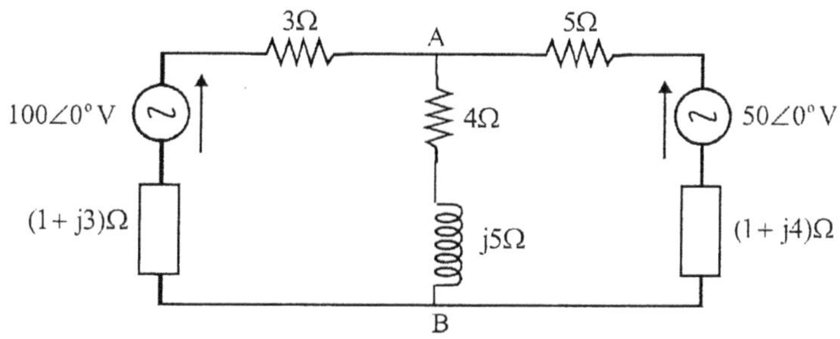

[**Ans.** $I_{AB} = 8.6164\angle -47.15°\,A$]

11. Using Thevenin's theorem, determine the current in a 1Ω resistor connected to terminals A,B of the network shown in figure.

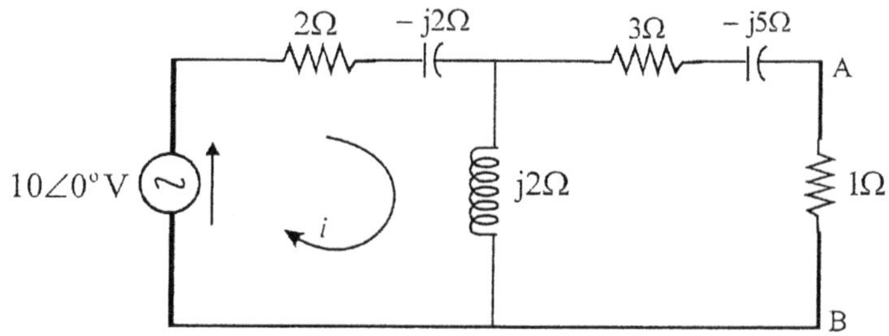

[**Ans.** $I = 1.4907\ \angle 116.565°\,A$]

12. The network of the given figure shows a voltage source through a T-circuit. What should be the value of a pure resistance load to be connected across the terminals A,B so that maximum power is received? What is the maximum power?

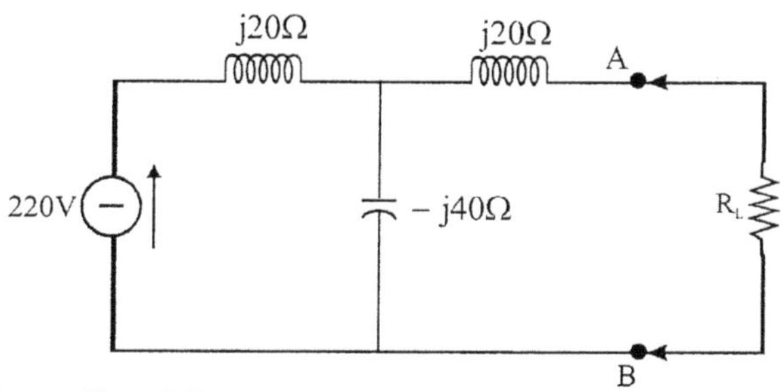

Hint : For Maximum Power Transfer, if R_L is Pure Resistance then $R_L + j0 = |\ Z_{Th}\ |$

[**Ans.** $P_{max} = 1613$ Watts.]

13. Obtain the current in 1 Ohm resistance across AB in the above problem by applying Norton's Theorem instead of Thevenin's Theorem.

[Ans. $I = 1.4907\ \angle 116.565°\,A$]

14. Using Thevenin's Theorem, find the current through $(5 + j4)\,\Omega$ impedance in Fig. 7

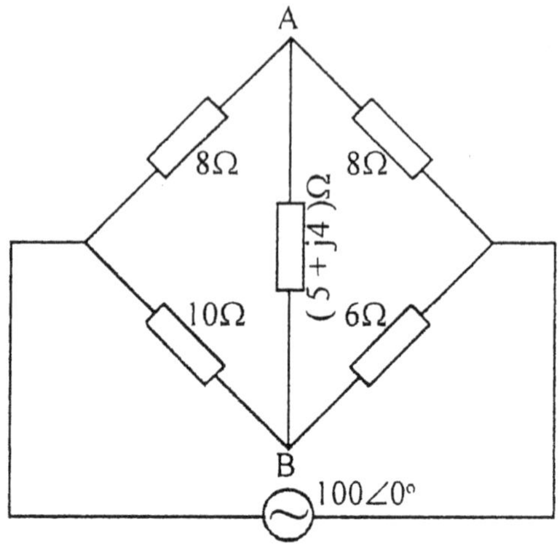

Fig. 7

[Ans: 0.998 ∠−17.42°A]

15. Using Norton's Theorem, Solve Problem No. 15

16. In the network shown in Fig. 8 two voltage sources act on the load impedance connected to the resistance, what load Z_L will receive maximum power? What is the value of Maximum Power?

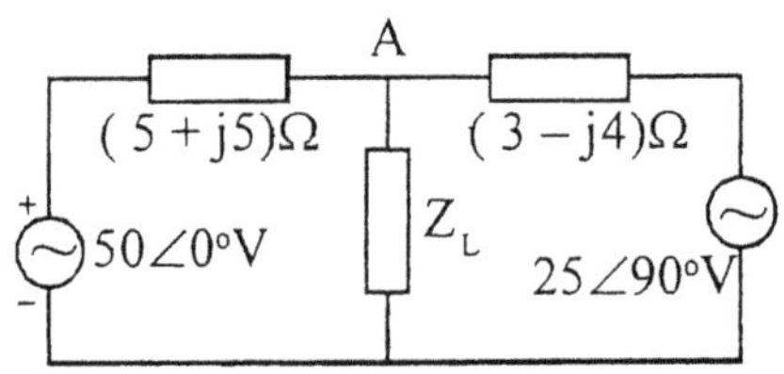

Fig. 8

[Ans: (4.23 + j 1.15)Ω, 5.68W]

17. A load of $(20 - JX_c)$ is supplied from a source of 10V RMS and internal impedance (10 + j20). Determine the value of X_c for maximum power supplied to the load and determine this power?

[Ans: 20Ω, 2.22W]

18. Verify the Reciporocity Theorem for the network given in Fig. 9 regarding the source current and the current in branch BC.

[Ans: I_{AB} = 5.71A; I_{BC} = 1.43A]

19. Verify the Reciprocity Theorem for the network given in Fig.10 regarding the source current and the load the current in branch AB.

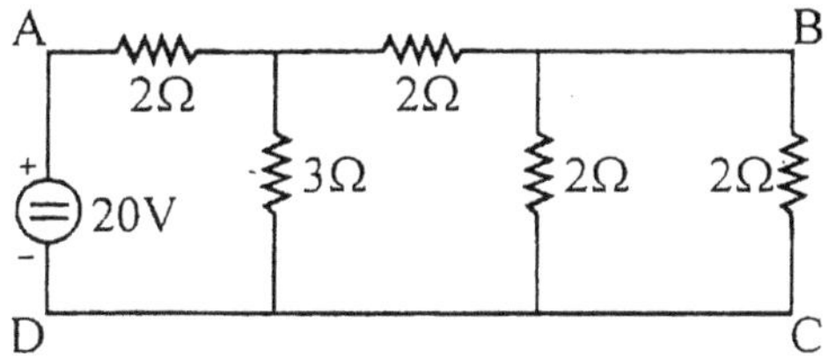

Fig. 9

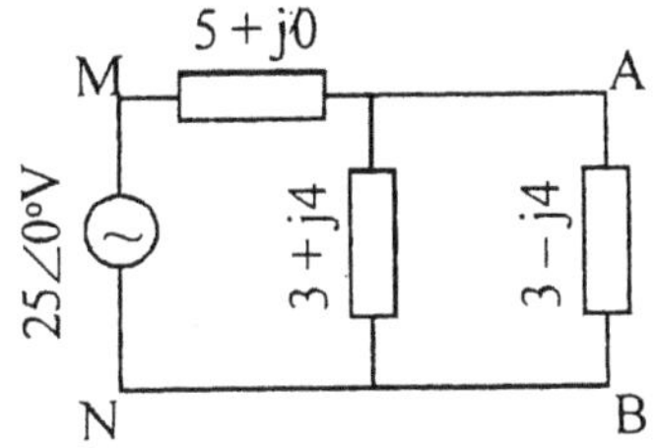

Fig. 10

[Ans: 2.7273 ∠0o A; 2.273 ∠53.13° A]

20. Verify Tellegen's Theorem for the network of Fig. 11.

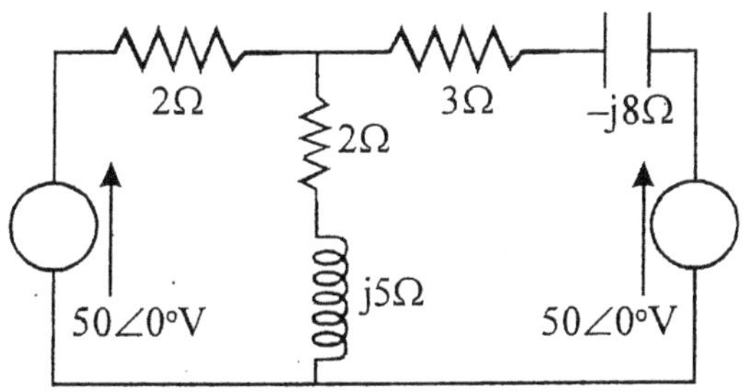

Fig. 11

21. Find the output voltage E_0 for the circuit shown in Fig, 12 using Millman's Theorem.

[Ans: 19V]

22. Using Millman's Theorem, find the current through R_L in the circuit shown in Fig. 13.

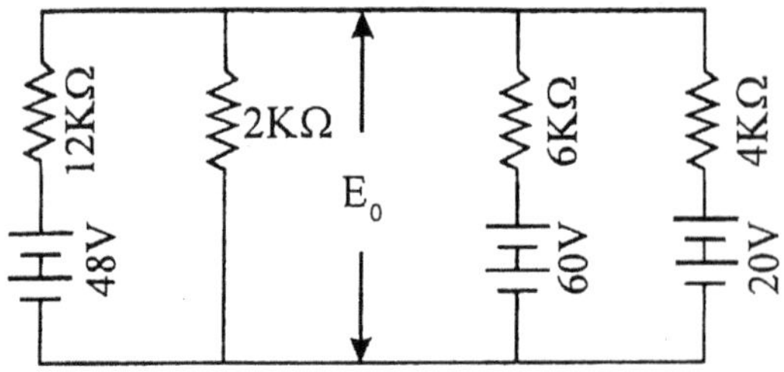

Fig. 12

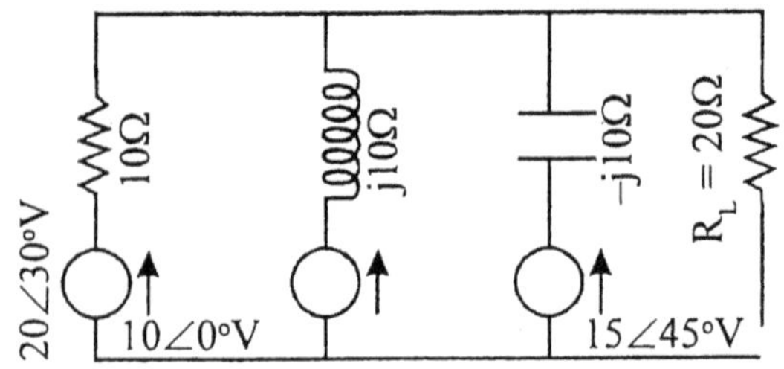

Fig. 13

[Ans: 0.4185 $\angle$ 57.69°A]

23. In the network of the Fig. 14, the 20 V source contains the current I. When the upper 10 Ω resistor is changed into 12 Ω the current output of the source change to I'. Find the change in the current? $\Delta I = (I' - I)$ by using Compensation Voltage Source as shown in Fig. 15.

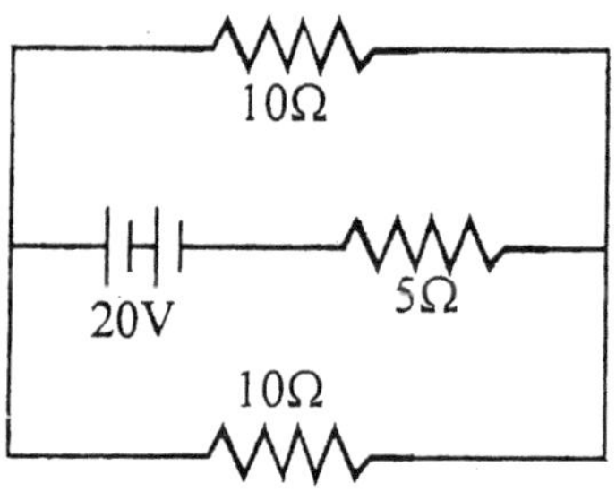

Fig. 14

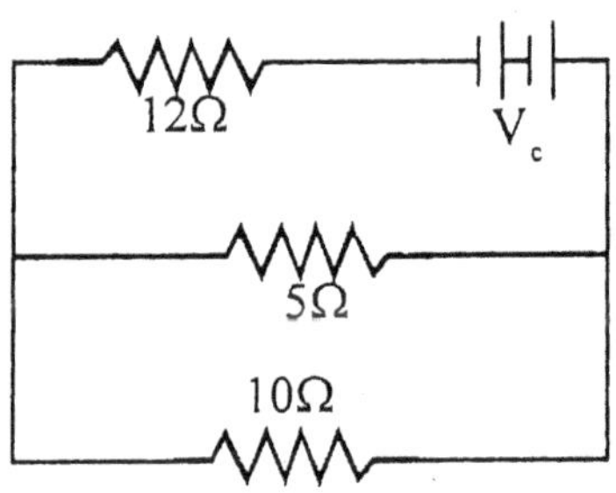

Fig. 15

[Ans. V_C = + 20, 000 V

V_C compensates additional $I\Delta Z$ drop to have the original I]

* * * * *

6

SEMICONDUCTOR DIODES, RECTIFIERS AND FILTERS

6.1 INTRODUCTION

If a junction is formed using *p-type* and *n-type* semiconductors, a diode is realised and it has the properties of a rectifier. In this chapter, the volt ampere characteristics of the diodes, electron-hole currents as a function of distance from the junction and junction capacitances will be studied.

6.1.1 Theory of p-n Junction

Take an intrinsic Silicon or Germanium crystal. If donor (*n-type*) impurities are diffused from one point and acceptors impurities from the other, a p-n junction is formed. The donor atoms will donate electrons. So they loose electrons and become positively charged. Similarly, acceptor atoms accept an electron, and become negatively charged. Therefore in the p-n junction on the *p-side*, holes and negative ions are shown and on the *n-side* free electrons and positive ions are shown. To start with, there are only *p-type* carriers to the left of the junction and only *n-type* carriers to the right of the junction. But because of the concentration gradient across the junction, holes are in large number on the left side and they diffuse from left side to right side. Similarly electrons will diffuse to the right side because of concentration gradient.

Because of thedisplacement of these charges, electric field will appear across the junction. Since *p-side* looses holes, negative field exists near the junction towards

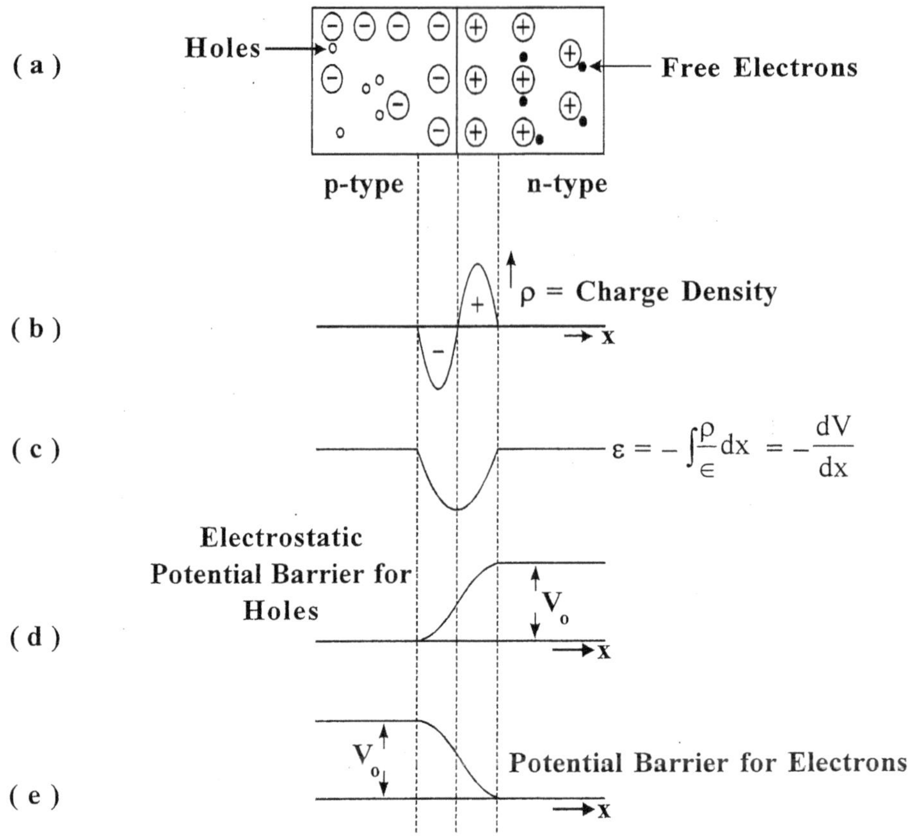

Fig. 6.1 Potential distribution in p-n junction diode

left. Since *n-side* looses electrons, positive electric field exists on the n side. But at a particular stage the negative field on *p-side* becomes large enough to prevent the flow of electrons from *n-side.* Positive charge on *n-side* becomes large enough to prevent the movement of holes from the *p-side*. The charge distribution is as shown in Fig. 6.1 (b). The charge density far away from the junction is zero, since before all the holes from *p-side* move to *n-side*, the barrier potential is developed. Acceptor atoms near the junction have lost the holes. But for this they would have been electrically neutral. Now these holes have combined with free electrons and disappeared leaving the acceptor atom negative. Donor atoms on *n-side* have lost free electrons. These free electrons have combined with holes and disappeared. So *the region near the junction is depleted of mobile charges. This is called Depletion Region, space charge region or transition region.* The thickness of this region will be of the order of few microns

$$1 \text{ micron} = 10^{-6} \text{ m} = 10^{-4} \text{cm}.$$

The electric field intensity near the junction is shown in Fig. 6.1 (c). This curve is the integral of the density function ρ. The electro static potential variation in the depletion region is shown in Fig. 6.1 (d). $\varepsilon = -\int \frac{dv}{dx}$. This variation constitutes a potential energy barrier against further diffusion of holes across the barrier.

When the diode is open circuited, that is not connected in any circuit, the hole current must be zero. Because of the concentration gradient, holes from the *p-side* move towards *n-side*. So, all the holes from *p-side* should move towards *n-side*. This should result in large hole current flowing even when diode is not connected in the circuit. But this will not happen. So to counteract the diffusion current, concentration gradient should be nullified by drift current due to potential barrier. Because of the movement of holes from *p-side* to *n-side*, that region (p-region) becomes negative. A *potential gradient* is set up across the junction such that *drift current flows* in opposite direction to the *diffusion current*. So the net hole current is zero when the diode is open circuited. The potential which exists to cause drift is called *contact potential* or *diffusion potential*. Its magnitude is a few tenths of a volt (0.01V).

6.2 p-n JUNCTION AS A DIODE

The p-n Junction shown here forms a semiconductor device called **DIODE**. Its symbol is A —▷|— K. **A** is anode. **K** is the cathode. It has two leads or electrodes and hence the name Diode. If the anode is connected to positive voltage terminal of a battery with respect to cathode, it is called ***Forward Bias***, (Fig. 6.2 (a)). If the anode is connected to negative voltage terminal of a battery with respect to cathode, it is called

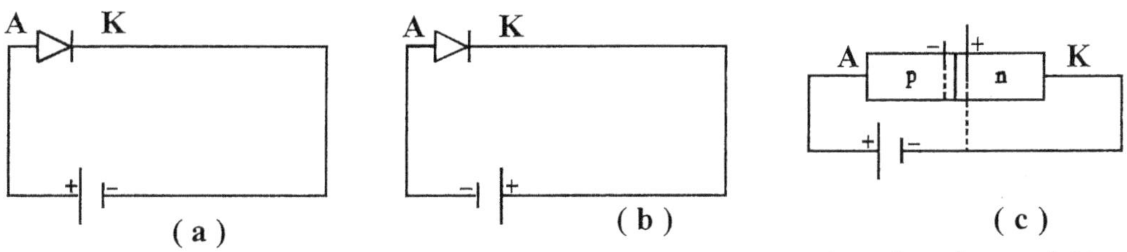

(a) *Forward bias* (b) *Reverse bias* (c) *p-n junction forward bias*

Fig. 6.2

Reverse Bias, (Fig. 6.2 (b)).

6.2.1 Reverse Bias

Because of the battery connected as shown, holes in (Fig. 6.2 (b)) *p-type* and electrons in *n-type* will move away from the junction. As the holes near the junction in *p-region* they will move away from the junction and negative charge spreads towards the left of the junction. Positive charge density spreads towards right. But this process cannot continue indefinitely, because to have continuous flow of holes from right to left, the holes must come from the *n-side*. But *n-side* has few holes. So very less current results. But some electron hole pairs are generated through of thermal agitation. The newly generated holes on the *n-side* will move towards junction. Electrons created on the *p-side* will move towards the junction. So there results some small current called ***Reverse Saturation Current***. It is denoted by I_0. I_0 will increase with the temperature. So the *reverse resistance* or *back resistance decreases with temperature*. I_0 is of the order of a few μA. *The reverse resistance of a diode will be of the order of MΩ. For ideal diode, reverse resistance is ¥.*

The same thing can be explained in a different way. When the diode is open circuited, there exists a barrier potential. If the diode is reverse biased, the barrier potential height increases by a magnitude depending upon the reverse bias voltage. So the flow of holes from *p-side* to *n-side* and electrons from *n-side* to *p-side* is restricted. But this barrier doesn't apply to the minority carriers on the *p-sides* and *n-sides*. The flow of the minority carriers across the junction results in some current.

6.2.2 Forward Bias

When a diode is forward biased, the potential barrier that exists when the diode is open circuited, is reduced. Majority carriers from *p-side* and *n-side* flow across the junction. So a large current results. *For ideal diode, the forward*

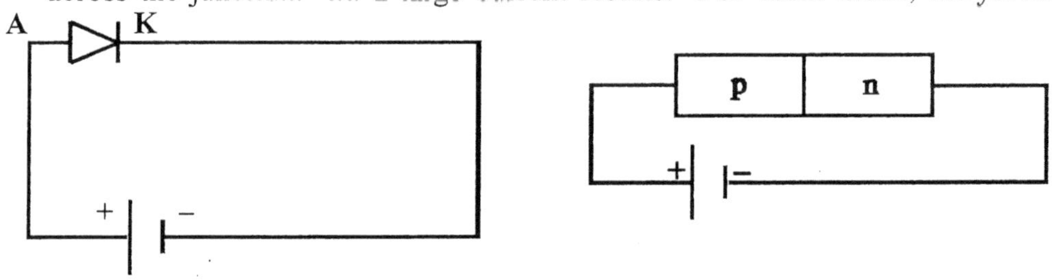

Fig. 6.3 Diode in forward bias

resistance $R_f = 0$. The forward current I_F, will be of the order of mA (milli-amperes).

6.2.3 Ohmic Contact

In the above circuits, external battery is connected to the diode. But directly external supply cannot be given to a semiconductor. So metal contacts are to be provided for *p-region* and *n-region*. A Metal-Semiconductor Junction is introduced on both sides of p-n junction. So these must be contact potentials across the metal-semiconductor junctions. But this is minimized by fabrication techniques and the contact resistance is almost zero. Such a contact is called ohmic contact. So the

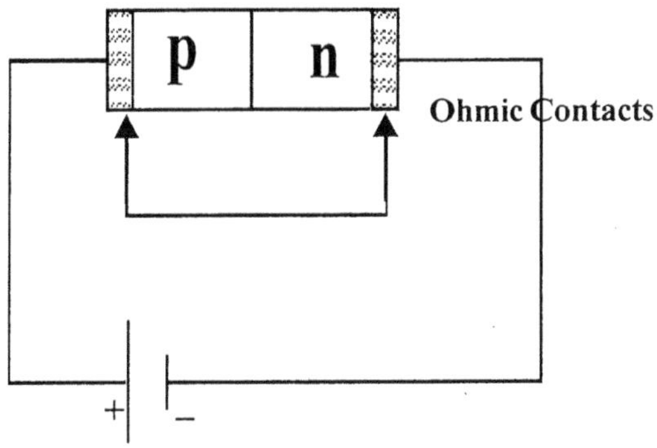

Fig. 6.4 Ohmic contacts

entire voltage appears across the junction of the diode.

6.2.4 Forward Current

If a large forward voltage is applied (Fig. 6.5), the current must increase. If the barrier potential across the junction is made zero, infinite amount of current should flow. But this is not practically possible since the bulk resistance of the crystal and the contact resistance together will limit the current. We may see in the other sections

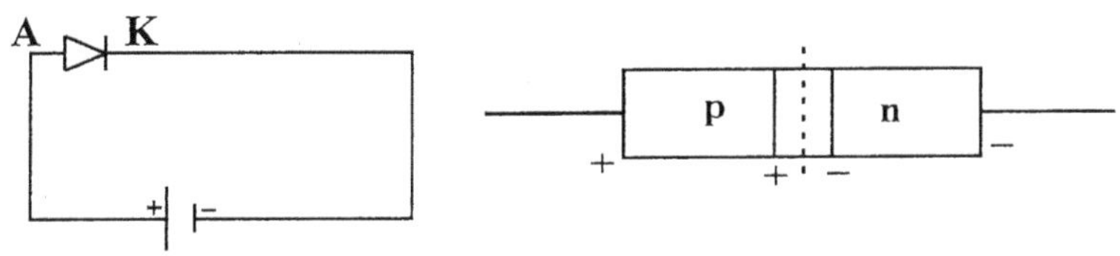

Fig. 6.5 Forward biasing

that when the diode is conducting, the voltage across it remains constant at V_γ *cut in voltage*. If the applied voltage is too large junction breakdown will occur.

6.3 ENERGY BAND DIAGRAM

When *p-type* and *n-type* semiconductors are brought into intimate contact *p-n junction* is formed. Then the fermilevel must be constant throughout the specimen. If it is not so, electrons on one side will have higher energy than on the other side. So the transfer of energy from higher energy electrons to lower energy electrons will take place till fermilevel on both sides comes to the same level. But we have already seen that in *n-type* semiconductors, E_F is close to conduction band E_{cn} and it is close to valence band edge E_{vp} on *p-side*. So the conduction band edge of *n-type*

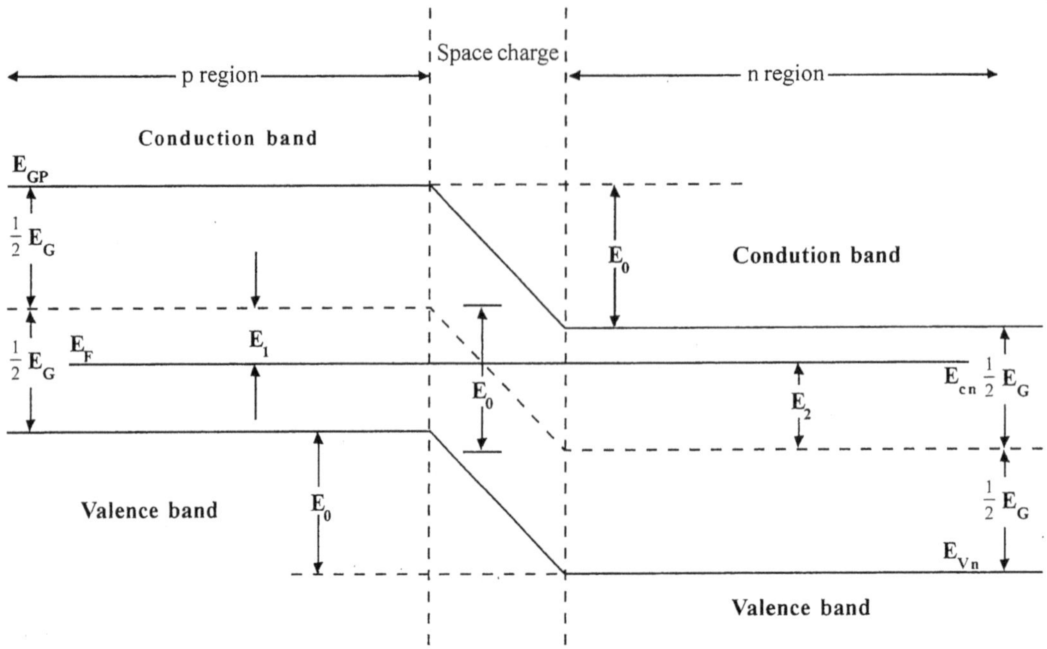

Fig. 6.6 Band Structure of Open Circuited Diode

semiconductor cannot be at the same level as that of *p-type* semiconductor. Hence, as shown, the energy band diagram for a *p-n junction* is where a shift in energy levels E_0 is indicated.

E_G = Energy gap in eV

E_F = Fermi energy level

E_O = Contact difference of potential

E_{cn} = Conduction band energy level on the *n-side*.

E_{cp} = Conduction band energy level on the *p-side*.

E_{vn} = Valence band energy level on the *n-side*.

E_{vp} = Valence band energy level on the *p-side*.

If a central line $\frac{E_G}{2}$ is taken, the shift in energy levels is the difference between the two central lines $\frac{E_G}{2}$ of the two semiconductors.

$$E_0 = E_{cp} - E_{cn} = E_{vp} - E_{vn}.$$

$$E_1 = \frac{E_G}{2} - (E_F - E_{vp})$$

$$E_2 = \frac{E_G}{2} - (E_{cn} - E_F)$$

$$E_1 + E_2 = E_G - E_F + E_{vp} - E_{cn} + E_F$$

But $E_G = E_{cp} - E_{vp}$

$\therefore$ $E_1 + E_2 = E_{cp} - E_{vp} - E_{cn} + E_{vp}.$

But $E_{cp} - E_{cn} = E_O$

$E_1 + E_2 = E_O.$

This energy E_0 represents the potential energy barrier for electrons. The contact difference of potential

$$E_F - E_{vp} = ½ (E_G) - E_1 = \left(\frac{E_G}{2}\right) - E_1 \quad .....(6.3.1)$$

and

$$E_{cn} - E_F = ½ (E_G) - E_2 = \left(\frac{E_G}{2}\right) - E_2 \quad .....(6.3.2)$$

Adding (6.3.1) and (6.3.2),

$$(E_F - E_{vp}) + (E_{cn} - E_F) = E_G - E_1 - E_2$$

or

$$(E_1 + E_2) = E_G - (E_{cn} - E_F) - (E_F - E_{vp})$$

$$(E_1 + E_2) = E_O$$

$$(E_{cn} - E_F) = KT \ln\left(\frac{N_C}{N_D}\right)$$

N_D = Donor atom concentration N_o/m^3.

N_A = Acceptor atom concentration N_o/m^3.

$$(E_F - E_{vp}) = KT \ln\left(\frac{N_V}{N_A}\right)$$

$$E_G = KT \ln\left(\frac{N_C.N_V}{n_i^2}\right)$$

$$\therefore \quad E_0 = KT \left(\ln\left(\frac{N_C - N_V}{n_i^2}\right) - \ln\left(\frac{N_C}{N_D}\right) - \ln\left(\frac{N_V}{N_A}\right)\right)$$

$$E_0 = KT \ln\left(\frac{N_C - N_V}{n_i^2} \times \frac{N_D}{N_C} \times \frac{N_A}{N_V}\right) = KT \ln\left(\frac{N_A.N_C}{n_i^2}\right) \quad .....(6.3.3)$$

The energy is expressed in electron volts eV.

K is Botlzman's Constant in eV / °K = 8.62×10^{-5} eV / °K

Therefore, E_0 is in eV and V_0 is the contact difference potential in volts V_0 is numerically equal to E_0. In the case of *n-type* semiconductors, $n_n = N_D$. (Subscript 'n' indicates electron concentration in *n-type* semiconductor)

$$n_i^2 = n_n \times p_n = N_D \times p_n$$

$$\because \quad n_n = N_D$$

n_i = Intrinsic concentration

n_p = Electron concentration in *p-type* semiconductor

n_n = Electron concentration in *n-type* semiconductor

p_p = Hole concentration in *p-type* semiconductor

p_n = Hole concentration in *n-type* semiconductor

$$p_n = \frac{n_i^2}{N_D} \quad \text{and} \quad N_D = \frac{n_i^2}{p_n}$$

$$n_p = \frac{n_i^2}{N_A} \quad \text{and} \quad N_A = \frac{n_i^2}{n_p}$$

$$n_i^2 = n_n \times p_n$$

Substituting all these value in

$$E_0 = KT \ln\left(\frac{N_A.N_D}{n_i^2}\right)$$

$$E_0 = KT \ln\left(\frac{n_i^2}{p_n} \times \frac{n_i^2}{n_p} \times \frac{1}{n_i^2}\right)$$

$$= KT \ln\left(\frac{n_i^2}{p_n.n_p}\right)$$

$$= KT \ln\left(\frac{n_n.p_n}{p_n.n_p}\right) = KT \ln\left(\frac{n_n}{n_p}\right)$$

$$E_0 = KT \ln\left(\frac{n_{n0}}{n_{p0}}\right) = KT \ln\left(\frac{p_{pn}}{p_{n0}}\right)$$

Taking reasonable values of $n_{n0} = 10^{16} / cm^3$

$$n_{p0} = 10^4 / cm^3$$

$$KT = 0.026 \text{ eV}$$

$$E_0 = 0.026 \times \ln \frac{10^{16}}{10^4} = 0.5 \text{ eV.}$$

6.3.1 The Current Components in a p-n Junction Diode

When a forward bias is applied to the diode, holes are injected into the *n-side* and electrons to the *p-side*. The number of this injected carriers decreases exponentially with distance from the junction. Since the diffusion current of minority carriers is proportional to the number of carriers, the minority carriers current decreases exponentially, with distance. There are two minority currents, one due to electrons in the *p-region* ***Inp***, and due to holes in the *n-region* ***Ipn***. As these currents vary with distance, they are represented as ***Ipn(x)***.

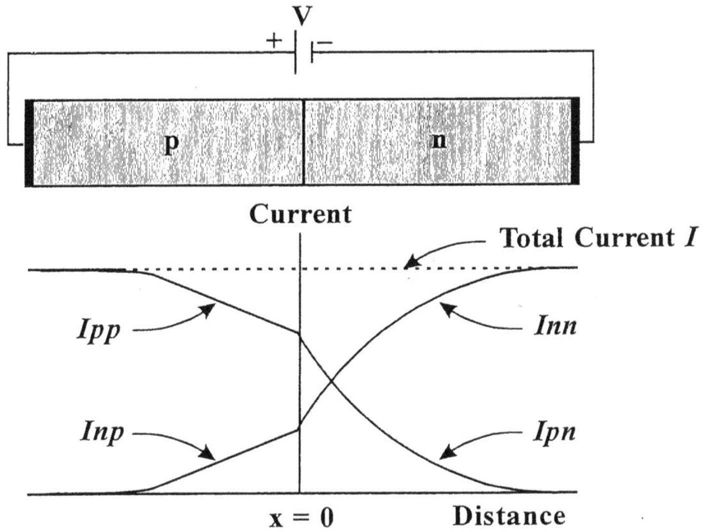

Fig. 6.7 *Current components in a p-n junction*

Electrons crossing from *n* to *p* will constitute current in the same direction as holes crossing from *p* to *n*. Therefore, the total current at the junction where x = 0 is

$$I = Ipn(0) + Inp(0)$$

The total current remains the same. The decrease in ***Ipn*** is compensated by increase in ***Inp*** on the *p-side*.

Now deep into the *p-region* (where x is large) the current is because of the electric field (since bias is applied) and it is drift current ***Ipp*** of holes. As the holes approach the junction, some of them recombine with electrons crossing the junction from *n* to *p*. So ***Ipp*** decreases near the junction and is just equal in magnitude to the diffusion current ***Inp***. What remains of ***Ipp*** at the junction enters the *n-side* and becomes hole diffusion current ***Ipn*** in the *n-region*. Since holes are minority carriers in the *n-side*, ***Ipn*** is small and as hole concentration decreases in the *n-region*, ***Ipn*** also exponentially decreases with distance.

In a forward biased *p-n junction* diode, at the edge of the diode on *p-side*, the current is hole current (majority carriers are holes). This current decreases at the junction as the junction approaches and at a point away from the junction, on the *n-side*, hole current is practically zero. But at the other edge of the diode, on the *n-side*, the current is electron current since electrons are the majority carriers. Thus in a *p-n junction* diode, the current enters as hole current and leaves as electron current.

6.4 DIODE EQUATION

p_{po} = Thermal equilibrium hole concentration on *p-side*.

p_{no} = Thermal equilibrium hole concentration on n side

$$p_{po} = p_{no}\, e^{V_o/V_T} \qquad .....(6.4.1)$$

where V_o is the electrostatics barrier potential that exists on both sides of the junction. But the thermal equilibrium hole concentration on the *p-side*

$$p_{po} = p_n(0)\, e^{(V_o-V)/V_T} \qquad .....(6.4.2)$$

where $p_n(0)$ = Hole concentration on *n-side* near the junction

V = Applied forward bias voltage.

This relationship is called Boltzman's Relationship.

∴ Equating (6.4.1) and (6.4.2).

$$p_n(0)\, e^{(V_o-V)/V_T} = p_{no} \times e^{V_o/V_T}$$

or
$$p_n(0) = p_{no} \times e^{\frac{V_O}{V_T} - \frac{V_O}{V_T} + \frac{V}{V_T}}$$

$$p_n(o) = p_{no} \times e^{V/V_T}$$

Therefore, the total hole concentration in 'n' region at the junction varies with applied forward bias voltage V as given by the above expression.

This is called the **Law of the Junction**.

$$p_n(0) = p_n(0) - p_{no}$$

$$= p_{no}\, e^{V/V_T} - p_{no}$$

$$\boxed{p_n(0) = p_{no}\left(e^{V/V_T} - 1\right)} \qquad .....(6.4.3)$$

The Forward Current

The hole current in the *n-side* ***Ipn(x)*** is given as

$$Ipn(x) = \frac{Ae \times Dp}{L_p}\, p_n(0)\, e^{-x/L_p}$$

But
$$p_n(0) = p_{no}\left(e^{V/V_T} - 1\right)$$

$$Ipn(0) = \frac{Ae \times Dp}{L_p} \times p_{no}\left(e^{V/V_T} - 1\right)$$

D_p = Diffusion coefficient of holes

D_n = Diffusion coefficient of electrons.

$\therefore$ e^{-x/L_p} at $x = 0$ is 1.

Similarly the electron current due to the diffusion of electrons from *n-side* to *p-side* is obtained from the above equation itself, by interchanging n and p.

$$\therefore \quad Inp(0) = \frac{Ae \times Dn}{L_n} \times n_{po}\left(e^{V/V_T} - 1\right)$$

The total diode current is the sum of ***Ipn (0)*** and ***Inp(0)***

or $$\boxed{I = I_o\left(e^{V/V_T} - 1\right)} \quad .....(6.4.4)$$

where $$I_0 = \frac{AeDp}{Lp} \times p_{no} + \frac{AeDn}{L_n} \times n_{po}$$

In this analysis we have neglected charge generation and recombination. Only the current that results as a result of the diffusion of the carriers owing to the applied voltage is considered.

Reverse Saturation Current

$$I = I_0 \times \left(e^{V/V_T} - 1\right)$$

This is the expression for current I when the diode is forward biased. If the diode is reverse biased, V is replaced by –V. V_T value at room temperature is ~ 26 mV. If the reverse bias voltage is very large, $e^{\frac{-V}{V_T}}$ is very small. So it can be neglected.

$$\therefore \quad I = -I_0$$

I_0 will have a small value and I_0 is called the ***Reverse Saturation Current.***

$$I_0 = \frac{AeD_p p_{no}}{L_p} + \frac{AeD_n n_{po}}{L_n}.$$

In *n-type* semiconductor,

$$n_n = N_D$$

But $$n_n \times p_n = n_i^2$$

$$\therefore \quad n_p = \frac{n_i^2}{N_A}$$

Substituting these values in the expression for I_0,

$$I_0 = Ae\left(\frac{D_p}{L_pN_D} + \frac{D_n}{L_n.N_A}\right) \times n_i^2$$

where $n_i^2 = A_0\, T^3\, e^{-E_G/KT}$

E_G is in electron volts = e.V_G, where V_G is in Volts.

$$\therefore \quad n_i^2 = A_0\, T^3\, e^{-\frac{E_{GO}}{KT}}$$

$$E_{G_O} = V_G \cdot e$$

But $\frac{KT}{e}$ = Volt equivalent of Temperature V_T.

For Germanium, D_p and D_n decrease with temperature and n_i^2 increases with T. Therefore, temperature dependance of I_0 can be written as,

$$I_0 = K_1 T^2\, e^{-V_G/V_T}$$

For Germanium, the current due to thermal generation of carriers and recombination can be neglected. But for Silicon it cannot be neglected. So the expression for current is modified as

$$I = I_0\left(e^{\frac{V}{\eta V_T}} - 1\right)$$

where n = 2 for small currents and n = 1 forM large currents.

Temperature Dependance Of p-n Junction Diode Characteristics

I_0 depends upon temperature. The expression is

$$I_0 = KT^m \cdot e^{-V_{G0}/\eta V_T}$$

$$ln(I_0) = lb\,(K) + m/n\,(T) - \left(T - \frac{V_{G0}}{\eta V_T}\right)$$

Reverse Saturation Current increases 7% /°C rise in temperature for both Silicon and Germanium. For a rise of 1°C in temperature, the new value of I_0 is,

$$I_0 = \left(1+\frac{7}{100}\right) I_0$$
$$= 1.07\ I_0.$$

For another degree rise in temperature, the increase is 7% of $(1.07\ I_0)$.

Therefore for 10 °C rise in temperature, the increase is $(1.07)^{10} = 2$.

Thus for every 10 °C rise in temp I_0 for Silicon and Germanium gets doubled.

6.5 VOLT-AMPERE CHARACTERISTICS

The general expression for I for a *p-n junction* diode is

$$I = I_0\left(e^{\frac{V}{\eta V_T}} - 1\right).$$

$\eta = 1$ for Germanium. 1 or 2 for Silicon. For Silicon, I will be less than that for Germanium.

$$V_T = 26\ \text{mV}.$$

If V is much larger than V_T, 1 can be neglected. So I increases exponentially with forward bias voltage V. In the case of reverse bias, if the reverse voltage –V >> V_T, then e^{-V/V_T} can be neglected and so reverse current is $-I_0$ and remains constant independent of V. So the characteristics are as shown in Fig. 6.8 and not

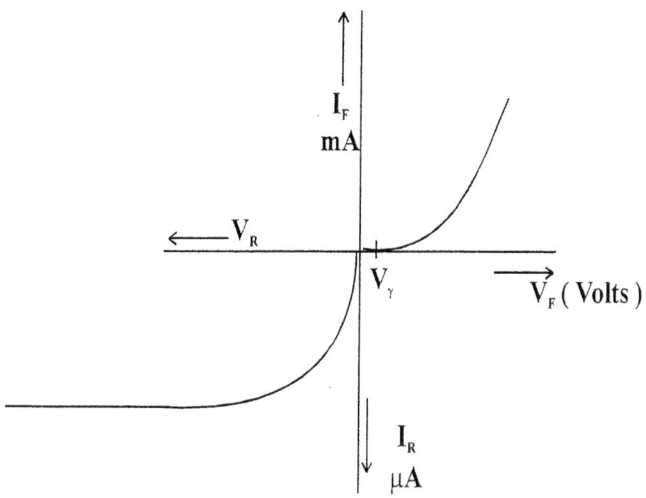

Fig. 6.8 *V-I Characteristics of p-n junction Diode*

like theoretical characteristics. The difference is that the practical characteristics are plotted at different scales. If plotted to the same scale, (reverse and forward) they may be similar to the theoretical curves. Another point is, in deriving the equations the breakdown mechanism is not considered. As V increases **Avalanche multiplication** sets in. So the actual current is more than the theoretical current.

6.5.1 Cut in Voltage V_γ

In the case of Silicon and Germanium, diodes there is a ***Cut In*** or ***Threshold*** or ***Off Set*** or ***Break Point Voltage***, below which the current is negligible. It's magnitude is 0.2V for Germanium and 0.6V for Silicon (Fig. 6.9).

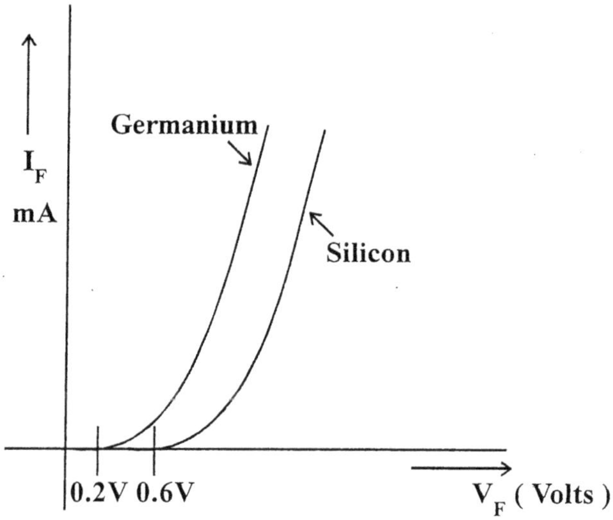

Fig. 6.9 *Forward characteristics of a diode*

6.6 TEMPERATURE DEPENDANCE

The expression for reverse saturation current I_0

$$I_0 = Ae\left(\frac{D_p}{L_pN_D} + \frac{D_n}{L_n.N_A}\right) \times n_i^2$$

$$n_i^2 \;\alpha\; T^3$$

$$\therefore \quad n_i^2 = A_0T^3\, e^{-E_{G0}/KT}$$

D_p decreases with temperature.

$$\therefore \quad I_0 \;\alpha\; T^2$$

$$\text{or} \quad I_0 = KT^m\, e^{-V_{G0}/\eta V_T}$$

where V_G is the energy gap in volts. (E_G in eV)

For Germanium, $\eta = 1, m = 2$

For Silicon, $\eta = 2, m = 1.5$

$$I_0 = KT^m e^{-V_{G0}/\eta V_T}$$

Taking ln, (Natural Logarithms) on both sides,

$$\ln I_0 = \ln(K) + m \times \ln(T) \frac{-V_{G0}}{\eta V_T}$$

Differentiating with respect to temperature,

$$\frac{1}{I_0} \times \frac{dI_0}{dT} = 0 + \frac{m}{T} - \left(-\frac{V_{G0}}{\eta V_T} \times \frac{1}{T} \right)$$

$$\frac{1}{I_0} \times \frac{dI_0}{dT} = \frac{m}{T} + \frac{V_{G0}}{\eta T V_T}$$

$\frac{m}{T}$ value is negligible 7

$$\therefore \boxed{\frac{1}{I_0} \times \frac{dI_0}{dT} \simeq \frac{V_{G0}}{\eta T V_T}} \qquad .....(6.6.1)$$

Experimentally it is found that reverse saturation current increases $\simeq$ 7% / °C for both Silicon and Germanium or for every 10° C rise in temperature, I_0 gets doubled. The reverse saturation current increases if expanded during the increasing portion.

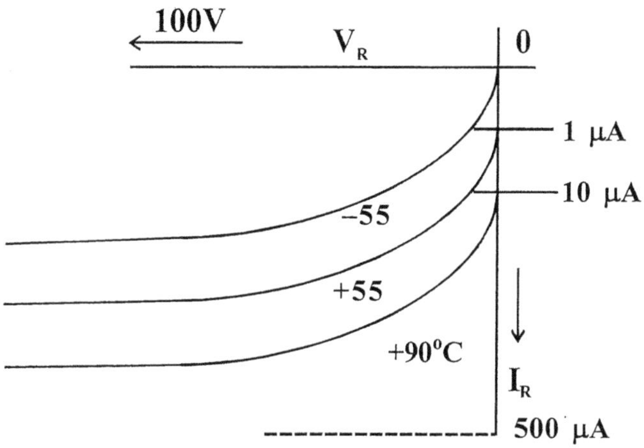

Fig. 6.10 Reverse Characteristics of a p-n junction *Diode*

6.6.1 Application of p-n Junction Diode

1. *As a rectifier*
2. *As a switch*
3. *In clipping circuits*
4. *In Clamping circuits*
5. *Modulator / Demodulator circuits*
6. *Bias stabilization circuits.*

6.7 IDEAL VS PRACTICAL DIODE

Ideal: The p-n junction diode allows flow of heavy current during forward bias condition and allows a small level of current when it is reverse biased. The direction of the arrow in the following Fig.6.11(a) depicts the direction of flow of current when the diode is forward biased. Similarly the direction of the arrow in Fig.6.11 (b) depicts the flow of current direction when the diode is reversed biased.

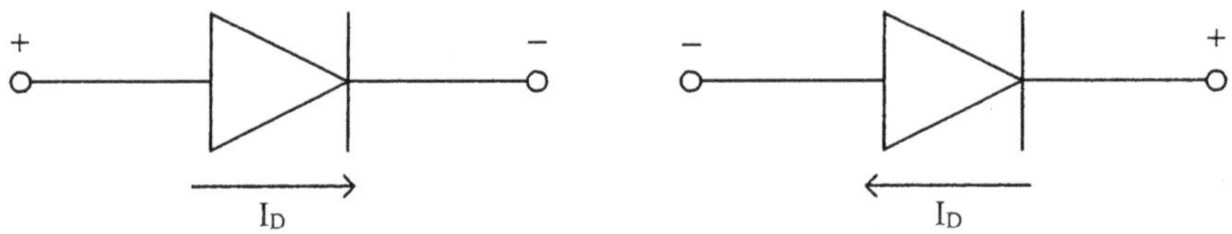

Fig. 6.11(a) Illustration of flow of current when the diode is forward biased

Fig. 6.11(b) Illustration of flow of current when the diode is Reverse biased

Analogy

A common analogy that can be employed to describe the behaviour of a p-n junction diode is Mechanical switch.

During the forward bias the p-n junction diode allows flow of heavy currents through the junction. And hence the amount of resistance offered by the p-n junction diode during forward bias condition is very small and assumed to be zero. Thus the pn junction diode acts as a closed switch when the diode is forward biased. On the other hand when the diode is reverse biased it offers a flow very small level of current across the junction which is assumed to be zero. In otherwords it offers very high amount of resistance during its reverse bias (prevents flow of current across the junction i.e., offer ∞ resistance). Thus the p-n junction diode acts on a open switch during the reverse bias.

The following Fig.6.12(a) shows the analogy of closed switch of p-n junction diode and Fig. 6.12(b) shows the analogy of open switch of p-n junction diode.

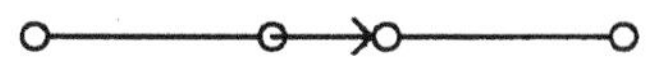

Fig. 6.12(a) Analogy of p-n junction diode as closed switch

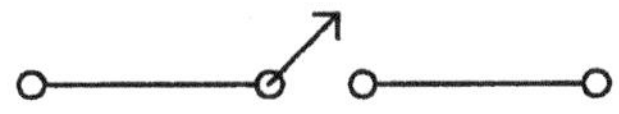

Fig. 6.12(b) Analogy of p-n junction diode as open switch

Practical

Even though the p-n junction diode offers a flow of heavy currents during forward bias, a small amount of current (which is due to presence of minority charge carriers across the junction) opposes the flow of forward biased current (I_D). In otherwords the p-n junction diode offers a small amount of forward resistance (r_f) which opposes the flow forward biased current (I_D).

Similarly when the diode is reverse biased the switch is assumed to be open but in practical the p-n junction diode allows the flow small amounts of currents through the junction which can be refered as leakage current (or) minority charge carrier current.

Ideal vs Practical

The following figure shows super imposition of practical characteristics in ideal characteristics. From the figure we can observe that the ideal diode does not offer any forward resistance and hence it does not require any additional voltage to make I_D to flow through the diode. In otherwords the forward bias current I_D starts flowing through the diode even when applied forward bias voltage is at 0V.

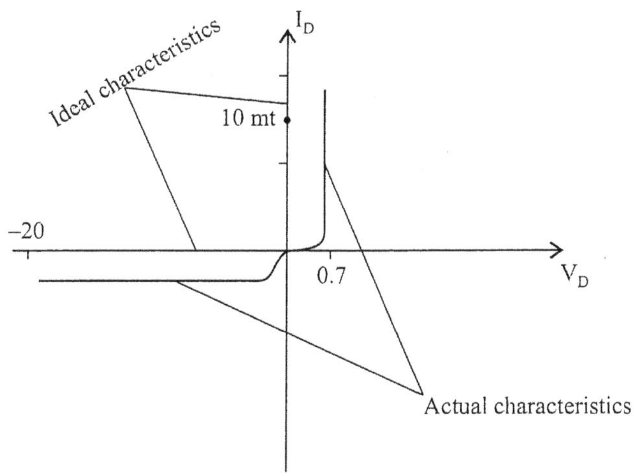

Where as in case of practical diode as it offers a small amount of forward resistance (r_f) it requires a knee voltage (or) threshold voltage of 0.6 in case of silicon (0.3 of germanium) to make forward current I_D to flow through the junction.

Analogy: Closed Switch

Ideal: When the switch is closed, the resistance between the contacts is assumed to be zero. In the plot a point chosen in vertical axis where the diode current 10 mA and voltage across diode is 0V. According to Ohm's law.

$$R_F = \frac{V_D}{I_D} = \frac{0V}{10 \text{ mA}} = 0\ \Omega$$

Practical: In practical diode the minimum voltage required to obtain I_D is 0.6 (for Si). Hence

$$R_F = \frac{V_D}{I_D} = \frac{0.6}{10 \text{ mA}} = 60\ \Omega$$

Analogy: Open switch

Ideal: For any value of voltage on horizontal axis (any point on X-axis has zero coordinate) the current is zero. Applying ohm's law. In plot V_D is taken as 20.

$$R_r = \frac{V_D}{I_D} = \frac{20}{0 \text{ mA}} = \infty$$

Practical: For practical diode a small level of current (usually in order of hA for si and mA for Ge) flows when the diode is reverse biased. Hence

$$R_r = \frac{20}{10 \text{ mA (assume)}} = \frac{20}{10 \times 10^{-9}} = 2000 \text{ M}\Omega \text{ for si}$$

	Ideal	Practical
When switch is closed	$R_f = 0$	$R_f = R_f$ (in order of ohms)
When switch is open	$R_r = \infty$	$R_r = R_r$ (in order of $M\Omega$)

6.8 STATIC AND DYNAMIC RESISTANCES

The static resisitance (R) of a diode is defined as the ratio of $\frac{V}{I}$ of the diode.

Static resistance varies widely with V and the dynamic resisitance or incremental resistance is defined as the reciprocal of the slope of the Volt-Ampere Characteristic $\frac{dV}{dI}$. This is also not a constant but depends upon V and I.

6.9 EQUIVALENT CIRCUITS

An equivalent circuit is a combination of elements that are chosen to represent actual characteristics of device for particular operating region.

Piecewise Linear Equivalent Circuit: The best method of obtaining equivalent circuit is to approximate its characteristics with straight line segments. The forward bias characteristics of p-n junction diode is shown in following Fig. 6.13.

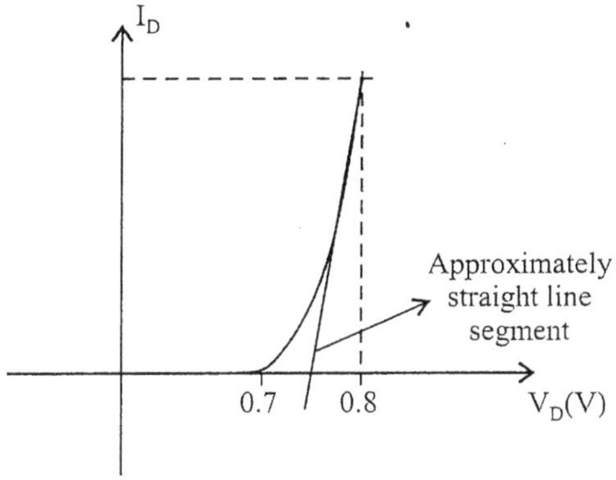

Fig. 6.13 *Approximation of diode characteristics*

The above Fig. 6.13 represent approximation of exact characteristics of p-n junction diode. From the figure it is clear that this approximation does not yield the exact duplication of actual characteristics of p-n junction diode. However these approximation yields sufficiently close to actual curve to represent the behaviour of the diode.

The actual characteristics have knee voltage of 0.7 V but the approximate model has knee voltage greater than 0.7 V. To achieve this 0.7 V of knee voltage connect a battery of 0.7 V to the p-n junction diode assuring its polarity is in opposition with the applying voltage as show in following Fig.6.14 and assuming the diode to be ideal, the ideal diode does not conduct any current when it is in reverse bias condition hence approximation of reverse bias characteristics of diode can be neglected.

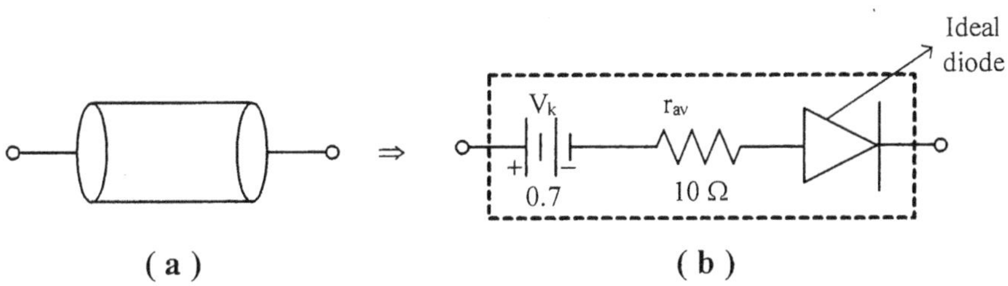

Fig. 6.14 *Equivalent circuit of diode*

The above Fig. 6.14 assures that the diode is not conducted until voltage across it reaches greater than its knee voltage i.e., 0.7. And once the diode starts conducting it offers a specified value of resistance r_{av}.

The approximate value of r_{av} can be obtained from the operating point of the diode. Assuming its operating point to be (10, 0.8) i.e., the current through the diode is 10 mA when applied voltage is 0.8 (show in Fig. 6.17(a)). We get

$$r_{av} = \left.\frac{\Delta V_d}{\Delta I_d}\right|_{\text{pt to pt}} = \frac{0.8 - 0.7}{10\text{ mA} - 0\text{ mA}} = \frac{0.1}{10\text{ mA}} = 10\ \Omega$$

$\therefore \qquad r_{av} = 10\ \Omega$

The 0.7 V is knee voltage i.e., minimum voltage required to make diode ready for conducting and at exact knee point the diode will be ready to conduct, but it will not be conducting hence I_D at knee voltage is 0 mA. When the voltage across the diode just crosses its knee voltage the diode starts conducting and hence I_D starts increasing. But at point when applied voltage is equal to knee voltage the diode will be in equilibrium resulting $I_D = 0$.

Simplified Equivalent Circuit

In general the value of r_{av} is very small in comparison with the other elements of the circuit and hence it can be neglected. Removing r_{av} from diode equivalent circuits give simplified equivalent diode circuit. The following shows the simplified equivalent circuit of the diode and its characteristics.

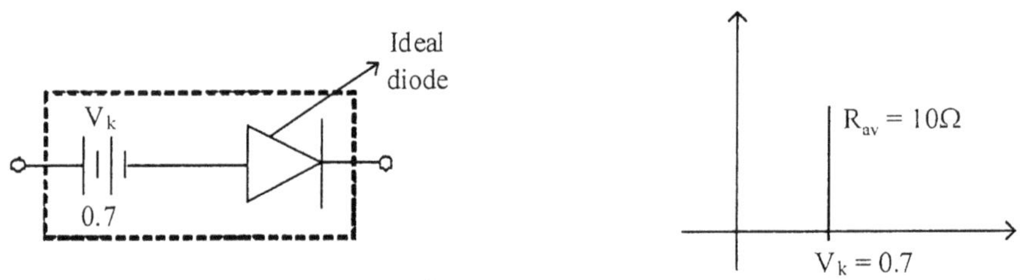

6.10 DIFFUSION CAPACITANCE

When a *p-n junction* diode is forward biased, the junction capacitance will be much larger than the transition capacitance C_T. When the diode is forward biased, the barrier potential is reduced. Holes from *p-side* are injected into the *n-side* and electrons into the *p-side*. Holes which are the minority carriers in the *n-side* are injected into the *n-side* from the *p-region*. The concentration of holes in the *n-side* decrease exponentially from the junction. So we can say that a positive charge is injected into the *n-side* from *p-side*. This injected charge is proportional to the applied forward bias voltage 'V'. So the rate of change of injected charge 'Q' with voltage 'V' is called the Diffusion Capacitance C_D. Because of C_D total capacitance will be much larger than C_T in the case of forward bias, (C_D is few mF (2mF.)) C_T value will be a few pico farads.

6.10.1 Derivation for C_D

Assume that, the *p-side* is heavily doped compared to *n-side*. When the diode is forward biased, the holes that are injected into the *n-side* are much larger than the electrons injected into the *p-side*. So we can say that the total diode current is mainly due to holes only. So the excess charge due to minority carriers will exist only on *n-side*. The total charge Q is equal to the area under the curve multiplied by the charge of electrons and the cross sectional area A of the diode. $P_n(0)$ is the Concentration of holes/cm^3. Area is in cm^2, x in cm,

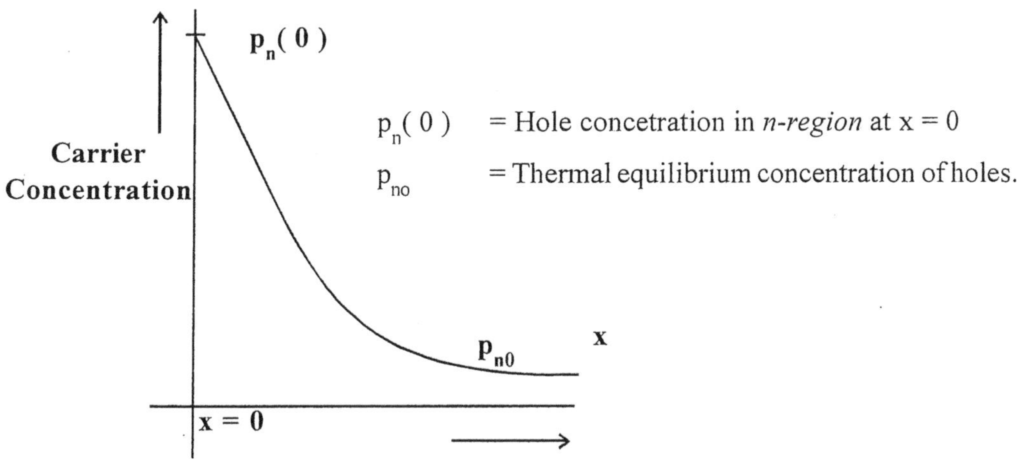

Fig. 6.15 Carrier concentration variation

$$\therefore \quad Q = \int_0^{\infty} AeP_n(0)e^{-x/Lp}\,dx$$

$$= Ae\,P_n(0)\int_0^{\infty} e^{-x/Lp}.dx$$

$$Q = AeP_n(0)\,[-Lp\,[0\text{-}1]\,]$$

$$= AeL_p p_n(0)$$

$$C_D = \frac{dQ}{dv} = AeL_p \times \frac{dp_n(0)}{dv} \qquad .....(6.10.1)$$

We know that

$$Ipn(x) = +\frac{e \times A \times Dp \times p_n(0) \times e^{-x/Lp}}{Lp}$$

or

$$Ipn(0) = I = \frac{AeD_p p_n(0)}{L_p}$$

$$p_n(0) = L_p \times I / AeD_p$$

$$\frac{dp_n(0)}{dv} = \frac{L_p}{AeD_p} \times \frac{dI}{dv}$$

$$= \frac{L_p}{AeD_p} \times g \qquad \text{..........(6.10.2)}$$

where g is the conductance of the diode.

Substitute equation (6.10.2) in (6.10.1).

$$\therefore \quad C_D = Ae \times L_p \times \left(\frac{L_p}{AeD_p}\right) \times g$$

$$= \frac{L_p^2}{D_p} \times g$$

The lifetime for holes $\tau_p = \tau$ is given by the eq.

$$\tau = \frac{L_p^2}{D_p}$$

where D_p = Diffusion coefficient for holes.

L_p = Diffusion length for holes.

D_p = cm^2/sec

$$\therefore \quad C_D = \tau \times g.$$

But diode resistance

$$r \simeq \frac{\eta . V_T}{I}$$

where $\eta = 1$ for Germanium

$\eta = 2$ for Silicon

$$\therefore \quad g = \frac{I}{\eta.V_T}$$

$$\therefore \quad C_D = \frac{I.\tau}{\eta.V_T} \qquad \text{..........(6.10.3)}$$

C_D is proportional to I. In the above analysis we have assumed that the current is due to holes only. So it can be represented as C_{DP}. If the current due to electron is also to be considered then we get corresponding value of C_{Dn}. The total diffusion capacitance = $C_{DP} + C_{Dn}$. Its value will be ***around 20mF***.

$$C_D = \tau \times g$$

or $r \times C_D = \tau$

$r \times C_D$ is called the time constant of the given diode. It is of importance in circuit applications. Its value ranges from nano-secs to hundreds of micro-seconds.

Charge control description of a diode :

$$Q = A \times e \times L_p \times p_n(0)$$

$$I = \frac{AeD_p p_n(0)}{L_p}$$

$$\therefore \quad \frac{Q}{L_p} = A \times e \times p_n(0)$$

$$I = Q \times D_p/L_p^2$$

But $L_p^2/Dp = \tau$

$$\therefore \quad I = \frac{Q}{\tau}$$

6.11 TRANSISITION CAPACITANCE

When a reverse bias is applied to a *p-n junction* diode, electrons from the *p-side* will move to the *n-side* and vice versa. When electrons cross the junction into the *n-region*, and hole away from the junction, negative charge is developed on the *p-side* and similarly positive charge on the *n-side*. Before reverse bias is applied, because of concentration gradient, there is some space charge region. Its thickness increases with reverse bias. So space charge Q increases as reverse bias voltage increases.

But $\quad C = \frac{Q}{V}$

Therefore, Incremental Capacitance,

$$C_T = \left|\frac{dQ}{dV}\right|$$

where |dQ| is the magnitude of charge increase due to voltage dV. It is to be noted that there is negative charge on the *p-side* and positive charge on *n-side*. But we must consider only its magnitude.

$$\text{Current } I = \frac{dQ}{dt}$$

Therefore, if the voltage dV is changing in time dt, then a current will result, given by

$$I = C_T \times \frac{dV}{dt}$$

This current exists for A.C. only. For D.C, Voltage is not changing with time. For D.C. capacitance is open circuit.

The knowledge of C_T is important in considering diode as a circuit element. *C_T is called Transition region capacitance or space charge capacitance or barrier capacitance, or depletion region capacitance.* This capacitance is not constant but depends upon the reverse bias voltage V. If the diode is forward biased, since space charge $\approx$ 0,(this doesn't exist). It will be negligible. C_T is of the order of 50 pf etc.

6.11.1 Alloy Junction

Indium is trivalent. If this is placed against *n-type* Germanium, and heated to a high temperature, indium diffuses into the Germanium crystal, a pn junction will be formed and for such a junction there will be abrupt change from acceptor ions on one side to donor ions on the other side. Such a junction is called ***Alloy Junction*** or ***Diffusion Junction***. In the figure, the acceptor ion concentration N_A and donor atom concentration, N_D is shown in Fig. 6.16. There is sudden change in concentration levels. To satisfy the condition of charge neutrality,

$$e \times N_A \times W_P = e \times N_D \times W_n.$$

If $N_A << N_D$, then $W_p >> W_n$. In practice the width of the region, W_n will be very small. So it can be neglected. So we can assume that the entire barrier potential appears across the *p-region* just near the junction.

Poisson's Equation gives the relation between the charge density and potential.

It is $\quad \frac{d^2V}{dx^2} = \frac{eN_A}{\epsilon}$

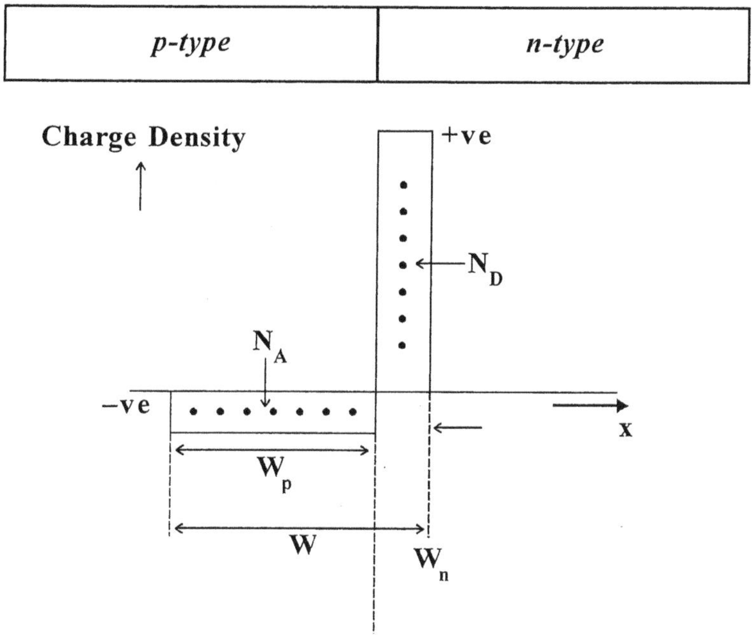

Fig. 6.16 Abrupt p-n junction

where $\in$ is the permitivity of the semiconductor

$$\frac{dV}{dx} = \frac{eN_A}{\in} x$$

$$V = \frac{e.N_A}{\in} \times \frac{x^2}{2}$$

At $x = w_p$, $V = V_B$ the barrier potential. $W_p = W$.

$$\therefore \quad V_B = \frac{e.N_A}{2\in} \times W^2 \qquad \text{..........(6.11.1)}$$

The value of W depends upon the applied reverse bias V. If V_0 is the contact potential, $V_B = V_o - V$ where V is the reverse bias voltage with negative sign.

So as V increases, W also increases and V_B increases

$$W \, \alpha \, \sqrt{V_B}$$

If A is the area of the junction, the charge in the width W is

$$Q = e \times N_A \times W \times A$$

where $\quad W \times A = \text{Volume}$

$$e \times N_A = \text{Total charge density}$$

$$C_T = \left|\frac{dQ}{dV}\right| = e \times N_A \times A \times \left|\frac{dW}{dV}\right|$$

$$Q = \sqrt{e \times N_A}\sqrt{2 \in V_B \times A}$$

But $$V_B = \frac{e.N_A}{2 \in} \times W^2$$

or $$W = \sqrt{\frac{2 \in V_B}{e \times N_A}}$$

$$\frac{dW}{dV} = \frac{1}{2}\sqrt{\frac{2 \in}{e \times N_A}}\, V_B^{-\frac{1}{2}}$$

But $$\sqrt{V_B} = W\sqrt{\frac{e \times N_A}{2 \in}}$$

$$\frac{dW}{dV} = \frac{1}{2}\sqrt{\frac{2 \in}{e \times N_A}} \times \frac{\sqrt{2 \in}}{\sqrt{e \times N_A} \times W}$$

$$= \frac{\in}{e \times N_A} \times \frac{1}{W}$$

$$= \frac{\in}{e \times N_A \times W}$$

$$\therefore \quad \boxed{C_T = \frac{\in \times A}{W}} \qquad \text{..........(6.11.2)}$$

This expression is similar to that of the Parallel Plate Capacitor.

Grown Junction

This is the second method of forming a *p-n junction*. This is obtained by drawing a single crystal from a molten Germanium. During this process first *p-type* and then *n-type* purities are added. For such a grown junction, charge density varies linearly.

RECTIFIERS AND FILTERS

6.12 p-n JUNCTION AS A RECTIFIER

The electronic circuits require a D.C. source of power. For transistor A.C. amplifier circuit for biasing, D.C supply is required. The input signal can be A.C. and so the output signal will be amplified A.C. signal. But without biasing with D.C. supply, the circuit will not work. So more or less all electronic A.C. instruments require D.C. power. To get this, D.C. batteries can be used. But they will get dried quickly and replacing them every time is a costly affair. Hence it is economical to convert A.C. power into D.C. Such circuits, their efficiency (η) etc., will be discussed in this Chapter.

Rectifier is a circuit which offers low resistance to the current in one direction and high resistance in the opposite direction.

Rectifier converts sinusoidal signal to unidirectional flow and not pure D.C.

Filter converts unidirectional flow into pure D.C.

6.13 HALF WAVE RECTIFIER

If the input to the rectifier is a pure sinusoidal wave, the average value of such a wave is zero, since the positive half cycle and negative half cycle are exactly equal.

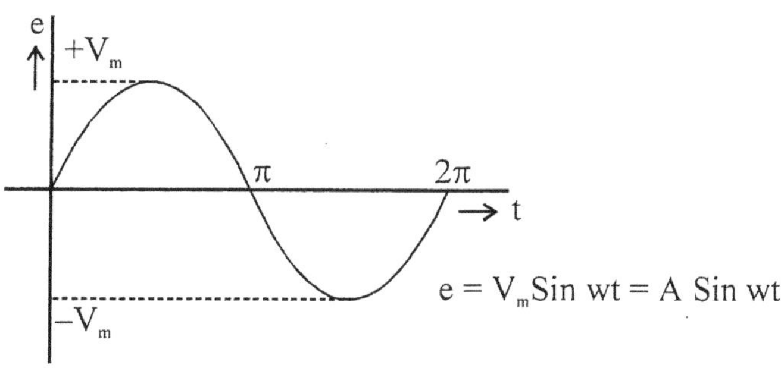

Fig. 6.17 AC input wave form

$$T_{av} = \frac{1}{2\pi} \int_0^{2\pi} A \, Sin(\omega t) \, dt$$

If this signal is given to the rectifier circuit, say Half Wave Rectifier Circuit, the output will be as shown in Fig. 6.18.

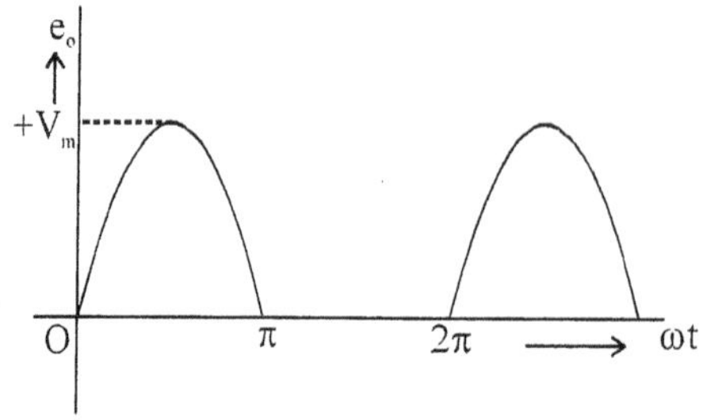

***Fig. 6.18** Half Wave Rectified Output (unidirectional flow)*

Now the average value of this waveform is *not zero*, since there is no negative half. Hence a rectifier circuit converts A.C. Signal with zero average value to a unidirectional wave form with non zero average value.

The rectifying devices are semiconductor diodes for low voltage signals and vacuum diodes for high voltage circuit. A basic circuit for rectification is as shown in Fig. 6.19.

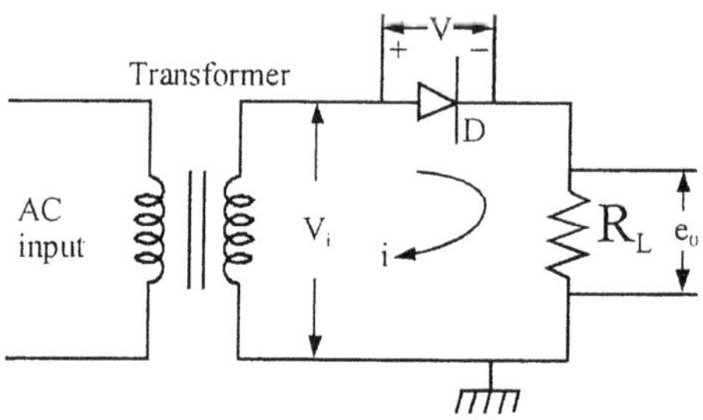

***Fig. 6.19** Halfwave Rectifier (HWR) circuit*

A.C input is normally the A.C. main supply. Since the voltage is 230V, and such a high voltage cannot be applied to the semiconductor diode, step down transformer should be used. If large D.C. voltage is required vacuum tubes should be used. Output voltage is taken across the load resistor R_L. Since the peak value of A.C. signal is much larger than V_γ, we neglect V_γ, for analysis.

The output current waveform for half wave rectification is shown in Fig. 6.21

$$\therefore \quad i = I_m \, Sin\, \alpha \qquad \alpha = \omega t \qquad 0 \le \alpha \le \pi$$

$$i = 0 \qquad \pi \le \alpha \le 2\pi$$

$$I_m = \frac{V_m}{R_f + R_L}$$

$i = I_m \sin\alpha$ between $0 \le \alpha \le \pi$
$i = 0$ between $\pi \le \alpha \le 2\pi$

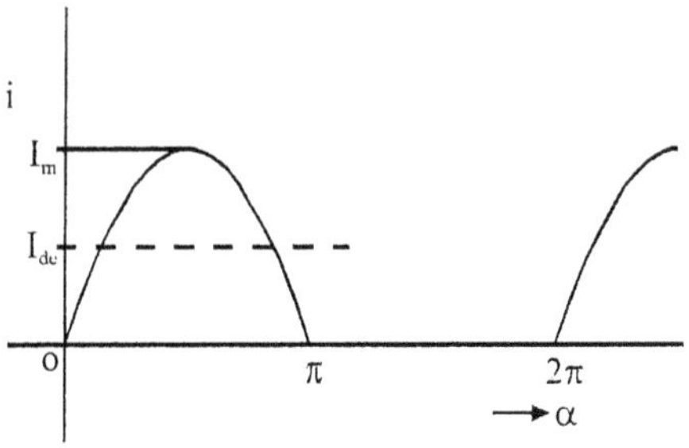

Fig. 6.20 Half Wave Rectified Output

R_f is the forward resistance of the diode. R_L is the load resistance

If a D.C. Ammeter is connected in the rectifier output circuit, what reading will it indicate? Is it the peak value or will the needle oscillate from O to maximum and then to O and so on, or will it indicate average value? The meter is so constructed that it reads the average value.

By definition, average value $= \dfrac{\text{Area of the curve}}{\text{Base}}$

∴ For Half wave rectified output, base value is 2π for one cycle,

$$I_{DC} = \frac{1}{2\pi}\int_0^{\pi} I_m \sin\alpha \, d\alpha = \frac{I_m}{2\pi}\cdot\left|-\cos\alpha\right|_0^{\pi}$$

$$I_{DC} = \frac{I_m}{2\pi}[1+1] = \frac{I_m}{\pi}$$

Upper limit is only π, because the signal is zero from π to 2π. The complete cycle is from 0 to 2π.

6.14 READING OF A.C AMMETER

An A.C. ammeter is constructed such that the needle deflection indicates the effective or RMS current passing through it.

Effective or RMS value of an A.C. quantity is the equivalent D.C. value which produces the same heating effect as the alternating component. If some A.C. current is passed through a resistor, during positive and negative half cycles, also because of the current, the resistor gets heated, or there is some equivalent power dissipation. What is the value of D.C. which produces the same heating effects as the A.C. quantity? The magnitude of this equivalent D.C. is called the RMS or Effective Value of A.C.

By definition $$I_{rms} = \sqrt{\frac{I}{2\pi}\int_0^{2\pi}(I_m \sin\alpha)^2 d\alpha} = \sqrt{\frac{I_m^2}{2\pi}\int_0^{2\pi}\sin^2\alpha . d\alpha}$$

$$= \frac{I_m}{\sqrt{2\pi}}\sqrt{\int_0^{2\pi}\left\{\frac{1-\cos 2\alpha}{2}\right\}d\alpha} = \frac{I_m}{\sqrt{2\pi}}\sqrt{\left.\frac{\alpha}{2}\right|_0^{2\pi} + \left.\frac{\sin 2\alpha}{4}\right|_0^{2\pi}}$$

$$\therefore \quad \text{RMS value of a sine wave} = \frac{I_m}{\sqrt{2\pi}} \times \sqrt{\frac{2\pi}{2} + 0}$$

$$= \frac{I_m\sqrt{\pi}}{\sqrt{2}\sqrt{\pi}} = \frac{I_m}{\sqrt{2}} = 0.707\, I_m$$

$$\text{Form Factor} = \frac{\text{RMS Value}}{\text{Average Value}}$$

For a sine wave, $I_{average} = 0.636 I_m = 2 \times \frac{1}{2\pi} \int_0^{\pi} (I_m \operatorname{Sin}\alpha) d\alpha$

$$I_{rms} = \frac{I_m}{\sqrt{2}} = 0.707 I_m$$

$$\text{Form factor} = \frac{\frac{I_m}{\sqrt{2}}}{} = \frac{0.707 I_m}{0.636 I_m} = 1.11$$

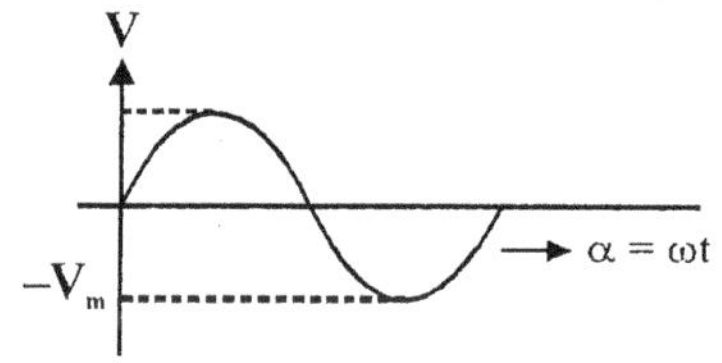

Fig. 6.21

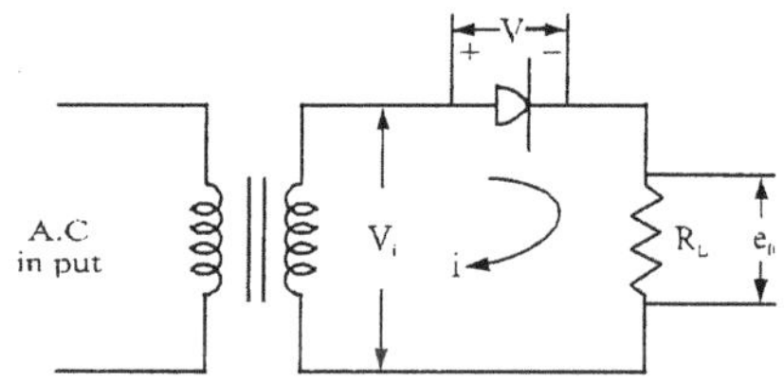

Fig. 6.22 Half Wave Rectifier Circuit

6.15 PEAK INVERSE VOLTAGE

For the circuit shown, the input is A.C. signal. Now during the positive half cycle, the diode conducts. The forward resistance of the diode R_f will be small.

$\therefore$ The voltage across the diode $v = i \times R_f$

$i = I_m \operatorname{Sin}\alpha \qquad 0 \le \alpha \le \pi.$

$\therefore\ v = I_m R_f \operatorname{Sin}\alpha \qquad 0 \le \alpha \le \pi.$

Since R_f is small, the voltage across the diode V during positive half cycle will be small, and the waveform is as shown. But during the negative half cycle, the diode will not conduct. Therefore, the current i through the circuit is zero. So the voltage across the diode is not zero but the voltage of the secondary of the transformer V_i will appear across the diode ($\because$ effectively the diode is across the secondary of the transformer.)

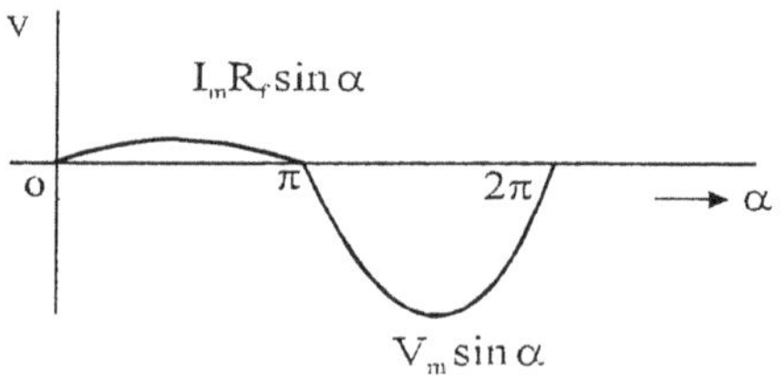

Fig. 6.23 Voltage Waveform across the Diode

$\therefore \quad v = V_m \text{ Sin } \alpha \qquad \pi \le \alpha \le 2\pi.$

The waveform across the diode is Fig. 6.23.

So the D.C. Voltage that is read by a D.C. Voltmeter is the average value.

$$V_{DC} = \frac{1}{2\pi}\left[\int_0^{\pi} I_m R_f.\text{Sin}\alpha \, d\alpha + \int_{\pi}^{2\pi} V_m \text{Sin}\alpha \, d\alpha\right]$$

$$= \frac{1}{2\pi}\left[(2)I_m R_f - (2)I_m (R_f + R_L)\right]$$

$$\because \quad V_m = I_m (R_f + R_c)$$

$$\therefore \quad \boxed{V_{DC} = \frac{-I_m R_L}{\pi}} \qquad .....(6.15.1)$$

If we connect a CRO across the diode, this is the wave form that we see is as shown in Fig. 6.24. So in the above circuits the diode is being subjected to a maximum voltage of V_m. It occurs when the diode is not conducting. Hence it is called the ***Peak Inverse Voltage*** (**PIV**)

6.15.1 Regulation

The variation of D.C output voltage as a function of D.C load current is called '*regulation*'.

$$\% \text{ Regulation} = \frac{V_{\text{No Load Voltage}} - V_{\text{Full Load}}}{V_{\text{Full Load}}} \times 100\,\%$$

For an ideal power supply output voltage is independent of the load or the output in voltage remains constant even if the load current varies, like in Zener diode near breakdown. Therefore, ***regulation is zero or it should be low for a given circuit***.

6.15.2 Expression for V_{DC} the Output DC Voltage

For half wave rectifier circuit, I_{DC} the average value is :

The waveform is shown is Fig. 6.24.

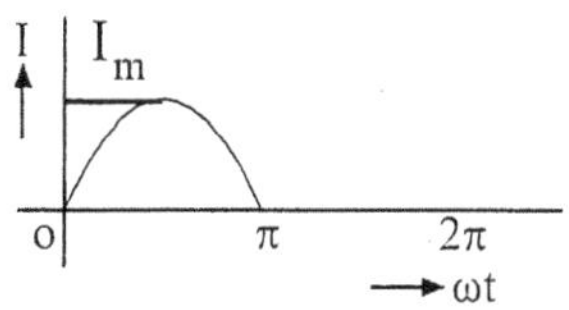

***Fig. 6.24** HWR current output*

$$I_{DC} = \frac{1}{2\pi}\int_0^{2\pi} i.d\alpha$$

$$= \frac{1}{2\pi}\int_0^{\pi} I_m \sin\alpha.d\alpha = \frac{I_m}{\pi}$$

$$\boxed{I_{DC} = \frac{I_m}{\pi}} \qquad .....(6.15.2)$$

But $$I_m = \frac{V_m}{R_f + R_L}$$ $R_L =$ Load Resistance

$R_f =$ Forward Resistance of the Diode.

$$V_{DC} = I_{DC}\, R_L$$

$$\therefore \quad I_{DC} = \frac{V_m / \pi}{R_f + R_L}$$

But $$I_{DC} = \frac{V_m}{\pi(R_f + R_L)}$$

$$\therefore \quad V_{DC} = \frac{V_m . R_L}{\pi(R_f + R_L)}$$

Adding and Subtracting R_f,

$$V_{DC} = \frac{V_m .(R_L + R_f - R_f)}{\pi(R_f + R_L)}$$

$$= \frac{V_m}{\pi} - \frac{V_m . R_f}{\pi(R_f + R_L)}$$

$$I_{DC} = \frac{V_m}{\pi(R_f + R_L)}$$

I_{DC} is determined by R_L. Hence V_{DC} depends upon R_L.

$$\boxed{V_{DC} = \frac{V_m}{\pi} - I_{DC} \times R_f} \qquad .....(6.15.3)$$

This expression indicates that V_{DC} is $\frac{V_m}{\pi}$ at ***no load or when the load current is zero***, and it decreases with increase in I_{DC} linearly since R_f is more or less constant for a given diode. The larger the value of R_f, the greater is the decrease in V_{DC} with I_{DC}. But, the series resistance of the secondary winding of the transformer R_s should also be considered.

For a given circuit of half wave rectifier, if a graph is plotted between V_{DC} and I_{DC} the slope of the curve gives ($R_f + R_s$) where R_f is the forward resistance of diode and R_s the series resistance of secondary of transformer.

Suppose for a given rectifier circuit, the specifications are 15V and 100 mA, i.e., the no load voltage is 15V and max load current that can be drawn is 100 mA. If the value of ($R_f + R_s$) = 25 Ω then the percentage regulation of the circuit is :

No Load Voltage = 15V

Drop across Diode = $I_m (R_f + R_s)$

R_S = Transformer Secondary Resistance

Max. Voltage with load = $15V - (I_m \times (R_f + R_s))$

$= 15 - 100 \text{ mA} \times 25\Omega$

$= 15 - 2.5 \text{ volts} = 12.5V$

$\therefore$ Percentage Regulation $= \dfrac{15-12.5}{12.5} \times 100 = \dfrac{2.5}{12.5} \times 100 \simeq 20\%.$

6.15.3 Power Supply Specifications

The input characteristics which must be specified for a power supply are :

1. *The required output D.C voltage*
2. *Regulation*
3. *Average and peak currents in each diode*
4. *Peak inverse voltage (**PIV**)*
5. *Ripple factor.*

6.16 RIPPLE FACTOR

The purpose of a rectifier circuit is to convert A.C. to D.C. But the simple circuit shown before will not achieve this. Rectifier converts A.C. to unidirectional flow and not D.C. So filters are used to get pure D.C. Filters convert unidirectional flow into D.C. Ripple factor is a measure of the fluctuating components present in rectifier circuits.

$$\text{Ripple factor, } \gamma = \frac{\text{RMS Value of alternating components of the waveform}}{\text{Average Value of the waveform}}$$

$$\gamma = \frac{I'_{rms}}{I_{DC}} = \frac{V'_{rms}}{V_{DC}}$$

I'_{rms} and V'_{rms} denote the value of the A.C components of the current and voltage in the output respectively. While determining the ripple factor of a given rectifier system experimentally, a voltmeter or ammeter which can respond to high frequencies (greater than power supply frequency 50 Hz) should be used and a capacitor should be connected in series with the input meter in order to block the D.C. component. Ripple factor should be small. (Total current $i = I_m \sin \omega t$ according to Fourier Series, only A.C. is sum of D.C. and harmonics).

We shall now derive the expression for the ripple factor. The instantaneous current is given by $i' = i - I_{DC}$.

i is the total current. In a half wave rectifier, some D.C components are also present. Hence A.C component is,

$$i' = (i - I_{DC})\ [\ (\text{Total current} - I_{DC})\ ]$$

$$\text{RMS value is } I'_{rms} = \sqrt{\frac{1}{2\pi}\int_0^{2\pi}(i - I_{DC})^2 d\alpha}$$

$$\therefore \quad I'_{rms} = \sqrt{\frac{1}{2\pi}\int_0^{2\pi}\left(i^2 - 2I_{DC}.i + I_{DC}^2\right) d\alpha}$$

Now, $\frac{1}{2\pi}\int_0^{2\pi} i^2 d\alpha$ = Square of the rms value of a sine wave by definition.

$$= (I_{rms})^2$$

$\frac{1}{2\pi}\int_0^{2\pi} i.d\alpha$ = The average value or D.C value I_{DC}.

I_{DC} is constant. So taking this term outside,

$$\frac{I_{DC}^2}{2\pi}\int_0^{2\pi} d\alpha = \frac{I_{dc}}{2\pi}[2\pi]$$

$$\therefore \quad I'_{rms} = \sqrt{(I_{rms})^2 - 2I_{DC}^2 + I_{DC}^2}$$

The rms ripple current is

$$I'_{rms} = \sqrt{(I_{rms})^2 - I_{DC}^2}$$

$$\text{Ripple factor, } \gamma = \frac{I'_{rms}}{I_{dc}} = \frac{\sqrt{(I_{rms})^2 - I_{DC}^2}}{I_{DC}}$$

$$\gamma = \sqrt{\left(\frac{I_{rms}}{I_{DC}}\right)^2 - 1} \qquad \text{.......... (6.16.1)}$$

This is independent of the current waveshape and is not restricted to a half wave configuration. If a capacitor is used to block D.C. and then I_{rms} or V_{rms} is measured,

Then, $\gamma = \frac{I_{rms}}{I_{DC}}$

$\because \quad I'_{rms} = \sqrt{I_{rms}^{\ 2} - I_{DC}}$

and $\quad I_{DC} = 0$ (blocked by Capacitor)

output waveform

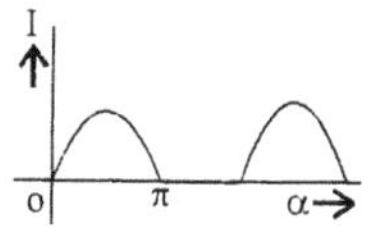

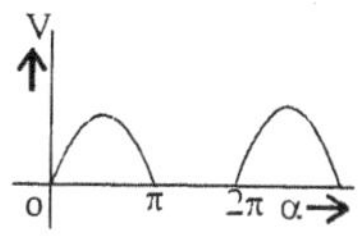

6.17 HALF WAVE RECTIFIER CIRCUIT (HWR)

$$I = I_m \sin \alpha$$

$$I_{av} = I_{DC} = \frac{1}{2\pi} \int_0^{\pi} I_m \sin \alpha . d\alpha = \frac{I_m}{\pi}$$

$$I_{rms} = \sqrt{\frac{1}{2\pi} \int_0^{\pi} I_m^2 \text{Sin}^2(\alpha)\, d\alpha} = \frac{I_m}{2}$$

$$I_{DC} = \frac{I_m}{\pi}; \; V_{DC} = \frac{V_m}{\pi} \quad \text{Peak Inverse Voltage (PIV)} = V_m$$

$$\text{Ripple Factor } \gamma = \frac{\text{RMS value of ripple current}}{\text{Average value of the current}} = \frac{I'_{rms}}{I_{DC}}$$

Total current $\quad I = I_{DC} + I'(\text{ripple})$

$I'(\text{ripple}) = (I - I_{DC})$

$$I'_{rms} = \sqrt{\frac{1}{2\pi} \int_0^{2\pi} (I - I_{DC})^2 \, d\alpha}$$

$$= \sqrt{\frac{1}{2\pi} \int_0^{2\pi} \left(I^2 - 2I.I_{DC} + I_{DC}^{\ 2}\right).d\alpha}$$

$$= \sqrt{(I_{rms})^2 - 2I_{DC}^{\ 2} + I_{DC}^{\ 2}}$$

$$\therefore \; I'_{rms} = \sqrt{(I_{rms})^2 - I_{DC}^{\ 2}}$$

$$\gamma = \frac{I'_{rms}}{I_{DC}} = \sqrt{\left(\frac{I_{rms}}{I_{DC}}\right)^2 - 1}$$

for Half Wave Rectifier, $\gamma = \sqrt{\left(\frac{I_m \times \pi}{2 \times I_m}\right)^2 - 1} = \sqrt{\left(\frac{\pi}{2}\right)^2 - 1} = \mathbf{1.21}$

6.17.1 Ratio of Rectification (η) : $\frac{P_{DC}}{P_{AC}}$

$$P_{DC} = I_{DC}^2 \times R_L$$

$$P_{AC} = (I_{rms})^2(R_f + R_L)$$

$$\eta = \frac{P_{DC}}{P_{AC}} = \frac{I_m^2 . R_L \times 4}{\pi^2 \times I_m^2 . R_L} = \frac{4}{\pi^2} = 0.408 \simeq \mathbf{40.8\%}$$

6.17.2 Ratio of Rectification for Half Wave Rectifier

$$= \frac{\text{DC power delivered to the load}}{\text{AC input power from transfer secondary}}$$

$$= \frac{P_{DC}}{P_{AC}}$$

$$P_{DC} = I_{DC}^2 \times R_L \qquad \text{But } I_{DC} \text{ for HWR} = \frac{I_m}{\pi}$$

$$\therefore \qquad P_{DC} = \left(\frac{I_m}{\pi}\right)^2 \times R_L$$

P_{AC} is what a Wattmeter would indicate if placed in the HWR circuit with its voltage terminal connected across the secondary of the transformer.

From $\pi - 2\pi$ the diode is not conducting. Hence $I_{ac} = 0$.

$$P_{ac} = (I_{rms})^2 \ (R_f + R_L)$$

Q from $\pi - 2\pi$ the AC is not being converted to D.C. So this power is wasted, as heat across diode and transformer.

$$\therefore \qquad I_{rms}(\ \pi - 2\pi\ ) \text{ portion of AC is} = \sqrt{\frac{1}{2\pi}\int_{\pi}^{2\pi} I_m^2 \sin^2 \alpha d\alpha} = \frac{I_m}{2}$$

I_{rms} for HWR is $I_m/2$. During negative half cycle, diode is not conducting and current I in the loop is zero even though Voltage V is present.

$$\therefore \qquad P_{AC} = \left(\frac{I_m}{2}\right)^2 (R_f + R_L)$$

$$= \left(\frac{I_m}{2}\right)^2 \times R_L$$

$$\therefore \quad \frac{P_{dc}}{P_{ac}} = \frac{I_m^2 . R_L \times 4}{\pi^2 I_m^2 \times R_L} = \frac{4}{\pi^2} = 0.406$$

$\therefore$ Ratio of rectification for HWR = **0.406**

Considering ideal diode. If we consider R_f also, the expression = $\dfrac{0.406}{1+\left(\dfrac{R_f}{R_L}\right)}$

The AC input power is not converted to D.C. Only part of it is converted to D.C and is dessipated in the load. The balance of power is also dessipated in the load itself as AC power. We have to consider the rating of the secondary of the transformer. In ratio of rectification we have considered only the A.C output as the secondary of the transformer.

$$\text{Transformer Utilization Factor (TUF)} = \frac{\text{D.C power delivered to the load}}{\text{AC rating of trnasformer secondary}}$$

$$= \frac{P_{DC}}{P_{AC} \text{ rated}}$$

This term TUF is not ratio of rectification, because all the rated current of the secondary is not being drawn by the circuit.

$$P_{ac} = I_{DC}^2 . R_L - \left(\frac{I_m}{\pi}\right)^2 . R_L$$

$$P_{ac} = V_{rms} \times I_{rms}$$

$$V_{rms} = \frac{V_m}{\sqrt{2}} = \text{Rated voltage of (Secondary Transformer)}$$

$$I_{rms} = I_m/2$$

$$P_{ac} = V_{rms} \times I_{rms}$$

But $\quad V_m = I_m (R_f + R_L) \simeq I_m R_L$ (R_f is small)

$$TUF = \frac{I_m^2}{\pi^2} \times R_L \Big/ \frac{I_m R_L}{\sqrt{2}} \times \frac{I_m}{2} = \frac{2\sqrt{2}}{\pi^2} = 0.287$$

Transformer Utilization Factor for Half Wave Rectifier is **0.287**

6.17.3 Disadvantages of Half Wave Rectifier

1. High Ripple Factor (1.21)
2. Low ratio of rectification (0.406)
3. Low TUF (0.287)
4. D.C saturation of transformer.

For half wave configuration,

$$I_{rms} = \sqrt{\frac{1}{2\pi}\int_0^{\pi} I_m^2 \sin^2 \alpha . d\alpha} = \frac{I_m}{2}$$

Because during negative half cycle the diode will not conduct hence i = 0 in the loop from $\pi - 2\pi$. So integration is from $0 - \pi$ only. Even though V is present, i = 0 for one half cycle. Therefore, Power is zero ($V \times I = P = 0$, since, $I = 0$)

$$I_{DC} = \frac{1}{2\pi}\int_0^{\pi} I_m \sin\alpha . d\alpha = \frac{I_m}{2\pi}\left|-\cos\alpha\right|_0^{\pi} = \frac{I_m}{2\pi}|1+1| = \frac{I_m}{\pi}$$

$\therefore$ Ripple Factor for Half Wave Rectification $\frac{I_{rms}}{I_{DC}} = \frac{I_m/2}{I_m/\pi} = \frac{\pi}{2}$

$$\gamma = \sqrt{(1.57)^2 - 1} = 1.21$$

$\because$ $\frac{\pi}{2} = 1.57$

$\therefore$ $\gamma > 1$

So the ripple voltage exceeds D.C voltage. Hence HWR is a poor circuit for converting AC to DC.

6.18 FULL WAVE RECTIFIER (FWR)

Since half wave rectifier circuit has poor ripple factor, for ripple voltage is greater than DC voltage, it cannot be used. So now analyze a full wave rectifier circuit.

The circuit is as shown in Fig. 6.25.

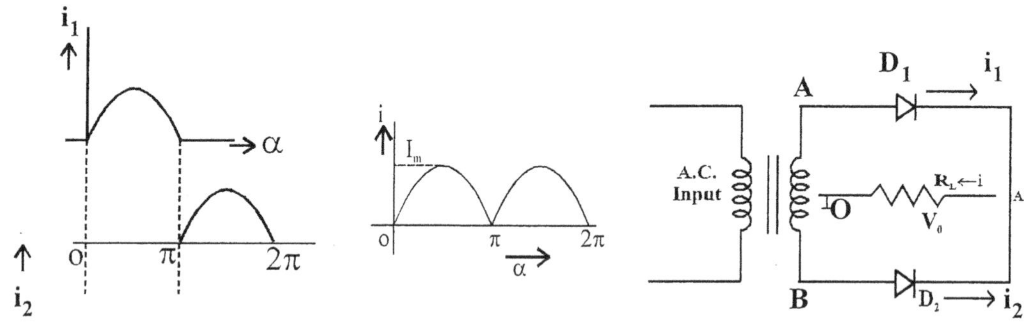

Waveforms *Fig. 6.25* **FWR Circuit**

During the +ve half cycle, D_1 conducts and the current through D_2 is zero

During the – ve half cycle, D_2 conducts and the current through D_1 is zero

A centre tapped transformer is used.

The total current i flows through the load resistor R_L and the output voltage V_0 is taken across R_L.

For half wave rectifier, circuit, $I_{DC} = \frac{1}{2\pi}\int_0^{2\pi} I_m \sin\alpha d\alpha = \frac{I_m}{\pi}$

For full wave rectifier, circuit, I_{DC} = Twice that of half wave rectifier circuit

$\therefore \quad I_{DC} = 2\ I_m/\pi$

For half wave rectifier circuit , $I_{rms} = \sqrt{\frac{1}{2\pi}\int_0^{\pi} I_m^2 \sin\alpha d\alpha} = \frac{I_m}{2}$

For full wave rectifier circuit, $I_{rms} = \sqrt{2 \times \frac{1}{2\pi}\int_0^{\pi} I_m^2 \sin^2\alpha d\alpha}$

$$I_{rms} = \sqrt{2} \times \frac{I_m}{2} = \frac{I_m}{\sqrt{2}}$$

$$I_m = \frac{V_m}{R_f + R_L}$$

R_f is the forward resistance of each diode.

A centre tapped transformer is essential to get full wave rectification. So there is a phase shift of 180^0, because of centre tapping. So D_1 is forward biased during the input cycle of 0 to π, D_2 is forward biased during the period π to 2π since the input to D_2 has a phase shift of 180^0 compared to the input to D_1. So positive half cycle for D_2 starts at a full wave rectifier circuit, while the D.C. current starts through the load resistance R_L is twice that of the Half Wave Rectifier Circuit.

Therefore, for Half Wave Rectifier Circuit,

$$I_{DC} = \frac{I_m}{\pi}$$

For Full Wave Rectifier Circuit,

$$I_{DC} = \frac{2I_m}{\pi}$$

Hence, Ripple Factor is improved.

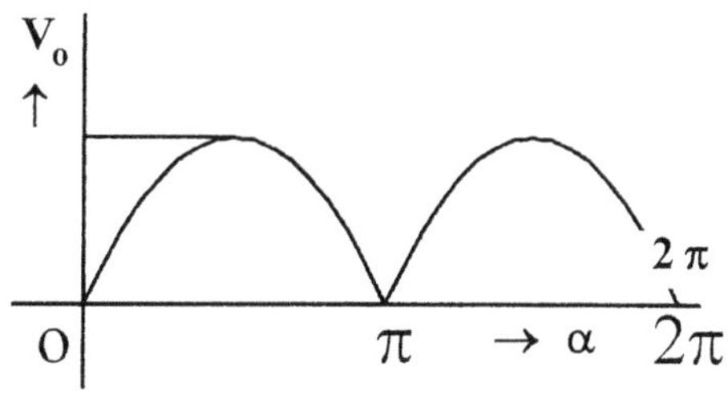

Fig. 6.26 HWR voltage output

6.18.1 Ripple Factor

$$\frac{I'_{rms}}{I_{DC}} = \frac{I_m/\sqrt{2}}{2I_m/\pi} = \frac{\pi}{2\sqrt{2}} = 1.11$$

$$I'_{rms} = \sqrt{I_{rms}^2 - I_{DC}^2}\ ; \quad \gamma = \frac{\sqrt{I_{rms}^2 - I_{DC}^2}}{I_{DC}}$$

Therefore, $\gamma = 1.21$ for HWR, and it is is 0.482 for FWR.

6.18.2 Regulation for FWR

$$V_{DC} = \frac{2V_m}{\pi} - I_{DC}R_F.$$

6.18.3 Ratio of Rectifications

$$= \frac{\text{D.C. power delivered to Load}}{\text{A.C. input power from transformer secondary}}$$

$$p_{DC} = I_{DC}^2 \times R_L \text{ for FWR, } I_{DC} = \frac{2I_m}{\pi}$$

$$= \frac{4.I_m^2}{\pi^2} R_L.$$

P_{AC} is what a Wattmeter would indicate if placed in the FWR Circuit with its voltage terminals connected across the transformer secondary.

$$P_{ac} = (I_{rms})^2. (R_F + R_L);\ I_{rms} \text{ for FWR} = \frac{I_m}{\sqrt{2}}$$

$$= \left(\frac{I_m}{\sqrt{2}}\right)^2 .(R_F + R_L).$$

$$= \frac{I_m^2}{2}(R_F + R_L).$$

If we assume that R_f is the forward resistance of the diode is very small, compared to R_L.

$$R_f + R_L \cong R_L.$$

$$\therefore \quad P_{AC} = \frac{I_m^2}{2}(R_L).$$

$$\text{Ratio of Rectification} = \frac{p_{DC}}{p_{AC}} = \frac{4I_m^2 R_L \times 2}{\pi^2 I_m^2 \times R_L} = \frac{8}{\pi^2} = 0.812$$

For HWR it is 0.406. Therefore, for FWR the ratio of rectification is twice that of HWR.

Transformer Utilization Factor : TUF

$$\text{TUF} = \frac{\text{D.C. power delivered to Load}}{\text{A.C. rating of the transformer secondary}}$$

$$= \frac{P_{DC}}{P_{AC}\text{rated}}.$$

This is not the same as the ratio of rectification because, the current delivered to the load is not the same as the actual rated current of the transformer (It is always less). For the transformer to be utilized best, the rated current of the secondary of the transformer should be drawn by the load.

$$P_{DC} = I_{DC}^2 . R_L = \left(\frac{2I_m}{\pi}\right)^2 . R_L.$$

$$P_{AC} \text{ rated} = 2V_{AC} \times I_{AC}$$

$$V_{AC} = \frac{2V_m}{\sqrt{2}}; \quad I_{AC} = I_{rms} = \frac{I_m}{\sqrt{2}}$$

$$V_m = I_m R_L \text{ (neglecting } R_F \text{)}$$

$$\therefore \quad TUF = \frac{4I_m^2 \times R_L \times \sqrt{2}}{\pi^2 \times I_m^2 \times R_L} = \frac{4\sqrt{2}}{\pi^2} = 0.693 = \mathbf{69.32\%}$$

6.18.4 Peak Inverse Voltage (PIV) for Full Wave Rectifier

With reference to the FWR circuit, during the -ve half cycle, D_1 is not conducting and D_2 is conducting. Hence maximum voltage across R_L is V_m, since voltage is also present between A and O of the transformer, the total voltage across $D_1 = V_m + V_m = 2V_m$.

6.18.5 D.C. Saturation

In a FWR, the D.C. currents I_1 and I_2 flowing through the diodes D_1 and D_2 are in opposite direction and hence cancel each other. So there is no problem of D.C. current flowing through the core of the transformer and causing saturation of the magnetic flux in the core.

PIV is 2 V_m for each diode in FWR circuit because, when D_1 is conducting, the drop across it is zero. Voltage delivered to R_L is V_m. D_2 is across R_L. During the same half cycle, D_2 is not conducting. Therefore, peak voltage across it is V_m from the second half of the transformer. The total voltage across D_2 is $V_m + V_m = 2V_m$.

6.19 BRIDGE RECTIFIERS

The circuit is shown in Fig. 6.27. During the positive half cycle, D_1 and D_2 are forward biased. D_3 and D_4 are open. So current will flow through D_1 first and then through R_L and then through D_2 back to the ground. During the –ve half cycle D_4 and D_3 are forward biased and they conduct. The current flows from D_3 through R_L to D_4. Hence the direction of current is the same. So we get full wave rectified output.

In Bridge rectifier circuit, there is no need for centre tapped transformer. So the transformer secondary line to line voltage should be one half of that, used for the FWR circuit, employing two diodes.

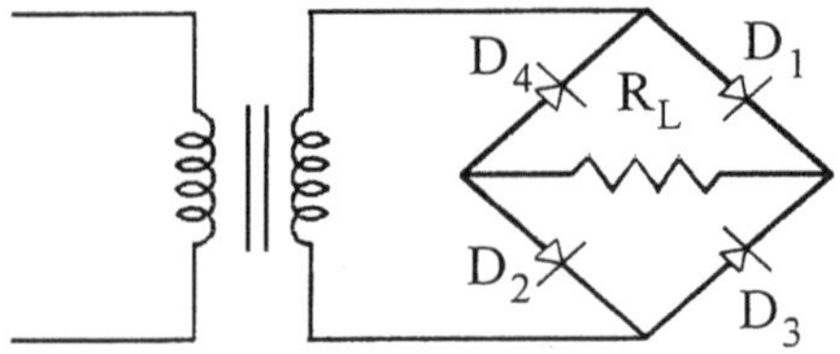

Fig. 6.27 Bridge Rectifier Circuit

$$I_{DC} = \frac{2I_m}{\pi}$$

where $$I_m = \frac{V_m}{R_S + 2R_f + R_L}$$

$$V_{DC} = \frac{2V_m}{\pi} - I_{DC}(R_S + R_F)$$

R_S = Resistance of Transformer Secondary.

R_f = Forward Resistance of the Diode.

R_L = Load Resistance.

$2R_f$ should be used since two diodes in series are conducting at the same time. The ripple factor and ratio of rectification are the same as for Full Wave Rectifier.

6.19.1 Advantages of Bridge Rectifier

1. *The peak inverse voltage (PIV) across each diode is V_m and not $2V_m$ as in the case of FWR. Hence the Voltage rating of the diodes can be less.*
2. *Centre tapped transformer is not required.*
3. *There is no D.C. Current flowing through the transformer since there is no centre tapping and the return path is to the ground. So the transformer utilization factor is high.*

6.19.2 Disadvantages

1. *Four diodes are to be used.*
2. *There is some voltage drop across each diode and so output voltage will be slightly less compared to FWR. But these factors are minor compared to the advantages.*

Bridge rectifiers are available in a package with all the 4 diodes incorporated in one unit. It will have two terminals for A.C. Input and two terminals for DC output. Selenium rectifiers are also available as a package.

Comparison of Rectifier Circuits

	HWR	FWR	Bridge Rectifier Circuit
Peak Inverse Voltage	V_m	$2V_m$	V_m
No Load Voltage	$\frac{Vm}{\pi}$	$\frac{2Vm}{\pi}$	$\frac{2V_m}{\pi}$
Ripple factor	1.21	0.482	0.482
Number of Diodes required	1	2	4
Ratio of Rectification $\frac{P_{dc}}{P_{AC}}$	0.406	0.812	0.812
TUF = $\frac{\text{DC Power delivered to the load}}{\text{AC ratings of transformer secondary}}$	0.287	0.693	0.812

6.20 THE HARMONIC COMPONENTS IN RECTIFIER CIRCUIT

The analytical representation of output current wave of halfwave rectifier using fourier series is given by

$$i = I_m \left[\frac{1}{\pi} + \frac{1}{2}\sin\omega t - \frac{2}{\pi}\sum_{k=even}\frac{\cos\omega t}{(k+1)(k-1)}\right] \quad .....(6.20.1)$$

The lowest angular frequency present in eq. (6.20.1) is of a.c source. And all other terms present in ab e equation are of even harmonic of power source.

Similarly in case of full wave rectifier in which one diode conducts for positive half cycle and the other c de conducting for negative half cycle and the net output current is difference of the currents conducted in positive half cycle ($i_1(\alpha)$) and the current conducted in negative half cycle ($i_2(\alpha)$). i.e.,

$$\text{N curr t } i = i_1(\alpha) + i_2(\alpha) \qquad (i_2(\alpha) = i_2(\alpha + \pi)) \quad .....(6.20.2)$$

Here both $i_1(\alpha)$ and $i_2(\alpha)$ are same in magnitude but opposite in polarity. The difference $i_1(\alpha) + i_2(\alpha)$ results in cancellation of lowest angular frequency of power source present in eq. (6.20.2). And hence the total current.

$i = i_1 + i_2$ attains of form

$$i = I_m \left[\frac{2}{\pi} - \frac{4}{\pi} \sum_{\substack{k\ \text{even} \\ k \neq 0}} \frac{\cos k\omega t}{(k+1)(k-1)} \right] \qquad .....(6.20.3)$$

Here in eq. (6.20.3) the fundamental frequency 'ω' has been eliminated from the equation. And hence lowest frequency in the output being 2ω, i.e., the second hormonic term. This offers an definite advantage in the effectiveness of filtering the output.

The other advantage of fullwave rectifier is that the direction of the current during two halfs of cycle is of opposite in nature, thus the magnetic cycle through which iron core is take is essentially that of the altering current and this eliminates d.c. saturation of the transformer, which would yield additional hormonics in the output.

6.21 FILTERS

A power supply must provide ripple free source of power from an A.C. line But the output of a rectifier circuit contains ripple components in addition to a D.C. term. It is necessary to include a filter between the rectifier and the load in order to eliminate these ripple components. Ripple components are high frequency A.C. Signals in the D.C output of the rectifier. These are not desirable, so they must be filtered. So filter circuits are used.

6.22 INDUCTOR FILTER CIRCUITS

Flux linkages per ampere of current $L = \frac{N\phi}{I}$. The ability of a component to develop induced voltage when alternating current is flowing through the element is the property of the inductor. Types of Inductors are :

1. *Iron Cored*
2. *Air Cored*

An inductor opposes any change of current in the circuit. So any sudden change that might occur in a circuit without an inductor are smoothed out with the presence of an inductor. In the case of AC, there is change in the magnitude of current with time.

Inductor is a short circuit for DC and offers some impedance for A.C. ($X_L = j\omega L$). So it can be used as a filter. AC voltage is dropped across the inductors, where as D.C. passes through it. Therefore, A.C is minimized in the output.

Inductor filter is used with FWR circuit. Therefore, HWR gives 121% ripple and using filter circuit for such high ripple factor has no meaning . FWR gives 48% ripple and by using filter circuit we can improve it. According to Fourier Analysis, current, I in HWR Circuit is

$$i = \frac{I_m}{\pi} + \frac{I_m}{2}\text{Sin}\,\omega t - \frac{2I_m}{\pi}\frac{\text{Cos}\,4\omega t}{3}$$

Current in the case of FWR is $i \cong \frac{2I_m}{\pi} - \frac{4I_m}{3\pi}\cos 2\omega t - \frac{4I_m \cos 4\omega t}{15\pi}$

The reactance offered by inductor to higher order frequencies like $4\omega t$ can be neglected. So the output current is

$$i \cong \frac{2I_m}{\pi} - \frac{4I_m}{3\pi}\cos 2\omega t$$

the fundamental harmonic ω is eliminated.

The circuit for FWR with inductor filter is as shown in Fig. 6.28.

One winding of transformer can be used as 'L'. For simplification, diode and

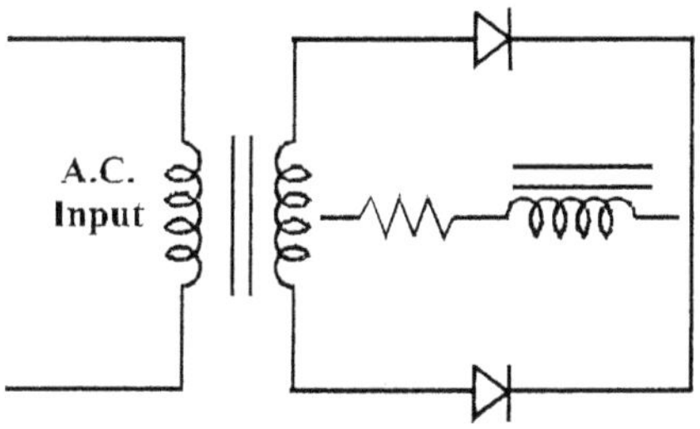

Fig. 6.28 *Inductor filter circuit*

choke (inductor) resistances can be neglected, compared with R_L. The D.C. component of the current is

$$I_m = \frac{V_m}{R_L}$$

Impedance due to L and R_L in Series $|Z| = \sqrt{R_L^{\,2} + (2\omega L)^2}$

For second Harmonic, the frequency is 2ω.

Therefore, A.C. Component of current $I_m = V_m / \sqrt{R_L^{\,2} + 4\omega^2 L^2}$. So substituting these values in the expression, for current,

$$I = \frac{2I_m}{\pi} - \frac{4I_m}{3\pi}\text{Cos}\,2\omega t$$

by inductor to higher order frequencies like $4\omega t$ etc., we get,

$$i = \frac{2V_m}{\pi R_L} - \frac{4V_m}{3\pi\sqrt{R_L^{\,2} + 4\omega L^2}}\cos(2\omega t - \theta)$$

where θ is the angle by which the load current lags behind the voltage and is given by

$$\phi = \text{Tan}^{-1}\frac{2\omega L}{R_L}.$$

6.22.1 Ripple Factor

$$\gamma = I_{r'rms}/I_{D.C.} = I'_{rms}/I_{DC} \quad I_{DC} = 2V_m/\pi R_L$$

$$I_{rrms} = \frac{I_m}{\sqrt{2}} = \frac{4V_m}{3\sqrt{2}\pi\sqrt{R_L^2 + 4\omega^2L^2}}$$

$$\therefore \quad \text{Ripple factor} = \frac{4v_m \times \pi R_L}{3\sqrt{2}\pi\sqrt{R_L^2 + 4\omega^2L^2} \times 2V_m}$$

$$= \frac{2}{3\sqrt{2}} \times \frac{1}{\sqrt{1 + \frac{4\omega^2L^2}{R_L^2}}}$$

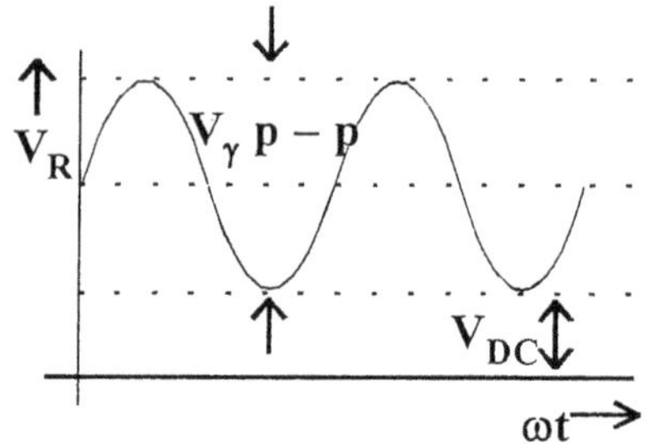

Fig. 6.29 Ripple Voltage V_V

If $\frac{4\omega^2L^2}{R_L^2} >> 1$, then

$$\text{Ripple Factor } r = \frac{R_L \times 2}{3\sqrt{2} \times 2\omega L}$$

$$\therefore \quad \boxed{\text{Ripple factor} = \frac{R_L}{3\sqrt{2}\omega L}} \quad .....(6.22.1)$$

Therefore, for higher values of L, the ripple factor is low. If R_L is large, then also γ is high. Hence inductor filter should be used where the value of R_L is low. Suppose the output wave form from FWR supply is as shown in Fig. 6.29, then,

$V_{\gamma p-p}$ is the peak to peak value of the ripple voltage. Suppose

$V_{DC} = 300V$, and $V_{\gamma p-p}$ is 10V.

then V_R maximum $= \frac{10}{2} = 5V$. Therefore, $V_\gamma rms = \frac{V_\gamma \max}{\sqrt{2}} = \frac{5}{\sqrt{2}} = 3.54V$

$$\therefore \quad \text{Ripple Factor } \gamma = \frac{V_r rms}{V_x} = \frac{3.54}{300} = 0.0118.$$

$$\% \text{ ripple} = \text{Ripple factor} \times 100\ \% = 1.18$$

6.22.2 Regulation

$V_{DC} = I_{DC}.R_L$; I_{DC} for F W R is $\frac{2I_m}{\pi}$.

$$V_{DC} = \frac{2I_mR_L}{\pi} = \frac{2V_{ms}}{\pi}$$

$$I_m.R_L = V_m.$$

Therefore, V_{DC} is constant irrespective of R_L. But this is true if L is ideal. In practice

$$I_{DC} = \frac{2V_m}{\pi} - I_{DC}.R_f$$

where R_f is the resistance of diode.

An inductor stores magnetic energy when the current flowing through it is greater than the average value and releases this energy when the current is less than the average value.

Another formula that is used for Inductor Filter for

$$\boxed{\text{Ripple Factor} = 0.236\ \frac{R_C + R_L}{\omega_L}} \qquad \text{.......... (6.22.2)}$$

where R_L is load resistance R_c is the series resistance of the inductors. This is the same as the above formula $\frac{R_L}{3\sqrt{2}\omega_L}$

$$\because \qquad \frac{1}{3\sqrt{2}} = 0.236.$$

Example 6.1:

A FWR is used to supply power to a 2000W Load. Choke Inductors of 20 H inductance and capacitors of 16mf are available. Compute the ripple factor using **1**. One Inductor filter **2**. One capacitor filter **3**. Single L type filter.

Solution:

1. One Inductor Filter :

$$I = \frac{2I_m}{\pi} - \frac{4I_m}{3\pi}\cos(2\omega t - \phi) \text{ where } \phi = \text{Tan}^{-1}\left(\frac{2\omega L}{R}\right)$$

$$I_{DC} = \frac{2V_m}{\pi R_L}$$

$$V = \frac{I \text{ ripple}}{I_{DC}} = \frac{R_L}{3\sqrt{2}\omega_L} = \frac{100}{3\times 1.414\times 2}$$

$$= 0.074$$

2. Capacitor filter :

The ripple voltage for a capacitor filter is of Triangular waveform approximately.

The rms value for Triangular wave is $\frac{E_d}{2\sqrt{3}}$ or $\frac{V_{rms}}{2\sqrt{3}}$.

$$\therefore \quad E_{rms} = \frac{E_d}{2\sqrt{3}}$$

Suppose the discharging of the capacitor will cut one for one full half cycle then the change last by the capacitor

$$Q = \omega t = I_{DC} \times \frac{1}{2f}.$$

$$\therefore \quad \text{voltage} = \frac{Q}{c}$$

$$\therefore \quad E_d = \frac{I_{DC}}{2fc}$$

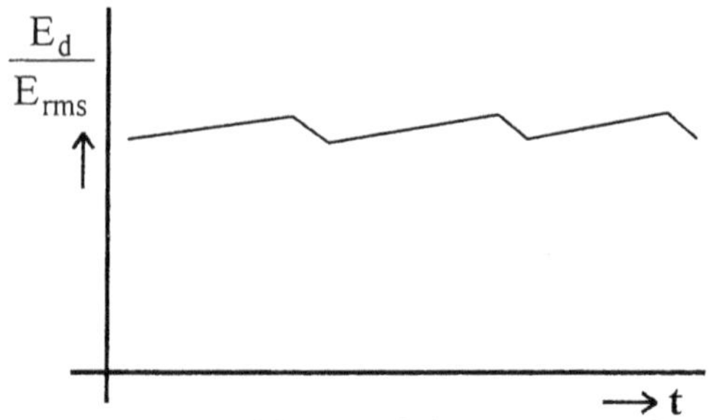

Fig. 6.30 V_0 of Capacitor Filter

$$\gamma = \frac{I_{DC}}{2fc \times 2\sqrt{3} \times I_{DC}} = \frac{1}{4fc\sqrt{3}R_L}$$

$$= 4 \times \sqrt{3} \times 50 \times 16 \times 10^{-6} \times 2000 = 0.09$$

$$3.\ \text{L Type factor}: \ \gamma = \frac{\sqrt{2}}{3} \times \frac{1}{4\omega^2 LC} = 0.0037$$

Example 6.2:

A diode whose internal resistance is 20 Ω is to supply power to a 1000Ω load from a 110V ((rms) source of supply. Calculate **(a.)** The peak load current. **(b.)** The DC load current **(c.)** AC Load Current **(d.)** The DC diode voltage. **(e.)** The total input power to the circuit. **(f.)** % regulation from no load to the given load.

Solution:

Since only one diode is being used, it is for HWR Circuit.

(a) $$I_m = \frac{V_m}{R_f + R_L}. = \frac{110\sqrt{2}}{1020} = 152.5\text{mA}.$$

(b) $$I_{DC} = \frac{I_m}{\pi} = \frac{152.5}{\pi} = 48.5\text{mA}.$$

(c) $$I_{rms} = \frac{I_m}{2} = \frac{152.5}{2} = 76.2\text{mA}.$$

(d) $$V_{dc} = \frac{-I_m.R_L}{\pi} = -48.5 \times 1 = -48.5\text{V}.$$

(e) $P_i = I_{rms} \times (R_f + R_L) = 5.92\Omega.$

(f) $$\% \text{ regulation} = \frac{V_{NL} - V_{FL}}{V_{FL}} \times 100\%$$

$$= \frac{\frac{V_m}{\pi} - I_{DC}.R_L}{I_{DC}.R_L} = 2.06\%$$

Example 6.3:

Show that the maximum DC output power $P_{DC} = V_{DC}\, I_{DC}$ in a half wave single phase circuit occurs when the load resistance equals the diode resistance R_f.

Solution:

$$P_{DC} = I_{DC}^2.R_L = \frac{V_m^{\ 2}.R_L}{\pi^2(R_F + R_L)}$$

when R_L is very large, $V_{DC} \cong \frac{V_m}{\pi}$; $I_{DC} = \frac{V_{DC}}{R_L} = \frac{V_m}{\pi.R_L}$

for P_{DC} to be maximum, $\frac{dp_{DC}}{dR_L} = 0.$

or $$\frac{V_m^{\ 2}}{\pi^2}\left(\frac{(R_F + R_L)^2 - (R_F + R_L).R_L}{(R + R_L)^4}\right) = 0.$$

$R_F + R_L = 2R_L$ or $R_L = R_F$.

Therefore, P_{DC} is maximum when R_L is equal to R_F.

Example 6.4:

A 1mA meter whose resistance is 10Ω is calibrated to read rms volts when used in a bridge circuit with semiconductor diodes. The effective resistance of each element may be considered to be zero in the forward direction and ∞ in the reverse direction. The sinusoidal input voltage is applied in series with a 5 - KΩ resistance. What is the full scale reading of this meter?

Solution:

$I_{DC} = \frac{2I_m}{\pi}$ for bridge rectifier circuit.

$$I_m = \frac{V_m}{R_L}; V_m = \sqrt{2}V_{rms}.$$

$$\therefore \quad I_{DC} = \frac{2\sqrt{2} \times V_{rms}}{\pi R_L};$$

$$R_L = 5K\Omega + 10\ \Omega = (5000 + 10)\Omega$$

$$= 5010\ \Omega$$

$$1mA = \frac{2\sqrt{2} \times V_{rms}}{\pi \times 5010.}$$

$$\therefore \quad V_{rms} = \frac{1 \times 10^{-3} \times \pi \times 5010}{2\sqrt{2}} = 5.56v.$$

6.22.3 Voltage Doubler Circuit

The circuit is shown in Fig. 6.31. C_1 charges to V_m, through D_1, in the +ve half cycle when D_1 conducts and D_2 is not conducting. During the –ve half cycle, the voltage across D_1 is

$$V_m + V_m = 2V_m,$$

with anode –ve this is the PIV across D1. During –ve half cycle, D_2 conducts therefore, it is Forward Biased and C_2 is charged to the peak value of 2 V_m, which shows that this circuit is a doubler at no load.

C_1 gets charged to V_m, C_2 gets charged to 2 V_m. By connecting two half wave voltage doubler, a voltage quadrupler can be built.

During +ve half cycle, D_1 is forward biased and so no current flows through C_2 current flows through C_2 during second half cycle only. So it is Half Wave Voltage doubler.

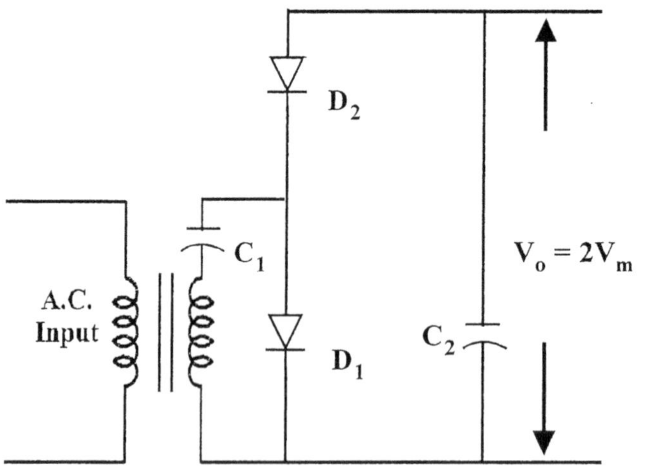

Fig. 6.31 Half Wave Voltage Doubler Circuit

6.23 CAPACITOR FILTER

Here X_c should be smaller than R_L. Because, current should pass through C and C should get charged. If C value is very small, X_C will be large and hence current flows through R only (Fig. 6.32) and no filtering action takes place. During +ve half cycle for a H W R circuit, with C filter, C gets charged when the diode is conducting and gets discharged (when the diode is not conducting) through R_L. When the input voltage $e = E_m \sin \omega t$ is greater than the capacitor voltage, C gets charged. When the input voltage is less than that of the capacitor voltage, C will discharged through R_L. The stored energy in the capacitor maintains the load voltage at a high value for a long period. The diode conducts only for a short interval of high current. The waveforms are as shown in Fig. 6.33. Capacitor opposes sudden fluctuations in voltage across it. So the ripple voltage is minimised.

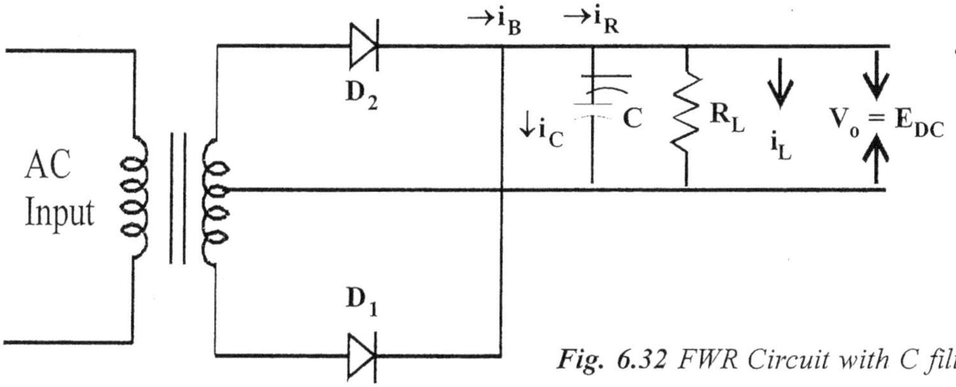

Fig. 6.32 FWR Circuit with C filter

So the voltage across R_L is more or less constant and hence ripple is reduced, as shown in the graph, the voltage V_o is closer to DC wave form. At θ_1 'C' gets charged and at θ_2 C starts discharging.

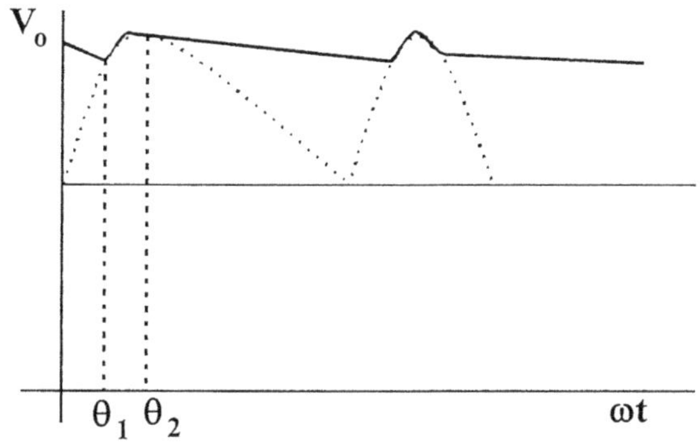

Fig. 6.33 Output of Capacitor filter circuit

The point at which diode starts conducting and C gets charged is known as **cut in point** θ_1. The point at which the diode stops conducting or the capacitor gets discharged is known as ***Cut Out Point*** θ_2.

C will not start charging at π itself. Therefore, even though the second diode is forward biased, e_c is almost $\simeq V_m$. So the first diode conducts from θ_1 to θ_2 and the second diode conducts from $\pi + \theta_1$ to $\pi + \theta_2$. Diode conducts only for a short of time when input current is high and charges C to the peak voltage.

Considering one diode, $i_B = i_C + i_R$ (Fig. 6.32)

$$i_C = C.\frac{dV}{dt} = C.\frac{de_c}{d_t}.$$

$$i_R = \frac{e_c}{R}$$ where e_c is the voltage to which c gets charged.

$$i_B = C.\frac{de_c}{d_t} + \frac{e_c}{R}.$$

But $e_c = E_m$ Sin wt. Therefore, at that point, diode conducts and C gets charged.

$$\theta_1 = \omega t_1.$$

$$\therefore \quad \frac{de_c}{dt} = \omega E_m. \text{ Cos } \omega t.$$

General expression for ib $= \omega C\, E_m \cos \omega t + \frac{E_m}{R} \text{ Sin } \omega t \qquad \theta_1 < \omega t < \theta_2.$

By some mathematical manipulation, i_b can be written as

$$i_b = \frac{E_m}{R}\sqrt{1+\omega^2 R^2 C^2}\,\text{Sin}(wt + \phi)$$

where $\phi = \text{Tan}^{-1}(\omega RC)$

At an ω instant θ_2, when the capacitor fully discharges, $i_B = 0$

$$\therefore \quad \omega C\, E_m \cos \theta_2 + \frac{E_m}{R}\text{Sin}\theta_2 = 0.$$

or $\theta_2 = \tan^{-1}(-\Omega RC)$

Thus the value of conduction angle depends upon the values of R and C, and also the value of i_b. During the non conducting interval, the capacitor discharges into the load R supplying load current.

$$-i_C = i_R.$$

The circuit equation is $-C \times \frac{de_c}{d_t} = \frac{e_c}{R}$

or $e_C = A\,e^{-t/R_L}$

At $\omega t = \theta_2 \qquad e_C = E_m \sin \theta_2.$

The final expression for E_{DC} seems a little complicated. So a simplified expression which is widely used is $E_{DC} = \left(E_m - \frac{E_R}{2} \cdot \right)$, where E_R is the peak to peak ripple voltage. The expression for Ripple Factor,

$$\gamma = \frac{E'_{AC}}{E_{DC}} = \frac{\pi + (\theta_1 - \theta_2)}{2\sqrt{3}\omega RC}.$$

where $(\theta_1 - \theta_2)$ is the angle of conduction θ_C. For large ωRC typical value 100), $\theta_c = 14.6^o$. The larger the value of R, the smaller the γ.

6.23.1 Ripple Factor for C Filter

Initially, the capacitor charges to the peak value V_m when the diode D_1 is conducting. When the capacitor voltage is equal to the input voltage, the diode stops conducting or the current through the diode D_1 is zero. Now the capacitor starts discharging through R_L depending upon the time constant $C.R_L$. Therefore, R_L value is large. Rate of charging is different from rate of discharging. When the voltage across the capacitor falls below the input voltage and when the diode D_2 is forward biased, the capacitor will again charge to the peak value Vm and the current through D_2 becomes zero when $V_c = V_{in}$. Thus the diodes D_1 and D_2 conduct for a very short period θ_1 to θ_2 and $\pi + \theta_1$ to $\pi + \theta_2$ respectively (Fig. 6.34).

T_2 = Time period of the discharging of the capacitor

T_1 = Time period of the charging of the capacitor

The amount of charge lost by the capacitor when it is discharging = $I_{DC}.T_2$. Because, I_{DC} is the average value of the capacitor discharge current. This charge is replaced during a short interval T_1 during which the voltage across the capacitor changes by an amount = peak to peak voltage of the ripple $V'r_{p-p}$.

$$Q = V \times C.$$

$$\therefore \quad Q \text{ charge} = V'_{p-p} \times C$$

$$Q \text{ charge} = Q \text{ discharge}.$$

$$\therefore \quad V'_{p-p} \times C = I_{DC} \times T_2.$$

$$\text{or} \quad V'_{p-p} = \frac{I_{DC} \times T_2}{C}$$

The output waveform can be assumed to be a triangular wave $T_2 >> T_1 (T_1 + T_2) = T/2$ when the diode is conducting (0 to π). ($T_1 + T_2$ corresponds to one half cycle).

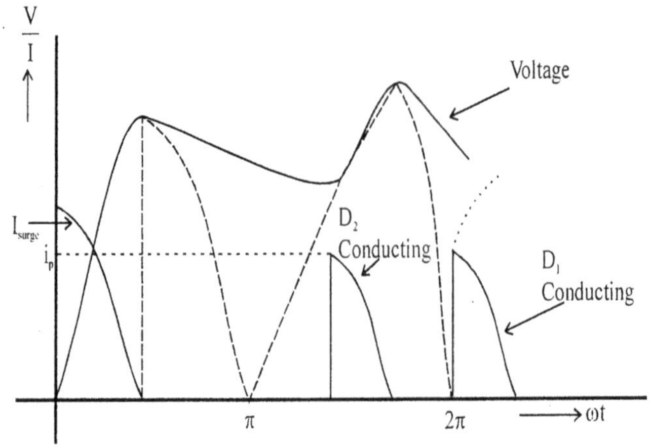

Fig. 6.34 Output waveforms of a capacitor filter

$$\therefore \quad T_2 = \frac{T}{2} = \frac{1}{2f} . \because T_2 = \frac{1}{2f}.$$

$$\therefore \quad V' \text{p-p} = \frac{I_{DC} \times 1}{C2f} \because T_2 = \frac{1}{2f} \cdot \text{ for the triangular wave, form factor} = \frac{1}{\sqrt{3}}.$$

$$\therefore \quad V'_{rms} = \frac{V'_{p-p}}{2\sqrt{3}}$$

$$I_{DC} = \frac{V_{DC}}{R_L}.$$

$$V' \text{rms} = \frac{V_{DC}}{4\sqrt{3}fCR_L}.$$

$$\text{Ripple Factor} \quad \boxed{\gamma = \frac{V'_{rms}}{V_{DC}} = \frac{1}{4\sqrt{3}fCR_L}.}$$

Therefore, Ripple factor may be decreased by increasing C or R_L or both.

Expression for V_{DC} :

$$V_{DC} = V_m - \frac{V'p-p}{2}.; \quad V' \text{P-P} = \frac{I_{DC}}{2fc}.I_{DC} = \frac{V_{DC}}{R_L}$$

$$V_{DC} = V_m - \frac{V_{DC}}{4fC.R_L}$$

$$\text{or} \quad V_{DC}\left(1 + \frac{1}{4fcR_L}\right) = V_m$$

$$\because \quad V_{DC} = V_m\left(\frac{4fR_LC}{1 + 4fR_LC}\right)$$

6.24 LC FILTER

In an inductor filter, ripple increases with increase in R_L. In a capacitor filter, ripple decreases with increase in R_L. A combination of these two into a choke ***input or L-Section Filter*** should then make the ripple independent of load resistance.

If L is small, the capacitor will be charged to V_m, the diodes will be cut off allowing a short pulse of current. As L is increases, the pulses of current are smoothed and made to flow for a larger period, but at reduced amplitude. But for a critical value of inductance L_c, either one diode or the other will be always conducting, with the result that the input voltage V and I to the filter are full wave rectified sine waves.

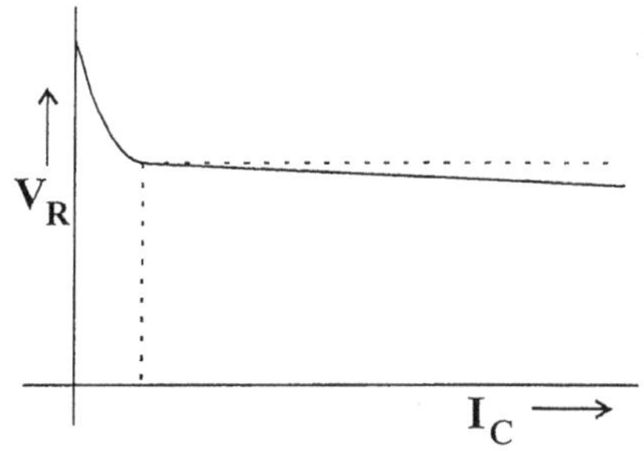

Fig. 6.35 *LC filter characteristic*

The graph of DC output voltage for LC Filter is as shown in Fig. 6.35.

$$V_{DC} = \frac{2V_m}{\pi}$$

for Full Wave Rectifier Circuit. considering ideal elements, conduction angle is increased when inductor is placed because there is some drop across L. So C cannot charge to V_m. Therefore, the diode will be forward biased for a longer period.

The ripple circuit which passes through L, is not allowed to develop much ripple voltage across R_L, if X_c at ripple frequency is small compared to R_L. Because the current will pass through C only. Since X_c is small and Capacitor will get charged to a constant voltage. So V_0 across R_L will not vary or ripple will not be there. Since for a properly designed LC Filter,

$$X_c << R_L$$

and $$X_L >> X_c \quad \text{at } \omega = 2\pi f$$

X_L should be greater than X_c because, all the AC should be dropped across X_L itself so that AC Voltage across C is nil and hence ripple is low.

6.24.1 Ripple Factor in LC Filter

AC Component of current through L is determined by X_L.

$$X_L = 2\omega L$$

RMS value of ripple current for Full Wave Rectifier with L Filter is,

$$I'rms = \frac{4V_m}{3\pi\sqrt{2}X_L} = \text{RMS Value of Ripple Current for L Filter}$$

$$I'rms = \frac{2}{3\sqrt{2}X_L} = \frac{2V_m}{\pi}$$

$$\therefore \quad I'rms = \frac{2}{3\sqrt{2}X_L} \cdot V_{DC} = \frac{\sqrt{2}V_{DC}}{3X_L}$$

The ripple voltage in the output is developed by the ripple current flowing through X_C.

$$\therefore \quad V'rms = I'rms \cdot X_C$$

$$= \frac{\sqrt{2}}{3} \cdot \frac{X_C}{X_L} \cdot V_{DC}$$

$$\text{Ripple factor} = \frac{V'rms}{V_{DC}}$$

$$\gamma = \frac{\sqrt{2}.X_C}{3X_L}$$

$$X_C = \frac{1}{j2\omega fc} \cdot X_L = jL\omega L = 2\pi fL$$

$$\therefore \quad \boxed{\gamma = \frac{\sqrt{2}}{3} \cdot \frac{1}{2\omega C} \cdot \frac{1}{2\omega L}}$$

$X_L = 2wL$ since ripple is being considered at a frequency, twice the line frequency.

If f = 60 Hz, $\boxed{\gamma = \frac{0.83}{LC}}$

6.24.2 Bleeder Resistance

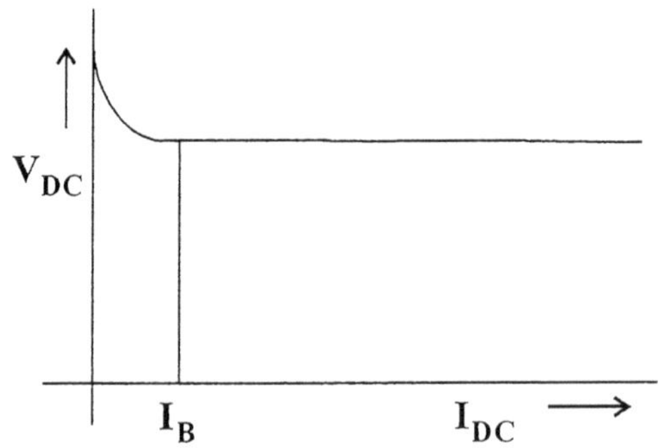

Fig. 6.36 LC filter characteristic

For the LC filter, the graph between I_{DC} and V_{DC} is as shown in Fig. 6.39. For light loads, i.e., when the load current is small, the capacitor gets charged to the peak value and so the no load voltage is $\frac{2V_m}{\pi}$. As the load resistance is decreased, I_{DC} increases so the drop across other elements V_{iz} diodes and choke increases and so the average voltage across the capacitor will be less than the peak value of $\frac{2V_m}{\pi}$. The out put voltage remains constant beyond a certain point I_B and so the regulation will be good. The voltage V_{DC} remains constant even if I_{DC} increases, because, the capacitor gets charged every time to a value just below the peak voltage, even though the drop across diode and choke increases. So for currents above I_B, the filter acts more like an inductor filter than C filter and so the regulation is good. Therefore, for LC filters, the load is chosen such that, the $I_{DC} \geq I_B$. The corresponding resistance is known as ***Bleeder Resistance*** (R_B). So Bleeder Resistance is the value for which $I_{DC} \geq I_B$ and good regulation is obtained, and conduction angle is 180^0. When $R_L = R_B$, the conduction angle = 180^0, and the current is continuous. So just as we have determined the critical inductance L_C, when $R_L = R_B$, the Bleeder Resistance, I_{DC} = the negative peak of the second harmonic term.

$$\therefore \quad I_{DC} = \frac{2V_m}{\pi R_B}$$

$$I' \text{ peak} = \frac{4V_m}{3\pi} \cdot \frac{1}{XL}.$$

$$\therefore \quad \frac{2V_m}{\pi R_B} = \frac{24V_m}{3\pi} \cdot \frac{1}{X_L}$$

$$R_B = \frac{3.X_L}{2}; \qquad R_B \text{ is Bleeder Resistance.}$$

I_{DC} should b_e equal to or greater than I peak, since, X_L determines the peak value of the ripple component. If X_L is large I' peak can be $\leq I_{DC}$ and so the ripple is negligible or pure D.C. is obtained, where $X_L = 2\omega L$ corresponding to the second harmonic. Therefore, R should be at least equal to this value of R_B, the Bleeder Resistor.

6.24.3 Swinging Choke

The value of the inductance in the LC circuits should be > L_C the minimum value so that conduction angle of the diodes is 180^0 and ripple is reduced. But if the current I_{DC} is large, then the inductance for air cored inductors as $L = \frac{N\phi}{I}$ (flux linkages per ampere of current), as I increases, L decreases and may become below the critical inductance value L_C. Therefore, Iron cored inductors or chokes are chosen for filters such that the value of L varies within certain limits and when I is large, the core saturates, and the inductance value will not be $< L_C$. Such chokes, whose inductance varies with current within permissible limits, are called ***Swinging Chokes***.

(In general for inductors $\frac{\phi}{I}$ remains constant so that L is constant for any value of current I) This can be avoided by choosing very large value of L so that, even if current is large, Inductance is large enough $> L_C$ But this increases the cost of the choke. Therefore, swinging choke are used.

$$L_C \geq \frac{R_L}{3\omega}$$

$R_L =$ Load Resistance

$L_C =$ Critical Inductance

$\omega = 2\pi f$

At no load $R_L = \infty$. Therefore, L should be ∞ which is not possible. Therefore,

Bleeder Resistance of value $= \frac{3X_L}{2}$ is connected in parallel with R_L, so that even

when R_L is ∞, the conduction angle is 180^0 for each diode.

The inductance of an iron - cored inductor depends up on the D.C. Current flowing through it, L is high at low currents and low at high currents. Thus its L varies or swings within certain limits. This is known as ***Swinging Choke***.

Typical values are

L = 30 Henrys at I = 20mA and

L = 4 Henrys at I = 100mA.

Example 6.5:

Design a full wave rectifier with an LC filter (single section) to provide 9V DC at 100mA with a maximum ripple of 2%. Line frequency f = 60 HZ.

Solution:

$$\text{Ripple factor } \gamma = \frac{\sqrt{3}}{2} \times \frac{1}{2\omega C} \times \frac{1}{2\omega L} = \frac{0.83}{LC} \mu$$

$$\therefore \qquad 0.02 = \frac{0.83}{LC} \text{or} LC = \frac{0.83}{0.02} = 42.\mu$$

$$R_L = \frac{V_{DC}}{I_{DC}} = \frac{9V}{0.1} = 90\Omega.$$

$$LC \geq \frac{R_L}{3\omega} \geq \frac{R_L}{1130}.$$

But LC should be 25% larger.

$\therefore$ for f = 60 Hz,the value of LC should be $\geq \frac{R_L}{900}$.

$$LC \geq \frac{R_L}{900}. \geq \frac{90}{900} = 0.1 \text{ Henry.}$$

If L = 0.1H, then $C = \frac{42}{0.1} = 420\mu f$. This is high value

If L = 1H, then $C = 42\ \mu f$.

If the series resistance of L is assumed to be 50 Ω, the drop across L is

$$I_{DC} \times R = 0.1 \times 50 = 5V.$$

Transformer Rating :

$$V_{DC} = 9V + 5V = 14V$$

$$\therefore \qquad V_m = \frac{\pi}{2}(9 + 5) = 22V$$

$$\therefore \qquad \text{rms value is } \frac{22}{\sqrt{2}} = 15.5V$$

Therefore, a 15.5 – 0 –15.5 V, 100mA transformer is required. PIV of the diodes is $2V_m$ Because it is FWR Circuit.

$$\therefore \qquad PIV = 44V$$

So, diodes with 44V, 100mA ratings are required.

Example 6.6:

In a FWR with C filter circuit, V_rp-p = 0.8v and the maximum voltage is 8.8 volts (Fig. 6.37). $R_L = 100\Omega$ and C = 1050 μf. Power line frequency = 60 Hz. Determine the Ripple Factor and DC output voltage from the graph and compare with calculated values.

Solution:

$$V_r' \text{ p-p} = 0.8\text{V}.$$

$$\therefore \quad V_{rms}' = \frac{0.8}{2\sqrt{3}} = 0.231\text{V}$$

$$V_{DC} = \text{Vm} - \frac{V_r'\text{p}-\text{p}}{2} = 8.8 - \frac{0.8}{2} = 8.4\text{V}.$$

$$\therefore \quad \gamma = \frac{\text{V'rms}}{V_{DC}} = \frac{0.2}{8.4} = 0.0238 \text{or} 2.38\%$$

Theoretical values,

$$\gamma = \frac{1}{4\sqrt{3}fR_L.C}$$

$$= \frac{1}{4\sqrt{3} \times 60 \times 100 \times 1050\mu f}$$

$$= 0.023 \text{ or } 2.3\%$$

$$V_{DC} = \frac{4fR_LC}{1 + 4fR_LC} \times V_m = 8.46\text{V}$$

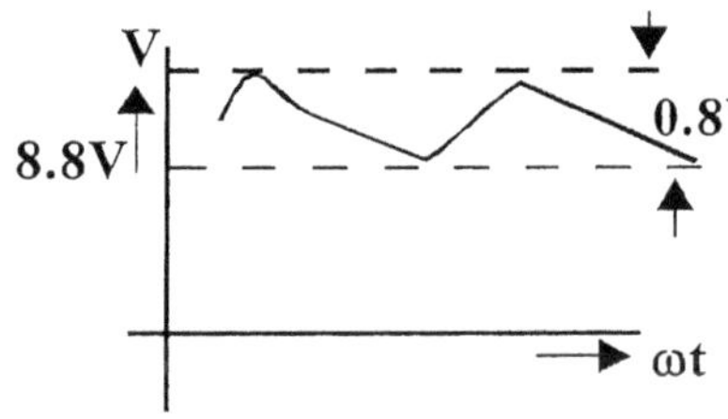

Fig. 6.37 For Example 6.6

6.25 CLC OR π FILTER

In the LC filter or L section filter, there is some voltage drop across L. If this cannot be tolerated and more D.C. output voltage V_0 is desired, π filter or CLC filter is to be used. The ripple factor will be the same as that of L section filter, but the regulation will be poor. It can be regarded as a L section filter with L_1 and C_1, before which there is a capacitor C,i.e. the capacitor filter. The input to the L section filter is the output of the capacitor filter C. The output of capacitor filter will be a triangular wave superimposed over D.C.

Now the output V_0 is the voltage across the input capacitor C less by the drop across L_1. The ripple contained in the output of 'C' filter is reduced by the L section filter L_1C_1.

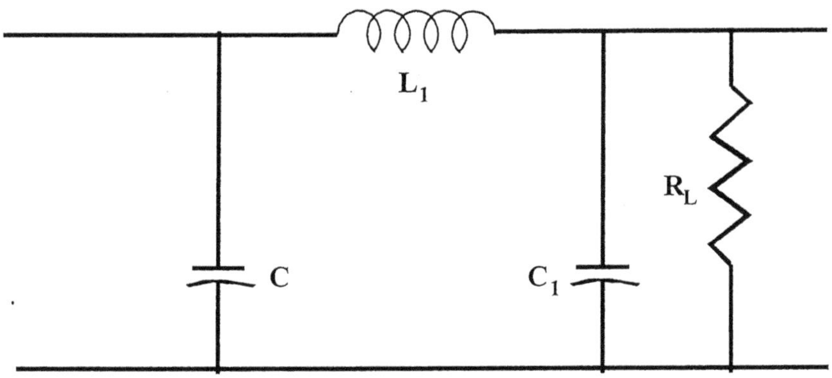

Fig. 6.38 CLC or π Section Filter

6.26 π SECTION FILTER WITH A RESISTOR REPLACING THE INDUCTOR

Consider the circuit shown here. It is a π filter with L replaced by R. If the resistor R is chosen equal to the reactance of L (the impedance in Ω should be the same), the ripple remains unchanged.

V_{DC} for a single capacitor filter = $\left(V_m - \frac{I_{DC}}{4fc} \cdot \right)$ If you consider C_1 and R, the out put across C_1 is $\left(V_m - \frac{I_{DC}}{4fc} \cdot \right)$. There is some drop across R. Therefore, the net output is

$$V_m - \frac{I_{DC}}{4fC} - I_{DC} \times R$$

The Ripple Factor for π sector is

$$\gamma = \frac{\sqrt{2} X_c X_{c_1}}{R_L . X_{L_1}} . \qquad .....(6.26.1)$$

So by this, saving in the cost, weight and space of the choke are made . But this is practical only for low current power supplies.

Suppose a FWR output current is 100 mA and a 20 H choke is being used in π section filter. If this curve is to be replaced by a resistor, $X_L = R$. Taking the ripple frequency as 2f = 100 Hz,

$$X_L = \omega L = 2\pi(2f).L = 4\pi f L$$

$$= 4 \times 3.14 \times 50 \times 20 = 12{,}000\Omega$$

$$\simeq 12\ K\Omega$$

Voltage drop across R = 12,000 × 0.1 = 1200V, which is very large.

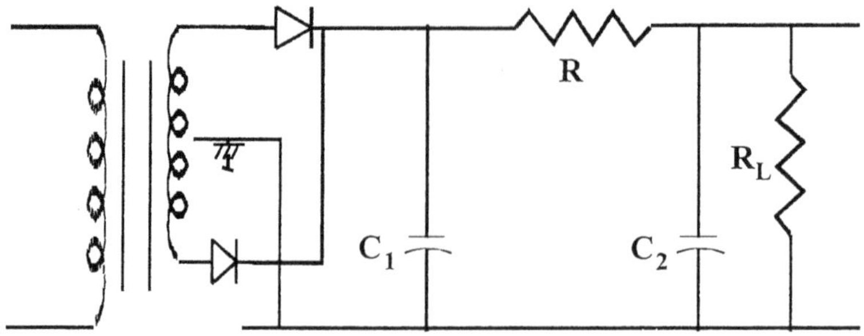

Fig. 6.39 **FWR Circuit π Section / CRC Filter**

6.27 FILTER CIRCUITS

Nearly pure DC can be obtained for a π-filter.

The output waveforms of a π-filter circuit is shown in Fig. 6.40(a)

The output waveforms of a C-filter circuit is shown in Fig. 6.40(b)

The output waveforms of a Inductor Filter Circuit is shown in Fig. 6.40(c)

The filter is used when the desired output voltage is large. But like in capacitor filter the regulation is poor.

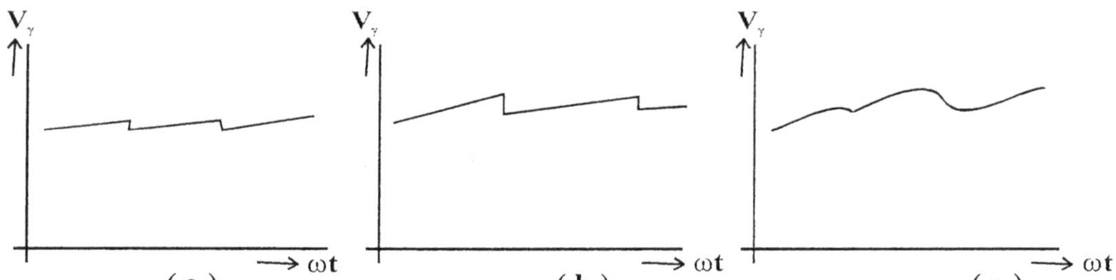

(a) π Filter

(b) C Filter

(c) L Filter

Fig. 6.40

6.27.1 Expression for Ripple Factor for p Filter

The output of the capacitor in the case of a capacitor filter is a Triangular wave. So assuming the output V_o across C to be a Triangular wave, it can be represented by Fourier series as

$$V = V_{DC} - \frac{V'P - P}{\pi}(\text{Sin } \omega 2\omega t - \frac{\sin 4\omega t}{2} + \frac{\sin 6\omega t}{3} ...)$$

But the ripple voltage peak to peak $V'\text{ p-p} = \frac{I_{DC}}{2fc}$. (for Δ wave)

Neglecting 4th and higher harmonic, the rms voltage of the Second Harmonic Ripple is

$$\frac{v'p-p}{\pi\sqrt{2}} = \frac{I_{DC}}{2\sqrt{2}\pi fc}$$

$$= \frac{I_{DC}\times 2}{4\sqrt{2}\pi fc} = \frac{\sqrt{2}I_{DC}}{4\pi fc}$$

$$\frac{1}{4\pi fc} = X_c$$

It is the capacitive reactance corresponding to the second harmonic.

$$V'_{rms} = \sqrt{2}I_{DC}X_C. \qquad \text{(across 'C' filter only)}$$

The output of C filter is the input for the L section filter of L_1C_1. Therefore, as we have done in the case of L section filter,

$$\text{Current through the inductor (harmonic component)} = \frac{\sqrt{2}I_{DC}X_C}{X_{L_1}}$$

$$\text{Output Voltage} = \text{Current} (X_C) \qquad (V_o = I.X_C)$$

$$V'_{rms} = \frac{\sqrt{2}I_{DC}X_C}{X_{L_1}}.X_{C_1} \qquad \text{(after L section)}$$

$$\text{Ripple factor,} \qquad \gamma = \frac{v'_{rms}}{V_{DC}} = \frac{\left(\sqrt{2}.I_{DC}.X_C\right)}{V_{DC}}\frac{X_{C_1}}{X_{L_1}}$$

$$V_{DC} = I_{DC}.R_L$$

$$\therefore \qquad \gamma = \frac{\left(\sqrt{2}.X_C\right)}{R_L}.\left(\frac{X_{C_1}}{X_{L_1}}\right).$$

where all reactances are calculated at the 2nd harmonic frequency $\omega = 2\pi f$.

DC output voltage = (The DC voltage for a capacitor filter) – (The drop across L_1).

6.27.2 Half Wave Voltage Doubler Circuit

During the first –ve half cycle, D_1 is forward biased. So C_1 will get charged, to the peak of the input voltage V_p, with the polarity as shown in Fig. 6.41. During the +ve half cycle, D_2 is forward biased. Therefore, C_2 will get charged to $V_p + V_P = 2V_P$. Capacitor C_1 will charge C_2 to V_p and from the source, through D_2, C_2 will get charged to a further value of V_p. So after several cycles, the voltage across C_2 will be $2V_p$. Here R_L should be large. Other wise, C_2 will be discharging quickly and the voltage may not be maintained constant at $2V_p$. This is half wave doubler circuit, since the output capacitor C_2 is charged only once during the full cycle.

Such circuits are used where large output voltage is required. Large V_o can be obtained by using a bigger transformer. But its cost will be more.

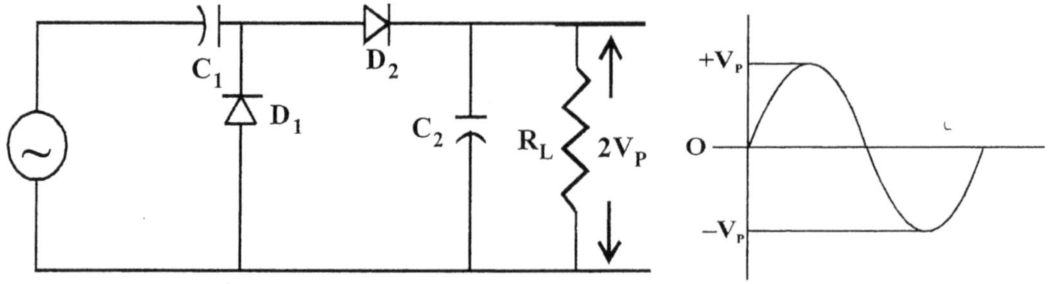

Fig. 6.41 Half wave voltage doubler circuit

6.27.3 Full Wave Voltage Doubler

During the +ve half cycle, D_1 is forward biased C_1 gets charged to V_p. During the negative half cycle, D_2 is forward biased. So C_2 gets charged to V_p. In the steady state, the voltage across R_L is $2V_p$. Here centre tapped transformer is not used. It is called full wave, since one of the diodes is conducting in each half cycle (similar to a full wave rectifier circuit.)But the problem is, since no centre tapped transformer is used, there is no common ground between the input and the output. If the bottom of R_L is grounded, input is floating. This is not desirable in electronic circuits. So in F W R circuits, centre tapped transformer is used In this circuit ripple will be less as shown in Fig. 6.42.

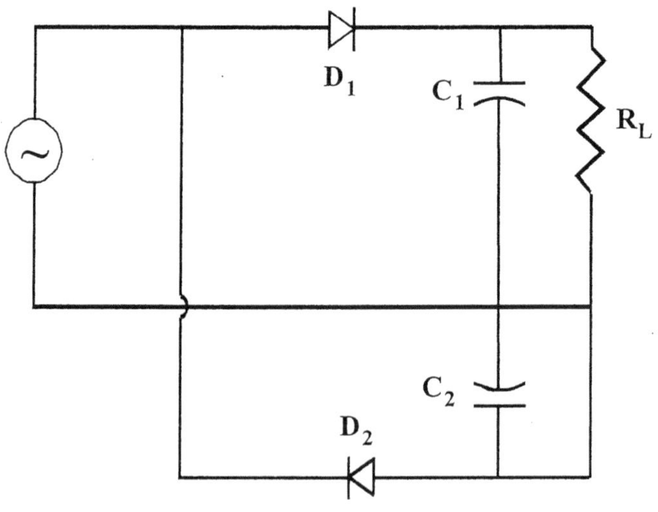

Fig. 6.42 Voltage Doubler Circuit

6.27.4 Voltage Tripler

The circuit is as shown in Fig. 6.43. Working is similar to the previous circuit.

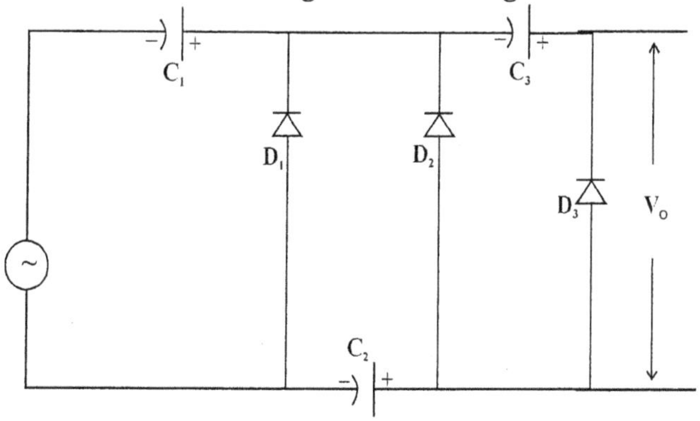

Fig. 6.43 Voltage Tripler Circuit

6.27.5 Voltage Quadrupler

The Circuit is shown in Fig. 6.44. Is there any limit to the number of sections that can be added to get large voltage? Theoretically it is possible. But the ripple voltage will get worse as additional sections are added. (A diode is being put across or in series. So ripple will increase). So these circuits are used only where very high voltage doubling is to be done. The input voltage itself will be higher.

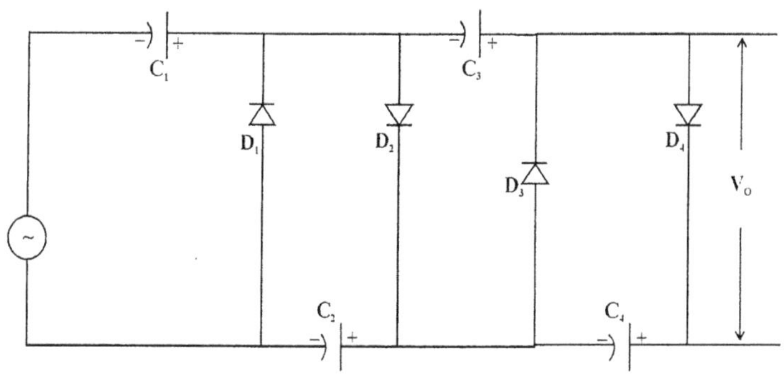

Fig. 6.44 Voltage Quadrupler circuit

6.27.6 Peak to Peak Detector

The Circuit is as shown in Fig. 6.45.

The input sine wave is +ve ly clamped. C_1 — D_1 is a +ve clamper circuit. D_2 — C_2 is a peak detector circuit. When D_1 is forward biased, C_1 gets charged to V_p. When D_2 is forward biased, C_2 gets charged to $2V_p$. So output will measure peak to peak value. R_L should be large. If the input is not symmetrical, +ve peak and

–ve peak are not same, the average value or rms value measured will not be correct. In such cases, first peak to peak detector is connected and then the DC meter or AC meter are connected. In certain measurements, peak to peak value is to be determined. In such cases this type of circuits are used.

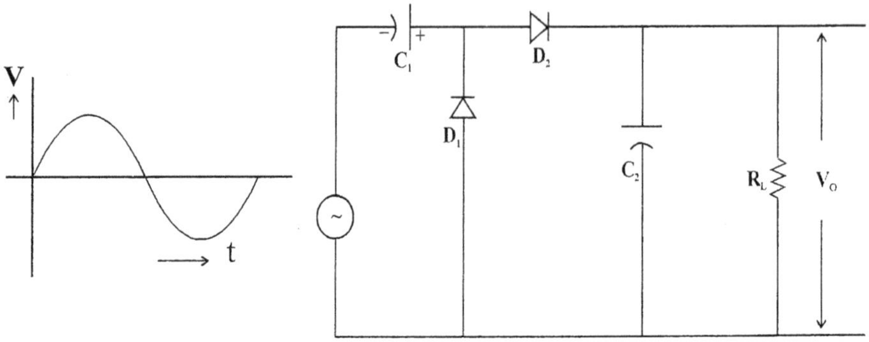

Fig. 6.45 Peak to Peak detector

Ripple Factors :

γ Ripple Factor for 'C' filter $= \dfrac{1}{4\sqrt{3}fCR_L}$

γ Ripple Factor for 'L' filter $= \dfrac{R_L}{4\sqrt{3}\,\omega L}$

γ Ripple factor LC filter $= \dfrac{\sqrt{2}}{3} \times \dfrac{1}{2\omega C} \times \dfrac{1}{2\omega L}$

Critical Inductance $L_C \geq \dfrac{R_L}{3\omega}$

Bleeder Resistor $R_B = \dfrac{3X_L}{2}$

γ for π-filter $\dfrac{\sqrt{2}X_C}{R_L} \cdot \left(\dfrac{X_{C1}}{X_{L1}}\right)$

Example 6.7:

Design a power supply using a π-filter to give DC output of 25V at 100mA with a ripple factor not to exceed 0.01%. Design of the circuit means, we have to determine L,C, diodes and transformers.

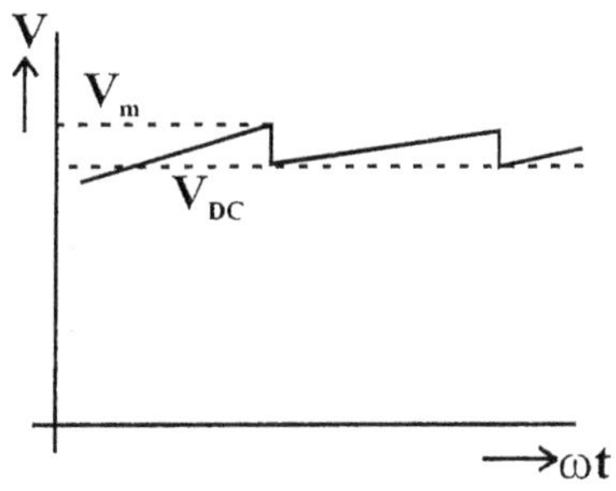

Fig. 6.46 *For Example 6.7*

Solution:

Design of the circuit means, we have to determine L, C, Diodes and Transformer

$$R_L = \frac{V_{DC}}{I_{DC}} = \frac{25V}{I_{DC}} = \frac{25V}{100mA} = 250\Omega.$$

$$\text{Ripple factor } \gamma = \sqrt{2}.\frac{X_c}{R_L}.\frac{X_{C1}}{X_{L1}}.$$

X_C can be chosen to be $= Xc_1$.

$$\therefore \qquad \gamma = \sqrt{2}.\frac{X_C^2}{R_L.X_{L1}}.$$

This gives a relation between C and L.

$$C^2 L = y$$

There is no unique solution to this.

Assume a reasonable value of L which is commercially available and determine the corresponding value of capacitor. Suppose L is chosen as 20 H at 100mA with a D.C. Resistance of 375Ω (of Inductor).

$$\therefore \qquad c^2 = \frac{y}{L}$$

$$\text{or} \qquad c = \sqrt{\frac{y}{L}}$$

$$V_{DC} = Vm - \frac{V_\gamma}{2}$$

$$V_\gamma = \frac{I_{DC}}{2fc}.$$

Now the transformer voltage ratings are to be chosen.

The voltage drop across the choke = choke resistance × I_{DC}.

$= 375 \times 100 \times 10^{-3} = 37.5V.$

$V_{DC} = 25V.$

Therefore, voltage across the first capacitor C in the π-filter is

$V_c = 25 + 37.5 = 62.5V.$

The peak transformer voltage, to centre tap is

$$V_m = (V_C) + \frac{V_\gamma}{2} \text{ (for C filter)}$$

$$V_\gamma = \frac{I_{DC}}{2fc}.$$

$$\therefore \quad V_m = 62.5 \text{ V} + \frac{0.1}{2 \times 50 \times c}$$

$$V_{rms} = \frac{V_m}{\sqrt{2}}. \cong 60v$$

Therefore, a transformer with 60 - 0 -60V is chosen . The ratings of the diode should be, current of 125mA. and voltage = PIV = $2V_m$ = 2 × 84.6V = 169.2V

Example 6.8:

A full wave rectifier with LC filter is to supply 250 v at 100m.a. D.C. Determine the ratings of the needed diodes and transformer, the value of the bleeder resistor and the ripple, if R_C of the choke = 400Ω. L = 10H and C = 20 μF.

Solution:

$$R_L = \frac{V_{DC}}{I_{DC}}$$

$$R_L = \frac{250v}{0.1} = 2{,}500\Omega.$$

For the choke input resistor,

$$E_{DC} = \frac{2E_m}{\pi(1+\frac{R_C}{R})}$$

and $\quad I_{rms} \cong I_{DC}.$

$$E_m = \frac{\pi E_{DC}}{2}(1+\frac{R_C}{R_L})$$

$$= \frac{\pi \times 250}{2}(1+\frac{400}{2500}) = 455V$$

$\therefore$
$$E_{rms} = \frac{455}{\sqrt{2}} = 322V$$

Therefore, the transformer should supply 322V rms on each side of the centre tap. This includes no allowance for transformer impedance, so that the transformer should be rated at about 340 volts 100 mA, D.C

The Bleeder Resistance $R_B = \frac{3X_L}{2}$

$$L = 10000\Omega.$$

$$I_B = \frac{2E_m}{3\pi\omega L} = \frac{2 \times 455}{3\pi \times 377 \times 10} = 0.256A$$

Ripple factor $\gamma = \frac{0.47}{4\omega^2 LC - 1} = \frac{0.47}{4 \times 377^2 \times 20 \times 10^{-6}}$

The current ratings of each diode = 0.00413. Should be 50mA.

6.27.7 Multiple LC Filters

Better filtering can be obtained by using two or more LC sections as shown in the Fig. 6.47.

Ripple factor $\gamma = \frac{\sqrt{2}}{3} \cdot \frac{XC_1}{XL_1} \cdot \frac{XC_2}{XL_2}$.

If we take $L = L_1 = L_2$

$C = C_1 = C_2$ and f = 60 Hz.

$$\gamma = \frac{1.46}{L^2C^2}$$

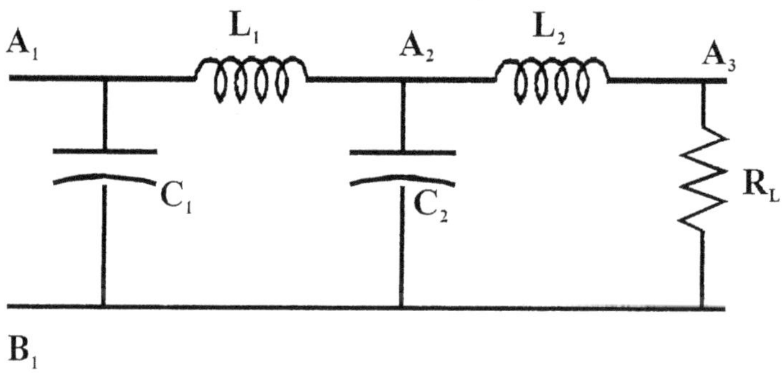

Fig. 6.47 Multiple LC filter sections.

Inductor value L in Henrys and Capacitor value C in μF(microfarads)

So ripple factor will be much smaller or better filtering is achieved.

$$\therefore \quad I_1 = \frac{\sqrt{2}}{3} \times \frac{V_{DC}}{L_1} \text{ (RMS Value of the ripple component of current.)}$$

$$V_{A_2B_2} = I_1 \times C_1 \frac{\sqrt{2}}{3} \cdot \frac{V_{DDC}}{X_{L_1}} . XC_1$$

$$I_2 = \frac{V_{A_2B_2}}{X_{L_2}}$$

The value of XL_2 should be much larger than XC_2. Therefore, A.C. current should get dropped across C_2.

$\therefore \qquad I_2 = V_{A_2B_2} / XL_2$ and not $\dfrac{V_{A_2B_2}}{X_{L_2} + X_{L_2}}$.

$\therefore \qquad V_o \cong I_2\, Xc_2 / R_L$ is in parallel with C_2.

$$= \frac{\sqrt{2}}{3}.V_{DC}.\frac{X_{C_1}}{X_{L_1}}.\frac{X_{C_2}}{X_{L_2}}$$

Ripple factor $\qquad \gamma = \dfrac{V_o}{V_{DC}} = \dfrac{\sqrt{2}}{3}.\dfrac{X_{C_1}}{X_{L_1}}.\dfrac{X_{C_2}}{X_{L_2}}$

But the actual value of L_c should be 25% more than this because of the approximately made in the series used to represent V.

6.27.8 Specification of a Power Supply

1. *The required D.C. output voltage.*
2. *Regulation.*
3. *The average and peak currents in each diode.*
4. *PIV of each diode.*
5. *Ripple factor.*

Example 6.9:

A full wave rectifier circuit with C-type capacitor filter filter is to supply a D.C. Current of 20 mA at 16V. If frequence is 50 Hz ripple allowed is 5% Calculate :

a. *Required secondary voltage of the transformer.*

b. *Ratio of I peak/I $_{max}$ through diodes.*

c. *The value of C required.*

Solution:

$$I_{DC} = 20\text{mA}\ V_{DC} = 16\text{V}.$$

$$\therefore \qquad R_L = \frac{V_{DC}}{I_{DC}} = \frac{16}{20\text{mA}} = 0.8\text{K}\Omega = 800\Omega.$$

$$\gamma = 5\% = \frac{1}{4\sqrt{3}fcR_L}$$

$$\therefore \quad 0.05 = \frac{1}{4\sqrt{3} \times 50 \times C \times 800}$$

$$\therefore \quad C = \frac{1}{4 \times 1.732 \times 50 \times 0.05 \times 800}$$

$$= \frac{1}{4\sqrt{3} \times 2.5 \times 800}$$

$$= \frac{1}{4\sqrt{3} \times 2000}$$

$$C = \frac{\sqrt{3}}{8000 \times 3} = 72\mu 8$$

$$V_{DC} \text{ with load } I_{DC} = \frac{V_m}{1 + \dfrac{1}{4fcR_L}} \quad V_m = ?$$

$$V_{DC} = V_m \left(\frac{4fCR_L}{1 + 4fCR_L} \right)$$

$$V_{DC} = V_m \left(\frac{L_1 f_C R_L}{1 + L_1 f_C R_L} \right)$$

$$V_{DC} = 16V \text{ at } I_{DC} = 20 \text{ mA.}$$

$$V_{DC} = V_m \left(\frac{0.05\sqrt{3}}{1 + 005\sqrt{3}} \right) 16$$

$$= \frac{V_m}{1 + \dfrac{1}{4 \times 50 \times 72 \times 10^{-6} \times 800}}$$

$$V_m = \frac{16 \times 1.086}{0.086}$$

$$16 = \frac{V_m}{1+0.05\sqrt{3}}$$

$$V_m = 180V$$

Example 6.10:

A half wave rectifier allows current to flow from time $t_1 = \pi/6\omega$ to $t_2 = 5\pi/6\omega$, the A.C. Voltage being given by $V_s = 100$ Sin ωt. Calculate the RMS and DC output voltage.

Solution:

$$V_{DC} = \frac{1}{2\pi}\int_{\pi/6}^{5\pi/6} 100\ \text{Sin}\ \omega t.\ d\omega t$$

$$= \frac{1}{2\pi} \times 100 \times [1 - \cos\omega t]_{\pi/6}^{5\pi/6}$$

$$= \frac{100}{2\pi}\left[-\cos\frac{5\pi}{6} + \cos\frac{\pi}{6}\right]$$

$$= \frac{100}{2\pi}\left[-\cos(150^0) + \cos(30^0)\right]$$

$$= \frac{50}{\pi}\left[+\frac{\sqrt{3}}{2} + \frac{\sqrt{3}}{2}\right]$$

$$= \frac{50\times\sqrt{3}}{3.14} = \frac{50\times 1.732}{3.14}$$

$$= 27.6V$$

$$V_{rms} = \sqrt{\frac{1}{2\pi}\int_{\pi/6}^{5\pi/6}(100\ .\ \text{Sin}\omega t)^2 . d\omega t}$$

$$= 48.5V.$$

Example 6.11:

A full wave rectifier operating at 50 Hz is to provide D.C. Current of 50mA at 30V. with a 80μF. C-type filter system used, calculate

(a) V_m the peak secondary voltage of transformer.

(b) Ratio of surge to mean currents of diodes.

(c) The ripple factor of the output.

Solution:

(a) $$R_L = \frac{V_{dc}}{I_{dc}} = \frac{30\text{ V}}{50\text{mA}} = 600\Omega.$$

$$V_{DC} = \left(\frac{V_m}{1 + \frac{I_{DC}}{4fcV_{DC}}} \right) = \frac{V_m}{1 + \frac{1}{4fcR_L}}$$

$$\frac{I_{DC}}{V_{DC}} = \frac{1}{R_L}.$$

$$V_m = V_{DC} + \frac{V_{DC}}{4fcR_L} = V_{DC} + \frac{I_{DC}}{4fc}$$

$$V_m = 30 + \frac{50 \times 10^{-3}}{4 \times 50 \times 80 \times 10^{-6}}$$

$$= 33.125\text{V}.$$

$$V_{rms} = \frac{V_m}{\sqrt{2}} = \frac{33.125}{\sqrt{2}} = 23.4\text{V}$$

(b) Surge Current through the diode, I_{diode} surge is I_m.

$$\frac{V_m}{X_c} = V_m \omega C$$

$$= 33.125 \times 2\,p \times 50 \times 80 \times 10^{-6} = 0.833\text{ A}$$

$$I_{DC} = 50\text{ mA}$$

$= I_{diode\ mean}$ is the average or mean current through the diode.

$$\frac{I_{diode\ swing}}{I_{diode\ mean}} = \frac{0.833}{50 \times 10^{-3}} = 16.65.$$

(c) $$R_L = \frac{V_{DC}}{I_{DC}} = 600\Omega.$$

$$\gamma = \frac{1}{4\sqrt{3}fcR_L} = 0.06.$$

Example 6.13:

For a FWR circuit, the A.C. Voltage input to transformer primary = 115V. Transformer secondary voltage is 50V. R_L = 25Ω. Determine

1. *Peak DC component, RMS and AC component of load voltage.*
2. *Peak DC component, RMS and AC component of load current.*

Solution:

Since, the transformer develops 50V between secondary terminals, there must be 25V across each half of the secondary winding.

The peak value across each half of secondary is $25\sqrt{2} = 35.4V$

Assuming ideal diodes, the rectified voltage across R_L also has a peak value of 35.4V(V_m).

∴ Average value V_{DC} for FWR circuit is

$$= \frac{2V_m}{\pi} = 0.636V_m$$

$$V_{DC} = 0.36 \times (35.4) = 22.5V.$$

A.C.Component load voltage is

$$V'_{rms} = \sqrt{V_m^{\ 2} - V_{DC}^{\ 2}}$$

$$= \sqrt{(25)^2 - (22.5)^2} = 10.9V.$$

$$I_m = \frac{V_m}{R_L} = \frac{35.4}{25} = 1.41A.$$

$$I_{DC} = \frac{2I_m}{\pi} = 0.636I_m = 0.636(1.41) = 0.897A.$$

$$I_L(rms) = 0.707\ I_m$$

$$= 1.41(0.707) = 1A$$

$$I'_{rms} = \sqrt{(I_{rms})^2 - (I_{DC})^2}$$

$$= \sqrt{1^2 - (0.897)^2}$$

$$= 0.441\ A.$$

SUMMARY

- Expression for current through a *p-n junction.*

$$\text{Diode is } I = I_O\left(e^{\frac{V}{\eta V_T}} - 1\right)$$

- Forward Resistance R_f of a diode will be small of the order of few Ω. Reverse Resistance R_γ will be very high, of the order of MΩ
- Cut in voltage V_γ for Germanium diode is 0.1V and for Silicon diode it is 0.5V at room temperature. It decreases with increase in temperature.
- I_O the reverse saturation current of a diode will be of the order of a few μA. It increases with temperature. I_O gets doubled for every 10°C rise in temperature.
- When the diode is reverse biased, transition capacitance C_T results when the diode is forward biased, diffusion capacitance C_D is exhibited.
- The three types of breakdown mechanisms in semiconductor diodes are
  1. Avalanche Breakdown
  2. Zener Breakdown
  3. Thermal Breakdown.
- Rectifier circuit converts AC to Unidirectional Flow.
- Filter Circuit converts Unidirectional flow to DC. It minimises Ripple.
- The different types of filter circuits are
  1. Capacitor Filter
  2. Inductor Filter
  3. L-Section (LC) Filter.
  4. π-Section Filter
  5. CRC and CLC Filters
- Ripple Factor $= \dfrac{I'_{rms}}{I_{DC}}$. For Half Wave Rectifier, $\gamma = 1.21$, for Full Wave

Rectifier $\gamma = 0.482$

- Expression for Ripple Factor for C Filter,

$$\gamma = \frac{1}{4\sqrt{3} f CR_L}$$

- Expression for Ripple Factor for L Filter,

$$\gamma = \frac{R_L}{4\sqrt{3}\omega L}$$

- Expression for Ripple Factor for LC Filter,

$$\gamma = \frac{\sqrt{2}}{3} \times \frac{1}{2\omega C} \times \frac{1}{2\omega}$$

- Expression for Ripple Factor for π Filter,

$$\gamma = \frac{\sqrt{2}C}{R_L}\left(\frac{X_{C_1}}{X_{L_1}}\right)$$

- Critical Inductance $L_C \geq \dfrac{R_L}{3\omega}$
- Bleeder Resistance $R_B = \dfrac{3X_L}{2}$

OBJECTIVE TYPE QUESTIONS

1. General expression for E_o, the contact difference of potential in an open circuited *p-n junction* in terms of N_C, N_A and n_i is ....
2. Typical value of E_o = ...eV
3. The equation governing the law of the junction is ....
4. The expression for the current in a forward biased diode is ....
5. The value of cut in voltage in the case of Germanium diode .... and Silicon Diode .... at room temperature.
6. Expression for V_B the barrier potential in terms of depletion region width W is V_B = ...

7. Expression for I_0 in terms of temperature T and V_T is ....
8. Zener breakdown mechanism needs relatively .... electric field compared to Avalanche Breakdown.
9. Expression for the diffusion capacitance C_D in terms of L_P and D_P is ....
10. Expression for Volt equivalent of temperature V_T in terms of temperature T is $V_T =$ ....
11. The salient feature of a Tunnel diode is, it exhibits .... characteristics.
12. Rectifier Converts .... to ....
13. Filter Circuit Converts .... to ....
14. Ripple Factor in the case of Full Wave Rectifier Circuit is ....
15. The characteristics of a swinging choke is ....
16. In the case of LC Filter Circuits, Bleeder Resistance ensures
    (1) ....
    (2) ....

ESSAY TYPE QUESTIONS

1. Derive the expression for E_0 in the case of open circuited *p-n junction* diode.
2. Qualitatively explain the forward and reverse characteristic of *p-n junction* diode.
3. Derive the expression for Transistion Capacitance C_T in the case of *p-n junction* diode.
4. How junction capacitances come into existance in *p-n junction* diode.
5. Derive the expression for the diffusion capacitance C_D in the case of *p-n junction* diode.
6. Obtain the expression for ripple factor in the case of Full Wave Rectifier Circuit with Capacitor Filter.
7. Explain the terms Swinging Choke and Bleeder Resistor.
8. Compare C, L, L-Section, π-Section (CLC and CRC) Filters in all respects.

MULTIPLE CHOICE QUESTIONS

1. The force of electron 'F' between nucleus and electron with charge 'e' and radius 'r' is, proportional to, F α

(a) $\frac{e^2}{r^2}$ (b) $\frac{r^2}{e^2}$ (c) $e^2 r^2$ (d) $\frac{e}{r}$

2. The energy possesed by the electron, W orbitting round the nucleus is,

(a) $\frac{-e}{8\pi \varepsilon_0 r}$ (b) $\frac{-e^2}{8\pi \varepsilon_0 r}$ (c) $\frac{-e.r}{8\pi \varepsilon_0}$ (d) $\frac{-e^2}{8\pi \varepsilon_0 r^2}$

3. The expression for the Kinetic Energy E of free electron in terms of momentum of the electron 'p' and mass of the electron m is,

(a) $\frac{p}{2m}$ (b) $\frac{p}{2m^2}$ (c) $\frac{p^2}{2m^2}$ (d) $\frac{p^2}{2m}$

4. The expression for radius of stable state 'e' r, is

(a) $\frac{n^2 h^2 \varepsilon_o}{\pi me^2}$ (b) $\frac{n h \varepsilon_o}{\pi^2 m^2 e^2}$ (c) $\frac{n^2 h \varepsilon_o^2}{\pi m e^2}$ (d) $\frac{n^2 h^2 \varepsilon_o^2}{\pi m e^2}$

5. The value of the radius of the lowest state or ground state is, given, h = 6.626 × 10^{-34} J - Secs $\varepsilon_o = 10^{-9}/36\pi$

(a) 0.28 A° (b) 0.98 A° (c) 0.18 A° (d) 0.58 A°

6. The energy required to detach an electron from its parent atom is called,

(a) Ionization potential (b) Eletric potential
(c) Kinetic energy (d) Threshold potential

7. Electron collision without transfer of energy in collision is called,

(a) Null collision (b) Impact collision
(c) Elastic collision (d) Stiff collision

8. Wave mechanics in electron theory is also known as

(a) Eienstein theory (b) Quantum mechanics
(c) Bohr mechanics (d) Classical theory

9. The expression for threshold frequency f_T to cause photoelectric emission is, with usual notation is,

(a) $\frac{e\phi^2}{h}$ (b) $\frac{e\phi}{h}$ (c) $\frac{eh}{\phi}$ (d) $\frac{e^2\phi^2}{h^2}$

10. The conductivity of a good conductor is typically

(a) 10^{-3} Ω/cm (b) 10^3 Ω/cm (c) 10^3 ℧/cm (d) 10 ℧/m

11. The free electron concentration of a good conductor is of the order of

(a) 10^{10} electrons/m3 (b) 10^{28} electron/m^3

(c) 10^2 electrons/cm^3 (d) 10^{10} electrons/m^3

12. Fermi level in Intrinsic semiconductor lies

(a) close to conduction band (b) close to valence band

(c) In the middle (d) None of the these

13. The potential which exists in a p-n junction to cause drift of charge carriers is called

(a) contact potential (b) diffusion potential

(c) ionisation potential (d) threshold potential

14. The rate of increase of reverse saturation current for Germanium diode is,

(a) 5% (b) 4% (c) 1% (d) 7%

15. Special types of diodes in which transition time and storage time are made small are called ...

(a) Snap diodes (b) Rectifier diodes (c) Storage diodes (d) Memory diodes

16. The Circuit which converts undirectional flow to D.C. is called

(a) Rectifier circuit (b) Converter circuit

(c) filter circuit (d) Eliminator

17. For ideal Rectifier and filter circuits, % regulations must be ...

(a) 1% (b) 0.1 % (c) 5% (d) 0%

18. The value of current that flows through R_L in a 'π' section filter circuit at no load is

(a) ∞ (b) 0.1 mA (c) 0 (d) few mA

CONCEPTUAL QUESTIONS (Interview Questions)

1. When it is said that an electron moves from valence band to conduction band, will there be physical movement of the electron ?

   Ans : No, The energy band diagram represents the energy levels of the electrons. When an electron "moves" from valence band to conduction band, it only means that the energy possessed by the electron has increased from Ev electron volts to Ec+ electron volts. The energy diagram is only Pictorial representation of the energy possessed by the electrons. The energy levels of different electrons will be different.

2. What does the arrow in the symbol of a P-n junction diode (—▷|—) indicate?

Ans : The arrow mark A▷|K (p) (n) indicates the direction of current flow when the p - n juction is forward biased.

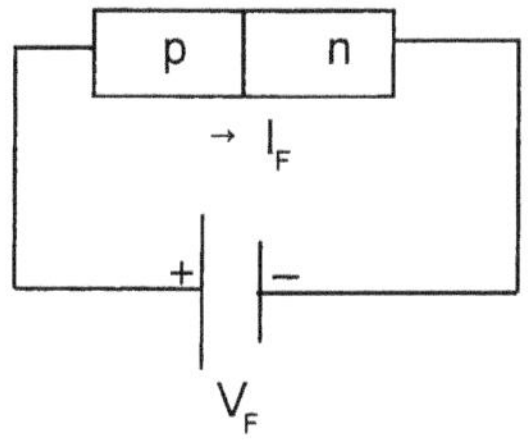

Direction of I_F is the same as that of the direction of arrow.

3. When a Semiconductor device 'breaks down'; what does it mean?

   Ans : It does not mean physical breakdown. The shape of the device remains the same. But the device looses its Semiconductor properties. The device may become a short or open circuited.

4. What is the difference between Avalanche and zener breakdown mechanisms?

   Ans : In Zener breakdown mechanism, due to sudden increase in the electric field strength ε, when it reaches a critical value, covalent bonds break and due to this there will be sudden increase in the number of free electrons and hence, the reverse current which was small till that time increases suddenly. But in Avalanche breakdown the field strength is not that high. A free electron with lot of energy collides with another electron, transfers its energy and makes it a free electron. This in turn generates some more free electrons. Thus there is multiplication of carriers. This will occur usually in a forward biased device. Avalanche breakdown precedes Zener breakdown, for comparision, though both cannot occur in the same device at a time.

5. How for a low voltage like 6.8 V, in the case of Fz 6.8 Zener doide zener breakdown occurs, which requires high electric field strength?

   Ans : Electric field strength $\varepsilon = \frac{V}{d}$. Even if V is small, if 'd' is very small, ε can be very high. This is what happens in a Zener diode. Due to heavy doping depletion, region width near the junction becomes very small. So, when the voltage V_Z reaches the required value, electric field 'ε' reaches the critical value, covalent bonds will be broken and Zener breakdown occurs.

Some of the Semiconductors Devices and IC Manufacutring Companies

BEL	:	Bharat Elecronics Ltd., Bangalore, India
CDIL	:	Continental Devices India Ltd., New Delhi, India.
SCL	:	Semiconductors complex Ltd., Chandigarh, India.
CEL	:	Control electronics Ltd., Ghaziabad, Uttar Pradesh.
ITI	:	Indian Telephone Industries, Bangalore, India.
TALS	:	Team Asia Lakhi Semiconductor Ltd., Hyderabad, India.
AD	:	Analog Devices Inc. MA 02062, USA
BB	:	Burr - Brown Research Corporation, USA
Intel	:	Intel Corporation, Santaclara, CA95051 USA
M	:	Motorola Semiconductor Products. Phoenix, AZ85036
TI	:	Texas Instruments, Semiconductor Group Dallas, Tx 75222.

Specifications for Silicon Diode

Sl.No.	Parameter	Symbol	Typical Value	Units
1.	Reverse breakdown voltage	V_{br}	75	V
2.	Static Reverse Current	I_R	5	μA
3.	Static Forward voltage (for silicon)	V_F	0.5	V
4.	Total Capacitance	C_T	2	pf
5.	Reverse Recovery Time	t_{rr}	4	ns
6.	Continuous Power Dissipation	P	500	mw
7.	Max. For Ward Current	I_F	50	mA

Some type numbers of Diodes

1. 0A79 : 0 → Semiconductor Device.
   A → Denotes Germanium device.

2. BY 127 : B → Denotes Silicon Device.
   Y → Rectifying Diode.
   Number 127 is owing a type no and has significance.

3. 1 N 4153 : N → Bipolar Device.
   1 → Single Polar function.

4. BY 100 : 800V, 1 A Silicon Diode.

7

BJT AND TRANSISTOR CONFIGURATIONS

7.1 TRANSISTOR CONSTRUCTION

Two commonly employed methods for the fabrication of diodes, transistors and other semiconductor devices are :

1. Grown type **2.** Alloy type

7.1.1 Grown Type

It is made by drawing a single crystal from a melt of Silicon or Germanium whose impurity concentration is changed during the crystal drawing operation.

7.1.2 Alloy Type

A thin wafer of ***n-type*** Germanium is taken. Two small dots of indium are attached to the wafer on both sides. The temperature is raised to a high value where indium melts, dissolves Germanium beneath it and forms a saturator solution. On cooling, indium crystallizes and changes Germanium to ***p-type***. So a PNP transistor formed. The doping concentration depends upon the amount of indium placed and diffusion length on the temperature raised.

7.1.3 Manufacture of Grown Junction Transistor

Grown Junction Transistors are manufactured through growing a single crystal which is slowly pulled from the melt in the crystal growing furnace Fig. 7.1. The purified polycrystalline semiconductor is kept in the chamber and heated in an atmosphere of N_2 and H_2 to prevent oxidation. A seed crystal attached to the vertical shaft which is slowly pulled at the rate of (1mm / 4hrs) or even less. The seed crystal initially makes contact with the molten semiconductors. The crystal starts growing as the seed crystal is pulled. To get ***n-type*** impurities, to the molten polycrystalline solution, impurities are added. Now the crystal that grows is ***n-type*** semiconductor. Then ***p-type*** impurities are added. So the crystal now is ***p-type***. Like this NPN transistor can be fabricated.

1. Furnace
2. Molten Germanium
3. Seed Crystal
4. Gas Inlet
5. Gas Outlet
6. Vertical Drive

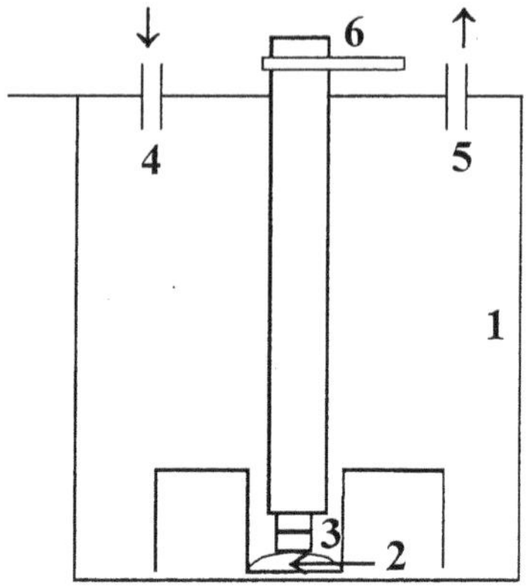

Fig. 7.1 Crystal growth.

7.1.4 Diffusion Type

To fabricate NPN transistor, ***n-type*** 'Si' (or 'Ge') wafer is taken. This forms the collector. Base-collector junction areas is determined by a mask and diffusion length. ***p-type*** impurity (Boron) is diffused into the wafer by diffusion process. The wafer to be diffused and the Boron wafer which acts as ***p-type*** impurity, are placed side by side. Nitrogen gas will be flowing at a particular rate, before the Boron is converted to Boron Oxide and this gets deposited over Siliconwafer. Baron is driven inside by drive in diffusion process. So the base-collector junction will be formed. Now again ***n-type*** impurity (Phosphorus) is diffused into the Base Area in a similar way. The different steps in process are :

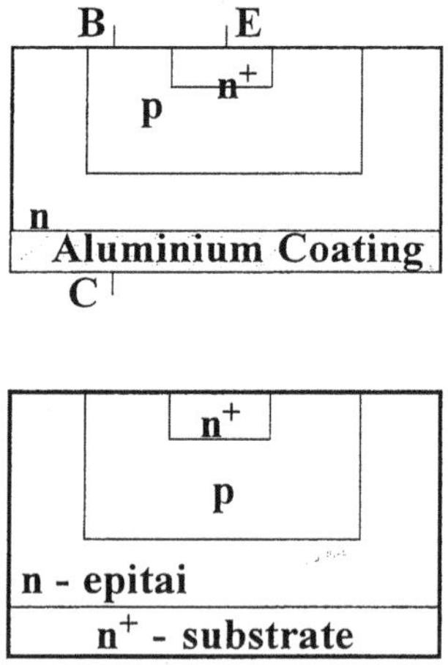

Fig 7.2(a) Epitaxial type transistor

1. *Oxidation*
2. *Photolithography*
3. *Diffusion*

4. *Metallization*
5. *Encapsulation*

7.1.5 Epitaxial Type

In this process, a very thick, high purity (high resistance), single crystal layer of Silicon or Germanium is grown as a heavily doped substrate of the same material. (n on n^+) or (p on p^+). This forms the collector. Over this base and emitter are diffused. ***Epitaxi*** is a greek word. *Epi* mean 'ON' *Taxi* means 'ARRANGEMENT'. This technique implies the growth of a crystal on a surface, with properties identical to that of the surface. By this technique abrupt step type p-n junction can be formed. The structure is shown in Fig 7.2.

7.2 BIPOLAR JUNCTION TRANSISTOR (BJT)

The device in which conduction takes place due to two types of carriers, electrons and holes is called a Bipolar Device. As *p-n junctions* exist in the construction of the device, it is a junction device. When there is transfer of resistance from input side which is Forward Biased (low resistance) to output side which is Reverse Biased (high resistance), it is a **Trans Resistor or Transistor Device**. There are two types of transistors NPN and PNP. In NPN Transistor, a ***p-type*** Silicon (Si) or Germanium (Ge) is sandwiched between two layers of ***n-type*** silicon.

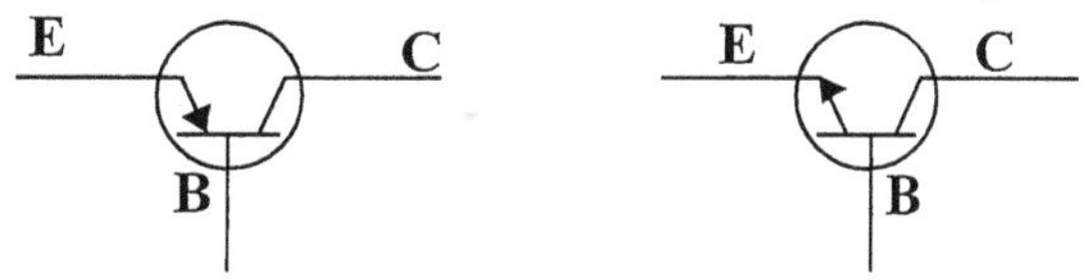

(a) **PNP** (b) **NPN**

Fig. *7.3 Transistor symbols.*

The symbol for PNP transistor is as shown in Fig. 7.3(a) and for NPN transistor, is as shown in Fig 7.3 (b).

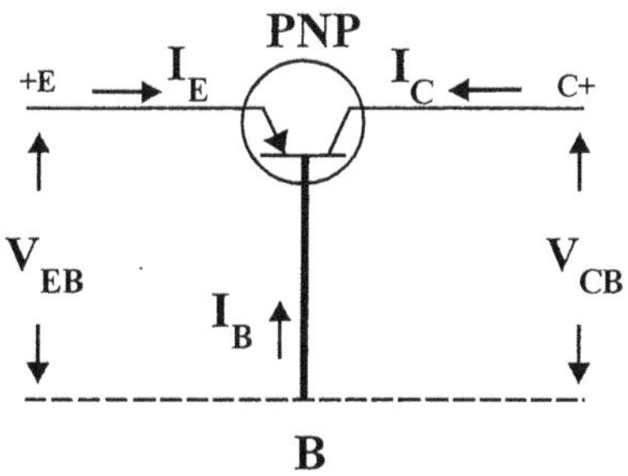

Fig. *7.4 Current components in a pnp transistor.*

The three sections of a transistor are Emitter, Base and Collector. If the ***arrow mark is towards the base,*** it is PNP transistor. If it is away then it is NPN transistor. The arrow mark on the emitter specifies the direction of current when the emitter base junction is forward biased. When the PNP transistor is forward biased, holes are injected into the base. So the holes move from emitter to base. The conventional current flows in the same direction as holes. So arrow mark is towards the base for PNP transistor. Similarly for NPN transistor, it is away. DC

Emitter Current is represented as I_E, Base Current as I_B and Collector Current as I_C. These currents are assumed to be positive when the currents flow into the transistor. V_{EB} refers to Emitter - Base Voltage. Emitter (E) Voltage being measured with reference to base B (Fig.7.4). Similarly V_{CB} and V_{CE}.

7.2.1 Principle of Operation

Fig. 7.5 shows a circuit for a PNP transistor. Emitter Base Junction is Forward Biased Collector - Base Junction is reverse biased. Fig. 7.6 shows potential distribution along the transistor (x). When the transistor is open circuited, there is no voltage applied. The potential barriers adjust themselves to a height V_o. V_o will be few tenths of a volt. It is the barrier potential. So when the transistor is open circuited holes will not be injected into the collector. (Potential barrier exists all along the base width, since the holes from emitter have to reach collector and not just remain in base).

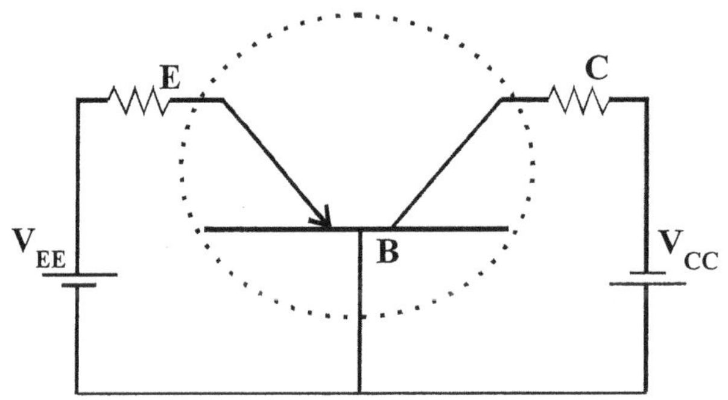

***Fig. 7.5** PNP transistor circuit connections*

But when transistor bias voltages are applied, ***emitter base junction is forward biased*** and the ***collector base junction is reverse biased***. The base potential almost remains constant. But if the reverse bias voltage of collector - base junction is increased, ***effective base width decreases***. Since EB junction is forward biased, in a PNP transistor, the emitter base barrier is lowered. So holes will be injected into the base. The injected holes diffuse into the collector across the ***n-type*** material (base). When EB junction is forward biased emitter - base potential is increased by $|V_{EB}|$. Similarly collector base potential is reduced by $|V_{CB}|$. Collector base junction is reverse biased. So collector extends into the base or depletion region width increases. Emitter - Base Junction is to be Forward Biased because, the carriers must be injected and Collector - Base Junction must be reverse biased, because the carriers must be attracted into the Collector Region. Then only current flow results through the transistor.

7.3 TRANSISTOR CURRENT COMPONENTS

Consider a PNP transistor. When Emitter - Base junction is forward biased, holes are injected into the base. So this current is I_{pE} (holes crossing form Emitter to base). Also electrons can be injected from base to emitter (which also contribute for emitter current I_E). So the resulting current is I_{nE}. Therefore the total emitter current $I_E = I_{pE} + I_{nE}$.

The ratio of $\frac{I_{pE}}{I_{nE}}$ is proportional to the ratio of conductivity of the 'p' material to that of 'n' material. In commercial transistor the ***Doping of the Emitter is made much large than the Doping of the Base***. So 'p' region conductivity will be much larger than 'n' region conductivity. So the emitter current mainly consists of holes only. (Such a condition is desired since electron hole recombination can take place at the emitter junction). ***Collector is lightly doped. Basewidth is narrow*** (Fig. 7.6).

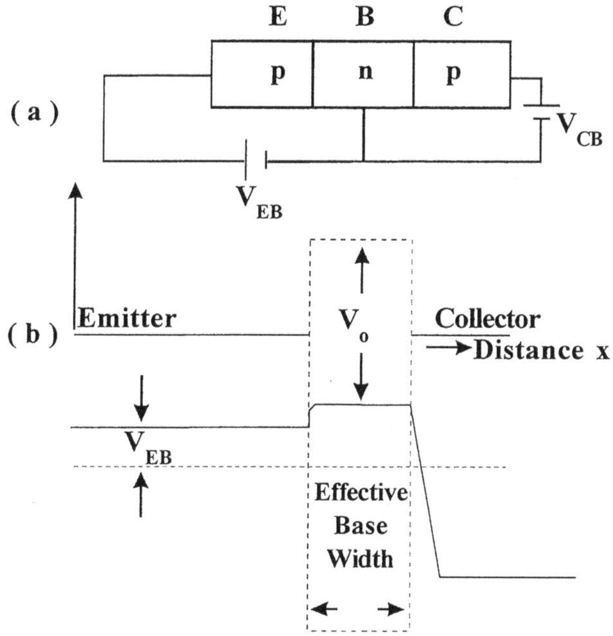

Fig. 7.6 Potential barrier levels in a transistor

All the holes injected from emitter to base will not reach collector, since some of them recombine with electrons in the base and disappear.

If I_{PE} is the current due to holes at the Emitter Junction, $I_{pE} > I_{pC}$ since some holes have recombined with electrons in the base. The recombination current is equal to ($I_{PE} - I_{PC}$). If the emitter is open circuited, so that $I_E = 0$, then I_{pC} would be zero. Under these conditions, junction acts as a reverse biased diode. So the Collector Current is approximately equal to reverse saturation Current I_{CO}. If I_E ¹ 0, then $I_C = I_{CO} + I_{pC}$.

For a PNP transistor, I_{CO} consists of holes moving from left to right (base to collector) and electrons crossing in the opposite direction. In the base, which is ***n-type***, holes are the minority carriers. In the collector, which is 'p' type, electrons are the minority carriers. I_{co} is due to minority carriers. Currents entering the transistor are positive. For NPN transistor, I_{CO} consists of electrons, the minority carriers in the base moving to collector and holes from collector moving to base. So the direction of I_{CO} for NPN transistor is the same as the conventional current entering into the transistor. ***Hence I_{CO} is positive for NPN transistor.***

Emitter Efficiency (γ) :

$$\gamma = \frac{\text{Current due to Injected at the Emitter - Base Junction}}{\text{Total Emitter Current}}$$

For a PNP transistor,

$$\gamma = \frac{I_{pE}}{I_{pE} + I_{nE}} = \frac{I_{pE}}{I_E}$$

Value of γ is always less than 1.

I_{pE} = Current due to injected holes from emitter to base

I_{nE} = Current due to injected electrons from base to emitter.

Transportation Factor (β^*) :

$$\beta^* = \frac{\text{Current due to injected carriers reaching B - C junction}}{\text{Injected carrier current at emitter - base junction}}$$

For a PNP transistor,

$$\beta^* = \frac{I_{pc}}{I_{pE}} = \frac{I_p \text{in the Collector}}{I_p \text{in the Emitter}} = \frac{\text{Hole Current in the Collector}}{\text{Hole Current in the Emitter}}$$

Large Signal Current Gain (α) :

$$\alpha = \frac{\text{Collector Current}}{\text{Emitter Current}}$$

$$\alpha = \frac{I_{pC}}{I_E}$$

Multiplying and dividing by I_{pE}.

$$\alpha = \frac{I_{pC}}{I_{pE}} \times \frac{I_{pE}}{I_E}$$

$$\frac{I_{pc}}{I_{pE}} = \beta^* ; \qquad \frac{I_{pE}}{I_E} = \gamma$$

$$\therefore \quad \boxed{\alpha = \beta^* \times \gamma}$$

7.4 AMPLIFYING ACTION

A small change DV_i of Emitter - Base Voltage causes relatively large Emitter - Current Change DI_E. a'is defined as the ratio of the change in collector current to the change in emitter current. Because of change in emitter current DI_E and consequent change in collector current, there will be change in output voltage DV_o.

$$\Delta V_0 = \alpha' \times R_L \times \Delta I_E \qquad \text{Since } \alpha' = \frac{\Delta I_C}{\Delta I_E}$$

$$\Delta V_i = r_e' \times \Delta I_E$$

R_L is load resistance. Therefore, the voltage amplification is,

$$A_V = \frac{\Delta V_0}{\Delta V_i} = \frac{\alpha' R_L . \Delta I_E}{r_e' \Delta I_E}$$

$$A_V = \frac{\alpha' . R_L}{r_e'} ; \alpha' = \left.\frac{\Delta I_C}{\Delta I_E}\right|_{V_{CB}=K}$$

$$R_L > r_e'$$

$$A_V > 1$$

α' = Small signal forward - circuit current gain.

r_e' = Emitter junction resistance.

The term, r_e'is used since this resistance does not include contact or lead resistances. r_e'value is small, because emitter - base Junction is forward biased.

So a small change in emitter base voltage produces large change in collector base voltage. Hence, the transistor acts as an amplifier. (This is for common base configuration. Voltages are measured with respect to base).

The input resistance of the circuit is low (typical value 40W) and output resistance is high (3000W). So current from low resistance input circuit is transferred to high resistance output circuit (R_L). ***So it is a transfer resistor device and abbreviated as transistor.*** The magnitude of current remains the same ($\alpha' \approx 1$).

7.5 TYPES OF TRANSISTOR CONFIGURATION

Transistors (BJTs) are operated in three configuration namely,

1. Common Base Configuration (C.B)
2. Common Emitter Configuration (C.E)
3. Common Collector Configuration (C.C)

7.6 COMMON EMITTER CONFIGURATION (C.E)

In this circuit Fig 7.7, emitter is common to both base and collector. So this is known as CE configuration or grounded emitter configuration. The input voltage V_{BE}, and output current I_C are taken as the dependent variables. These depend upon the output voltage V_{CE} and input current I_B.

$$\therefore \quad V_{BE} = f_1 (V_{CE}, I_B)$$

$$I_C = f_2(V_{CE}, I_B)$$

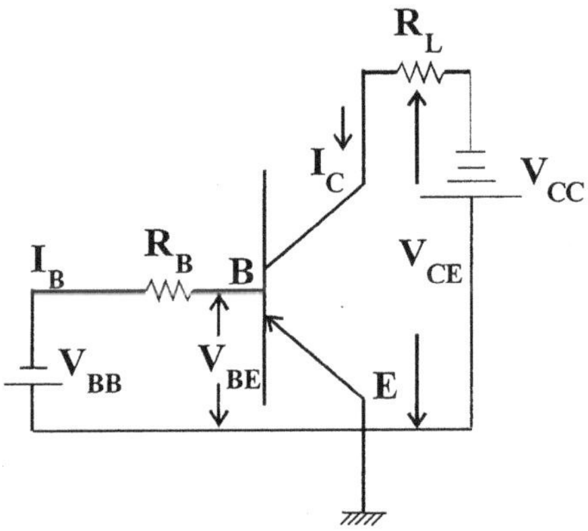

Fig. 7.7 Common Emitter pnp transistor amplifier circuit

7.6.1 Input Characteristics

If the collector is shorted to the emitter, the transistor is similar to a Forward Biased Diode. So I_B increases as V_{BE} increases. The input characteristics are as shown in Fig. 7.8. I_B increases with V_{BE} exponentially, beyond cut-in voltage V_g. The variation is similar to that of a Forward Biased Diode. If $V_{BE} = 0$, $I_B = 0$, since emitter and collector junctions are shorted. If V_{CE} increases, base width decreases (by Early effect), and results in decreased recombination. As V_{CE} is increased, from –1 to –3 I_b decreases. V_g is the cut in voltage.

$$I_E + I_B + I_C = 0$$

$$I_E = -(I_B + I_C)$$

As V_{CE} increases, I_C increases, therefore I_B decreases. Therefore, base recombination will be less. The input characteristics are similar to a forward biased p-n junction diode.

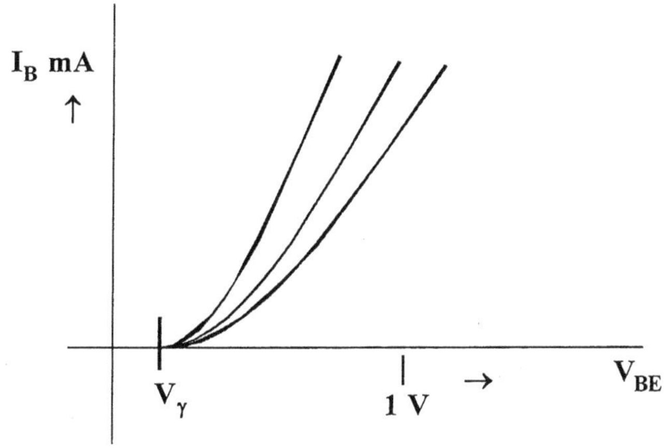

Fig. 7.8 *Input characteristics in C.E configuration*

$$I_B = I_o\left(e^{\frac{V_{BE}}{nV_T}} - 1\right)$$

The input characteristics I_B vs V_{BE} follow the equation,

$$I_B = I_o\left(e^{\frac{V_{BE}}{nV_T}} - 1\right)$$

7.6.2 Output Characteristics

The transistor is similar to a Collector Junction Reverse Biased Diode. The output characteristics is divided into three regions namely,

1. Active Region.
2. Saturation Region
3. Cut-Off Region.

These are shown in Fig 7.9. If I_B increases, there is more injection into the collector. So I_C increases. Hence characteristics are as shown.

The output characteristics are similar to a reverse biased *p-n junction*. (Collector Base junction is reverse biased). They follow the equation.

$$I_C = I_o\left(e^{\frac{V_{CE}}{nV_T}} - 1\right)$$

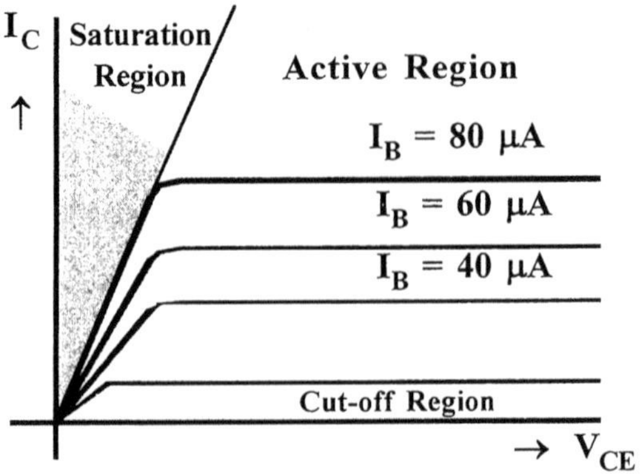

Fig. 7.9 Output characteristics in C.E Configuration

The three different regions are

1. Active Region
2. Saturation Region
3. Cut-off Region.

$$I_B + I_C + I_E = 0$$

$$\therefore \quad I_E = -(I_B + I_C)$$

$$-I_C = +I_{CO} - \alpha . I_E$$

But $$I_E = -(I_B + I_C)$$

$$\therefore \quad I_C = +I_{CO} + \alpha . I_B + \alpha . I_C$$

$$I_C (1 - \alpha) = -I_{CO} + \alpha . I_B$$

If the transistor were to be at cut off, I_B must be equal to zero.

7.6.2.1 Test for Saturation

1. $|I_B| \geq \left| \frac{I_C}{\beta} \right|$

2. If V_{CB} is positive for PNP transistor and negative for NPN transistor, the transistor is in saturation.

7.6.2.2 Common Emitter Cut Off Region

If a transistor was to be at cut off, $I_C = 0$. To achieve this, if the emitter-base junction is open circuited, there cannot be injection of carriers from E to C, through B. Hence $I_C = 0$. (Any reverse saturation current flowing because of thermal agitation is very small. So the transistor can not be operated). But in the case of C.E configuration, even if I_B is made 0, the transistor is not cut off. I_C will have a considerable value even when $I_B = 0$. The circuit is shown in Fig. 7.10.

Because, $I_C = -a\, I_E + I_{CO}$.

a = Current gain ;

I_{CO} = Reverse saturation current in the collector.

But $I_E = -(I_B + I_C)$.

If $I_B = 0,\ I_E = -I_C$

$\therefore$ $I_C = -a(-I_C) + I_{CO}$

or $I_C\,(1 - a) = I_{CO}$

or $$I_C = \frac{I_{CO}}{1-\alpha} = I_{CBO}$$

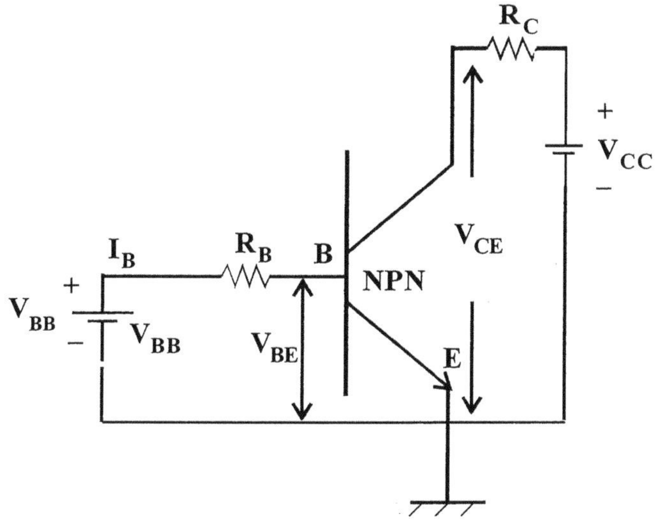

***Fig. 7.10** Common Emitter Cut-off region*

In order to achieve cut-off in the Common Emitter Configuration, Emitter Base Junction should be reverse biased. a is constant for a given transistor. So the actual collector current with collector junction reverse biased and base open circuited is designated by I_{COB}. If a $\approx$ 0.9, then $I_{CBO} \approx 10\, I_{CO}$. So the transistor is not cut off. Therefore, in the C.E configuration to cut of the transistor, it is not sufficient, if I_B = 0. The Emitter Base Junction must be reverse biased, so that I_E is equl to 0. Still I_C will not be zero, but will be I_{CO}. Since I_{CO} is small, the transistor can be regarded as at cut off.

7.7 REVERSE COLLECTOR SATURATION CURRENT I_{CBO}

I_{CBO} is due to leakage current flowing not through the junction, but around it. The collector current in a transistor, when the emitter current is zero, is designated by I_{CBO}. $|I_{CBO}|$ can be $> I_{CO}$. I_{CO} is collector reverse saturation current (when collector-base junction is reverse biased).

7.7.1 Saturation Resistance

For a transistor operating in the saturation region, saturation resistance for Common Emitter configuration is of importance. It is denoted by R_{CES} or R_{CE}'s (saturation) or R_{CS}

$$R_{CS} = \frac{V_{CE}(sat)}{I_C}$$

The operating point has to be specified for R_{CS}.

Saturation Voltage: Manufacturers specify this for a given transistor. This is done in number of ways. R_{CS} value may be given for different values of I_B or they may supply the output and input characteristic for the transistor itself. $V_{CE}(Sat)$ depends upon the operating point and also the semiconductor material, and on the type of transistor construction. Alloy junction and epitaxial transistors give the lowest values of V_{CE}. Grown junction transistor yield the highest V_{CE}.

7.7.2 DC Current Gain b_{dc}

b_{dc} is also designated as h_{FE} (DC forward current transfer ratio).

$$b = \frac{I_C}{I_B}$$

$$I_C = \frac{V_{CC}}{R_L}$$

So if b is known, $I_B = \left(\frac{I_C}{\beta}\right)$

Therefore, $\frac{I_C}{\beta}$ gives the value of base current to operate the transistor in saturation region. b varies with I_C for a given transistor.

7.7.3 Test for Saturation

To know whether a transistor is in saturation or not,

1. $|I_B| \geq \dfrac{I_C}{h_{fe}(\text{or}\beta)}$

2. V_{CB} should be positive for PNP transistor and negative for NPN transistor.

In the C.E configuration,

$$I_C = -(I_B + I_E)$$

$$I_C = -\alpha I_E + I_{CO}$$

$$\therefore \quad I_E = \frac{I_{CO} - I_C}{\alpha}$$

$$\therefore \quad I_C = \frac{I_{CBO}}{1-\alpha} + \frac{\alpha . I_B}{1-\alpha}$$

But, $$\beta = \frac{\alpha}{1-\alpha}$$

$$\therefore \quad (1 - \alpha) = \frac{\alpha}{\beta}$$

If we replace I_{CO} by I_{CBO},

$$I_C = I_{CBO} \times (\beta + 1) + \frac{\alpha . I_B . \beta}{\alpha}$$

$$I_C = (1 + \beta)\, I_{CBO} + \beta\, I_\beta$$

Transistor is cut off when $I_E = 0$, $I_C = I_{CBO}$ and $I_B = -I_{CBO}$.

So at cut-off $\beta = 0$.

7.8 COMMON BASE CONFIGURATION (C.B)

The base is at ground potential. So this is known as ***Common Base Configuration*** or ***Grounded Base Configuration.*** Emitter and Collector Voltages are measured with respect to the base. The convention is currents, entering the transistor are taken as positive and those leaving the transistor as negative. For a PNP transistor, holes are the majority carriers. Emitter Base Junction is Forward Biased. Since holes are entering the base or I_E is positive. Holes through the base reach collector. I_C is flowing out of transistor. Hence I_C is negative; similarly I_B is negative. The

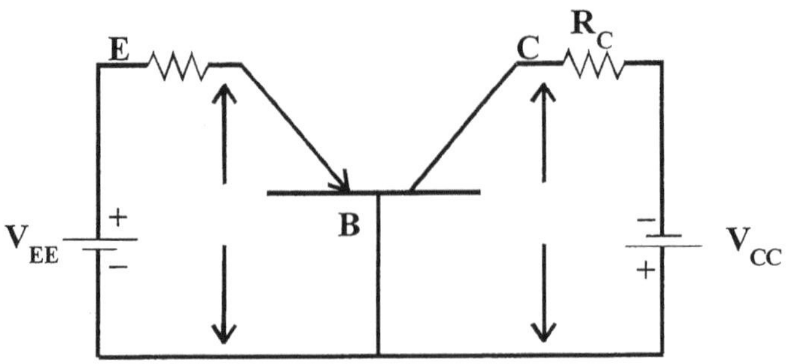

Fig. 7.11 Common Base Amplifier Circuit

characteristics of the transistor can be described as,

$$V_{EB} = f_1(V_{CB}, I_E)$$

Dependent variables are input voltage and output current.

$$I_C = f_2\,(V_{CB}, I_E)$$

Independent Variables are Input Current and Output Voltage. For a Forward Biased Junction V_{EB} is positive, for reverse bias collector junction V_{CB} is negative.

Transistor Characteristics in Common Base Configuration

The three regions in the output characteristics of a transistor are

1. *Active region*
2. *Saturation region*
3. *Cut off region*

The input and output characteristics for a transistor in C.B configuration are as shown in Fig. 7.12 (a) and (b) When $I_E = 0$, the emitter is open circuited. So the transistor is not conducting and the I_{CBO} is not considered. Generally the transistor is not used in this mode and it is regarded as cut off ; Even though I_{CBO} is present it is very small for operation. During the region OA, for a small variation of V_{BC} there is very large, change in current I_C with I_E. So current is independent of voltage. There is no control over the current. Hence the transistor can not be operated in this region. So it is named as Saturation region. The literal meaning of saturation should not be taken. The region to the right is called active region. For a given V_{CB}, with I_E increased, there is no appreciable change in I_C. It is in saturation. When transistor is being used as a switch, it is operated between cut off and saturation regions.

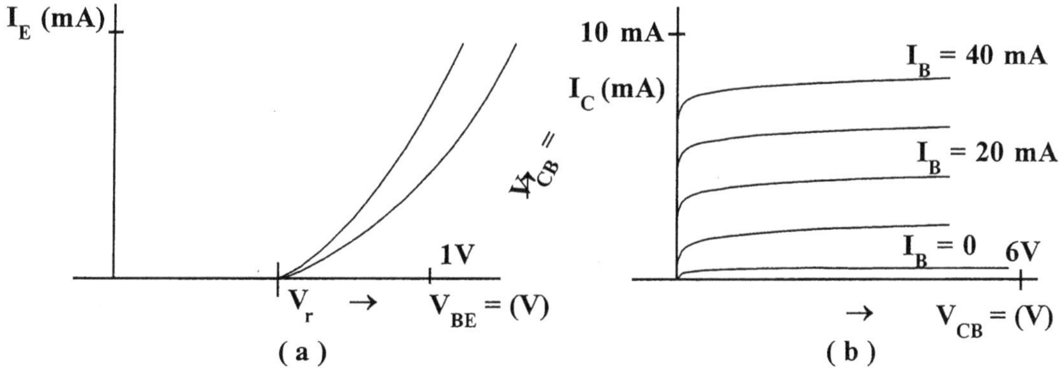

Fig. 7.12 (a) Input characteristics in C.B configuration
(b) Output characterisitics in C.B configuration

7.8.1 Early Effect

The collector base junction is reverse biased. If this reverse bias voltage is increased, the space charge width at the *n-p region* of Base - Collector Junction of PNP Transistor increases. So effective base width decreases (Fig. 7.13)

$$V_B = \frac{e.N_A}{2\epsilon}.W^2$$

W = Space Charge Region Width.

The decrease in base width with increase in collector reverse bias voltage is known as Early Effect. When the base width decreases, the probability of recombination of holes and electrons in the base region is less. So the transportation factors β^* increases.

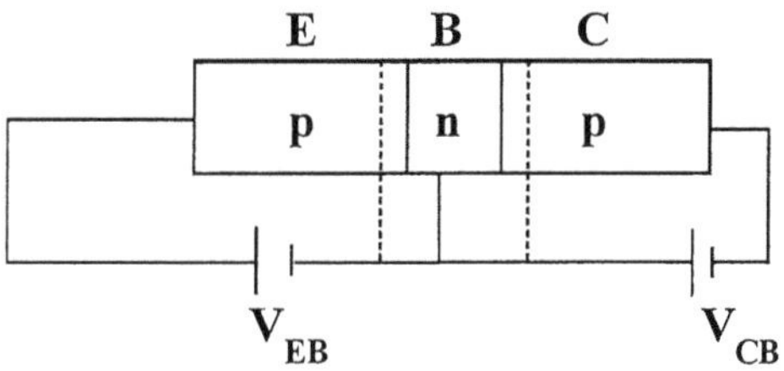

Fig. 7.13 Early effect

7.9 THE COMMON COLLECTOR CONFIGURATION (C.C)

Here the load resistor R_L is connected in the emitter circuit and not in the collector circuit. Input is given between base and ground. The drop across R_L itself acts as the bias for emitter base junction. The operation of the circuit similar to that of Common Emitter Configuration. When the base current is I_{CO}, emitter current will be zero. So no current flows through the load. Base current I_B should be increased so that emitter current is some finite value and the transistor comes out of cut-off region.

Input characteristics I_B vs V_{BC}

Output characteristics I_C vs V_{EC}

The circuit diagram is shown in Fig. 7.14 (a) and the characteristics are shown in Fig. 7.14 (b) and (c).

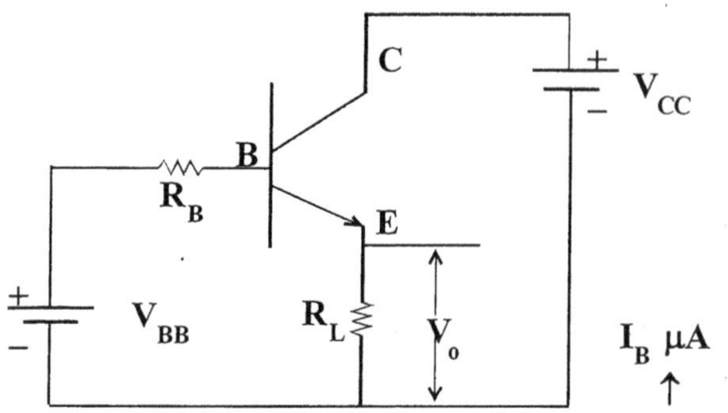

(a)

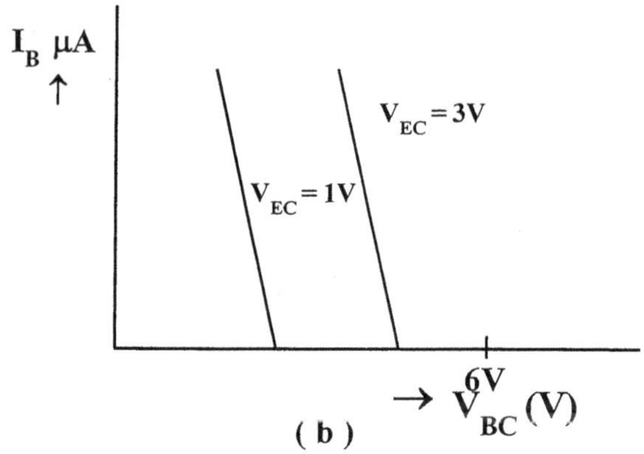

(b)

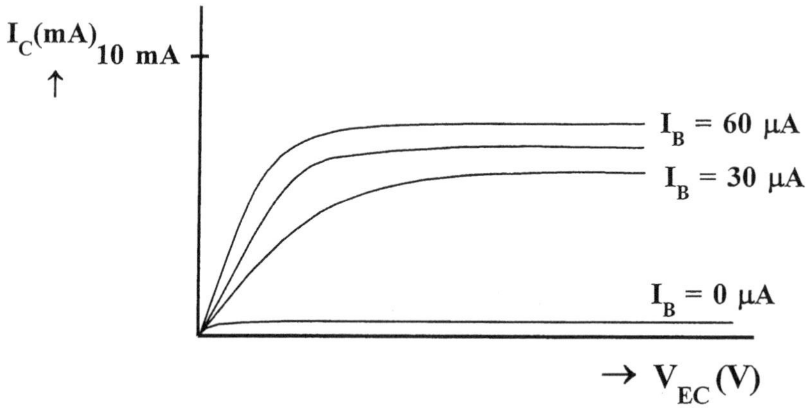

(c)

Fig. 7.14 (a) Cirucuit for C.C configuration

(b) Input characteristics in C.C configuration

(c) Output characteristics in C.C configuration

7.9.1 Parameters

γ = Emitter Efficiency

= I_{pE}/I_E for PNP transistor.

β^* = Transportation factor

$$= \frac{I_{pC}}{I_{pE}} = \frac{\text{No. of holes reaching collector}}{\text{No. of holes emitter}}$$

α = Small signal current gain

$$= \frac{I_C}{I_E} \; ; \alpha = \beta^*\gamma \qquad \alpha < 1$$

α' = Small signal current gain

$$= \partial I_C / \partial I_E = \frac{\Delta I_C}{\Delta I_E} \qquad \alpha' < 1$$

$h_{FE} = \beta$ = Large signal current gain

$$= \frac{I_C}{I_B} \qquad \beta >> 1$$

$h_{FE} = \beta'$ = Small signal current gain

$$= \frac{\partial I_C}{\partial I_B} = \frac{\Delta I_C}{\Delta I_B} \; \beta' >> 1$$

I_{CBO} = Reverse saturation current when E-B junction is open circuited.

I_{CO} = Normal reverse saturation current

$$\alpha = \frac{\beta}{1+\beta}$$

$$\beta = \frac{\alpha}{1-\alpha}$$

7.10 ANALYTICAL EXPRESSIONS FOR TRANSISTOR CHARACTERISTICS

We know that $I_C = (-\alpha I_E + I_{CO})$. I_{CO} is the reverse saturation current since Collector Base Junction is reverse biased. The above expression holds good only when the reverse bias voltage applied to Collector Base Junction is very large. Suppose, the reverse bias voltage is being varied from 0 to some value. Then the value of I_{CO} also varies. Hence to generalize the above expression,

$$I'_{CO} = I_C\left(e^{\frac{V_C}{V_T}} - 1\right)$$

where I_{CO} is the final steady state value.

I'_{CO} is the value at any voltage V_C.

$$\therefore \quad I_C = -\alpha \cdot I_E + I_{CO}\left(e^{\frac{V_C}{V_T}} - 1\right)$$

Now a transistor can be connected in normal mode with the same E, B and C. So in such a case α is represented as α_N. But the same transistor can also be used in the inverse mode by forward biasing the collector junction and reverse biasing the emitter junction. Then collector acts as emitter and vice versa. In such a case α is represented as α_I, I, indicating inverted mode. Sometimes the same are also expressed as α_F = Forward α and α_R = Reverse α.

Base Spread Resistance : r_{bb}'

The base region is narrow. So the resistance of the base region will be large, and it is called Base Spreading Resistance. V_{CB} represents the voltage drop between collector terminal and base terminal. Considering the base spread resistance. V_C represents the drop across the collector junction as shown in Fig 7.15. So the currents and voltages in a transistor PNP or NPN are as shown.

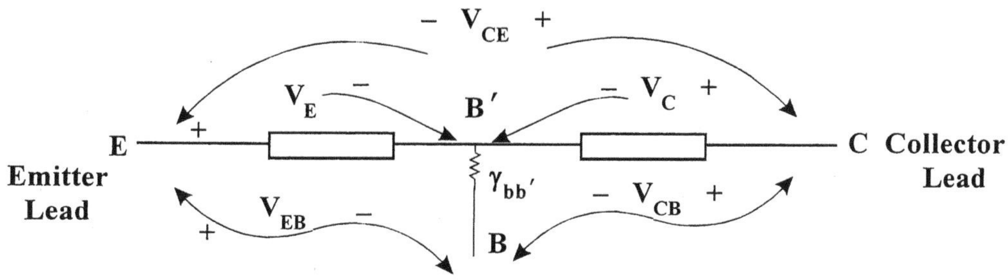

Fig. 7.15 Base spread resistance

7.11 EBERS - MOLL MODEL OF A TRANSISTOR

A transistor can be represented by two diodes connected back to back. Such a circuit for a PNP transistor shown is called ***Eber - Moll Model***. The transistor is represented by the diodes connected back to back. I_{EO} and I_{CO} represent the reverse saturation currents. $\alpha_N I_E$ is the current source. $\alpha_I I_C$ represents the current source in inverted mode (Fig. 7.16).

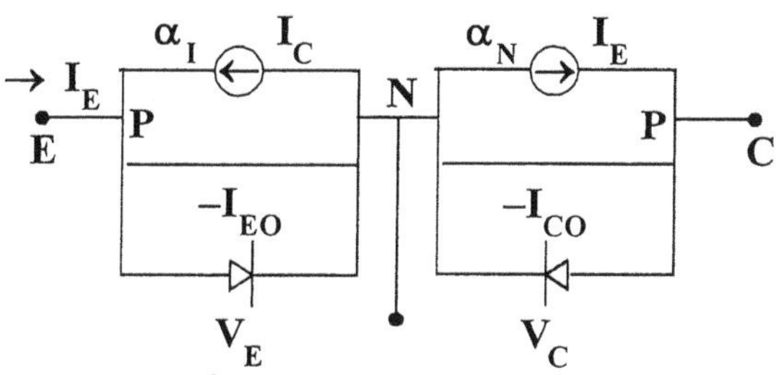

Fig. 7.16 Eber-Moll Model of PNP transistor

A transistor cannot be formed by simply connecting two diodes since, the base region should be narrow. Otherwise the majority carriers emitted from the emitter will recombine in the base and disappear. So transistor action will not be seen.

7.12 TRANSISTOR IN CUT-OFF REGION UNDER C.E. CONFIGURATION

A transistor is said to be cut off when $I_C = 0$ and $I_B = 0$. But actually I_C cannot be zero since when common base junction is reverse biased, there is sum reverse saturation current = I_{CO}. Therefore, a transistor is said to be cut off when, (Fig. 7.17)

$$I_B = 0,$$

$$I_C = I_{CO}$$

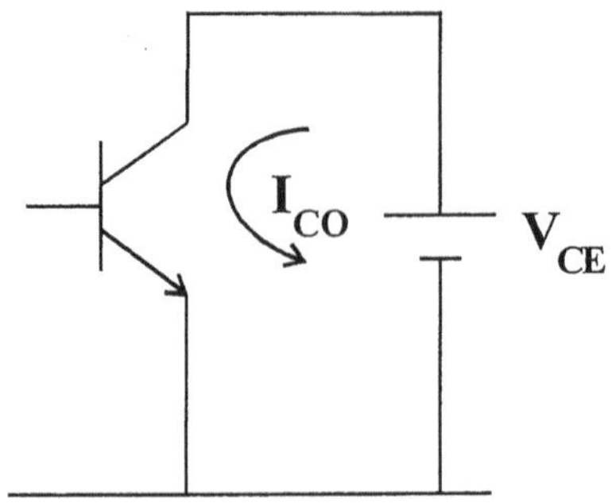

Fig. 7.17 Transistor in cut-off

I_{CO} is very small

$$\therefore \qquad I_C \approx I_{CO}.$$

Typical values are $\alpha_N = 0.9$;

$$\alpha_I = 0.5$$

To get cut off, if the E – B junction is left open, the reverse saturation current that flows $\approx$ 10 I_{CO}, and it is represented as I_{CBO}. I_{CBO} is the reverse saturation current when E – B junction is open circuited. The reason for I_{CBO} being so large is, when E – B junction is open circuited, the voltage that exists $\approx$ 60mV. This acts as emitter is forward biased slightly. Hence I_{CBO} is large. So to get cut off, we have to reverse bias the emitter base junction. For Germanium transistor it should be 0.1V. (Reverse Bias Voltage)

7.12.1 Short Circuit Base

Suppose instead of reverse biasing the emitter junction, we simply short the base to emitter terminal. Then $V_E = 0$, $I_C \approx I_{CO}$ and $I_E = 0$, because the transistor is almost cut off.

The transistor is cut off when $I_B = 0$, and $I_C = I_{CO}$.

Example 7.1:

For the transistor AF 114 connected as shown in Fig. 7.18, determine the values of I_B, I_C and V_{BC} given that $V_{CE} = -0.07$V and $V_{BE} = -0.21$V. PNP transistor collector is reverse biased. E-B junction is open circuited. So the current flowing is the reverse saturation current.

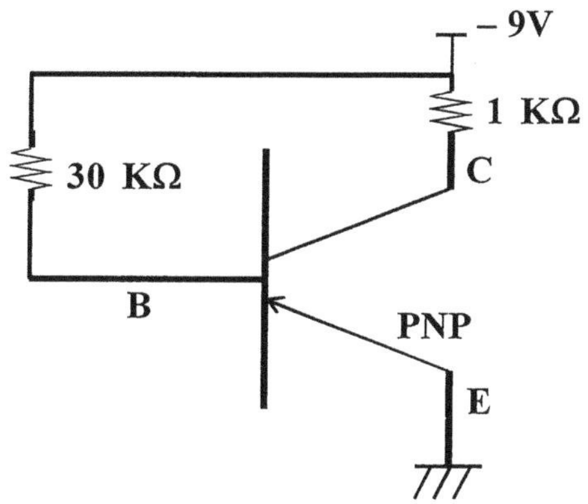

Fig. 7.18 For Example 7.1

Solution:

$$I_C = \frac{-9V - V_{CE}}{1K} = -8.93\ \mu A$$

$$I_B = \frac{-9 - V_{BE}}{30K}$$

$$= \frac{-9 + 0.21}{30K} = -0.293\ \mu A.$$

$$V_{BC} = V_{BE} - V_{CE} = -0.21 + 0.07 = -0.14V$$

Example 7.2:

If $\alpha = 0.98$, and $V_{BE} = 0.6$ V, find R_1 in the circuit shown in Fig. 7.19 an Emitter Current $I_E = -2$ mA; Neglect reverse saturation current.

Solution:

I_E is negative because it is NPN transistor. So actually I_E is flowing out of transistor. The convention is currents, entering into the transistor are positive. So I_E is negative here. Though there is no separate bias for the emitter, the voltage drop between emitter and ground itself acts as the forward bias voltage.

$$I_E = -2\mu A\ ;\ I_C = -\alpha\ .\ I_E = 0.98 \times 2 = 1.96\ \mu A$$

I_C is positive since it is entering into transistor.

$$I_B = (1 - \alpha)\ I_E = (1 - 0.98)(-2) = 40\ \mu A$$

The voltage between base and ground $V_{BN} = -V_{BE} + I_E\ R_E$ and (considering only magnitude)

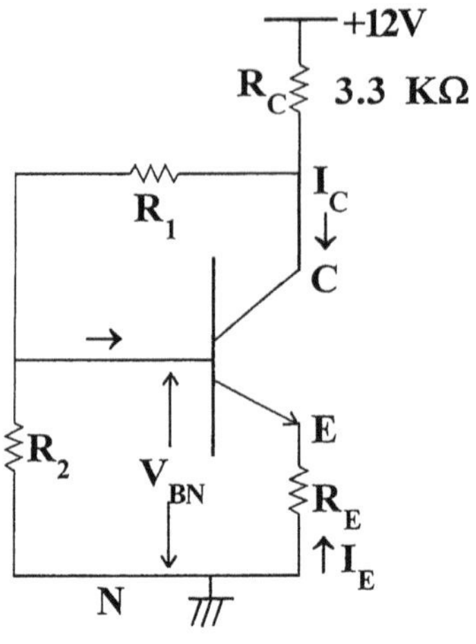

Fig. 7.19 For Example 7.2

$$V_{BE} = +0.6 : I_E = 2\mu A : R_E = 100\Omega$$

$$\therefore \quad V_{BN} = 0.6 + (2 \times 0.1) = 0.8V$$

V_{BN} is $> V_{BE}$. So it is $(V_{EB} + I_E R_E)$

$$\therefore \quad I_{R2} = \frac{V_{BN}}{R_2} = \frac{0.8}{20k} = 40\mu A$$

$$I_{R1} = I_B + I_{R_2} = 40 + 40 = 80\mu A$$

$$I_{3.3}K = I_{R_1} + I_C = I_{R_1} = 80\mu A\ ;$$

$$I_C = 1.96\mu A$$

$$I_{3.3}K = 80 \times 10^{-3} + 1.96 = 2.04\mu A$$

$$V_{R_1} = V_{CC} - V_{3.3}K - V_{BN} = 12 - (2.09 \times 3.3) - 0.8 = 4.47V$$

$$\therefore \quad R_1 = \frac{V_{R_1}}{I_{R_1}} = \frac{4.47}{0.08} = 56K$$

Example 7.3:

Show that for an NPN silicon transistor of the alloy junction type, in which the resistivity ρ_B of the base is much larger than that of the collector, the voltage V is given by

$$V = \frac{6.34 \times 10^3 . W^2}{\rho_B},$$

where ρ_B is in Ω – cm and base width W in *mills*.

Solution:

$$\in = \frac{12}{36\pi \times 10^{11}} \text{ F/m}$$

$$\mu_p = 500 \text{ cm}^2/\text{V - sec.}$$

The expression for $W^2 = 2\in . \dfrac{V_B}{eN_A},$

where $\in = 12/36\pi \times 10^{11}$ F/cm.

In NPN transistor, the base is 'P' type.

$$\sigma_p = \sigma_B \text{ (base)}$$

$$\therefore \quad \sigma_B = e . N_A . \mu_p$$

or $$e . N_A = \frac{\sigma_B}{\mu_p} = \frac{1}{\rho_B . \mu_p}$$

Thus $$e . N_A = \frac{1}{\rho_B . \mu_p}$$

$\therefore \quad V_B = \dfrac{e.N_A.W^2}{2\in}$ V_B is called the Barrer Potential or Contact Potential.

Substitute the value of $\in.N_A$ in the expression for V_B.

$$V_B = \frac{W^2}{2\in.\mu_p\rho_B}$$

Substitute the value of ε of Silicon as $\dfrac{12}{36\pi \times 10^{11}}$ F/cm.

Since W is in mills,

$$1 \text{ mil} = \frac{1}{1000''} \quad (\text{ one thousand of an inch })$$

To convert this to centimeters, multiply by 2.45

∴ Barrier Potential,

$$V_B = \frac{W^2(\text{mills}) \times (2.54)^2 \times \left(\frac{1}{1000}\right)^2}{\frac{2 \times 12}{36\pi \times 10^{11}} \times \frac{500}{V - \sec} \rho_B}$$

$$\mu_p = 500 \text{ cm}^2/\text{V - sec} = \text{Mobility of Holes}$$

$$V_B = 6.1 \times 10^3 \times \frac{W^2}{\rho_B}$$

Example 7.4:

A transistor having $\alpha = 0.96$ is placed in Common Base Configuration with a load resistance of 5 KΩ. If the emitter to base junction resistance is 80Ω, find the values of amplifier current, voltage and power gain.

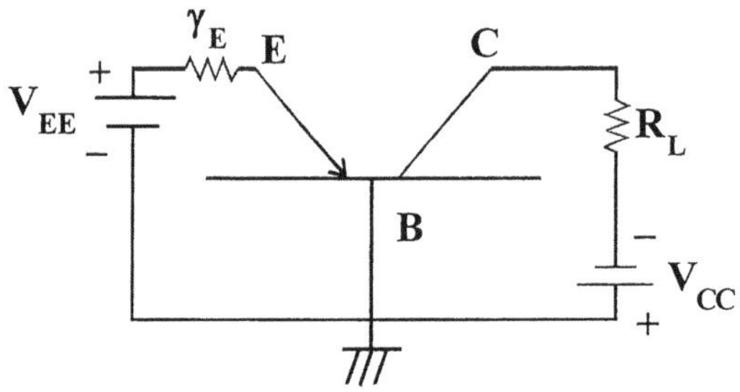

Fig. 7.20 *For Example 7.4*

Solution:

$$A_i = \text{current gain} = \frac{I_C}{I_E} = \alpha = 0.96$$

$$A_V = \frac{\alpha . R_L}{\gamma_{BE}}$$

$$= \frac{0.9 \times 5 \times 10^3}{80} = 60$$

$$A_V = \frac{\alpha' R_L}{\gamma e}$$

$\therefore$ Power Gain $A_P = A_V \times A_I = 60 \times 0.96 = 57.6$

Example 7.5:

If a transistor, with $\alpha = 0.96$ and emitter to base resistance 80Ω is placed in Common Emitter Configuration, find A_I, A_V and A_P.

Solution:

$$A_I = \frac{I_C}{I_B} = \beta$$

$$\therefore \quad \beta = \frac{\alpha}{1-\alpha}$$

$$= \frac{0.96}{1-0.96} = 24$$

$$A_V = \frac{-\beta R_L}{\gamma_{BE}}$$

$$= \frac{24 \times 7.5 \times 10^4}{80} = -22{,}500$$

$$A_P = A_I \times A_V = 54{,}000.$$

Example 7.6:

A transistor is operated at a forward current of $2\mu A$ and with the collector open circuited. Calculate the junction voltages V_C and V_E, the collector to emitter voltage V_{CE} assuming $I_{CO} = 2\mu A$, $I_{EO} = 1.6\mu A$ and $\alpha_N = 0.98$.

Solution:

$I_C = 0$ Since Collector Junction is open circuited.

$$I = I_0 \left(e^{V_N/V_T} - 1\right)$$

$$I_E = I_{EO}\left(e^{\frac{V_E}{nV_T}} - 1\right)$$

$$\frac{I_E}{I_{EO}} = e^{\frac{V_E}{nV_T}} - 1 \text{ or } e^{\frac{V_E}{nV_T}} = 1 + \frac{I_E}{I_{EO}}$$

$$\therefore \quad \frac{V_E}{nV_T} = ln\left(1+\frac{I_E}{I_{EO}}\right)$$

$$\therefore \quad V_E = \eta \cdot V_T \, ln\left(1+\frac{I_E}{I_{EO}}\right)$$

$$\therefore \quad V_E = 0.1853V$$

$$V_C = V_T \, ln \, . \left(1+\frac{I_C}{I_{CO}}\right)$$

$$= V_T \, ln\left(1-\frac{\alpha . I_E}{I_{CO}}\right)$$

$$= 0.179V$$

$$V_{CE} = V_C - V_E = -\,0.006 \text{ V}$$

Example 7.7:

A load draws current varying from 10 to 100 mA at a nominal voltage of 100V. A regulator consists of R_S = 200Ω and zener diode represented by 100 V, and R_Z = 20Ω. For I_L = 50 mA. Determine variation in V_i corresponding to a 1 % variation in V_L.

Solution:

The zener diode is represented by a voltage source with it. V_L can vary with in 1 % so what is the allowable range in which V_i can be permitted to vary ? In order that the zener diode is operated near the breakdown region, the current through the zener diode is operated near the breakdown region, the current through the zener should be at least 0.1 times the load current.

$$I_{Zmin} = 0.1\,(I_{Lmax})$$

$$I_{Lmax} = 100 \text{ mA.}$$

$$\therefore \quad I_Z = 0.1 \times 100 = 10 \text{ mA.}$$

$$V_L = V_Z + I_Z R_Z$$

$$= 100 + 0.01 \times 20 = 100.2V$$

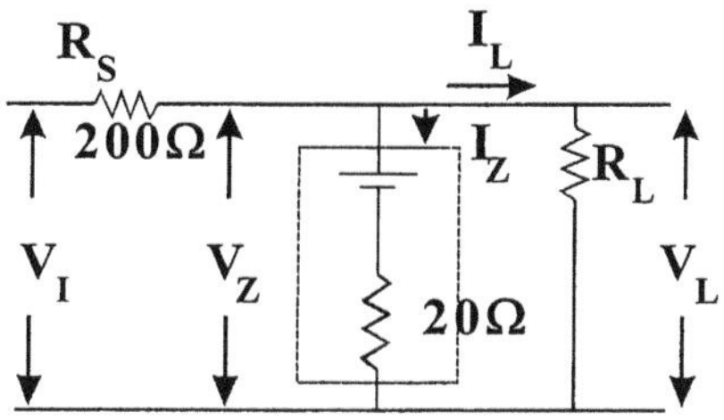

Fig. 7.21 For Example 7.15

$$V_1 = V_L + (I_L + I_Z)\,R_S$$
$$= 100.2 + (0.05 + 0.01)\,200$$
$$= 112.2\text{ V.}$$
$$V_L = 100.2\text{V}$$

1% increase is $V_L \approx 1\text{V}$.

$$V_L' = 100.2 + 1 = 101.2\text{V}$$

Therefore, the increase in

$$I_Z, I_Z' = \frac{V_C' - V_Z}{R_Z}$$
$$= \frac{101.2 - 100}{20}$$
$$= \frac{1.2}{20}\,\frac{V}{\Omega} = 0.06\text{A}$$

Total Current = 0.05 + 0.06 = 0.1 A

$\therefore$ Total Voltage $V_1 = V_Z + R_S \times (I_Z + I_L)$

$$= 101.2 + 2000 \times (0.05 + 0.06)$$
$$V_1 = 123.2\text{V}$$

Change in $V_1 = 123.2\text{V} - 112.2\text{V} = 11\text{V}$

A change of 11V in V_I on the input side produces a change of 1V on the output side due to zener diode action.

7.13 TRANSISTOR BIASING

7.13.1 Operating Point

The transistor output characteristics in Common Emitter Configuration are as shown in Fig. 7.22. The point in the characteristics of the transistors indicated by V_{CE}, I_C is called ***Operating Point*** or ***Quiescent Point*** **Q**. This is shown in Fig. 7.22. The operating point must be in the active region for transistor to act as an amplifier.

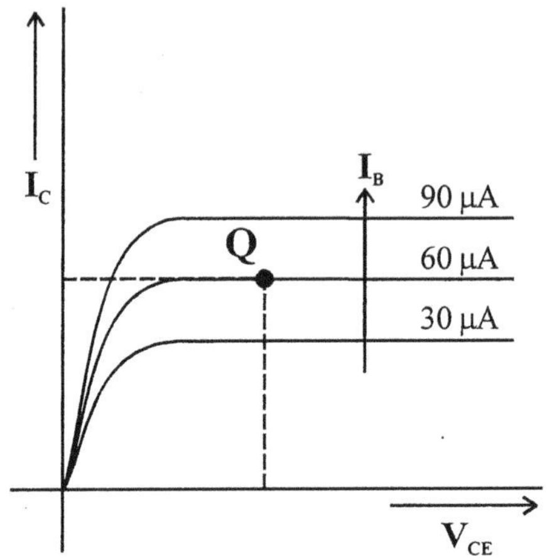

Fig. 7.22 Output characteristics of BJT

Transistor when being used as an amplifier is operated in the active region. As I_B (in μA) increases, I_C (in mA) increases. So for a small change in I_B, there is corresponding large change in I_C and thus transistor acts as an amplifier. The operating point or the quiescent point has to be fixed when any transistor is being used as an amplifier.

$$A_V = \frac{\alpha' R_L}{r_C} \qquad R_L << r_c.$$

R_L = Load Resistance

r_c = Collector resistance

α' = Small signal current gain (less than 1)

A_V = Voltage gain

Therefore, transistor acts as an amplifier.

$$\alpha' = \frac{\Delta I_C}{\Delta I_E}$$

$$\Delta V_C = \alpha' \times R_L \times \Delta I_E$$

$$\Delta V_i = r_C \, \Delta I_E$$

$$\therefore \quad A_V = \frac{\Delta V_C}{\Delta V_i} = \frac{\alpha' R_L \times \Delta I_E}{r_C \times \Delta I_E} = \frac{\alpha' R_L}{r_C}$$

$R_L >> r_c$. $S_o A_V > 1$

The quiescent point changes with temperature T, because it depends on the parameters β, I_{CO} or V_{BE} etc. β and I_{CO} depend upon temperature. Hence operating point changes with temperature. So by suitably biasing the transistor circuit, the quiescent point may be made stable or not to change with temperature. Compensation techniques are also used in order to make the operating point stable with temperature.

It is very essential to choose the operating point properly so that the transistor is operated in the active region and that the operating point does not change with temperature. Since if the input is sinusoidal wave, the output will also be a true replica of the input or it will also be a perfect sine wave. Otherwise, the output will be distorted, clipped etc., if the quiescent point is not chosen correctly or it is changing with temperature. Hence it is essential to choose the operating point correctly. Operating point is fixed with respect to the output characteristic of a transistor only.

So now the question is how to choose the operating point? Before this let us consider as to how the quiescent point is fixed. Consider the circuit as shown in Fig. 7.23 and 7.24. It is Common Emitter Transistor circuit with a load resistance R_L. PNP transistor is being considered.

From the above circuit,

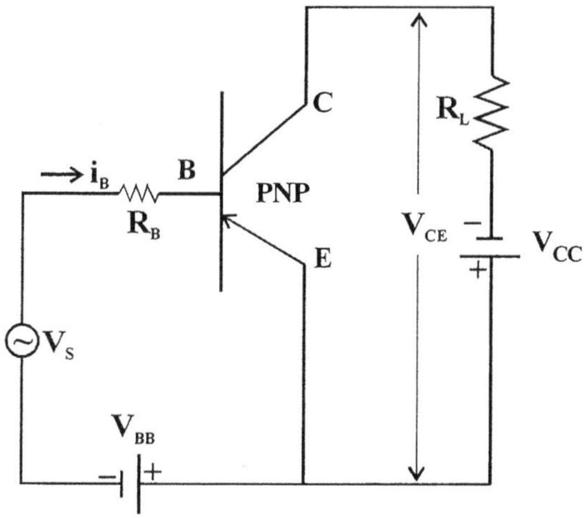

Fig 7.23 Amplifier circuit

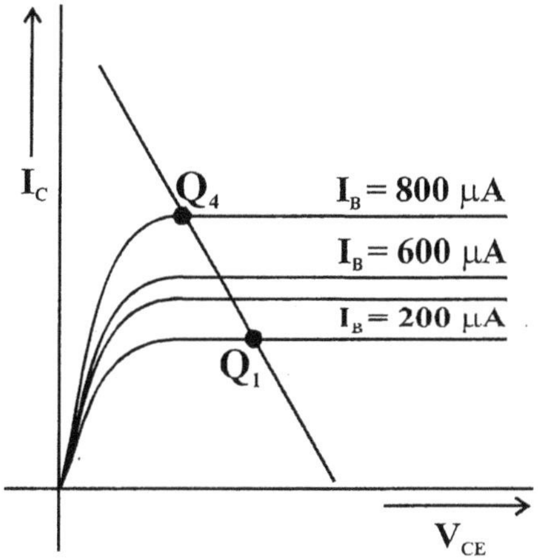

Fig 7.24 Variation of operating point

$$V_{CE} = V_{CC} - i_C \times R_L \qquad \text{.......... (7.13.1)}$$

V_{CC} is DC Bias Supply. i_C is instantaneous value of AC Collector Current.

V_{CE} = Instantaneous Voltage

From the Equation (6.1), the load line can be drawn. To draw the load line,

Let $i_C = 0$ then $V_{CE} = V_{CC}$. This is one point of the load line.

Let $V_{CE} = 0$ then $i_C = \dfrac{V_{CC}}{R_L}$. This another point of the load line.

The value of r_C is very small compared to R_L. Hence it can be neglected.

$$i_C = \left(-\frac{V_{CC}}{R_L + r_C}\right)$$

Neglecting, r_C, $\quad i_C = \left(-\dfrac{V_{CC}}{R_L}\right)$

So by setting $i_C = 0$, $V_{CE} = 0$, the load line is drawn. It is a straight line independent of the transistor parameters. The slope of the line depends on R_L the load resistance only. V_{CC} is a fixed quantity. For a given value of I_B, I_C is also fixed when the transistor is being operated in the active region. Therefore for $I_B = 200\mu A$, the quiescent point is Q_1. For $I_B = 800\ \mu A$, the operating point is Q_4 and so on. Now, when the input signal applied between base and emitter is a symmetrical signal, varying both on positive and negative sides equally, within the range of the transistor characteristics, then the operating point is chosen to lie in the centre of the output characteristics (i.e., corresponding to $I_B = 300$ mA), so that the output signal is a true replica of the input without distortion (V_{CC} and V_{BB} are bias voltages. Input signal is the a.c signal between B and E). When the input signal is symmetrical, the ***quiescent point is chosen at the centre of the load line***.

7.13.2 Static Load Line or DC Load Line

The load line as explained above, if drawn, is called as the ***dynamic (AC) load line***. If the output impedance in the circuit is reactive, then the load line will be elliptical. Since the voltage and current are shifted in phase, and the resulting equation will be that of an ellipse. For a resistance the load line is a straight line.

While drawing a load line, if only the collector resistance R_C is considered and load resistance R_L is infinity (∞), then the load line is drawn with the points $\left(\frac{V_{CC}}{R_C} \text{ and } V_{CE}\right)$. It is called as ***Static Load Line*** or ***DC Load Line***

But if the load line is drawn considering a finite load resistance R_L, and the resistance considered is R_L in parallel with R_C, it is called ***AC Load Line or Dynamic Load Line***, with AC Input.

But if the load line is drawn considering a finite load resistance R_L, and the resistance considered is R_L in parallel with R_C, it is called ***AC Load Line or Dynamic Load Line***, with AC Input.

7.14 BIASING CIRCUITS

There are three types of Biasing Circuits

1. Fixed Bias circuit or Base Bias circuit
2. Collector to Base Bias circuit
3. Self Bias or Emitter Bias circuit (Universal bias or Voltage divider bias circuit).

7.15 NEED FOR BIASING/IMPORTANCE OF BIASING

In order that amplification is achieved, the BJT must be operated in the active region. To ensure thus, E-B junction of the transistor (BJT) must be forward biased and collector-Base junction must be reverse biased. The operating point or quiescant point must be in the centre of the active region. To ensure thus, biasing must be done. Similarly to use the BJT as on-off switch, the operation point must change between cut-off and saturation region. So biasing is to be done.

7.16 FIXED BIAS CIRCUIT OR (BASE BIAS CIRCUIT)

Consider the NPN Transistor Circuit as shown in Fig. 7.25. R_C is collector resistance. C_C is coupling capacitor. It blocks DC since capacitor is open circuit for D.C. So the output signal will be pure a.c signal. R_C. limits the current I_C . R_B provides the bias voltage for the base. Since this is NPN transistor, V_{CC} should be positive, because the collector should be reverse biased.The drop across R_b, ($V_{CC} - I_B \times R_B$) is negative and it provides positive bias for the base. Since R_L is connected, dynamic load line has to be considered. Suppose the output characteristics of the transistor are as shown in Fig. 7.26.

Suppose Q_1 is the centre point, and it is chosen as the operating point. But if the input signal is greater than 40 μA on the negative side, with Q_1, the transistor will be cut off since for that swing $I_B = 0$, and the transistor is cut off, so another operating point is to be chosen. Choosing the operating point means, a correct value of I_B, I_C, V_{CE}. (if the transistor is in Common Emitter Configuration) so that we get undistorted output.

Since from the below circuit,

$$I_B = \frac{V_{CC} - V_{BE}}{R_B}$$

$$I_C = \frac{V_{CC} - V_{CE}}{R_C}$$

V_{BE} is the base emitter voltage for a forward biased EB Junction. This will be 0.2V for Germanium and 0.6 for Silicon. In order to get a large value of I_B or the change in the operating point, V_{CC} or R_B has to be changed. Since once V_{CC}, R_B etc., are fixed, I_B is fixed. So the above circuit is called as ***Fixed Bias Circuit***.

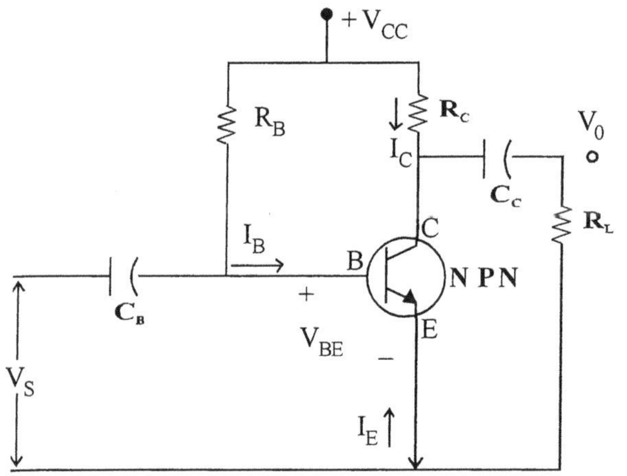

Fig. 7.25 *Fixed bias circuit*

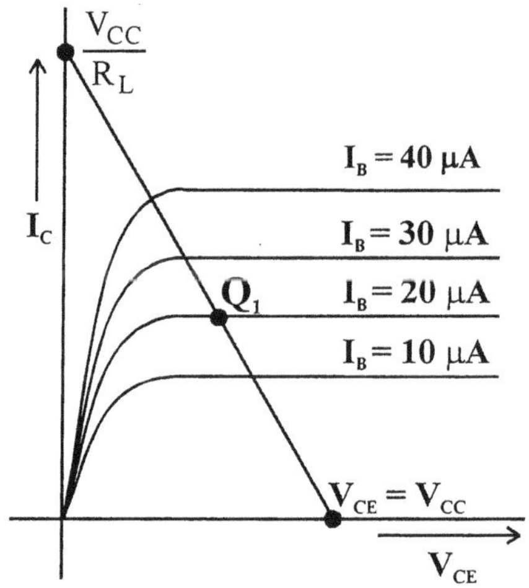

Fig. 7.26 *Operating point variation*

I_B is fixed when R_B and V_{CC} are fixed. Because in the expression, for I_B, β is not appearing. So once R_B and V_{CC} are fixed, the biasing point is also fixed. Hence the name Fixed Bias Circuit.

7.17 BIAS STABILITY

The circuit shown in Fig. 7.23 is called as a ***Fixed Bias Circuit***, because I_B remains constant for given values of V_{CC} and V_{BB} and R_B. ($I_B \simeq V_{BB}/R_B$). So the operating point must also remain fixed. Suppose the transistor in the circuit is AC128, and we replace this transistor by another AC 128 transistor. The characteristics of these two transistors will not be exactly the same. There will be slight difference. It is because of the limitations in the fabrication technology.

The doping concentration, diffusion length etc., in fabrication cannot be controlled precisely. If we say impurity concentration is 1 in $10^6/cm^3$, it need not be the same. The diffusion length will also be not the same since the temperature inside the furnace during drive in diffusion may vary. The furnace may not have constant temperature profile along the length of the furnace. Because of these reasons, transistors of the same type may not have exactly the same characteristics. Hence in the above circuit, if one AC 128 transistor is replaced by another transistor, the operating point will change, since the characteristics of the transistor will be slightly different.

Hence for the same circuit when one transistor is replaced by the other, operating point changes. In some cases, because of the change in the operating point, the transistor may be cut off. In the above circuit, the operating point is not fixed, since I_B is fixed, and β changes. On the other hand I_B should be changed to account for the change in β so that the operating point is fixed or I_C and V_{CE} are fixed.

7.18 THERMAL INSTABILITY

Change in temperature also adversely effects the operating point. I_{CO} doubles for every 10^0C rise in temperature.

$$I_C = (\beta + 1) I_{CO} + \beta \times I_B \qquad \text{.......... (7.18.1)}$$

$$I_C = \frac{I_{CO}}{1-\alpha} + \frac{\alpha . I_B}{1-\alpha} \qquad \text{.......... (7.18.2)}$$

The collector current causes the collector junction temperature to rise. Hence I_{CO} increases. So I_C also increases. This in turn, increases the temperature and so I_{CO} still increases, and then I_C. So the transistor output characteristics will shift upwards (because I_C increases). Then the operating point changes. So in certain cases, even if the operating point is fixed in the middle of active region, because of the change in temperature, the operating point will be shifted to the saturation region.

7.19 STABILITY FACTOR 'S'

In the previous circuit, if I_B is fixed, operating point will shift with changes in the values of I_C with temperature.

$$I_C = \frac{I_{CO}}{1-\alpha} + \frac{\alpha . I_B}{1-\alpha}$$

I_{CO} gets doubled for every 10°C rise in temperature.

I_C also increases with temperature. So if I_B is fixed, the operating point will shift. In order to keep operating point fixed, I_C and V_{CE} should be kept constant. There are two methods to keep I_C constant.

1. ***Stabilization Technique :***

Here resistive biasing circuits are used to allow I_B to vary to keep I_C relatively constant, with variation in I_{CO}, β and V_{BE}. i.e., I_B decreases if I_{CO} increase, to keep I_C constant.

2. ***Compensation Technique :***

In this method, temperature sensitive devices such as diodes, transistor, thermistors etc., which provide compensating voltage and currents to maintain the operating point constant are used.

Stability Factor S *is defined as the rate of change of I_C with respect to I_{CO} the reverse saturation current, keeping β and V_{BE} constant.*

$$S = \left.\frac{\partial I_C}{\partial I_{CO}}\right|_{\beta,\, V_{BE} = K}$$

$$= \frac{\Delta I_C}{\Delta I_{CO}}$$

The ***value of S should be small.*** If it is large, it indicates that the circuit is thermally unstable.

$\frac{\partial I_C}{\partial \beta}, \frac{\partial I_C}{\partial V_{BE}}$ are also some times called as Stability factors.

Expression for I_C,

$$I_C = (1 + \beta)\, I_{CO} + \beta I_B \qquad \text{.......... (7.19.1)}$$

$$S = \left.\frac{\partial I_C}{\partial I_{CO}}\right|_{\beta = K,\ V_{BE} = K}$$

Therefore, differentiating Equation (7.19.1) with respect to I_C, with β as constant,

$$1 = (1 + \beta) \times \frac{\partial I_{CO}}{\partial I_C} + \beta \times \frac{\partial I_B}{\partial I_C}$$

$$1 = \frac{(1+\beta)}{S} + \beta \cdot \frac{\partial I_B}{\partial I_C}$$

I_{CO} cannot be measured directly. I_C and I_B can be measured directly. So I_C is not differentiated directly with I_{CO}.

$$\left(1-\beta\cdot\frac{\partial I_B}{\partial I_C}\right) = \frac{1+\beta}{S}$$

S = 0 implies that I_C is independent of I_{CO}. This is ideal case. S = 5 is a reasonable value for practical circuit. General expression for S is,

$$S = \frac{1+\beta}{1-\beta\left(\frac{\partial I_B}{\partial I_C}\right)} \qquad \text{..........(7.19.2)}$$

$$\because \qquad I_B = \frac{V_{CC}-V_{BE}}{R_B}$$

For a fixed bias circuit, I_B is independent of I_C.

$$\therefore \qquad \frac{\partial I_B}{\partial I_C} = 0.$$

$$\therefore \qquad S = 1+\beta$$

If $\beta = 40$, S = 41. It is a very large value or thermal instability is more. So some circuits are available which make I_C more independent of I_{CO} and hence S will be low.

7.20 SELF BIAS OR EMITTER BIAS CIRCUIT

In the fixed bias circuit, since I_C increases with temperature, and I_B is fixed, operating point changes. In collector to base bias circuit, though I_C changes with temperature, I_B also changes to keep operating point fixed. But since β changes with temp. and I_C depends on β, the assumption has to be made that $\beta R_C >> R_B$ to make I_C independent of β. But if in one circuit, R_C is small, the above assumption will not hold good. [In transformer coupled circuits R_C will be small].

So collector to base circuit is as bad as fixed bias circuit. A circuit which can be used even if there is zero DC resistance in services with the collector terminal, is the ***Self Bias on Emitter Bias Circuit***. The circuit is as shown. This circuit is also known as ***Voltage Divider Bias or Universal Bias Circuit.***

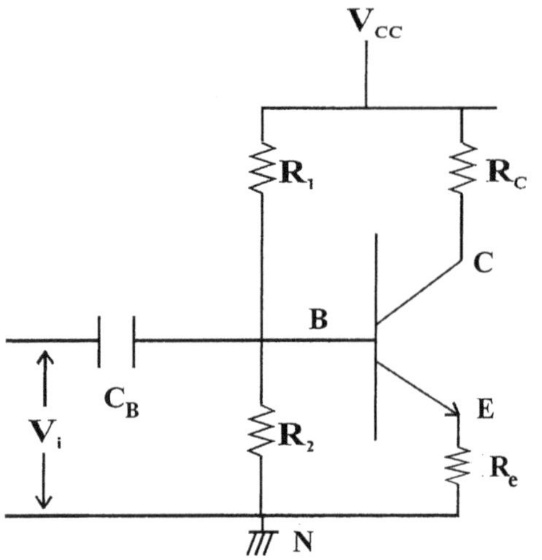

Fig. 7.27 Self Bias circuit

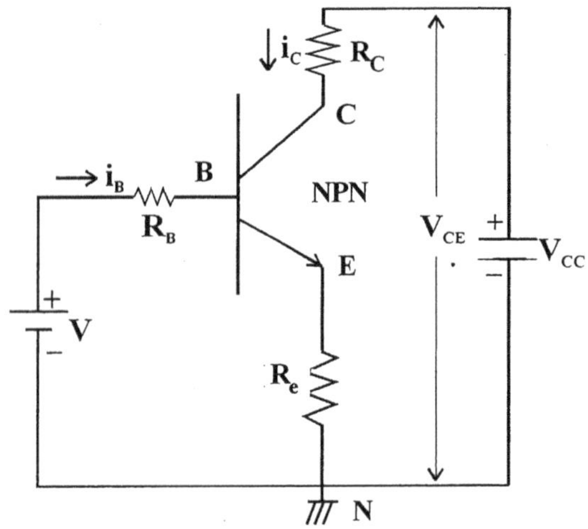

Fig. 7.28 Thevenin's Equivalent

The current in the emitter lead causes a voltage drop across R_e in such a direction, that it forward biases the emitter-base junction. It is NPN transistor emitter is *n-type*. Negative polarity should be at the emitter. If I_C increases due to increase in I_{CO} with temperature the current in R_C also increases.

$$I = I_C + I_B; \; V_{BE} = V_{BN} - V_{EN}$$

As I_B increases, I_C also increases.

So current I or emitter current increases. Therefore IR_e drop increases. Since the polarity of this voltage is to reverse bias the E – B junction, I_B decrease. Therefore I_C will increase less than it would have been, had there been no self biasing resistor.

[The voltage drop across R_C provides the self bias for the emitter]. Hence stability is good.

With reference to Fig. 7.28, it is NPN transistor. So the conventional current is flowing out of the transistor, from the emitter. That is why the symbol is with arrow mark pointing outwards. So emitter point is at positive with respect to N. Hence the drop $(I_B + I_C)\,R_C$ has the polarity such as to reverse bias or to oppose the forward bias voltage V_{BE} junction. R_B can be regarded as parallel combination of R_1 and R_2.

7.21 STABILIZATION AGAINST VARIATION IN I_{CO} AND V_{BE} AND β

S is calculated assuming that no AC signal is impressed and only DC voltages are present. In the Fig 7.28, the voltage V is the drop across R_2. Combination acts as a potential divider. So the drop across R_2 which is equal to V, is given by the expression,

$$V = \frac{V_{CC} \cdot R_2}{R_1 + R_2}.$$

R_B is the effective resistance looking back from the base terminal

$$\therefore \qquad R_B = \frac{R_1 R_2}{R_1 + R_2}$$

Applying Kirchhoff's Voltage Law around the base circuit,

$$V = I_B R_B + R_e (I_B + I_C) + V_{BE}$$

Assuming V_{BE} to be independent of I_C, differentiating I_B with respect to I_C, to get 'S'.

$$I_B, R_B + I_B R_e = + V - V_{BE} - I_C \times R_e$$

$$I_B (R_B + R_e) = V - V_{BE} - I_C \times R_e$$

$$\therefore \qquad I_B = \frac{(V - V_{BE} - I_C R_e)}{(R_B + R_e)}$$

$$\frac{\partial I_B}{\partial I_C} = -\frac{R_e}{R_e + R_B}$$

The expression for stability factor $S = \dfrac{1+\beta}{1-\beta\left(\dfrac{\partial I_B}{\partial I_C}\right)}$

$$\therefore \qquad S = \frac{1+\beta}{1+\beta\left(\dfrac{R_e}{R_e + R_B}\right)}$$

If $\dfrac{R_B}{R_e}$ is very small, $S \simeq 1$.

The fixed bias circuit, collector to base bias circuit and Emitter bias circuits provide stabilisation of I_C against variations in I_{CO}. But I_C also varies with V_{BE} and β. ***V_{BE} decreases at the rate of 2.5 mV/°C for both Germanium and Silicon transistors***. Because, with the increase in T, the potential barrier is reduced and more number of carriers can move from one side (p) to the other side (n). β also increases side with temperature. Hence I_C changes.

7.22 STABILITY FACTOR S'

The variation of I_C with V_{BE} is given by stability factors S'.

$$S' = \left.\frac{\partial I_C}{\partial V_{BE}}\right|_{\beta = K}$$

where both I_{CO} and β are considered constant.

$$V = I_B \times R_B + V_{BE} + (I_B + I_C)\, R_e \qquad \text{.......... (7.22.1)}$$

$$I_C = (1 + \beta)\, I_{CO} + \beta \,.\, I_B \qquad \text{.......... (7.22.2)}$$

Therefore, From equation (7.22.2),

$$I_B = \frac{I_C - (1+\beta)I_{CO}}{\beta}$$

From equation (7.22.1),

$$V_{BE} = V - I_B R_B \times (I_B + I_C)\, R_e \qquad \text{.......... (7.22.3)}$$

But $$I_B = \frac{I_C - (1+\beta)I_{CO}}{\beta}$$

Substituting this value in equation (7.22.3),

$$V_{BE} = V - \left\{\frac{I_C - (1+\beta)I_{CO}}{\beta}\right\} R_B - \left\{\frac{I_C - (1+\beta)I_{CO}}{\beta} + I_C\right\} R_e$$

$$V_{BE} = V - \frac{I_C . R_B}{\beta} + \frac{R_b.(1+\beta)}{\beta} \times I_{CO} + \left(\frac{1+\beta}{\beta}\right) R_e \times I_{CO} - \frac{I_C}{\beta} \times R_e - I_C \times R_e$$

$$V_{BE} = V - I_C \left\{\frac{R_B}{\beta} + \frac{R_e}{\beta} + R_e\right\} + I_{CO} \left\{\frac{R_B.(1+\beta)}{\beta} + \left(\frac{1+\beta}{\beta}\right) R_e\right\}$$

$$V_{BE} = V + (R_B + R_e) \left\{\frac{1+\beta}{\beta}\right\} I_{CO} - \left\{\frac{R_B + R_e(1+\beta)}{\beta}\right\} I_C \qquad \text{.......... (7.22.4)}$$

$$S' = \frac{\partial I_C}{\partial V_{BE}}$$

$$\frac{\partial V_{BE}}{\partial I_C} = -\left\{\frac{R_B + R_e(1+\beta)}{\beta}\right\}$$

or $$S' = \frac{\partial I_C}{\partial V_{BE}} = \frac{-\beta}{R_B + R_e(1+\beta)}$$

But $$S = \frac{(1+\beta)(R_e + R_B)}{R_B + R_e(\beta+1)}$$

$\therefore$ $$S' = \frac{-S.\beta}{(R_B + R_e)(\beta+1)}$$

$$\beta + 1 \simeq \beta$$

$\frac{R_B}{R_e}$ if it small can be neglected

$$\therefore \qquad S' = -\frac{S}{R_e}$$

S = 1, implies that the stability of the circuit is very good or to a better accuracy,

$$1 + \beta + \frac{R_B}{R_e} = 1 + \frac{R_B}{R_e} = 1 + \beta$$

when $\frac{R_B}{R_e}$ is small,

$$\therefore \qquad S = \frac{(1+\beta)\left(1+\frac{R_B}{R_e}\right)}{(1+\beta)} = 1 + \frac{R_B}{R_e}$$

If $\frac{R_B}{R_e}$ is very large, then S = 1 + β. Now for a fixed $\frac{R_B}{R_e}$, S increases with increases β. If β is small, S is almost independent of β. If a graph is plotted between S and $\frac{R_B}{R_e}$, then we get, the graph as

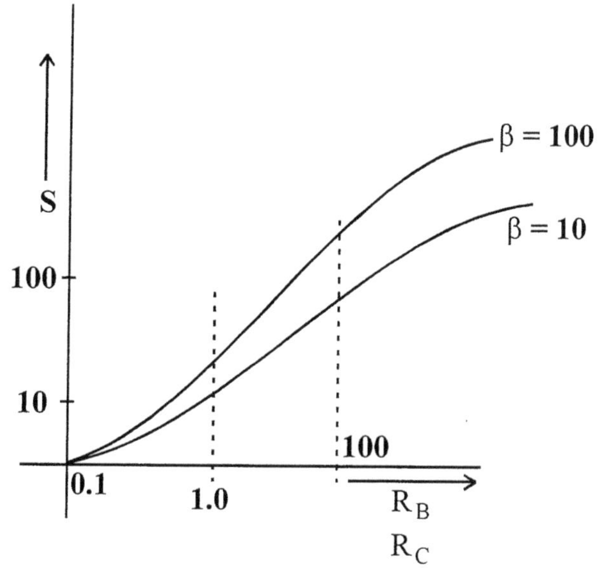

Fig. 7.29 Variation of stability factor

The smaller the value of R_B the better the stabilisation $R_B = R_1$ in parallel with R_2. So R_1 and R_2 should be small.

The smaller, the value of R_B, the better the stabilisation. Because of feedback through R_C, the AC gain is reduced. To prevent this, R_e is shunted with a capacitor whose reactance at the operating frequencies is very small so that AC signal passes through the capacitor only.

S' has the units of $\frac{1}{\Omega}$ or mhos. S' is negative indicate that as V_{BE} decreases, I_C increases.

$$S' \simeq \frac{-S}{R_E}$$

Therefore, if S is made unity, than S' will also be less. We get stability of I_C with respect to both V_{BE} and I_{CO}. For self bias circuit, variation of S'' with V_{BE} is also less. S'' is approximately equal to 1.

7.23 STABILITY FACTOR S''

The variation of I_C with respect to β is given by the stability factors S''.

$$S'' = \frac{\partial I_C}{\partial \beta}$$

where both I_{CO} and V_{BE} are assumed to be constant.

From equation (7.22.4),

$$V_{BE} = V + (R_B + R_e)\frac{\beta+1}{\beta} \times I_{CO} - \frac{R_B + R_e(1+\beta)}{\beta} \times I_C$$

$$V_{BE} - V - (R_B + R_e)\frac{\beta+1}{\beta} \times I_{CO} = -\frac{R_B + R_e(1+\beta)}{\beta} \times I_C \times (1+\beta)$$

$$I_C = \frac{-\beta.V_{BE}}{(R_B + R_e)(1+\beta)} + \frac{V.\beta}{(R_B + R_e)(1+\beta)} \frac{I_{CO}(R_B + R_e)(1+\beta)}{R_B + R_e(1+\beta)}$$

$$I_C = \frac{V.\beta - \beta V_{BE}}{(R_B + R_e)(1+\beta)} + I_{CO}$$

$$\frac{\partial I_C}{\partial \beta} = \frac{(R_B + R_e)(1+\beta)(V - V_{BE}) - \beta(V - V_{BE})(R_e + R_B)}{(R_B + R_e)^2(1+\beta)^2}$$

$$= \frac{(R_B + R_e)(V - V_{BE})}{(R_B + R_e) + (1+\beta)^2}$$

$$S'' = \frac{\partial I_C}{\partial \beta} = \frac{I_C.S}{\beta(1+\beta)}$$

7.24 PRACTICAL CONSIDERATIONS

Silicon transistors are superior to Germanium transistor as far as thermal stability is concerned. The variation in the value of I_C for Silicon transistor is 30% for a typical circuit when the temperature varies from –65°C to +175°C. But for Germanium transistor, the variation will be 30% when the temperature is varied from –65°C to +75°C. Therefore, ***the thermal stability is poor for Germanium transistor***. Silicon transistors are never used above 175°C and Germanium transistors above +75°C. Such high temperature will be encountered in the electronic control of temperature for furnaces, space applications etc.

7.25 BIAS COMPENSATION

The collector to base bias circuit and self bias circuit are used for stability of I_C with variation in I_{CO}, V_{BE} and β. But we said that there is feed back from the output to the input. Hence the amplification will be reduced, even though the stability is improved.

Example 7.8:

An NPN transistor with β = 50, is used in Common Emitter Circuit with V_{CE} = 10V and R_C = 2KΩ. The bias is obtained by connecting a 100 kΩ resistance from collector to base. Assume V_{BE} = 0. Find

(a) *The Quiescent Point*

(b) *The Stability Factor*

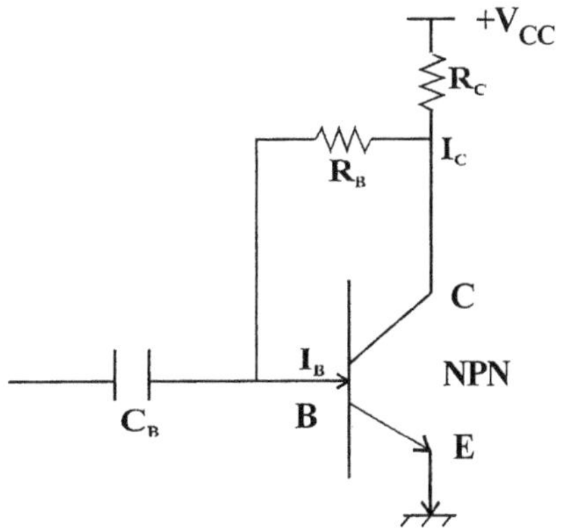

Fig. 7.30 *For Example 7.8*

Solution:

$$V_{CC} - (I_C + I_B)\, R_C - I_B\, R_B = 0$$

Neglecting I_{CO}, $\quad I_C = \beta \times I_B$

Then $\quad (\beta + 1)\, I_B\, R_C + I_B\, R_B = V_{CC}$

$$(50 + 1) \times 2 \times 10^3 + 100 \times 10^3 \times I_B = 10$$

$$\therefore \qquad I_B = \frac{10}{202 \times 10^3} = 49.5 \text{ mA}$$

$$I_C = 50 \times 49.5 \times 10^{-6} = 2.475 \text{ mA}$$

$$V_{CE} = I_B \times R_B = 4.95$$

$$S = \frac{\beta + 1}{1 + \dfrac{\beta . R_C}{R_C + R_B}} = \frac{51}{1 + \dfrac{50 \times 2}{102}} = 25.6$$

Example 7.9:

A transistor with $\beta = 100$ is to be used in Common Emitter Configuration with collector to base bias. $R_C = 1K\Omega$ $V_{CC} = 10V$. Assume $V_{BE} = 0$. Choose R_B so that quiescent collector to emitter voltage is 4V. Find Stability Factor.

Solution:

$$V_{CC} - (\beta + 1)\, I_B \times R_C - V_{CE} = 0$$

$$S = \frac{101}{1 + \dfrac{100 \times 1}{68.3}} = 41$$

$$10 - 101 \times 10^3\, I_B - 4 = 0$$

$$I_B = \frac{6}{101 \times 10^3} = 59.4 \mu A$$

$$V_{CE} = I_B\, R_B = 4V$$

$$\therefore \qquad R_B = \frac{4}{59.4 \times 10^{-6}} = 67.3K$$

Example 7.10:

One NPN transistor is used in the self biasing arrangement. The circuit components values are $V_{CC} = 4.5V$, $R_C = 1.5K$, $R_C = 0.27\ K\Omega$, $R_2 = 2.7\ K\Omega$ and $R_1 = 27K$. If $\beta = 44$. Find the Stability Factor. Also determine the Quiescent point $Q\ (V_{CE} \cdot I_C)$.

Solution:

The Quiescent Point

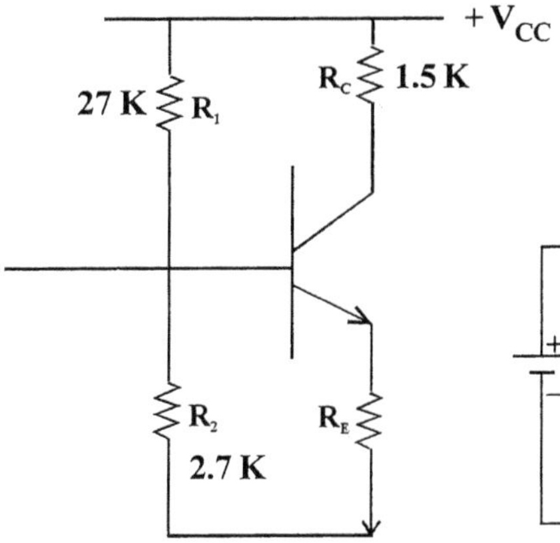

Fig. 7.31 For Example 7.10

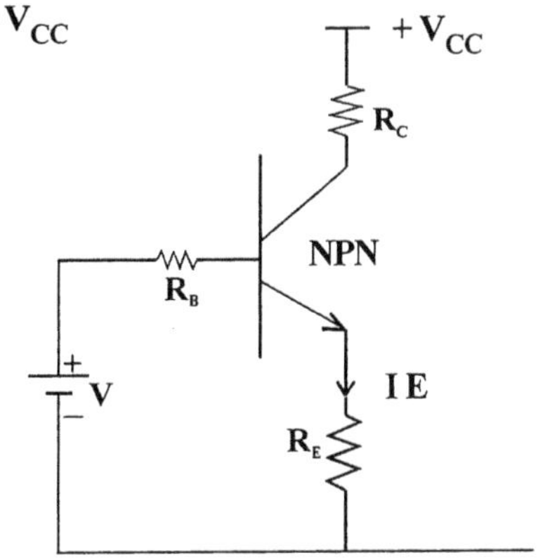

Fig. 7.32 For Example 7.10

$$V = \frac{R_2.V_{CC}}{R_1 + R_2} = \frac{2.2 \times 4.5}{27 + 2.7} = 0.409V$$

$$R_B = \frac{R_1 R_2}{R_1 + R_2} = \frac{27 \times 2.7}{27 + 2.7} = 2.46k\Omega$$

$$S = \frac{(1+\beta)1 + \frac{R_B}{R_e}}{1 + \beta + \frac{R_B}{R_e}} = 8.4$$

Applying KVL over the loop.

$$+V = I_B \times R_B + V_{BE} + (1 + \beta)\, R_C I_B$$

$$+0.409 = I_B \times (2.46 + 45 \times 0.27) + 0.2$$

or $$I_B = \frac{0.409 + 0.2}{14.6} = 1.1mA$$

$$I_C = \beta \times I_B = 44 \times 1.1 = 48.4 \text{ mA}$$

$$V_{CE} = V_{CC} - I_C \times R_C - (I_B + I_C)\, R_C = 2.4V$$

If the reduction in gain is not tolerable then compensation techniques are to be used. Thus both stabilization and compensation technique are used depending upon the requirements.

7.26 COLLECTOR TO BASE BIAS

The circuit shown in Fig. 7.25 is fixed bias circuit, since I_B is fixed, if V_{CC}, R_C and V_{BB} are fixed. But an improvement over this circuit is the circuit shown is Fig. 7.33.

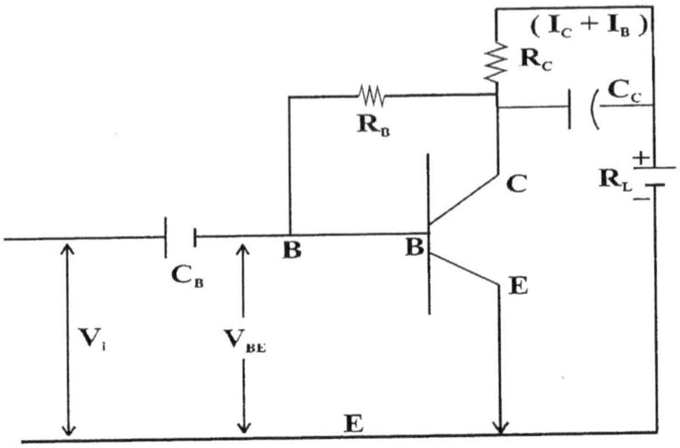

Fig. 7.33 Collector to Base Bias Circuit

In this circuit, R_B is connected to the collector point C, instead of V_{CC} as in Fig. 7.33. V_i is input, the capacitor will block DC input and allow only AC signal to the base. Since it is NPN transistor, collector junction is reverse biased. R_C is a resistor in the collector circuit, which controls the reverse bias voltage of the collector junction with respect to emitter.

$$\because \quad V_{CE} = V_{CC} - I R_C$$

$$I = I_C + I_B$$

This circuit is an improvement over the Fig. 7.25 fixed bias circuit because,

$$V_{CE} = V_{CC} - (I_C + I_B) R_C$$

$$I = I_C + I_B$$

With temperature I_C increases. Since V_{CC} is same, I is same and so I_B decreases.

As I_C increases with temperature because of the increases in I_{CO}, $(I_C + I_B)$ R_C product increase. So V_{CE} reduces. That means the reverse bias voltage of the collector junction is reduced or number of carriers collected by the collector reduces. Hence I_B also reduces. Therefore, because of the reduction in the bias current I_B, the increase in I_C will not be as much as it would have been considering only increase in I_{CO} with temperature.

We can calculate the stability factor for this circuit. Applying KVL,

$$V_{BE} - V_{CC} + (I_B + I_C)\, R_C + I_B \,.\, R_B = 0$$

or $$V_{CC} - V_{BE} = I_B\,(R_C + R_B) + I_C \,.\, R_C$$

$$I_B = \frac{V_{CC} - V_{BE} - I_C R_C}{R_C + R_B}$$

V_{BE} is the cut in voltage and it is 0.6V for silicon, and 0.2V for germanium. It is independent of I_C. So differentiate this expression with respect to I_C, we get

$$\frac{\partial I_B}{\partial I_C} = \frac{0 - R_C - 0}{R_C + R_B} = \frac{-R_C}{R_C + R_B}$$

The expression for stability factor S

$$S = \frac{1+\beta}{1-\beta\left(\frac{\partial I_B}{\partial I_C}\right)}$$

But $$\frac{\partial I_B}{\partial I_C} = \frac{-R_C}{R_C + R_b}$$

$$\therefore \quad \boxed{S = \frac{1+\beta}{1+\beta\times\frac{R_C}{R_C + R_B}}}$$

Therefore, the value of S is less than $(1 + \beta)$.

$(1 + \beta)$ is the maximum value for the fixed bias circuit. Therefore collector bias circuit, is an improvement over the fixed bias circuit. Now let us consider the effect of β on the stability of the circuit, discussed above.

$$I_C = (1+\beta)\, I_{CO} + \beta \times I_B \qquad \text{.......... (7.26.1)}$$

$$I_B = \frac{V_{CC} - I_C R_C - V_{BE}}{R_C + R_B} \qquad \text{.......... (7.26.2)}$$

From Equation (7.26.1), $I_B = \dfrac{I_C - (1+\beta)I_{CO}}{\beta}$ .... (7.26.3)

Equating (7.26.2) and (7.26.3) and transferring I_C to one side

$$\frac{I_C - (1+\beta)I_{CO}}{\beta} = \frac{V_{CC} - I_C R_C - V_{BE}}{R_C + R_B}$$

$$I_C (R_C + R_B) - (1 + \beta)(R_C + R_B) I_{CO} = \beta[V_{CC} - V_{BE}] - \beta I_C R_C$$

$$I_C (R_C + R_B) + \beta I_C R_C = \beta ([V_{CC} - V_{BE}] + (R_C + R_B) I_{CO} (1 + \beta))$$

$$\beta >> 1 = 1 + \beta \simeq \beta$$

$$\therefore \quad I_C[R_C + \beta R_C + R_B] = \beta[V_{CC} - V_{BE}] +)(R_C + R_B) I_{CO} \beta$$

$$I_C [(\beta + 1) R_C + R_B] \simeq \beta [V_{CC} - V_{BE} + (R_C + R_B) I_O]$$

Again $(\beta + 1) \simeq \beta$.

$$\therefore \quad I_C \simeq \frac{\beta[V_{CC} - V_{BE} + (R_C + R_B) I_{CO}]}{(\beta R_C + R_B)} \qquad \text{.......... (7.26.4)}$$

This expression, in this form is derived to get the condition to make I_C independent of β. Therefore, with the changes in the value of β, with temperature, I_C changes so the operating point also changes.

The above circuit, is an improvement over the fixed bias circuit, I_B is being decreased, with increase in I_C (Because I_{CO} increases with temperature) to keep operating point constant. Since I_C is dependent on the β from the above expression, and since β changes with temperature, I_C also changes and thereby operating point changes. Therefore from the above expression, it is essential to make I_C insensitive to β, so that the modified circuit works satisfactorily to keep the operating point fixed. The assumption to be made in the above expression (7.26.4) is,

$$\beta \times R_C >> R_B.$$

Therefore, the above expression becomes,

$$I_C \simeq \frac{\beta[V_{CC} - V_{BE} + (R_C + R_B) I_{CO}]}{R_C}$$

$$\therefore \quad \beta R_C >> R_b$$

$$\therefore \quad I_C \simeq \frac{V_{CC} - V_{BE} + (R_C + R_B) I_{CO}}{R_C}$$

So if we choose R_B very small compared to βR_C, the circuit will work satisfactorily. But in all cases, this may not work out. If R_B value is ***not so small***, then the above assumption is not valid and hence the circuit won't work satisfactorily.

For a given circuit, to determine the operating point, draw the load line on the output characteristics of the given transistor. Load line is drawn as given below :

$$V_{CC} = I_C R_L + V_{CE}$$

If $I_C = 0$, then, $\quad V_{CC} = V_{CE}$

$$\text{If } V_{CE} = 0, \text{ then } I_C = \frac{V_{CC}}{R_L}.$$

With these points, a straight. line is drawn on the output characteristics of the transistor. Corresponding to a given value of I_B, where this line cuts the output characteristic, it is called as the operating point.

In the collector to base bias circuit we said that the stability is good since, I_B is decreased with increase in I_C to keep operating point fixed. In that circuit, there is feedback from the output to the input through the resistor R_B. If the signal voltage causes an increase in I_B, I_C increases. But V_{CE} decreases. So because of this, the component of base current coming through R_B decreases. Hence the increase in I_B is less than what it would have been without feedback. In feedback circuits the amplification factor will be less than what it would have been without feedback. But the advantage is stability is improved. Such an amplifier circuit with feedback is called as ***Feedback Amplifier Circuit***. This type of circuits are discussed in subsequent chapters.

7.27 DIODE COMPENSATION FOR V_{BE}

The circuit shown in Fig. 7.34 employs self biasing technique for stabilization. In addition a diode is connected in the emitter circuit and it is forward biased by the source V_{DD}. Resistance R_d limits the current through the diode. The diode should be of the same material as the transistor. The negative voltage V_0 across the diode will have the same temperature coefficient as V_{BE}. Since the diode is connected as shown, if V_{BE} decreases with increase in temperature (V_B is cut in voltage), V_0 increases with temperature. Since with temperature, the mobility of carriers increases, so current through the diode increases and V_0 increases. Therefore, as V_{BE} decreases, V_0 increases, so that I_C will be insensitive to variations in V_{BE}.

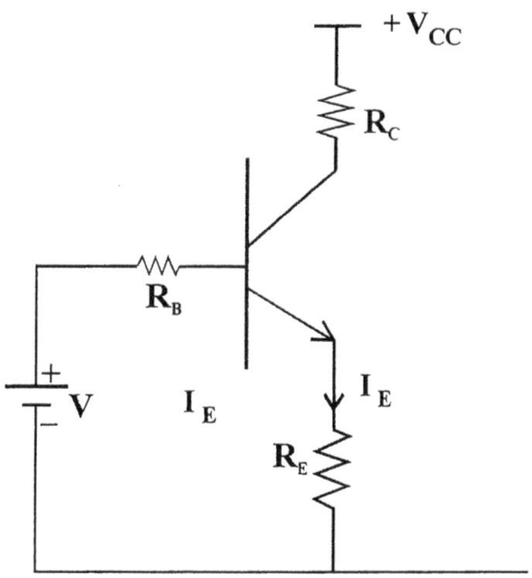

Fig. 7.34 Diode compensation for V_{BE}

The decrease in V_{BE} is nullified by corresponding increase in V_D. So the net voltage more or less remains constant. Thus the diode circuit compensates for the changes in the values of V_{BE}.

7.28 DIODE COMPENSATION FOR I_{CO}

I_{CO} the reverse saturation current of the collector junction increases with temperature. So I_C also increases with temperature. Therefore, diode compensation should also be given against variation is I_{CO}. A circuit to achieve this is as shown in the Fig. 7.35.

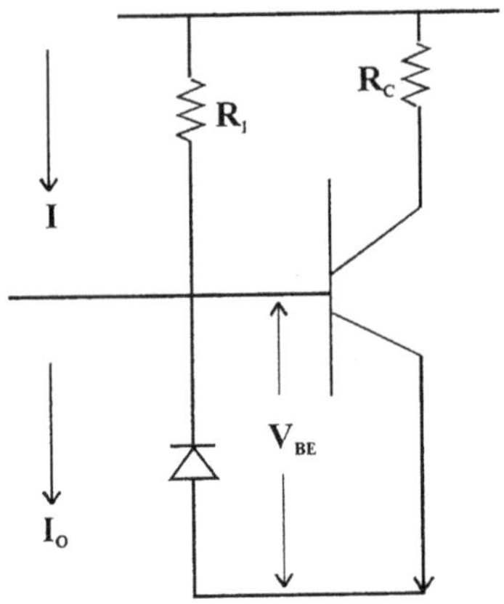

Fig. 7.35 Diode compensation for I_{CO}

The diode D is reverse biased since the voltage V_{BE} which also appears across the diode is so as to reverse bias the diode. If the diode and the transistor are of the same material, the reverse saturation current I_0 of diode will increase at the same rate as the transistor collector saturation current I_{CO}.

$$I = \frac{V_{CC} - V_{BE}}{R_1}$$

Therefore, V_{CC} is measured with respect to ground.

V_{BE} is small compare to V_{CC}.

$$V_{CC} \simeq 15V,$$

$$V_{BE} = 0.2 \text{ or } 0.6V.$$

$$\therefore \quad I \simeq \frac{V_{CC}}{R_1}$$

The diode D is reverse biased by an amount 0.2V for Germanium. Current through the diode is the reverse saturation current I_0.

$$\therefore \quad I_B = I - I_0$$

But we know that $I_C = (1+\beta)\, I_{CO} + \beta \times I_B$

Substituting the value of I_B,

$$I_C = \beta \times I - \beta \times I_0 + (1+\beta)\, I_{CO}$$

But $\beta >> 1,$

$$\therefore \quad \beta + 1 \simeq \beta$$

$$\therefore \quad I_C = \beta \times I - \beta \times I_0 + \beta \times I_{CO}$$

$-\beta \times I_0$ and $+\beta \times I_{CO}$ are of opposite signs. If the diode and the transistor are of the same material, the rate of increase of I_0 and I_{CO} will be the same. Therefore, I_C is essentially is constant. I_O and I_{CO} need not be the same. Only the changes in I_{CO} and I_O will be of the same order. So net value of I_C will remain constant.

7.29 BIASING CIRCUITS FOR LINEAR INTEGRATED CIRCUITS

With the advancements in semiconductor technology, the active components made with this technology are no more costlier than the passive components like resistors, capacitors etc. Moreover, incorporating the passive components particularly capacitors in ICs is difficult. With today's technology, in biasing and stability circuits, transistors and diodes can be used. Hence certain circuits incorporating transistors in biasing compensation are developed.

One such circuit is shown in Fig. 7.36. Why should transistor be used for compensation when it is being used as a diode only? It is because of convenience in fabrication technology. In IC fabrication for a diode, a transistor is being used as a diode. It is one step process and fabrication is easy.

Transistor Q_1 is used as a diode between base and emitter because, collector and base are shorted. Q_1 is connected as a diode across the emitter base junction of transistor Q_2. So effectively the circuit is same as circuit in Fig.7.36. Instead of a diode, a transistor is used. Now the collector current through transistor Q_1 is,

$$I_{C1} = \frac{V_{CC} - V_{BE}}{R_1} - I_{B1} - I_{B2}$$

$$V_{BE} << V_{CC}, \text{ and } (I_{B1} + I_{B2}) << I_{C1}$$

$$\therefore \quad I_{C1} = \frac{V_{CC}}{R_1} = \text{Constant.}$$

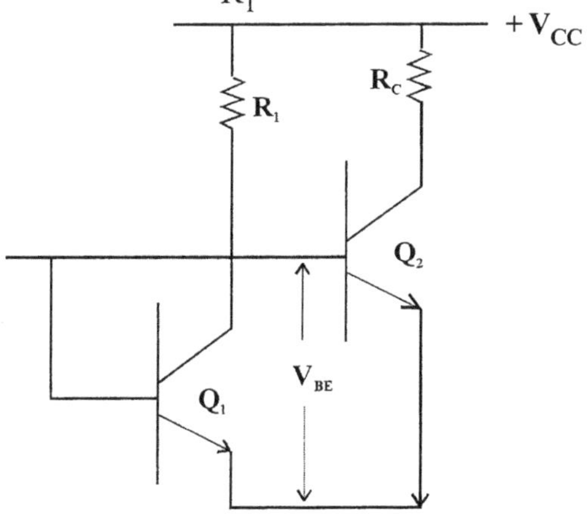

Fig. 7.36 Biasing circuit for ICs

If transistors Q_1 and Q_2 are identical, their V_{BE}, (cut in voltage) will be the same. Hence $I_{C1} = I_{C2}$ = Constant if $R_1 = R_C$. These two transistors are being driven by the base voltage V_{BE}. Since, V_{BE} is the same, I_C will not increase. V_{CC} and R_C are also fixed. Hence the stability will be good. Even if the two transistors are not exactly identical, practical experience has shown that, the stability factor will be around 5. Hence in ICs, such circuits are employed for biasing stability.

7.30 BJT MODELING USING h-PARAMETERS

In drawing the equivalent circuit of a transistor, we assume that the Transistor Parameters are constant. With the variation in operating points, transistor parameters will also vary. But we assume that the Quiescent or operating point variation is small.

The advantages of considering h-parameters for transistor are

1. *They are easy to measure.*
2. *Can be obtained from Transistor characteristics.*
3. *They are real numbers at Audio frequencies.*
4. *Convenient to use in circuit design and analysis.*

Manufactures specify a set of ***h-parameters*** for transistors.

7.31 SIMPLIFIED TRANSISTOR HYBRID MODEL

The analysis of transistors can be done using h-parameters which includes detailed calculation of all the four parameters related to transistor device such as voltage amplification, current amplification, input and output impedences. However in practice appropriate approximations can be allowed provided that if approximations are within tolerable limits of devices (considering all the parameters of the devices such as aging, temperature dependence of characteristics etc).

Out of available four parameters of transistors, two parameters i.e., h_{ie} and h_{fe} (CE configuration) are sufficient for simplified analysis of transistor for low frequency circuits provided that load resistance is small enough to satisfy the following condition.

$$h_{oe} R_L < 0.1$$

The simplified hybrid model for different configurations of transistors CB, CE and CC are shown in Fig. 7.38, Fig. 7.37, and Fig. 7.39 respectively.

Simplified CE Configuration

The Fig. 7.37 shows simplified hybrid model of CE transistor.

Current Gain: The current gain of CE transistor is given by

$$A_I = \frac{I_2}{I_1} = \frac{-h_{fe}}{1 + h_{oe} Z_L} \qquad .....(7.31.1)$$

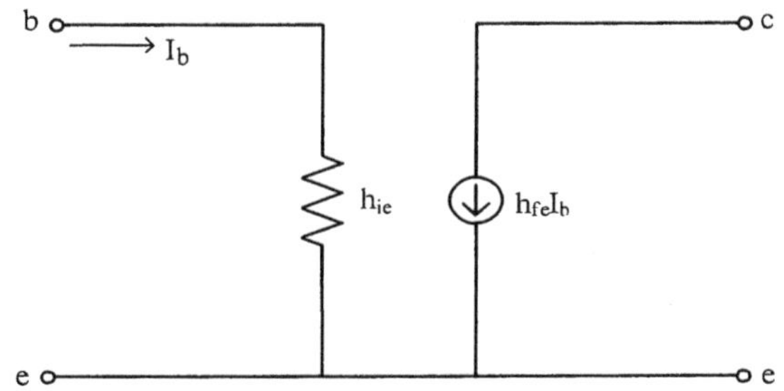

Fig. *7.37 Simplified hybrid model of CE configuration.*

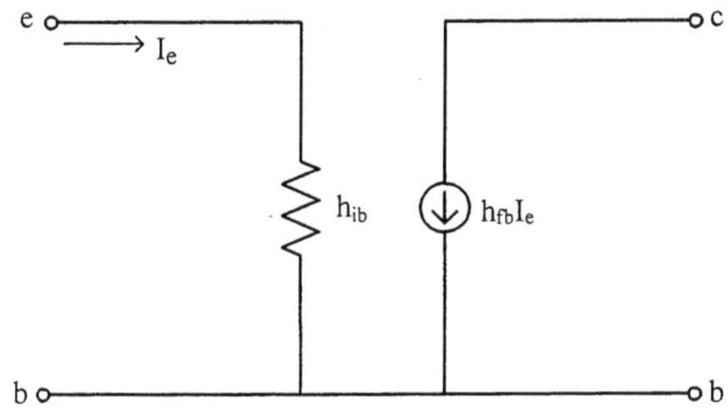

Fig. *7.38 Simplified hybrid model of CB configuration.*

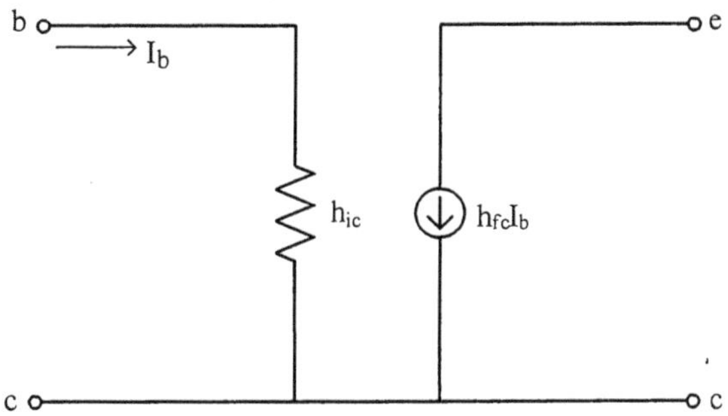

Fig. *7.39 Simplified hybrid model of CC configuration.*

By applying condition $h_{oe}R_L < 0.1$, the denominator term in eq.(7.31.1) becomes $\simeq 1$

Hence $$A_I = -h_{fe} \qquad .....(7.31.2)$$

Input impedance: The input impedance of CE transistor is given by

$$Z_i = h_i + h_r A_I Z_L \qquad .....(7.31.3)$$

$$= h_{ie}\left(1 - \frac{h_{re}h_{fe}|A_I|}{h_{ie}h_{oe}h_{fe}} h_{oe}R_L\right) \qquad .....(7.31.4)$$

For any transistor $h_{re} h_{fe}/h_{ie} h_{oe} = 0.5$, $|A_I| \simeq h_{fe}$ (from eq. (7.31.1)) and by applying the condition $h_{oe} R_L < 0.1$, the second term in brackets of eq. (7.31.4) becomes negligible. And hence

$$Z_i \simeq h_{ie} \qquad .....(7.31.5)$$

Voltage gain: The voltage gain (A_V) of CE configuration transistor is given by

$$A_V = A_I(R_2/R_i) \qquad .....(7.31.6)$$

Substituting values of A_I from eq. (7.31.2) and R_i from eq. (7.31.5) we get

$$A_V = -h_{fe} R_L/h_{ie} \qquad .....(7.31.7)$$

Output impedance: To calculate output impedance the input signal (Vs assume) is to be made zero and load resistance (assume 'RL' is made infinite) if Vs = 0, output impedence RL is infinite. There is one dependent ideal current source i.e., hfeIb, in the output circuit and so output impedence will be infinite and hence output impedence will be equal to load resistance i.e., $R_o = R_L$.

$$\text{Output impedance } (Z_o) = \text{Load Resistance } R_L \qquad .....(7.31.8)$$

By employing corresponding convertions of *h-parameters* the difference parameters like A_I, Z_i, A_V, Z_o etc., for different transistor configurations (CB, CE, CC) can be easily obtained.

7.32 DETERMINATION OF h-PARAMETERS FROM TRANSISTOR CHARACTERISTICS

The output characteristics of a Transistors in Common Emitter Configuration are as shown in Fig. 7.40.

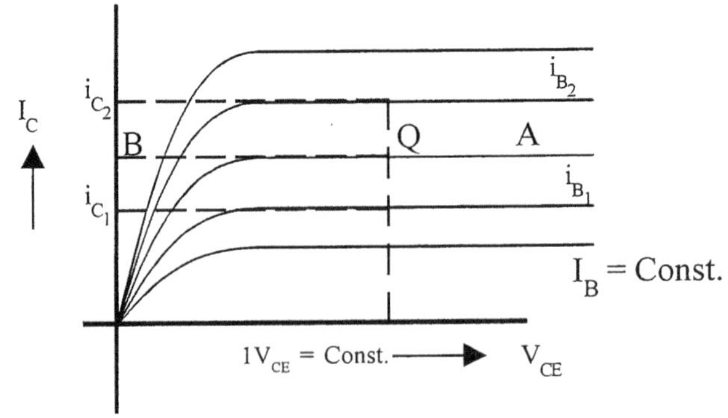

Fig. 7.40 Transistor Output characteristics (To determine h_{fe}, h_{oe} graphically).

$$h_{fe} = \left.\frac{\partial I_C}{\partial I_B}\right|_{V_C=K} = \left.\frac{\Delta I_C}{\Delta I_B}\right|_{V_C=0} \qquad V_{CE} = \text{Constant} = K$$

h_{fe}, h_{oe} can be determined from Output characteristics. h_{ie}, h_{re} can be determined from input characteristics.

Suppose Q is the Quiescent Point around the operating point h_{fe} is to be determined. For h_{fe}, we have to take the ratio of incremental change in I_B keeping V_{CE} constant. From the output characteristics two values i_{c1}, and i_{c2} are taken corresponding to i_{b_1} and i_{b_2}. Then,

$$\mathbf{h_{fe}} = \left(i_{c_2} - i_{c_1}\right)/\left(i_{b_2} - i_{b_1}\right)$$

h_{fe} is also known as ***Small-Signal Beeta*** (β) of the transistors. Since we are determining the incremental values of current ratio of I_C and I_B. h_{fe} is represented as β.

β is represented as h_{fe} and is called large signal β. $\beta = I_C / I_B$, since here no incremental values are being considered.

$$h_{oe} = \left.\frac{\partial I_C}{\partial v_C}\right|_{I_B = K}$$

Output characteristics of a transistor are plotted between i_c and v_c. Therefore, h_{oe} is the slope of the output characteristics, corresponding to a given value of I_b at the operating point. For Common Emitter Configuration, in exactly the same way we have drawn for amplifier system, the h-parameter equivalent circuit can be drawn. The Transistor in Common Emitter Model is shown in Fig. 7.41. The *h-parameter* equivalent is shown in Fig. 7.44.

Equations are $\quad v_b = h_{ie} i_b + h_{re} v_C$ .... (7.32.1)

$\quad i_C = h_{fe} i_b + h_{oe} v_C$ .... (7.32.2)

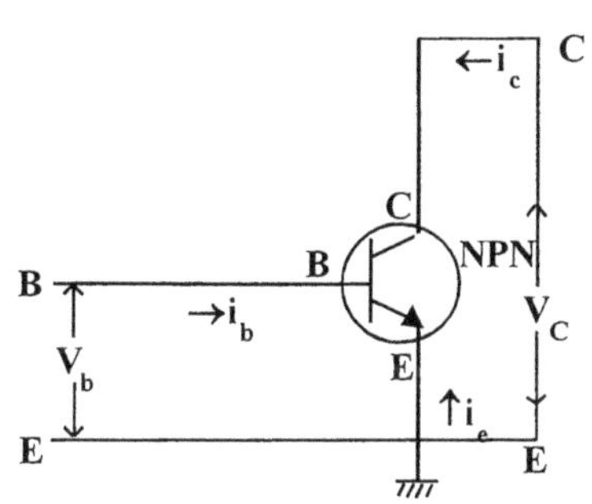

Fig. 7.41 Common Emitter configuration.

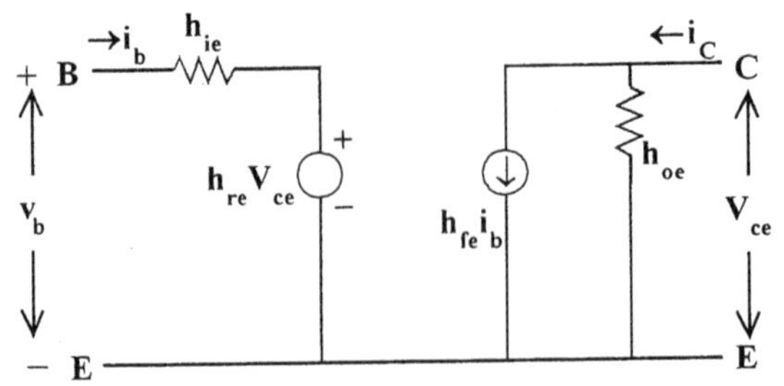

Fig. 7.42 Equivalent circuit in C.E. configuration.

7.33 MEASUREMENT OF h-PARAMETERS

In the circuit shown in Fig. 7.43 to determine the h-parameter, PNP transistors are taken into account. The Operating point is chosen by suitably adjusting R_2, V_{EE} and V_{CC}, there by adjusting the values of I_C and V_{CB}. The *h-parameters* depend upon the frequency of operation. So the signal generator is adjusted to be 1KHz, at which frequency normally all the transistor h parameters are specified by the manufactures. 1 KHz is in the audio frequency range. ***So the h-parameters are assumed to be real at audio frequency***. The tank circuit impedance will be very high at this frequency, of the order of 500kΩ. This is very high compared to the input resistance of the transistor. The tank circuit is tuned to the same frequency of the signal generator (1KHz). This will prevent any stray pick up. The stray signal will pass through the LC tank circuit, since it offers least impedance. So spurious signal will not affect the net voltages and currents of the transistor, while determining the h-parameters.

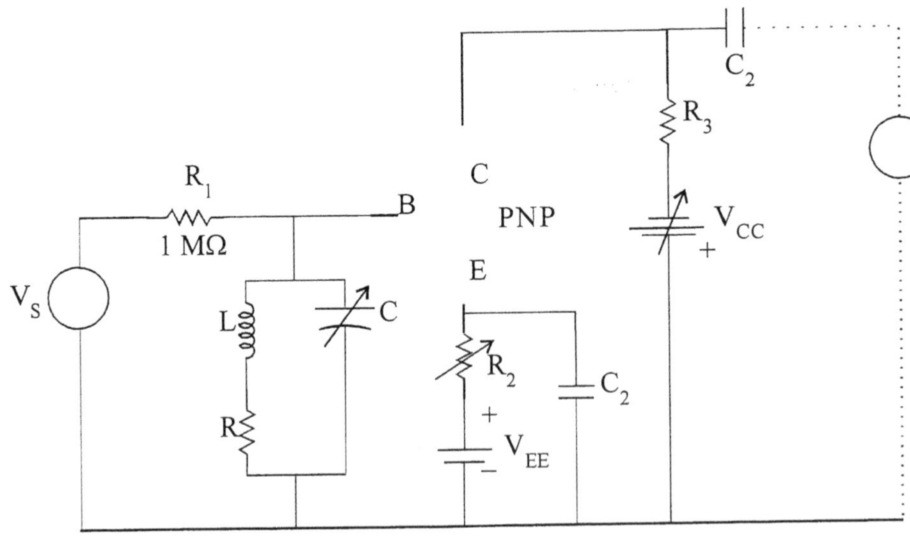

Fig. 7.43 Circuit for determining h - parameters.

The name, tank circuit, is used for LC Parallel circuit, because energy is stored in these elements, as water is stored in a tank.

$$h_{ie} = \left.\frac{V_b}{I_b}\right|_{V_c} \; ; \; I_b = \frac{V_S}{R_1}$$

Since the input resistance of the transistor is small compared to the Tank Circuit Resistance, all the current I_b flows into the transistors alone. R_S is negligible compared to R_1 (1MΩ).

$$\therefore \qquad I_b = \frac{V_S}{R_1}$$

Now h_{ie} and h_{fe} are determined when $V_c = 0$ or the collector is shorted to ground. But if the collector is directly connected to the ground, I_c will be different and V_{CE} will be different. The operating point will change h parameters. Depending on the Operating point *h - parameters* differ. Hence AC short circuit is to be done . C_2 will block D.C. Hence the D.C values of V_{CE}, I_c will not change at all. But the A.C. Voltage of V_c is = 0 or the collector is at ground potential for A.C. voltages . So this is called AC short circuit.

$$\text{Now} \qquad h_{ie} = \left.\frac{V_b}{I_b}\right|_{V_c = 0} \qquad \because \; I_b = \frac{V_S}{R_1}$$

$$h_{fe} = \left.\frac{I_c}{I_b}\right|_{V_c = 0} \qquad I_b = \frac{V_S}{R_1}$$

$$\therefore \qquad h_{fe} = \frac{I_C R_1}{V_S}$$

where V_o is the voltage across R_3.

h_{re} and h_{oe} are defined when $I_b = 0$ or input is open circuited. Therefore, the tank circuit impedance is very large. The input side can be regarded as open circuit. The signal generator is connected on the output side.(as shown by the dotted lines).

$$h_{re} = \left.\frac{V_b}{V_c}\right|_{I_b=0} = \frac{V_b}{V_c}$$

$$h_{oe} = \left.\frac{I_c}{V_c}\right|_{I_b=0} = \frac{I_c}{V_c} = \frac{V_O}{R_L V_C} \qquad \because \; I_C = \frac{V_O}{R_L}$$

7.34 ANALYSIS OF CE, CB AND CC CONFIGURATIONS USING h- PARAMETERS

To form transistor amplifier configuration, we connect a load impedance Z_L and a signal source as shown in Fig. 7.44.

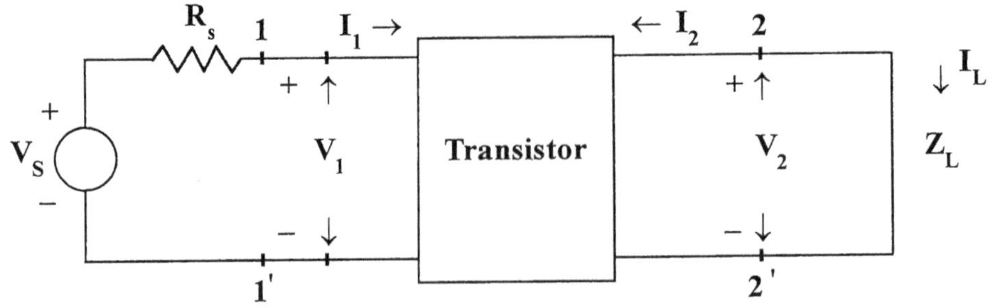

Fig. 7.44 Amplifier circuit.

V_s is the signal source, R_s is the resistance of the signal source, and Z_L is the load impedance. Transistor can be connected in C.E, C.B. and C.C. Configuration. To analyse these circuits i.e., to determine the current gain A_I, Voltage gain A_V, input impedance, output impedance etc, we can use the h parameters. So the equivalent circuit for the above transistor amplifier circuit in general form without indicating C.E, C.B, or C.C. Configuration, can be denoted as in the Fig. 7.45.

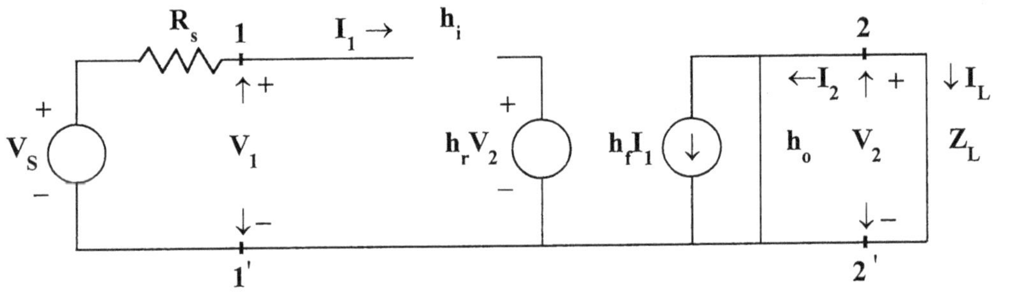

Fig. 7.45 h-parameter equipvalent circuit (general repersentation).

Current Amplification,

$$A_I = \frac{I_L}{I_1};$$

But $\quad I_L = -I_2$

$\therefore \quad A_I = -\frac{I_2}{I_1};$

Negative sign is because I_2 is always represented as flowing into the current source. For a transistor PNP or NPN type, if it is flowing out of the transistor, I_2 is represented as $-I_2$. Now voltage across Z_2 is taken with 2 as +ve and 2^1 as −ve The directions of I_L are as shown and $I_L = -I_2$.

From the above circuit we have,

$$I_2 = h_f I_1 + h_o V_2$$

But $V_2 = I_L Z_L = -I_2 . Z_L$ (Since $I_L = -I_2$)

∴ $I_2 = h_f I_1 - h_o . Z_L . I_2$

or $I_2 + h_o Z_L I_2 = h_f I_1 .$

$I_2 (1 + h_o Z_L) = h_f I_1$

∴ $$A_I = -\frac{I_2}{I_1} = \frac{-h_f}{1 + h_o Z_L} \quad \text{.......... (7.34.1)}$$

Negative sign indicates that I_1 and I_2 are out of phase by 180^0.

7.34.1 Input Impedance (Z_i)

Input impedance of any circuit is the impedance we measure looking back into the amplifier circuit. Now amplifier terminals are 1and 1^1. R_s is the resistance of the signal source. So to determine the output z of the amplification alone, it need not be considered. $Z_i = \frac{V_1}{I_1}$

But $V_1 = h_i I_1 + h_r V_2$

Hence $$Z_i = \frac{V_1}{I_1} = h_i + h_r \frac{V_2}{I_1} \quad \text{.......... (7.34.2)}$$

But $$V_2 = -I_2 . Z_L , \quad \therefore \frac{V_2}{I_1} = \left(\frac{-I_2}{I_1}\right) Z_L = A_I Z_L \quad \text{.......... (7.34.3)}$$

Substituting the values of V_2 in Equation (7.34.2),

$$Z_i = h_i + h_r . A_I . Z_L$$

But $$A_I = \frac{-h_f}{1 + h_o z_L}$$

∴ $$Z_i = h_i - \frac{h_f . h_r z_L}{1 + h_o z_L}$$

$$= h_i - \frac{h_f . h_r}{\frac{1}{z_L} + h_o}$$

$$Z_i = h_i - \frac{h_f . h_r}{Y_L + h_o} \quad \text{.......... (7.34.4)}$$

Y_L is load admittance. Therefore, Input impedance is a function of load impedance. If $Z_L = 0$, $Y_L = \infty$,

∴ $Z_I = h_i$.

7.34.2 Voltage Gain (A_V)

$$A_v = \frac{V_2}{V_1}$$

But $\quad V_2 = +I_L \cdot Z_L = -I_2 \cdot Z_L \qquad$ Since, $I_2 = -I_L$

But $\quad I_2 = -A_I \cdot I_1$

$\therefore \quad V_2 = +A_I \cdot I_1 Z_L$

$$\frac{V_2}{V_1} = \frac{A_I I_1 Z_L}{V_1}$$

But $\quad \frac{V_1}{I_1} = Z_i$

$$Z_i = h_i - \frac{h_f . h_r}{\frac{1}{R_L} + h_o}$$

$$\therefore \quad A_V = \frac{A_I . Z_L}{Z_i}$$

$$A_V = -\frac{-h_f . z_L}{h_i + z_L (h_i h_o - h_f h_r)} \qquad \text{.......... (7.34.5)}$$

7.34.3 Output Admittance (Y_o)

The output impedance can be determined by using two assumptions $Z_L = \infty$, and $V_s = 0$

Y_o is defined as $\frac{I_2}{V_2}$ with $Z_L = \infty$

But $\quad I_2 = h_f I_1 + h_o V_2$

Dividing by V_2,

$$\frac{I_2}{V_2} = Y_O = \frac{h_f . I_1}{V_2} + h_o. \qquad \text{.......... (7.34.6)}$$

From the equivalent circuit with $V_s = 0$,

$$R_s \cdot I_1 + h_i I_1 + h_r V_2 = 0.$$

Dividing by V_2 Through out, we get

$$\frac{R_S . I_1}{V_2} + \frac{h_i I_1}{V_2} + h_r = 0$$

or $$\frac{I_1}{V_2} = \frac{-h_r}{h_i + R_S} \qquad \text{.......... (7.34.7)}$$

Substitute this value in the equation for $Y_{o.}$

$$Y_o = h_f \left(\frac{-h_r}{h_i + R_s} \right) + h_o$$

$$Y_o = h_o - \frac{h_f . h_r}{h_i + R_s}$$

Therefore, $Z_o = 1/Y_o$

Therefore, Output admittance is a function of Source Resistance R_s where as Z_i is a function of Y_L. If $R_s = 0$, $Y_o \cong h_o$. Since ($h_f h_r / h_i$) is very small.

7.34.4 Voltage Gain (A_{vs}) considering Source Resistance R_s

$$\therefore \quad A_{vs} = \frac{V_2}{V_S} = \frac{V_2}{V_1} \times \frac{V_1}{V_S} = A_V . \frac{V_1}{V_S}.$$

This is the equivalent circuit of the input side of the amplifier circuit.

$$\therefore \quad V_1 = \frac{V_S . Z_i}{Z_i + R_S}$$

$$\therefore \quad A_{vs} = \frac{A_V . Z_i}{Z_i + R_S}$$

$$A_v = \frac{V_2}{V_1}$$

$$\therefore \quad A_V Z_i = \frac{V_2}{V_1} . Z_i$$

But $\quad V_2 = I_L . Z_L ; V_1 = I_1 Z_i$

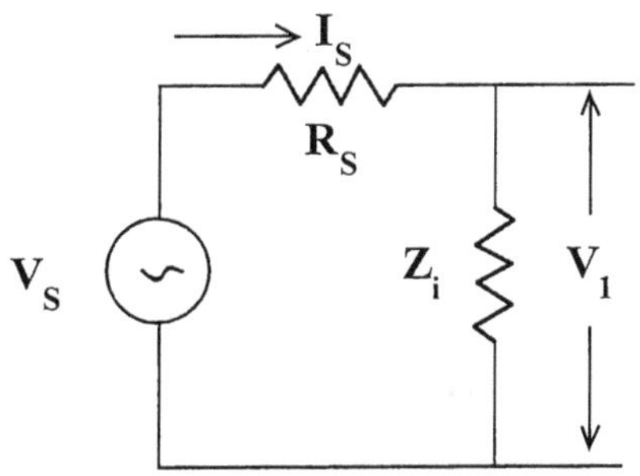

Fig. 7.46 Voltage source.

$$\therefore \quad A_v . Z_i = \frac{I_L . Z_L . Z_i}{I_1 . Z_L} = A_I . Z_L$$

$$\therefore \quad A_{vs} = \frac{A_I . Z_L}{R_S + Z_i} \qquad \text{If } R_s = 0, A_{vs} = A_v \qquad \text{....... (7.34.8)}$$

Hence A_V is the voltage gain of an ideal voltage source with zero terminal resistance.

7.34.5 Current Gain (A_{IS}) Considering Source Resistance

The input source can also be represented as a current source I_S in parallel with resistance R_s or Voltage source V_s in series with resistance R_s. Now let us consider the input source as a current source in parallel with R_s. The equivalent circuit is

$$A_{IS} = \frac{I_L}{I_S} ; I_L = -I_2$$

$$\therefore \quad A_{IS} = \frac{I_L}{I_S} = -\frac{I_2}{I_S}$$

$$A_{IS} = -\frac{I_2}{I_S} = -\frac{I_2}{I_1}.\frac{I_1}{I_S}$$

$$= A_I.\frac{I_1}{I_S} \qquad \text{.......... (7.34.9)}$$

$$\therefore \quad A_I = \frac{-I_2}{I_1}$$

From the Fig. 7.47,

$$I_1 = \frac{I_S.R_S}{R_S + Z_i}$$

$$\therefore \quad A_{IS} =$$

$$\frac{A_I \times R_S}{(R_S + Z_i)} = \frac{A_I.R_S}{(R_S + Z_i)}$$

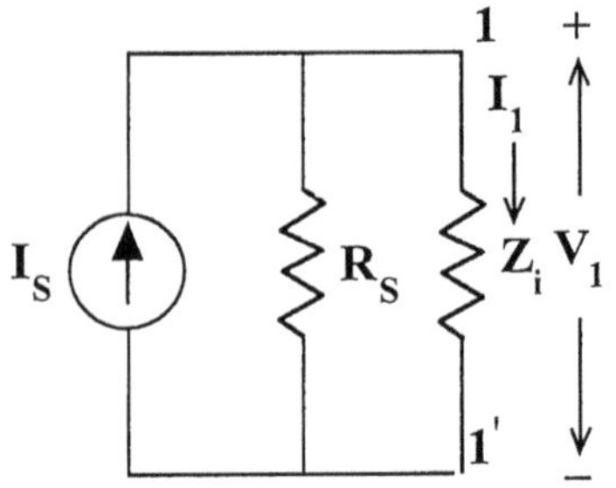

Fig. 7.47 Current source.

If $R_S = \infty$, $A_I = A_{IS}$. Therefore, A_I is the current gain for an ideal source

$$\text{Now} \qquad A_{VS} = \frac{A_I.Z_L}{Z_i + R_s} \qquad \text{.......... (7.34.10)}$$

$$A_{VS} = \frac{A_I.R_S}{Z_i + R_S} \qquad \text{.......... (7.34.11)}$$

$$\text{Dividing 1 and 2,} \simeq \frac{A_{vs}}{A_{IS}} = \frac{Z_L}{R_S}$$

$$\text{or} \qquad A_{vs} = A_{IS}.\frac{Z_L}{R_S} \qquad \text{.......... (7.34.12)}$$

This equation is independent of the transistor parameters. This is valid if the equivalent Current and Voltage sources have the same resistance.

7.34.6 Power Gain (A_P)

$$A_P = \frac{P_2}{P_1} = \frac{V_2 I_L}{V_1 I_1}$$

$$= A_v.A_I$$

$$\because \quad A_v = \frac{A_I.Z_L}{Z_i},$$

$$\therefore \quad A_P = \frac{A_I^2 Z_L}{Z_i} \qquad \text{........ (7.34.13)}$$

7.35 COMPARISON OF CE, CB, CC CONFIGURATIONS USING h-PARAMETERS

C.E : Of the three, it is the most versatile. Its voltage and current gain are > 1. Input and output resistance vary least with R_s and R_L. R_i and R_o values lie between maximum and minimum for all the three configurations. Phase shift of 180^0 between V_i and V_o. Power Gain is Maximum.

C.B : $A_I < 1$ $A_V > 1$. R_i is the lowest and R_o is the highest . It has few applications. Sometimes it is used to match low impedance source V_i and V_o. No phase shift.

C.C : $A_V < 1$. A_I is very high. It has very ***high input Z and low output Z. So it is used as a buffer*** between high Z source and low impedance load. It is also called as ***Emitter follower***.

To analyse circuit consisting of a number of Transistors, each Transistor should be replaced by its equivalent circuit in h-parameters. The emitter base and collector points are indicated and other circuit elements are connected without altering the circuit Configuration. This way the circuit analysis becomes easy.

Example 7.11:

Find the Common Emitter Hybrid parameters in terms of the Common Collector Hybrid parameters for a given transistor.

Solution:

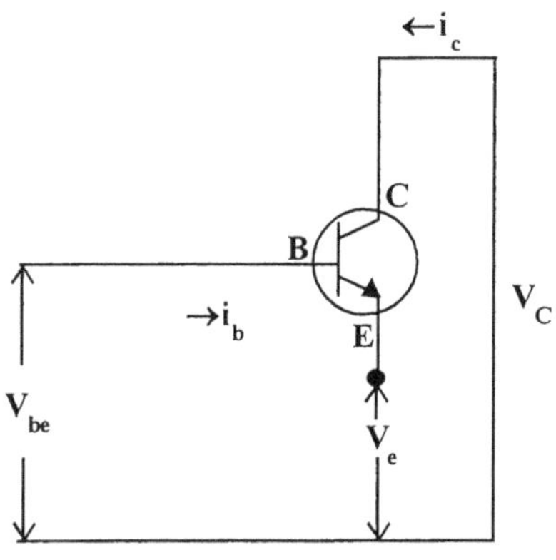

Fig. 7.48 *For Example 7.11.*

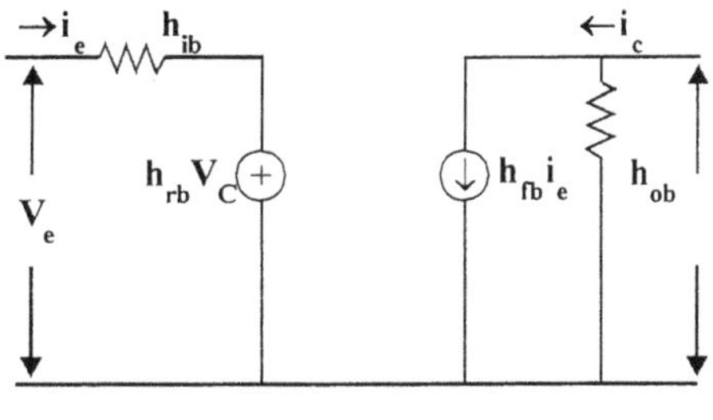

Fig. 7.49 *Equivalent circuit.*

We have to find h_{ie}, h_{re}, h_{fe} and h_{oe} in terms of h_{ic}, h_{rc}, h_{fc} and h_{oc}.

The transistor circuit in Common Collector Configuration is shown in Fig. 7.48. The h-parameter equivalent circuit is shown in Fig. 7.49.

C.C :

$$v_b = h_{ic} \cdot i_b + h_{rc} v_e$$
$$i_e = h_{fc} i_b + h_{oc} v_e$$
$$i_e = h_{fc} i_b - i_l$$
$$= (h_{fe} i_b - V_{ce} h_{oc})$$
$$V_{be} = i_b \cdot h_{ic} - h_{rc} v_{ce} + v_{ce} = i_b \cdot h_{ic} \qquad \text{Since } v_{ce} = 0$$

But $h_{ie} = \left.\dfrac{V_{be}}{I_b}\right|_{V_{ce}=0}$

$\therefore$ $h_{ie} = h_{ic}$ .... (7.35.1)

$$h_{fe} = \frac{i_c}{i_b}\bigg| V_{ce} = 0$$

$$i_c = - i_b - i_e = - i_b - h_{fc} \cdot i_b + h_{oc} v_{ce}$$

But $v_{ce} = 0$

$$= - i_b - (h_{fc} i_b - h_{oc} v_{ce})$$
$$= -i_b - h_{fc} i_b + h_{oc} v_{ce} \qquad v_{ce} = 0$$

$$h_{fe} = \frac{-i_b - h_{fc} i_b}{i_b} = -(1 + h_{fc})$$

$h_{fe} = -(1 + h_{fc})$ .... (7.35.2)

$$h_{re} = \frac{V_{be}}{V_{ce}}\bigg| i_{b=0}$$

$$V_{be} = i_b h_{ie} - h_{rc} V_{ce} + V_{ce}$$
$$i_b = 0$$

$\therefore$ $h_{re} = -\dfrac{h_{rc} V_{ce} + V_{ce}}{V_{ce}} = 1 - h_{rc}$

$h_{re} = 1 - h_{rc}$ .... (7.35.3)

$$h_{oe} = \frac{i_c}{V_{ce}}\bigg| i_b = 0$$

$$i_c = - i_e - i_b$$

But $i_b = 0$.

$i_c| = i_e$

$i_e = V_{ce}$ hoc

$\therefore \quad \frac{i_e}{V_{ce}} = \frac{i_e}{V_{ce}} = h_{oc}$

$\therefore \quad i_c = -i_e = -h_{fc} i_b + V_{ce} h_{oc}$

$\therefore \quad h_{oe} = h_{oc}$ .... (7.35.4)

Since, $\quad i_b = 0$

$\therefore \quad i_c = V_{ce} \times h_{oc}$

$\therefore \quad h_{oe} = h_{oc}$ is admittance in mhos (℧).

SUMMARY

- In order that for a transistor, Emitter-Base junction is forward biased and collector-Base junction is reverse biased, biasing circuit must be employed. The three types of biasing circuits are
    1. Fixed bias or Base bias circuit.
    2. Collector to base bias
    3. Self bias or Emitter bias
- Self bias circuit is commonly used. It ensures that the operating point is in the centre of the active region. The operating point is defined by I_C, V_{CE} and I_B. Usually V_{CE} is chosen to be equal to $\frac{V_{CC}}{2}$.
- Stability factor is a function of I_{Co}, β, V_{BE}

$$S = \left.\frac{\partial I_C}{\partial I_{co}}\right|_{\beta \text{ and } V_{BE} = \text{Constant}}$$

General Expression for,

$$S = \frac{1+\beta}{1-\beta\left(\frac{\partial I_B}{\partial I_C}\right)}$$

For a Self bias, circuit,

$$S = (1+\beta)\left[\frac{1+\frac{R_B}{R_e}}{1+\frac{R_B}{R_e}+\beta}\right]$$

- Quiescent point Q, (or operating point or biasing point) changes with temperature. So to nullify for the effect of temperature, compensating techniques must be used. The circuits are :
    1. Diode Compensation for V_{BE}
    2. Diode Compensation for I_{CO}
    3. Thermistor and Sensistor Compensation.

OBJECTIVE TYPE QUESTIONS

1. For identical construction, ... types of bipolar junction transistors (BJTs) have faster switching times.
2. The arrow mark in the symbols of BJTs indicates ....
3. For NPN Bipolar Junction Transistor Emitter Efficiency γ = ...
4. For PNP Transistor (BJT), the transportation factor β^* = ...
5. Relation between α, β^* and γ for a transistor (BJT). α = ...
6. The expression for α in terms of β for a BJT is ....
7. The expression for β in terms of α for a BJT is ....
8. Expression for I_C in terms of β, I_B and I_{CBO} is ....
9. Expression for I_{CEO} in terms of I_{CBO} and α is ....
10. β' of a BJT is defined as ....
11. β' of a transistor (BJT) is ....
12. Relation between β' and β is ....
13. β' is synonymous to .... and β is same as ....
14. β is .... (for DC Currents).
15. β' is .... (for AC Currents).
16. Base width modulation is ....
17. Operating point is also known as ....
18. Operating point depends upon the following BJT parameters ....
19. For a transistor to function as an amplifier, the operating point must be located at ....
20. For normal amplifier circuits, the operating point is chosen such that V_{CE} = ...V_{CC}.
21. Fixed bias circuit is also known as ....
22. Self bias circuit is also known as ....
23. Stability factor S is defined as S = ...
24. S' is defined as S' = ...
25. Stability factor S'' is defined as S'' = ...
26. General expression for S = ...
27. Expression for S for self bias circuit is.................. .
28. For both Silicon and Germanium, V_{BE} or V_r cut in voltage decreases at the rate of ....
29. Semiconductor materials having positive temperature coefficient of R (PTCR) are called.................. .

30. Materials used for Thermistors are ....
31. For most configurations, the D.C. analysis of the circuit begins with ...... current of the BJT.
32. A BJT circuit is most stable, and least sensitive to temperature changes if the circuit has .... stability factor.
33. 'β' of a BJT is very sensitive to .... and V_{BE} .... at the rate of 2.5 mv for each 1° increase in temperature.
34. If a transistor (BJT) is in active region, V_{CE} will be about .... % of V_{CC}.

ESSAY TYPE QUESTIONS

1. With the help of a neat graph qualitatively explain the Potential distribution through a transistor (BJT).
2. Explain about the different current components in a transistor.
3. Derive the relation between β and β′ .
4. Derive the relation between α, β* and γ.
5. Differentiate between the terms h_{FE} and h_{fe}. Derive the relationship between them.
6. Qualitatively explain the input and output.
7. Explain the V-I characteristics in Common Emitter Configuration.
8. Describe the V-I characteristics of a transistor in Common Collector Configuration and Explain.
9. Compare the input and output characteristics of BJT in the three configurations, critically.
10. Explain about the need for biasing in electronic circuits, what are the factors affecting the stability factor.
11. Draw the Fixed bias circuit and derive the expression for the stability factor S. What are the limitations of this circuit ?
12. Draw the Collector to Base bias circuit and derive the expression for the stability factors What are the limitations of this circuit ?
13. Draw the Self bias circuit and obtain the expression for the stability factor S. What are the advantages of this circuit ?
14. Compare all the three biasing circuits.
15. Explain the terms: Thermal Runaway and Thermal Resistance.
16. Explain the terms Bias Stabilization and Bias Compensation.
17. Draw the circuits and explain the principles of working of Diode Compensation for V_{BE} and I_{CO}.

MULTIPLE CHOICE QUESTIONS

1. As reverse bias voltage is increased, for a diode, the base width at the junction

(a) decreases (b) increases (c) remains same (d) none of these

2. Expression for 'α' of a BJT in terms of I_{pc}, I_E, I_c etc is

(a) $\frac{I_{pc}}{I_E}$ (b) $\frac{I_{pE}}{I_E}$ (c) $\frac{I_{nc}}{I_E}$ (d) $\frac{I_{nE}}{I_C}$

3. Expression for β^* is,

(a) $\frac{I\,pc}{I\,p_B}$ (b) $\frac{I\,pc}{In_B}$ (c) $\frac{I\,pc}{Ip_E}$ (d) $\frac{I\,p_E}{I\,pc}$

4. The relation between I_{CEO}, I_{CBO} and α is, I_{CEO} =

(a) $\frac{I_{CBO}}{\alpha}$ (b) $\frac{I_{CBO}}{(1+\alpha)}$ (c) $\frac{I_{CBO}}{(1-\alpha)^2}$ (d) $\frac{I_{CBO}}{(1-\alpha)}$

5. For a transistor in C.E configuration to be at cut-off, the condition is,

(a) $V_{BE} = -V_{BB} + R_B.I_{CBO} \leq -0.4$ (b) $V_{BE} = +V_{BB} - R_B.I_{CBO} \leq 0.4$

(c) $V_{BE} = -V_{BB} + R_B.I_{CBO} \leq -0.1$ (d) $V_{BE} = -V_{BB} - R_B.I_{CBO} \leq -0.1$

6. The condition for test for saturation in a BJT is

(a) $|I_B| \leq \left|\frac{I_C}{\alpha}\right|$ (b) $|I_B| \leq \left|\frac{I_C}{\beta}\right|$ (c) $|I_B| \geq \left|\frac{I_C}{\alpha}\right|$ (d) $|I_B| \geq \left|\frac{I_C}{\beta}\right|$

7. Expression for saturation resistance Rcs of a transistor (BJT) is ...

(a) $Rcs = \frac{V_{CE}}{I_C(sat)}$ (b) $Rcs = \frac{I_C(sat)}{V_{CE}}$

(c) $Rcs = \frac{V_{CE}\ (sat)}{I_C}$ (d) $Rcs = \frac{V_{CE}}{I_C}$

8. For high eletric field strengths $\varepsilon > 10^4$ V/cm, mobility of carriers is proportional to

(a) ε_x (b) $\frac{1}{(\varepsilon_x)^2}$ (c) $\frac{1}{\varepsilon_x}$ (d) None of these

9. When the input is symmetrical, to operate the BJT in active region, the quiescent point is chosen ....

(a) at the top edge of the load line

(b) at the bottom edge of the load line

(c) at the centre of the load line

(d) can be chosen any where on the load line

10. AC load line is also known as ....

(a) dynamic load line (b) variable load line

(c) quiescent load line (d) active load line

11. For better stability of the amplifier circuit, value of stability factor 'S' must be

(a) large (b) as small as possible (c) 1 (d) 0

12. For fixed bias circuit, the expression for stability factor 'S' is

(a) $1-\beta$ (b) $1+\beta^2$ (c) $\dfrac{1+\beta}{1-\beta\dfrac{(\partial I_C)}{(\partial I_B)}}$ (d) $1+\beta$

13. Voltage divider bias or universal bias circuit is also known as

(a) self bias circuit (b) collector bias circuit

(c) collector to base bias circuit (d) Fixed bias circuit

14. By definition, expression for stability factor S" is

(a) $S'' = \left.\dfrac{\partial I_C}{\partial \beta}\right|_{\substack{V_{BE}=K \\ I_{CO}=K}}$ (b) $S'' = \left.\dfrac{\partial \beta}{\partial I_C}\right| V_{BE}, I_{CO}=K$

(c) $S'' = \dfrac{\partial I_\beta}{\partial I_C}$ (d) $S'' = \dfrac{\partial I_\beta}{\partial \beta}$

15. The units of stability factor S are

(a) ohms (b) constant (c) mhos (d) volt-amperes

16. Compared to silicon BJT, thermal stability for germanium transistors is ...

(a) good (b) poor (c) same (d) can't be said

CONCEPTUAL QUESTIONS (Interview Questions)

1. Why transistor biasing is to be done?

Ans : In order that amplification is achieved, the BJT must be operated in the active region. To ensure thus, E-B junction of the transistor (BJT) must be forward biased and Collector - Base junction must be reverse biased. The operating point or quiescant point must be in the centre of the active region. To ensure thus, biasing must be done. Similarly to use the BJT as On - Off switch, the operating point must change between cut-off and saturation regions. So biasing is to be done.

8

JFET AND SPECIAL PURPOSE DEVICES

8.1 JUNCTION FIELD EFFECT TRANSISTOR (JFET)

The Field Effect Transistor is a semiconductor device which depends for its operation on the control of current by an electric field. Hence, the name FET.

There are two types of FETs :

1. Junction Field Effect Transistor (JFET) or simply FET.
2. Insulated Gate Field Effect Transistor IGFET. It is also called as Metal Oxide Semiconductor (MOS) transistor or MOST or MOSFET.

The principle of operations of these devices, their characteristic are given in this chapter.

Advantages of FET over BJT (Transistor)

1. JFET is unipolar device. Its operation depends upon the flow of majority carriers only. Vacuum tube is another example of unipolar device. Because the current depends upon the flow of electrons emitted from cathode. Transistor is a bipolar device. So recombination noise is more.
2. JFET has high input resistance (MΩ). BJT (Transistor) has less input resistance. So loading effect will be there.
3. It has good thermal stability.

4. Less noisy than a tube or transistor. Because JFET is unipolar, recombination noise is less.
5. It is relatively immune to radiation.
6. JFET can be used as a symmetrical bilateral switch.
7. By means of small charge stored or internal capacitance, it acts as a memory device.
8. It is simpler to fabricate and occupies less space in IC form. So packing density is high.

Disadvantage

1. Small gain - bandwidth product.

8.1.1 JFET

There are two types;

1. n - channel FET.
2. p - channel FET.

If n - type semiconductor bar is used it is n - channel FET.

If p - type semiconductor bar is used it is p - channel FET.

Ohmic contacts are made to the two ends of a semiconductors bar of n-type or p-type semiconductor. Current is caused to flow along the length of the bar, because of the voltage supply connected between the ends. If it is n - channel FET, the current is due to electrons only and if it is p-channel FET, the current is due to holes only.

The three lead of the device are 1. Gate 2. Drain 3. Source. They are similar to 1. Base 2. Collector 3. Emitter of BJT respectively.

8.2 CONSTRUCTION

FET will have gate junction on both sides of the silicon bar (as shown in Fig. 8.1).

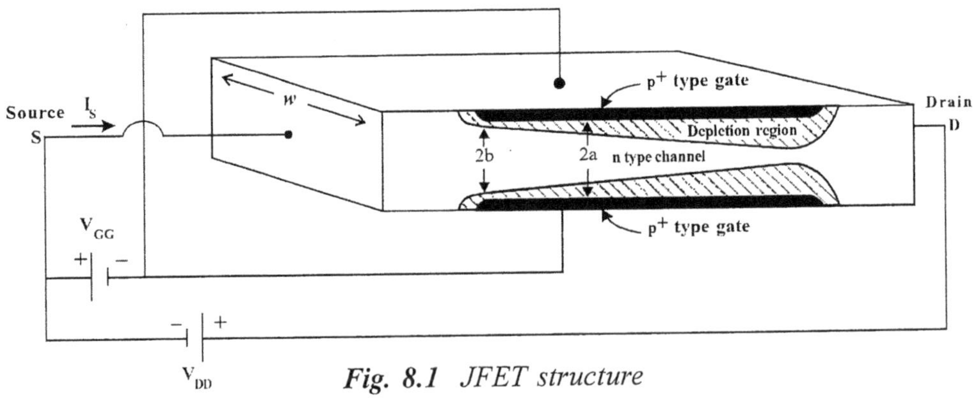

Fig. 8.1 JFET structure

But it is difficult to diffuse impurities from both sides of the wafer. So the normal structure that is adopted is, a p-type material is taken as substrate. Then n-type channel is epitaxially grown. A p-type gate is then diffused into the n-type channel. The region between the diffused p-type impurity and the source acts as the drain.

2b (x) = Actual channel width, at any point x, the spacing between the depletion regions at any point x from source end because the spacing between the depletion regions is not uniform, it is less towards the drain and more towards the source (see Fig.8.2). 2b depends upon the value of V_{GS}.

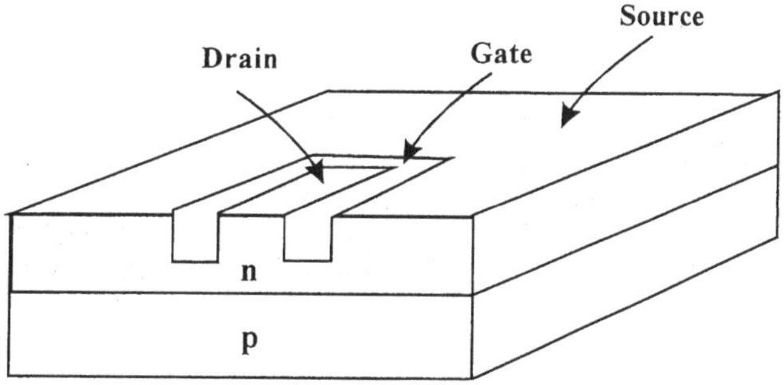

Fig. 8.2 Structure

2a = Channel width, the spacing between the doped regions of the gate from both sides. This is the channel width when $V_{GS} = 0$ (and the maximum value of channel width).

8.3 PRINCIPLE OPERATION

If a p-n junction is reverse biased, the majority carriers will move away from the junction. That is holes on the p-side will move away from the junction leaving negative charge or negative ions on the p-side (because each atom is deprived of a hole or an electron fills the hole. So it becomes negative by charge). Similarly, electrons on the n-side will move away from the junction leaving positive ions near the junction. Thus, there exists space charge on both sides of the junction in a reverse biased p-n junction diode. So, the electric field intensity, the lines of force originate from the positive charge region to the negative charge region. This is the source of voltage drop across the junction. As the reverse bias across the junction increases, the space charge region also increases, or the region of immobile uncovered charges increases (i.e., negative region on p-side and positive region on n-side increases as shown in Fig. 8.3). The conductivity of this region is usually zero or very small. Now in a FET, between gate and source, a reverse bias is applied. Therefore the channel

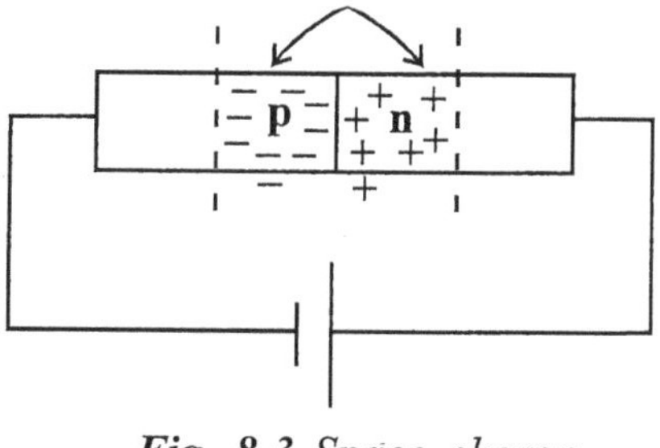

Fig. 8.3 Space charge in p-n junction

width is controlled by the reverse bias applied between gate and source. Space charge region exists near the gate region on both sides. The space between them is the channel. If the reverse bias is increased, the channel width decreases. Therefore, for a fixed drain to source voltage the drain current will be a function of the reverse biasing voltage across the junction. Drain is at positive potential (for n-type FET). Therefore, electrons tend to move towards drain from the source (1) Because, source is at negative potential, they tend to move towards the drain. But because of the reverse bias applied to the gate, there is depletion region or negative charge region near the gate which restricts the number of electrons reaching the drain. Therefore the drain current also depends upon the reverse bias voltage across the gate junction. The term field effect is used to describe this device because the mechanism of current control is the effect of the extensions, with increase reverse bias of the field associated with region of uncovered charges.

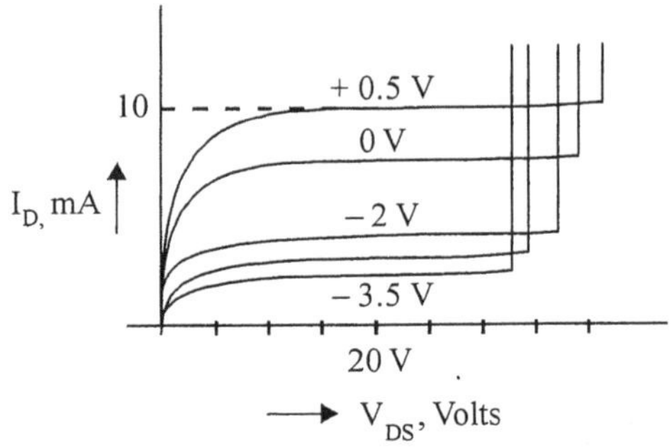

***Fig. 8.4** Drain characteristics*

The characteristics of n-channel FET between I_D, the drain current and V_{DS} the drain source voltage are as shown in Fig. 8.4, for different values of V_{GS}.

To explain these characteristics, suppose $V_{GS} = 0$. When $V_{DS} = 0$, $I_D = 0$, because the channel is entirely open. When a small V_{DS} is applied (source is forward biased or negative voltage is applied to n-type source), the n-type bar acts as a simple semiconductor resistor and so I_D increases linearly with V_{DS}. With increasing current, the ohmic voltage drop between the source and the channel region reverse biases the junction and the conducting portion of the channel begins to constrict. Because, the source gets reverse biased or the negative potential at the n-type source reduces. Because of the ohmic drop, along the length of the channel it self, the constrictions is not uniform, but is more ***pronounced*** at distances farther from the source. i.e., the channel width is narrow at the ***drain*** and wide at the source. Finally, the current will remain constant at a particular value and the corresponding voltage at which the current begins to level off is called as the ***pinch off voltage***. But the channel cannot be closedown completely and there by reducing the value of I_D to zero, because the required reverse bias will not be there.

Now if a gate voltage V_{GS} is applied in a direction to reverse bias the gate source junction, pinch off will occur for smaller values of (V_{DS}), and the maximum drain current will be smaller compared to when $V_{GS} = 0$. If V_{GS} is made + 0.5V, the gate source junction is forward biased. But at this voltage, the gate current will be very small because for Si FET, 0.5V is just equal to or less than the cut in voltage. Therefore, the characteristic for $V_{GS} = + 0.5$V, I_D value will be comparatively larger, (compared to $V_{GS} = 0$V or − 0.5V). Pinch off will occur early.

FET characteristics are similar to that of a pentode, in vacuum tubes.

The maximum voltage that can be applied between any two terminals of the FET is the lowest voltage that will cause avalanche breakdown, across the gate junction. From the FET characteristics it can be observed that, as the reverse bias voltage for the gate source junction is increased, avalanche breakdown occurs early or for a lower value of V_{DS}. This is because, the reverse bias gate source voltage adds to the drain voltage because drain though n -type (for n-channel FET), a positive voltage is applied to it. Therefore, the effective voltage across the gate junction is increased.

For n-channel FET, the gate is p-type, source and drain are n-type. The source should be forward-biased, so negative voltage is applied. Positive voltage is applied to the drain. ***Gate source junction should be reverse biased,*** and gate is p-type. Therefore,voltage or negative voltage is applied to the gate. Therefore, n-channel FET is exactly similar to a Vacuum Tube (Triode). Drain is similar to anode (at positive potential), source to cathode and gate to grid, (But the characteristics are similar to pentode).

For p-channel FET, gate is n-type, and positive voltage is applied, drain is at negative potential with respect to source.

Consider n-channel FET. The source and drain are n-type and p-type gate is diffused from both sides of the bar (See Fig. 8.5 below).

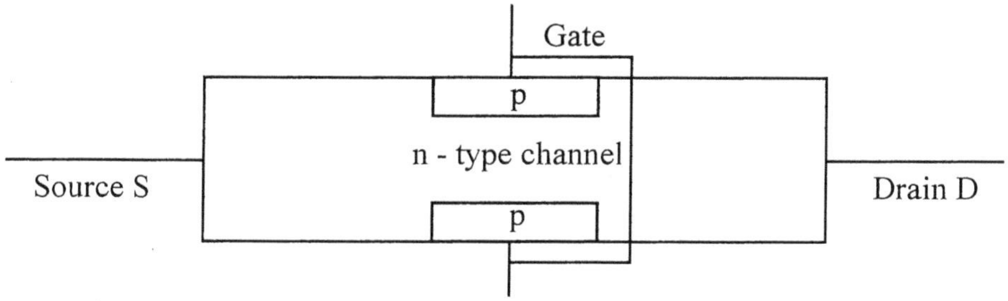

Fig. 8.5 *Structure of n-channel JFET.*

Suppose source and gate are at ground potential and small positive voltage is applied to the drain. Source is n-type. So it is Forward biased. Because drain is at positive potential, electrons from the source will move towards the drain. Negligible current flows between source and gate or gate and drain, since these p-n junctions are reverse biased and so the current is due to minority carriers only. Because gate is heavily doped, the current between S and G or G and D can be neglected. The current flowing from source to drain I_D depends on the potential between source and drain V_{DS}, resistance of the n-material in the channel between drain to source. This resistance is a function of the doping of the n-material and the channel width; length and thickness.

If V_{DS} is increased, the reverse bias voltage for the gate drain junction is increased since, drain is n-type and positive voltage at drain is increased. This problem is similar to that of a reverse biased p-n junction diode.

Consider a p-n junction which is reverse biased. The holes on the p-side remain near the negative terminal and electrons on the n-side reach near the positive terminal as shown in Fig.8.6. There are no mobile charges near the junction. So we call this as the depletion region since it is depleted of mobile carriers or charges. As the reverse voltage is increased, the depletion region width '*l*' increases.

This result is directly applicable for JFET. The depletion region extends more towards drain because points close to the drain are at higher positive voltage compared to points close to source. So the depletion region is not uniform $V_{DS} < V_{P0}$. But extends more towards drain than source. V_{P0} is the pinch off voltage.

As V_{DS} is increased, depletion region is increased (shaded portion). So channel width decreases and channel resistance increases. Therefore, the rate at which I_D increases reduces, eventhough there is positive potential for the electrons at the drain and hence they tend to move towards the drain and conventional current flows as shown by the arrow mark, in the Fig. 8.7.

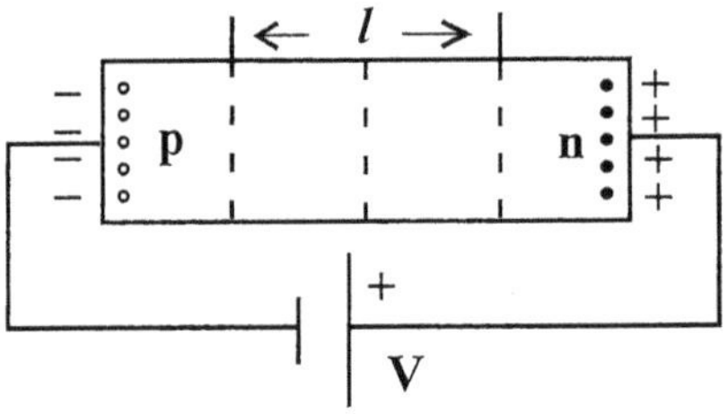

***Fig. 8.6** Reverse biased G - D junctions*

When $V_{DS} = V_{PO}$

When V_{DS} is further increased, the depletion region on each side of the channel join together as shown in Fig.8.8. ***The corresponding V_{DS} is called as V_{PD}, the pinch off voltage,*** because it pinches off the channel connection between drain and source.

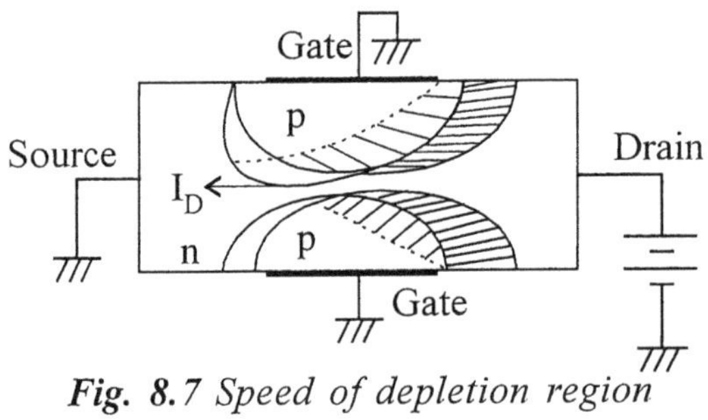

***Fig. 8.7** Speed of depletion region*

When $V_{DS} > V_{P0}$

If V_{DS} is further increased, the depletion region thickens. So the resistivity of the channel increases. Because drain is at more positive potential, more electrons tend to move towards the drain. Hence I_D should increase. But, because channel resistance increases, I_D decreases. Therefore, the net result is I_D levels off, for any V_{DS} above V_{P0}.

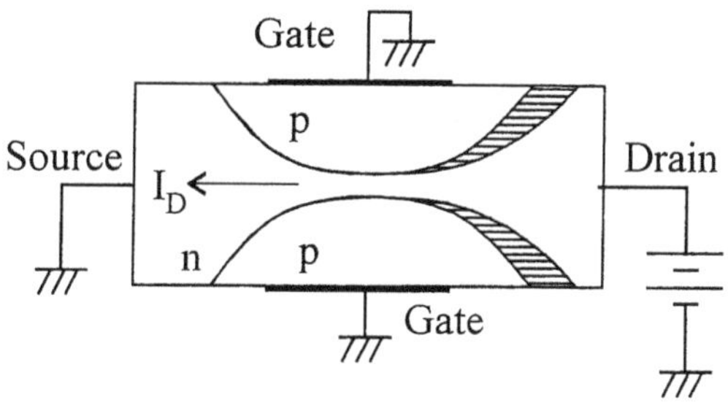

Fig. 8.8

$$I_D = \frac{V_{DS}}{r_{DS}}$$

In the linear region, r_{DS} is almost constant. So as V_{DS} is increased, I_D increases. When the two channels meet, as V_{DS} increases r_{DS} also increases. So I_D remains constant.(See Fig. 8.9)

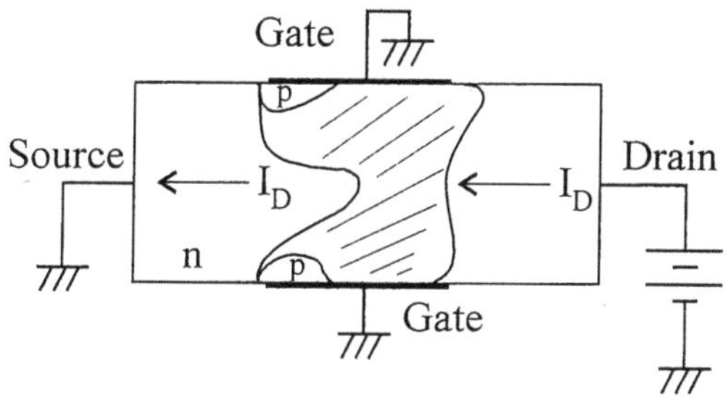

Fig. 8.9 Depletion region width variation in JFET.

8.4 SYMBOL

8.4.1 n-channel JFET

The arrow mark at the gate indicates the direction of current if the Gate source junction is forward biased.

On both sides of the n - type bar heavily doped (p^+) region of acceptor impurities have been formed by alloying on diffusion. These regions are called as Gates. Between the gate and source a voltage V_{GS} is applied in the direction to reverse bias the p-n junction.

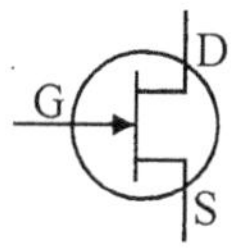

Fig. 8.10 (n-channel JFET).

8.4.2 p-channel JFET

The symbol for p-channel FET is given in Fig. 8.11. The semiconductor bar is p-type or source is p-type. Gate is heavily doped n-region. If G - S junction is forward biased, electrons from gate will travel towards source. Hence, the conventional current flows outside. So, the arrow mark is as shown in Fig. 8.11. indicating the direction of current, if G-S junction is forward biased.

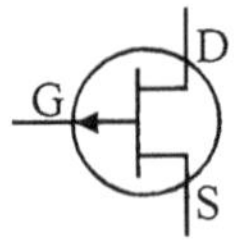

Fig. 8.11 (p-channel JFET).

The FET has three terminals, Source, drain and Gate.

Source

The source S is the terminal through which the majority carriers enter the bar (Fig. 8.12). Conventional current entering the bar at S is denoted as I_S.

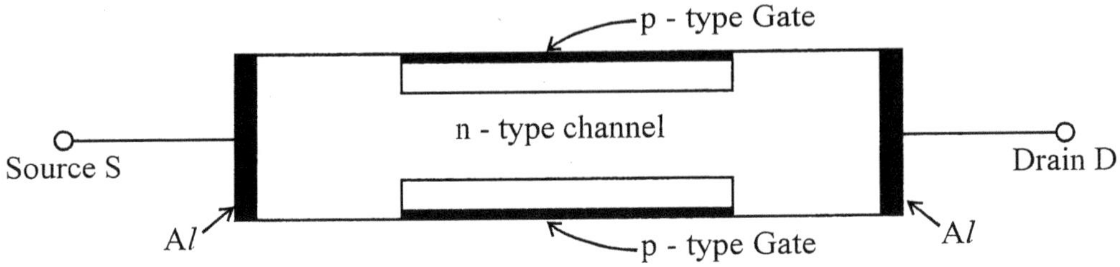

Fig. 8.12 *Structure of JFET.*

Drain

The drain D is the terminal through which the majority carriers leave the bar. Current entering the bar at D is designated by I_D. V_{DS} is positive if D is more positive than S, source is forward biased.

Gate

In the case of n - channel FET, on both sides of the bar heavily doped (p+) region of acceptor impurities have been formed by diffusion or other techniques. The impurity regions form as Gate. Both these region (p+ regions) are joined and a lead is taken out, which is called as the gate lead of the FET. Between the gate and source, a voltage V_{GS} is applied so as to reverse bias the gate source p-n junction. *Because of this reason JFET has high input impedance. In BJT Emitter Base junction is forward biased. So it has less Z.* Current entering the bar at G is designated as I_G.

Channel

The region between the two gate regions is the channel through which the majority carriers move from source to drain. By controlling the reverse bias voltage applied to the gate source junction the channel width and hence the current can be controlled.

8.5 PINCH OFF REGION

The voltage V_{DS} at which, the drain current I_D tends to level off is called the ***pinch off voltage***. When the value of V_{DS} is large, the electric field that appears along the x-axis, i.e., along the channel e_x will also be more. When the value of I_D is more, the drain end of the gate i.e., the gate region near the drain is more reverse biased than the source end. Because, the drain is at reverse potential, for n-channel FET, source is n-type, drain is n-type and positive voltage is applied for the drain, gate is also reverse biased. Therefore, the drop across the channel adds to the reverse bias voltage of the gate. Therefore, channel narrows more near the drain region and less near the source region. So, the boundaries of the depletion region are not parallel when V_{DS} is large and hence e_x is large.

As V_{DS} increases, e_x and I_D increase, whereas the channel width $b(x)$ narrows ($\because$ depletion region increases). Therefore the current density $J = I_D / 2b(x)\, w$ increases. If complete pinch off were to take place b = 0. But this cannot happen, because if b were to be 0, $J \to \infty$ which cannot physically happen.

Mobility m is a function of electric field intensity m remains constant for low electric fields when $\varepsilon_x < 10^3$V/cm.

$$\mu \,\alpha\, \frac{1}{\sqrt{\varepsilon_x}} \text{ when } \varepsilon_x \text{ is } 10^3 \text{ to } 10^4 \text{ V/cm.(for moderate fields).}$$

$$\mu \,\alpha\, \frac{1}{\varepsilon_x} \text{ for very high field strength} > 10^4 \text{ V/cm.}$$

$$I_D = 2bwe\, N_D\, \mu_n\, \varepsilon_x.$$

As V_{DS} increases, e_x increases but m_n decreases, b almost remains constant. Therefore I_D remains constant in the pinch off region.

Pinch off voltage (V_{P0} or V_P) is the voltage at which the drain current I_D levels off.

When $V_{GS} = 0$V, if $V_{GS} = -1$,V, the voltage at which the current I_D levels off decreases, (This is not pinch off voltage). If V_{GS} is large – 5, V_{DS} at which the current levels off may be zero. So the relationship between V_{DS}, V_{PD} and V_{GS} is

$V_{DS} = V_{P0} + V_{GS}$ V_{DS} is positive n-channel FET.

V_{P0} = P indicates pinch off voltage, zero indicate the voltage when V_{GS} = 0V.

V_{DS} is the voltage at which the current I_D levels off.(See Fig. 8.13).

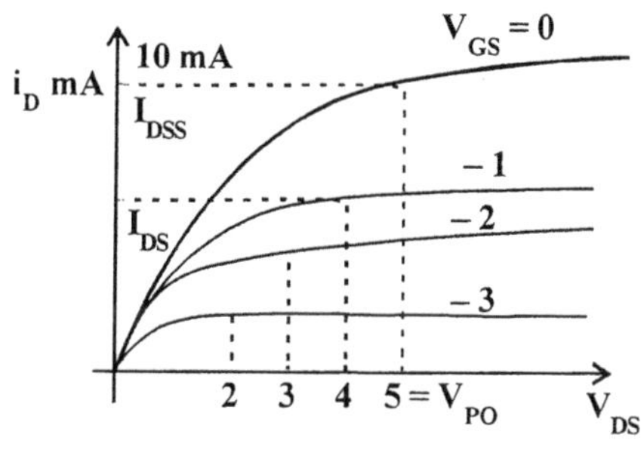

Fig. 8.13 JFET drain characteristics.

8.6 JFET VOLT-AMPERE CHARACTERISTICS

Suppose the applied voltage V_{DS} is small. The resulting small drain current I_D will not have appreciable effect on the channel profile. Therefore, the channel cross-section A can be assumed to be constant throughout, its length.

$$\therefore \quad A = 2b.w,$$

Where 2b is the channel width corresponding to negligible drain current and w is channel dimension perpendicular to the 'b' direction.

$$\therefore \quad I_D = Ae \cdot N_D \, \mu_n \in$$

$$A = 2b \cdot w$$

$$E = \frac{V_{DS}}{L}$$

$$= 2bwe \cdot N_D \, \mu_n \cdot \frac{V_{DS}}{L} \qquad .....(8.6.1)$$

Where L is the length of the channel. Eliminating 'b' which is unknown,

$$V_{GS} = \left(1-\frac{b}{a}\right)^2 \cdot V_P$$

$$\left(1-\frac{b}{a}\right) = \left(\frac{V_{GS}}{V_P}\right)^{1/2}$$

$$\frac{b}{a} = 1 - \left(\frac{V_{GS}}{V_P}\right)^{1/2}$$

$$b = a\left[1-\left(\frac{V_{GS}}{V_P}\right)^{1/2}\right]$$

Substituting, this value in equation (1),

$$\boxed{I_D = \frac{2awe\,N_D\,\mu_n}{L}\left[1-\left(\frac{V_{GS}}{V_P}\right)^{1/2}\right] V_{DS}}$$

This is the expression for I_D in terms of V_{GS}, V_P and V_{DS} because the value of b is not directly known.

$$V_P = \frac{e.N_D}{2\varepsilon} \cdot w^2$$

But, $w = a - b\ (x)$

$b = 0$, at pinch off,

$\therefore$ $w = a =$ The spacing between the two gate dopings.

$$\therefore \quad \boxed{|V_P| = \frac{e.N_D}{2\in} \cdot a^2}$$

V_{DS} controls the width of the depletion region (a – b) because for a given V_{GS}, as V_{DS} increases, the reverse potential between drain and gate increase. Therefore, depletion region width increases.

8.6.1 Expression for V_{Gs}

$$V = \frac{e.N_D x^2}{2\in}$$

We can get the expression for V_{GS} if we replace x by (a - b) and V by V_{GS}.

$$\therefore \quad V_{GS} = \frac{e.N_D(a-b)^2}{2\in}$$

But $$|V_P| = \frac{e.N_D}{2\in}.a^2$$

$$\therefore \quad \frac{e.N_D}{2\in} = \frac{|V_P|}{a^2}$$

Substituting this in expression for V_{Gs},

$$V_{GS} = \frac{V_P}{a^2}(a-b)^2$$

Channel width is controlled by V_{GS}. As V_{GS} increases, channel width reduces.

Therefore, channel width is controlled by V_{GS} and depletion region width by V_{DS}.

$$\boxed{V_{GS} = \left(1 - \frac{b}{a}\right)^2 .V_p}$$

8.6.2 The ON resistance r_{ds} (ON)

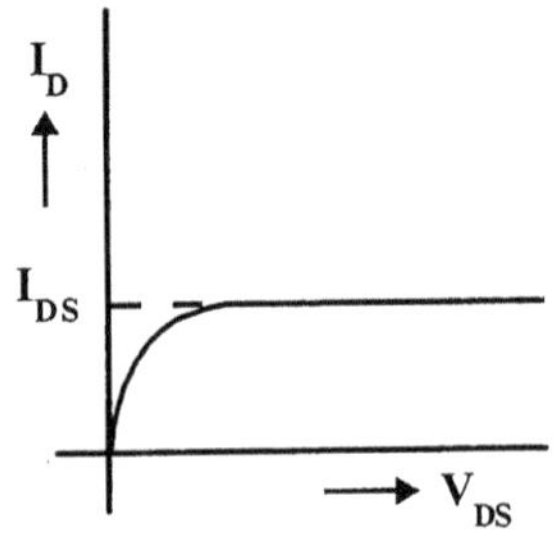

Fig. 8.14 Drain characteristics

When V_{DS} is small, the FET behaves like an ohmic resistance whose value is determined by V_{GS}.

The ratio $$\frac{V_{DS}}{I_D} = r_{ds(ON)}$$

$$I_D = 2bwe\, N_D\, \mu_n \cdot \frac{V_{DS}}{L}$$

$$\therefore \quad \frac{V_{DS}}{I_D} = \frac{L}{2a\,we\, N_D\, \mu n} = r_{ds(ON)}.$$

8.7 TRANSFER CHARACTERISTICS OF FET

I_{DS} = The saturation value of the drain current or the value at which I_D remains constant.

I_{DSS} = The saturation value of the drain current when gate is shorted to source that is $V_{GS} = 0$.

It is found that the transfer characteristics giving the relationship between I_{DS} and V_{GS} can be approximated by parabola.

$$I_{DS} = I_{DSS}\left(1 - \frac{V_{GS}}{V_p}\right)^2$$

Transfer characteristic follows the equation, I_D vs V_{GS}

$$I_D = (I_{DSS} + I_{GSS})\left(1 - \frac{V_{GS}}{V_p}\right)^2 - I_{GSS}$$

With negligible I_{GSS}, because I_{GS} is the leakage gate current between G and S when D is shorted, to S will be of the order of nano amperes, so it can be neglected.

$$\therefore \quad I_D = I_{DSS}\left(1 - \frac{V_{GS}}{V_p}\right)^2$$

This V_p is $V_{GS(off)}$ and not V_{DS} because when $V_p = V_{GS(off)}, I_D = 0$.

I_{DSS} = The saturation value of drain current when gate is shorted to source or $V_{GS} = 0$.

The transfer characteristics can be derived from the drain characteristics.

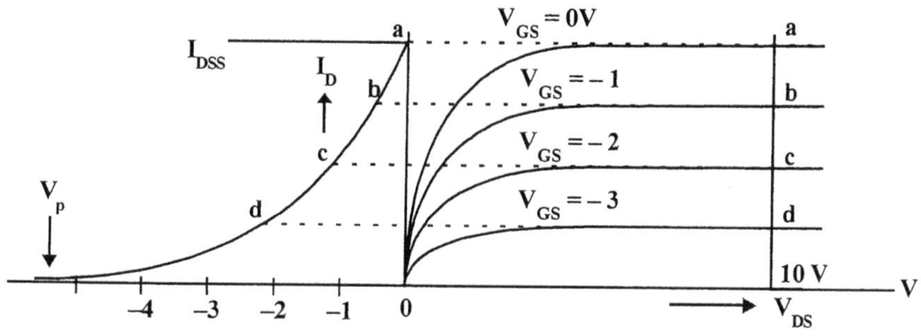

Fig. 8.15 Drain and Gate characteristics of JFET

8.8 COMPARISON BETWEEN BJT AND FET

The following table lists out the comparison of FET and BJT on different aspects.

BJT	FET
1. In BJT current flow is due to both majority and minority carriers.	1. Current flow is due to only majority carriers.
2. BJT has two junctions (E-B & B-C).	2. It has only one junction.
3. BJT is more noisy.	3. FET is less noisy.
4. BJT has less input impedance	4. FET offers high input impedance.
5. BJT has loading effect (as its input impedance is less).	5. FET has no loading effect.
6. It has poor thermal stability	6. FET has good thermal stability.
7. BJT is less immune to radiation	7. FET is more immune to radiation.
8. Difficult to fabricate as compared to FET.	8. Easy (or) simpler to fabricate as compared to BJT.
9. BJT occupies more space in the IC form. Hence its packing density is less.	9. FET occupies less space in IC form. Hence it offers high packing density.
10. It has more gain-band width product.	10. FET has less gain-bandwidth product.
11. There is no such facility available with BJT.	11. Conductivity of FET can varied by varying gate bias.
12. Reverse saturation current is more in BJT.	12. Reverse saturation current is less in FET and hence it can be operated over wide ranges of temperature.

8.9 SMALL SIGNAL MODEL

The expression for drain current I_d in the case of a FET is

$$I_d = g_m V_{gs} + \frac{1}{r_d} . V_{ds}$$

This is similar to the expression I_p of a triode. So the equivalent circuit can be drawn as a current source in parallel with a resistance. Between gate and source the capacitance is C_{gs} and C_{gd} is the barrier capacitance between gate and drain. C_{ds} represents the drain to source capacitance of the channel. (See Fig. 8.16).

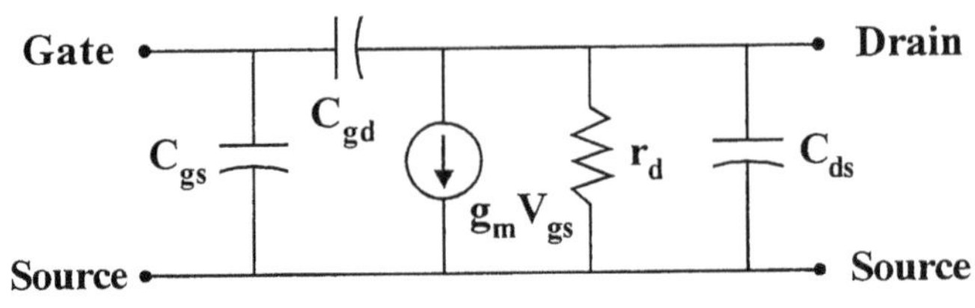

Fig. 8.16 JFET equivalent circuit

This is high frequency equivalent circuit because we are considering capacitances. In low frequency circuit, X_C will be very large so C is open circuit and neglected.

FET is a voltage controlled device; by controlling the channel width by voltage, we are controlling the current. If we neglect the capacitances the equivalent circuit of FET can be drawn as given in Fig. 8.17. This is similar of a transfer ***h-parameter*** equation circuit.

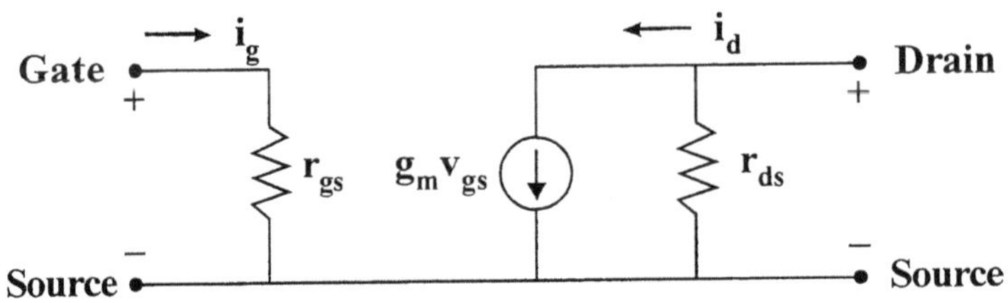

Fig. 8.17 Simplified equivalent circuit

Arrow mark is downwards in the current source $g_m V_{gs}$ because the current (conventional current) is flowing from drain to source for n-channel FET (Because electrons are the majority carriers).

8.10 FET BIASING FOR ZERO DRIFT CURRENT

As Temperature (T) increases, Mobility (μ) dercreases at constant Electric Field (ε), therefore I_D decreases. As T increases depletion region width decreases. So conductivity σ of the channel increases, I_D also increases.

I_D decreases by 0.7 % per degree centigrade.

I_D increase at a rate of 2.2 mV/^{0}C change in V_{GS}. Therefore change in V_{GS} = 2.2 mV/^{0}C.

$\therefore$ For zero drift current,

$$0.007\,|I_D| = 0.0022\,g_m$$

or $$\frac{|I_D|}{g_m} = 0.314\text{ V}$$

or $$\boxed{|V_p| - |V_{GS}| = 0.63\text{V}}$$

$$I_{DS} = I_{DSS}\left(1 - \frac{V_{GS}}{V_P}\right)^2$$

$$g_m = g_{mo}\left(1 - \frac{V_{GS}}{V_p}\right)$$

$$g_{mo} = \frac{-2I_{DSS}}{V_p}$$

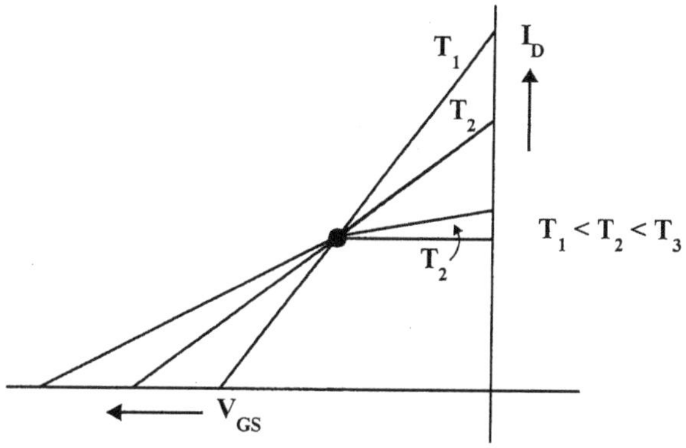

Fig. 8.18 Biasing for zero drift current

Example 8.1:

Design the bias circuit for zero drain current drift, for a FET having the following particulars :

$$V_p = -3\text{V}\ ;\ g_{mo} = 1.8\text{ mA/V}$$
$$I_{DSS} = 1.75\text{ mA, if } R_d = 5\text{k}\Omega$$

(a) Find I_D for zero drift current.

(b) V_{GS}

(c) R_S

(d) Voltage gain with R_s bypassed with a large capacitance,

Solution:

For zero drift current,

$$|V_P| - |V_{GS}| = 0.63$$
$$V_p = -3\text{V}$$
$$V_{GS} = V_P - 0.63$$

(a) $$I_D = I_{DSS}\left(1 - \frac{V_{GS}}{V_p}\right)^2 \simeq 0.08\text{ mA}$$

(b) $\therefore \quad V_{GS} = -3 - 0.63 = \mathbf{-3.63\ V}$

(c) $$R_S = \frac{-V_{GS}}{I_D} = \frac{3.63}{I_D} = \frac{3.63}{0.08\ \text{mA}} = 43\ \text{K}\Omega$$

(d) $$g_m = g_{mo}\left|\frac{V_{GS} - V_p}{V_p}\right| = 0.31\ \text{mA/V}$$

$$A_v = \mathbf{g_m \cdot R_d} = 12$$

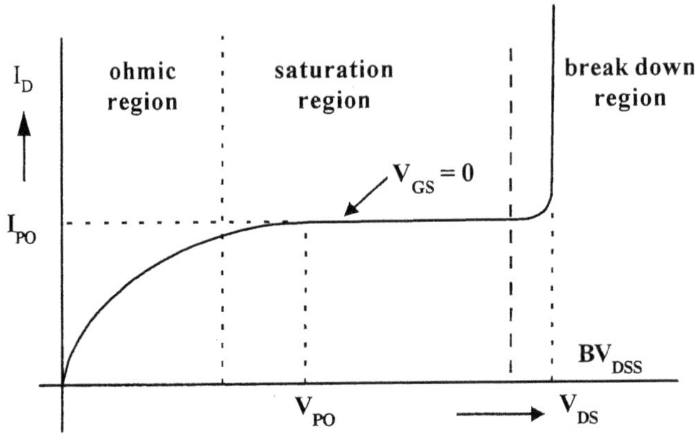

Fig. 8.19 *Drain characteristics*

The drain current I_D increases rapidly as V_{DS} increases, towards V_{PO}. Above V_{PO}. The current tends to level off at I_{PO} and then rises slowly. When V_{GS} = Breakdown voltage BV_{DSS}, breakdown (avalanche) occurs and the current rises rapidly. Current levels off because as V increases, resistance of the channel also rises. Therefore, I remains constant. As V increases, R increases, therefore V/R = constant. The region before V_{PO} is called ***ohmic region*** because ohms law is obeyed (see Fig.8.19). Just as a p-n junction breakdown when reverse voltage is very high FET also break down if V_{DS} is high due to high electric field.

Suppose V_{DS} is fixed and V_{GS} is varied. As V_{GS} is made negative, p-n junction is reverse biased and depletion region between gate and source increases. This decreases channel width and increases channel resistance. Hence I_D decreases. If gate voltage is made positive, depletion region decreases and hence I_D increases. The p-n junction, between gate and source becomes forward biased and current flows from gate to source. So n-type JFET is usually operated so that $V_{GS} = 0$ or negative.

V_{P0}	:	Pinch off voltage when $V_{GS} = 0$. 0 indicate $V_{GS} = 0$.
$I_{DSS}\ (I_{P0})$	:	Saturation value of the drain current when gate is shorted to source ($V_{GS} = 0$).
I_{DSS}	:	Saturation value of the drain current for $V_{GS} \neq 0. -1, -2$ etc.
BV_{DSS}	:	Breakdown voltage V_{DS}, when gate is shorted to source.
$I_{D(off)}$	:	Drain current when the JFET is in OFF state.
IGSS	:	Gate cut-off current when Drain is shorted to source.
g_{mo}	:	Mutual conductance g_m when $V_{GS} = 0$.

The expression for the pinch off voltage $|V_P| = \dfrac{\in . N_D}{2\varepsilon} . a^2$

where N_D = Donor atom concentration

$\in$ = Permittivity $\in_o - \in_r$

a = Effective width of Silicon bar

Example 8.2:

For a p - channel Silicon FET, with $a = 2 \times 10^{-4}$ cm and channel resistivity r = 10 W - cm. Find the pinch off voltage.

Solution:

$$V_P = \frac{e. N_A}{2 \in} . a^2 \quad \text{(for p - channel FET)}$$

For Si, $\in = 12\, e_o\ \mu_p = 500\ \text{cm}^2/\text{v} - \text{sec}$.

$$\sigma = \frac{1}{\rho} \simeq p\mu_p . e \simeq N_A\, \mu_p . e$$

or
$$e . N_A = \frac{1}{\rho \mu_p}$$

$$= \frac{1}{10 \times 500}$$

or
$$e . N_A = 2 \times 10^{-4}$$

$$\therefore \quad V_p = \frac{2 \times 10^{-4} \left(2 \times 10^{-4}\right)^2}{2 \times 12 \left(36\pi \times 10^{11}\right)^{-1}} = 3.78\text{V}$$

SPECIAL PURPOSE DEVICES

8.11 BREAK DOWN MECHANISMS IN SEMICONDUCTOR DIODES

There are three types of breakdown mechanisms in semiconductor devices.

1. Avalanche Breakdown **2.** Zener Breakdown **3.** Thermal Breakdown

8.11.1 Avalanche Breakdown

When there is no bias applied to the diode, there are certain number of thermally generated carriers. When bias is applied, electrons and holes acquire sufficient energy from the applied potential to produce new carriers by removing valence electrons from their bonds. These thermally generated carriers acquire additional energy from the applied bias. They strike the lattice and impart some energy to the valence electrons. So the valence electrons will break away from their parent atom and become free carriers. These newly generated additional carriers acquire more energy from the potential (since bias is applied). So they again strike the lattice and create more number of free electrons and holes. This process goes on as long as bias is increased and the number of free carriers gets multiplied. This is known as ***avalanche multiplication***, Since the number of carriers is large, the current flowing through the diode which is proportional to free carriers also increases and when this current is large, avalanche breakdown will occur.

8.11.2 Zener Breakdown

Now if the electric field is very strong to disrupt or break the covalent bonds, there will be sudden increase in the number of free carriers and hence large current and consequent breakdown. Even if thermally generated carriers do not have sufficient energy to break the covalent bonds, the electric field is very high, then covalent bonds are directly broken. This is ***Zener Breakdown***. A junction having narrow depletion layer and hence high field intensity will have zener breakdown effect. ($\cong 10^6$ V/m). If the doping concentration is high, the depletion region is narrow and will have high field intensity, to cause Zener breakdown.

8.11.3 Thermal Breakdown

If a diode is biased and the bias voltage is well within the breakdown voltage at room temperature, there will be certain amount of current which is less than the breakdown current. Now keeping the bias voltage as it is, if the temperature is increased, due to the thermal energy, more number of carriers will be produced and finally breakdown will occur. This is ***Thermal Breakdown***.

In zener breakdown, the covalent bonds are ruptured. But the covalent bonds of all the atoms will not be ruptured. Only those atoms, which have weak covalent bonds such as an atom at the surface which is not surrounded on all sides by atoms will be broken. But if the field strength is not greater than the critical field, when the applied voltage is removed, normal covalent bond structure will be more or less restored. This is Avalanche Breakdown. But if the field strength is very high, so that the covalent bonds of all the atoms are broken, then normal structure will not be achieved, and there will be large number of free electrons. This is ***Zener Breakdown***.

In Avalanche Breakdown, only the excess electron, loosely bound to the parent atom will become free electron because of the transfer of energy from the electrons possessing higher energy.

8.12 USE OF ZENER DIODE AS REGULATOR

Voltage Regulator Circuits are electronic circuits which give constant DC output voltage, irrespective of variations in *Input Voltage* V_i, *current drawn by the load* I_L from output terminals, and *Temperature T*. Voltage Regulator circuits are available in discrete form using BJTs, Diodes etc and in IC (Integrated Circuit) form. The term voltage regulator is used when the output delivered is DC voltage. The input can be DC which is not constant and fluctuating. If the input is AC, it is converted to DC by Rectifier and Filter Circuits and given to I.C. Voltage Regulator circuit, to get constant DC output voltage. If the input is A.C 230 V from mains, and the output desired is constant DC, a stepdown transformer is used and then Rectifier and filter circuits are used, before the electronic regulator circuit. The block diagrams are shown in Fig. 8.20 and 8.21.

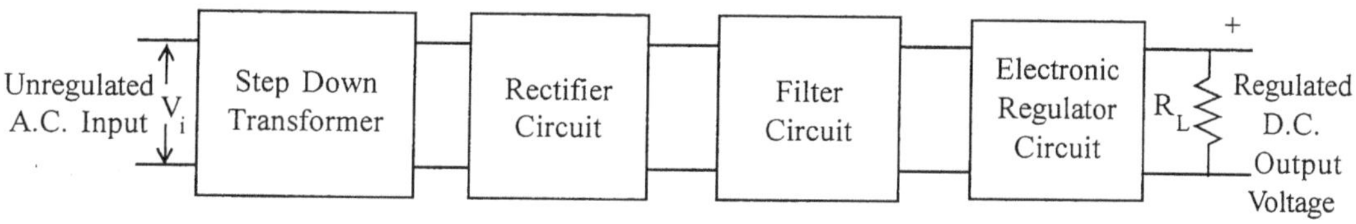

***Fig. 8.20** Block diagram of voltage regulator with A.C input*

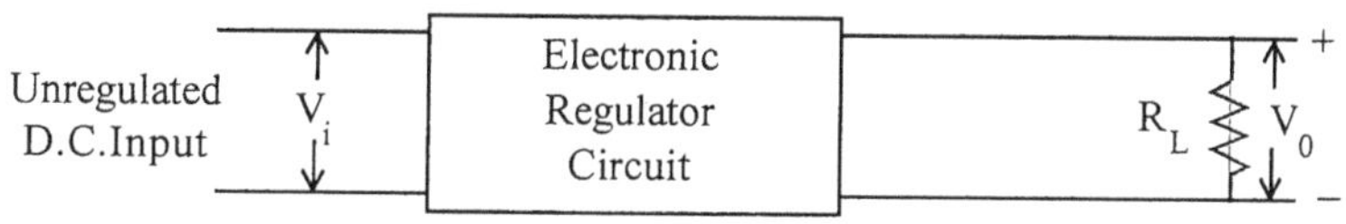

***Fig. 8.21** Block diagram of voltage regulator with D.C input*

The term *Voltage Stabilizer* is used, if the output voltage is AC and not DC. The circuits used for voltage stabilizers are different. The voltage regulator circuits are available in IC form also. Some of the commonly used ICs are, mA 723, LM 309, LM 105, CA 3085 A.

7805, 7806, 7808, 7812, 7815 : Three terminal positive voltage regulators.

7905, 7906, 7908, 7912, 7915 : Three terminal negative voltage regulators.

The Voltage Regulator Circuits are used for electronic systems, electronic circuits, IC circuits, etc.

The specifications and Ideal Values of Voltage Regulators are :

Specifications		Ideal Values
1. Regulation (S_V)	:	0 %
2. Input Resistance (R_i)	:	∞ ohms
3. Output Resistance (R_0)	:	0 ohms
4. Temperature Coefficient (S_T)	:	0 mv/oc.
5. Output Voltage V_0	:	-
6. Output current range (I_L)	:	-
7. Ripple Rejection	:	0 %

Different types of Voltage Regulators are

1. Zener regulator
2. Shunt regulator
3. Series regulator
4. Negative voltage regulator
5. Voltage regulator with foldback current limiting
6. Switching regulators
7. High Current regulator

8.12.1 Zener Voltage Regulator Circuit

A simple circuit without using any transistor is with a zener diode voltage regulator circuit. In the reverse characteristic voltage remains constant irrespective of the current that is flowing through zener diode. The voltage in the break down region remains constant (Fig. 8.22).

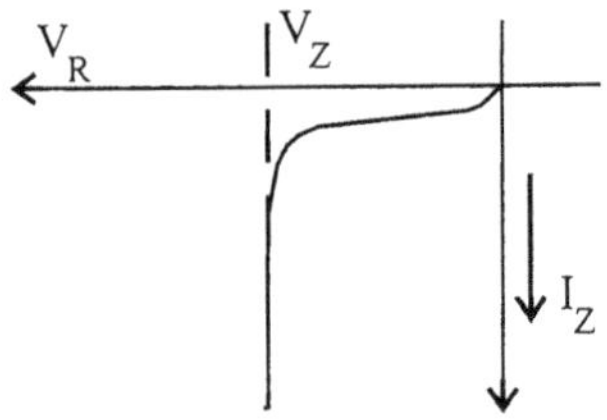

Fig. 8.22 *Zener diode reverse characteristic.*

Therefore in this region the zener diode can be used as a voltage regulator. If the output voltage is taken across the zener, even if the input voltage increases, the output voltage remains constant. The circuit as shown in Fig. 8.23.

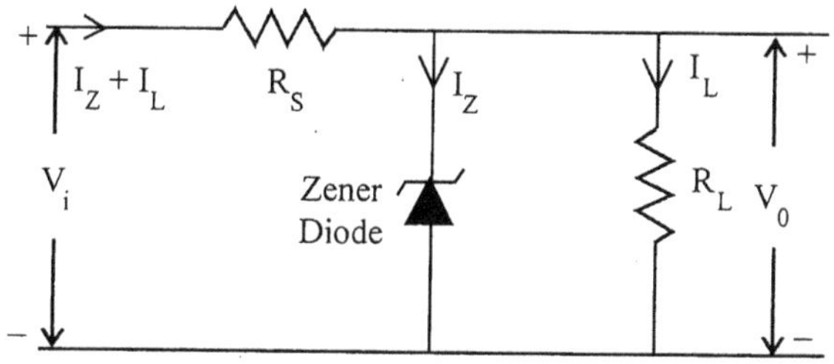

***Fig. 8.23** Zener regulator circuit*

The input V_i is DC. zener diode is reverse biased.

If the input voltage V_i increases, the current through R_s increases. This extra current flows, through the zener diode and not through R_L. Therefore zener diode resistance is much smaller than R_L when it is conducting. Therefore I_L remains constant and so V_0 remains constant.

The limitations of this circuits are

1. The output voltage remains constant only when the input voltage is sufficiently large so that the voltage across the zener is V_Z.
2. There is limit to the maximum current that we can pass through the zener. If V_i is increased enormously, I_Z increases and hence breakdown will occur.
3. Voltage regulation is maintained only between these limits, the minimum current and the maximum permissible current through the zener diode. Typical values are from 10 mA to 1 ampere.

8.12.2 Shunt Regulator

The shunt regulator uses a transistor to amplify the zener diode current and thus extending the Zener's current range by a factor equal to transistor h_{FE}. (Fig. 8.24)

Unregulated D.C.

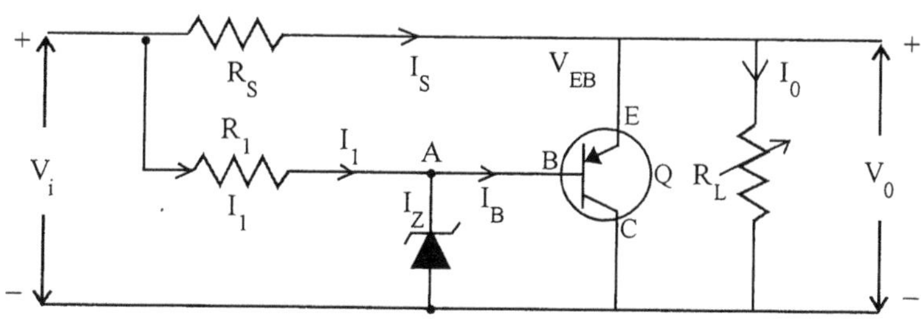

***Fig. 8.24** Shunt regulator circuit*

zener current, passes through R_1

nominal output voltage

$$= V_Z + V_{EB}$$

The current that gets branched as I_B is amplified by the transistor. Therefore the total current $I_0 = (b + 1)\ I_B$, flows through the load resistance R_L. Therefore for a small current through the zener, large current flows through R_L and voltage remains constant. In otherwords, for large current through R_L, V_0 remains constant. Voltage V_0 does not change with current.

Example 8.3:

For the shunt regulator shown, determine

1. The nominal voltage
2. Value of R_1,
3. Load current range
4. Maximum transistor power dissipation.
5. The value of R_S and its power dissipation.

V_i = Constant. Zener diode 6.3V, 200mW, requires 5mA minimum current.

Transistor Specifications :

$$V_{EB} = 0.2V,\ h_{FE} = 49,\ I_{CBO} = 0.$$

1. The nominal output voltage is the sum of the transistor V_{EB} and zener voltage.

$$V_0 = 0.2 + 6.3 = 6.5V = V_{Eb} + V_Z$$

2. R_1 must supply 5mA to the zener diode :

$$\therefore \qquad R_1 = \frac{8V - 6.3}{5 \times 10^{-3}} = \frac{1.7}{5 \times 10^{-3}} = 340\ \Omega$$

3. The maximum allowable zener current is

$$\frac{\text{Power rating}}{\text{Voltage rating}} = \frac{0.2}{6.3} = 31.8\ \text{mA}$$

The load current range is the difference between minimum and maximum current through the shunt path provided by the transistor. At junction A, we can write,

$$I_B = I_Z - I_1$$

I_1 is constant at 5 mA

$$\therefore \qquad I_B = I_Z - I_1$$

I_1 is constant at 5 mA

$$\therefore \qquad I_B = 5 \times 10^{-3} - 5 \times 10^{-3} = 0$$

$$I_B\ (\text{min}) = I_{Z\ Max} - I_2$$

$$= 31.8 \times 10^{-3} - 5 \times 10^{-3} = 26.8\ \text{mA}$$

The transistor emitter current $I_E = I_B + I_C$

$$I_C = \beta\ I_B = h_{FE}\ I_B$$

$\therefore \qquad I_E = (\beta + 1)\, I_B = (h_{FE} + 1)\, I_B$

I_B ranges from a minimum of 0 to maximum of 26.8 mA

$\therefore$ Total load current range is $(h_{FE} + 1)\, I_B$

$$= 50\,(26.8 \times 10^{-3}) = 1.34 \text{ A}$$

4. The maximum transistor power dissipation occurs when the current is maximum $I_E \simeq I_C$

$$P_D = V_o\, I_E = 6.5\,(1.34) = 8.7 \text{ W}$$

5. R_S must pass 1.34 A to supply current to the transistor and R_L.

$$R_S = \frac{V_i - V_0}{1.34} = \frac{8 - 6.5}{1.34} = 1.12\ \Omega$$

The power dissipated by R_S,

$$= I_S^2\, R_S$$

$$= (1.34)^2 \,.\, (1.12) = 3\text{W}$$

Regulated Power Supply

An unregulated power supply consists of a transformer, a rectifier, and a filter. For such a circuit regulation will be very poor i.e. as the load varies (*load means load current*) [*No load means no load current or 0 current. Full load means full load current or short circuit*], we want the output voltage to remain constant. But this will not be so for unregulated power supply. The short comings of the circuits are :

1. Poor regulation
2. DC output voltage varies directly as the A.C. input voltage varies
3. In simple rectifiers and filter circuits, the D.C. output voltage varies with temperature also, if semiconductors devices are used.

An electronic feedback control circuit is used in conjuction with an unregulated power supply to overcome the above three short comings. Such a system is called a "*regulated power supply*".

Stabilization

The output voltage depends upon the following factors in a power supply.

1. Input voltage V_i
2. Load current I_L
3. Temperature

$\therefore$ Change in the output voltage ΔV_0 can be expressed as

$$\Delta V_0 = \frac{\partial V_0}{\partial V_i} . \Delta V_i + \frac{\partial V_0}{\partial I_L} . \Delta I_L + \frac{\partial V_0}{\partial T} . \Delta T$$

$$\Delta V_0 = S_1\, \Delta V_i + R_0\, \Delta I_L + S_T\, \Delta T$$

Where the three coefficients are defined as,

(a) *Stability factors*

$$S_V = \frac{\Delta V_0}{\Delta V_i}\bigg|_{\Delta I_L = 0,\ \Delta_T = 0}$$

This should be as small as possible. Ideally 0 since V_0 should not change eve if V_i changes.

(b) ***Output Resistance***

$$R_0 = \frac{\Delta V_0}{\Delta I_L}\bigg|_{\Delta V_i = 0,\ \Delta_T = 0}$$

(c) ***Temperature Coefficient***

$$S_T = \frac{\Delta V_0}{\Delta_T}\bigg|_{\Delta V_i = 0,\ \Delta I_L = 0}$$

The smaller the values of the three coefficients, the better the circuit.

8.13 THE TUNNEL DIODE

In an ordinary p-n junction diode the doping concentration of impurity atoms is 1 in 10^8. With this doping, the depletion layer width, which constitutes barrier potential is 5mV. If the concentration of the impurity atoms is increased to say 1 in 10^3(This corresponds to impurity density of $\approx 10^{19}/m^3$), the characteristics of the diode will completely change. Such a diode is called ***Tunnel Diode***. This was found by ***Esaki*** in ***1958***.

8.13.1 Principle of Operation

Barrier Potential V_B :

$$V_B = \frac{e.N_A}{2\epsilon} \cdot W^2$$

or

$$W = \sqrt{\frac{2\epsilon.V_B}{e.N_A}}$$

where W = Width of Depletion Region in μ

N_A = Impurity Concentration N_O/m^3.

So, the width of the junction barrier varies inversely as the square root of impurity concentration. Therefore as N_A increases W decreases.

Therefore, in tunnel diodes, by increasing N_A, W can be reduced from 5m to 0.01m. According to classical mechanics, a particle must possess the potential which is at least equal to, or greater than, the barrier potential, to move from one side to the other. When the barrier width is so thin as 0.01m, according to Schrodinge equation, there is much probability that an electron will penetrate when a forward bias is applied to the diode, so that potential barrier decreases below E_0. The *n-side* levels must shift upward with respect to those on the *p-side*. So there are occupied states in the conduction band of the n material, which are at the same energy level as allowed empty states in the valence band of the *p-side*. Hence electrons will tunnel from the *n-side* to the *p-side* giving rise to forward current. As the forward bias is increased further, the number of electrons on *n-side* which occupy the same energy level as that ***vacant energy*** state existing on *p-side*, also increases. So more number of electrons tunnel through the barrier to empty states on the left side giving rise to peak current I_p. If still more forward bias is applied, the energy level of the electrons on the *n-side* increases, but the empty states existing on the *p-side*, reduces. So the tunneling current decreases. In addition to the ***Quantum Mechanical Tunneling Current***, there is regular *p-n junction* injection current also. The magnitude of this current is considerable only beyond a certain value of forward bias voltage.

Therefore, the current again starts beyond V_V. The graph is shown in Fig. 8.25.

I_p = Peak current

I_V = Valley current

V_F = Peak forward voltage

$V_p \approx$ 50 mV

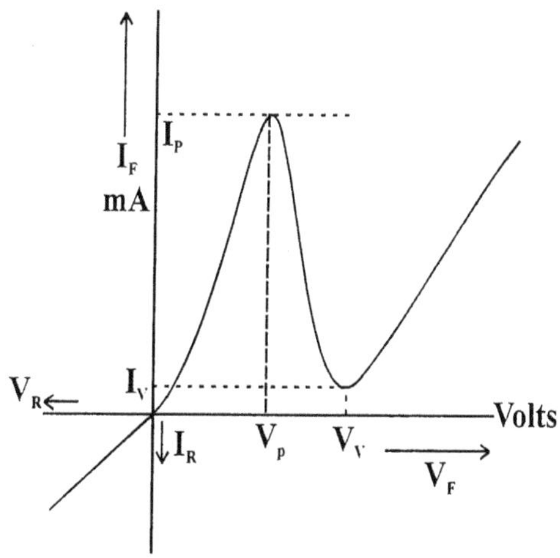

Fig. 8.25 V - I Characteristics of a tunnel diode.

8.13.2 Characteristics of a Tunnel Diode

For small forward voltage (ie., $V_p \approx$ 50mV for Ge), forward resistance is small $\approx 5\Omega$ and so current is large. At the ***Peak Current*** I_p, corresponding to voltage V_p, $\frac{dI}{dV}$ is zero. If V is beyond V_p, the current decreases. So the diode exhibits ***Negative Resistance***

Characteristics between I_p and I_v called the ***Valley Current***. The voltage at which the forward current again equals I_V is called as *peak* forward voltage V_V. Beyond this voltage, the current increases rapidly.

The symbol for tunnel diode is shown in Fig. 8.26(a). Typical values of a tunnel diode are :

$V_p = 50$ mV, $V_V = 350$ mV,

$V_F = 0.5$V, $\frac{I_P}{I_V} = 8$

$I_P = 10$ mA, Negative resistance $R_n = -30\Omega$.

Series Ohmic resistance $R_s = 1\Omega$. The series inductance L_s depends upon the lead length and the geometry of the diode package. $L_s \approx 5$ nH. Junction capacitance C= 20 pF. Its circuit equivalent is shown in Fig. 8.26(b).

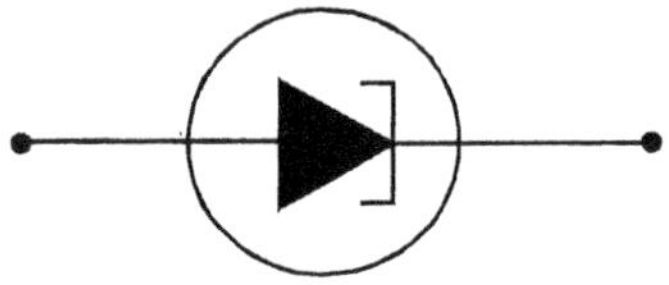

(a) *Symbol of tunnel Diode*

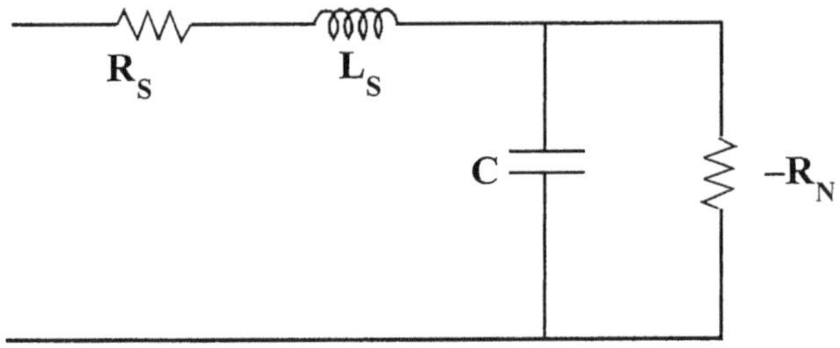

(b) *Equivalent circuit*

Fig. 8.26

Advantages :

1. Low Cost
2. Low Noise
3. Simplicity
4. High Speed
5. Low Input Power
6. Environmental Immunity.

Disadvantages :

1. Low output voltage swings, even for small voltage, while the current goes to large values. So the swing is limited.
2. Circuit design difficulty, since it is a two terminal device and there is no isolation between input and output.

Applications :

1. As a high frequency oscillator (GHz)

2. As a fast switching device. switching time is in nano-seconds. since tunneling is a quantum mechanical phenomenon, there is no time lag between the application of voltage and consequent current variation. So it can be used for high frequencies.

If the barrier width is $\approx$ 3A°, the probability that electrons will tunnel through the barrier is large. The barrier width will be $\approx$ 3A°, when impurity doping concentration is $\geq 10^{19}/cm^3$. If width $\approx 0.01\mu$ electrons will tunnel without the application of field. But if width $\approx$ 3A°, very small applied voltage is sufficient for the electron to tunnel.

The two conditions to be satisfied for tunneling phenomenon to take place are:

Necessary Condition :

1. The effective depletion region width near the junction must be small, of the order of 3A° by heavy doping.

Sufficient Condition :

2. There must be equivalent empty energy states on the *p-side* corresponding to energy levels of electrons on the *n-side*, for these electrons to tunnel from *n-side* to *p-side*.

When the tunnel diode is reverse biased, there will be some empty states on *n-side* corresponding to filled states on the *p-side*. So electrons will tunnel through the barrier from the p-side to n-side (since when the diode is reverse biased the energy levels on n side will decrease). As bias voltage is increased the number of electrons tunneling will increase, so forward current increases. When the diode is forward biased, electrons from *n-side* will increase, and tunnel through the barrier to *p-side*. This is called quantum mechanical current. Apart from this, normal p-n diode current will also be there. Beyond valley voltage V_V, it is normal diode forward current.

8.14 VARACTOR DIODE

Barrier of transition capacitance C_T varies with the value of reverse bias voltage. The larger the reverse voltage, the larger the W.

$$C_T = \frac{\in \times A}{W}$$

So C_T of a *p-n junction* diode varies with the applied reverse bias voltage.

Diodes made especially for that particular property of variable capacitance with bias are called ***Varactors***, ***Varicaps*** or ***Voltacaps***. These are used in LC Oscillator Circuits.

The Symbol is shown in Fig. 8.27

Varactor diodes are used in high frequency circuits.

Fig. 8.27 *Varactor diode*

In the case of abrupt junction,

$$C = \frac{\in A}{W}$$

But $W \alpha (V)^{1/2}$

$\therefore$ $C \alpha (V)^{-1/2}$

In the case of linearly graded junction,

$W \alpha (V)^{1/3}$

$\therefore$ $C \alpha (V)^{-1/3}$

The variation of impurity atom concentration with ω for different types of junctions is shown in Fig. 8.28.

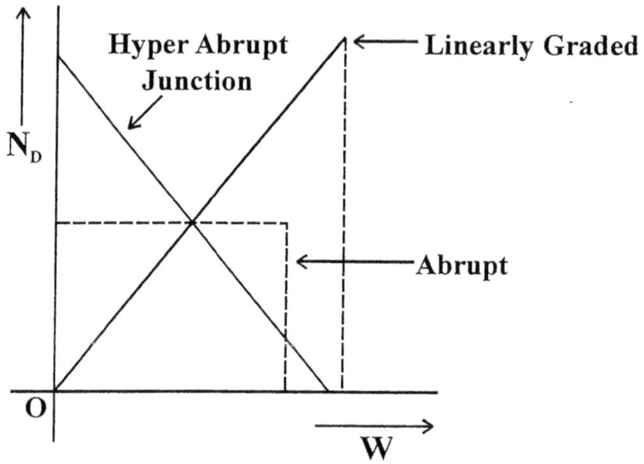

Fig. 8.28 *Impurity concentration variation in the junction region.*

But 'C' can be made proportional to V^2 by changing the doping profile. Such a junction is called ***Hyper Abrupt Junction***. This junction can be obtained by ***Epitaxial Process***. Therefore, C changes rapidly with voltage. Varactors are used in Reverse Bias condition, since C_T proportional to V and $C_D = \tau \times g$ No direct relation with V . So Varactors are used in Reverse Bias condition.

Capacitance, $$C = \frac{\in A}{W}$$

As the reverse bias voltage increases, width of the depletion region increases. So W increases $C \alpha \frac{1}{W}$. Hence C decreases. C is maximum under no reverse bias condition, since W is minimum C is minimum under maximum reverse bias, since W is maximum.

$$\text{Figure of merit} = \frac{C_{Max}}{C_{Min}}$$

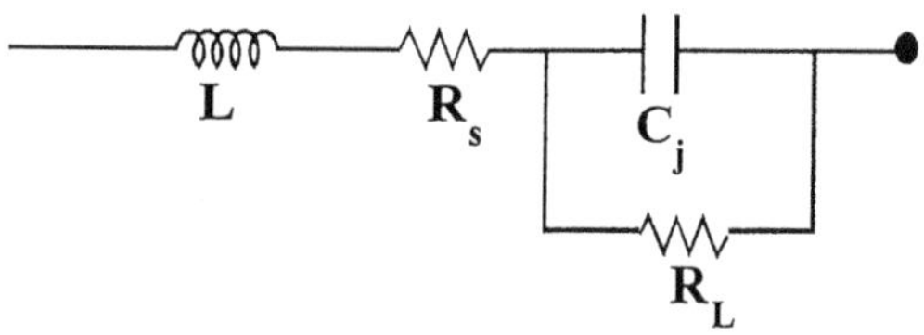

Fig. 8.29 Equivalent circuit of varactor diode

L : Lead inductance.

R_S : Series resistance of the diode due to the semiconductor material

C_j : Junction Capacitance

R_L : Leakage resistance of the diode since the capacitor can not be ideal one with out any leakage.

8.15 SILICON CONTROLLED RECTIFIER

Silicon controlled rectifier (SCR) is a 4 layer p-n-p-n device. In one type of construction, n-type material is diffused into a silicon pnp pellet to form alloy junction. From the diffused n region, the cathode connection is taken. From the p region on the other side anode is formed. Gate is taken from the p - region into which n - type impurity is diffused. So in the symbol for SCR, the gate is shown close to the cathode, projecting into it. This is known as planar construction.

8.15.1 Annular Type of Construction of SCR

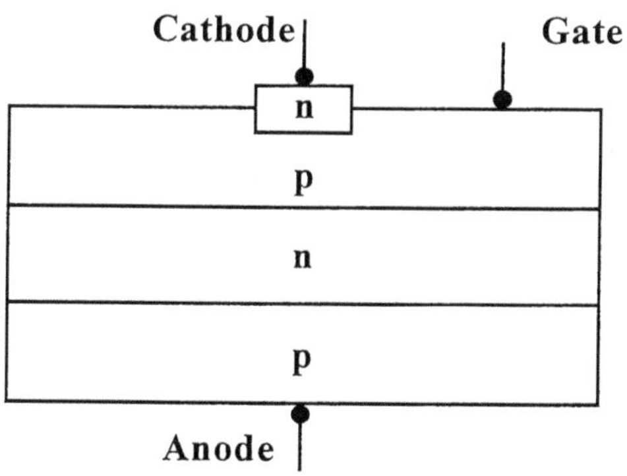

Fig. 8.30 SCR structure.

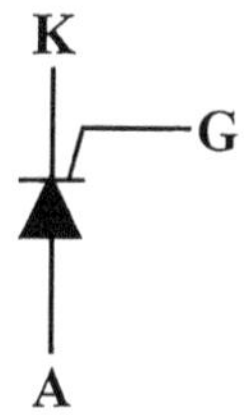

Fig. 8.31 SCR symbol.

In this type of construction gate is central to the device, with the cathode surrounding it. This is also called as shorted emitter construction. The advantages of this type construction are

(i) Fast turn on time

(ii) Improved high temperature capabilities.

8.15.2 Static Characteristics

The gate current I_G determines the necessary forward voltage to be applied between anode and cathode (Anode positive with reference to cathode) to cause conduction or to turn SCR 'ON'. So that current flows through the device.

As the gate current increases, the voltage to be applied between anode and cathode to turn the device ON decreases. If the gate is open, $I_G = 0$, the SCR is in the OFF state. This is also ***known as forward blocking region.*** The anode current that flows is only due to leakage. But if the voltage between anode and cathode is increased beyond a certain voltage, V_{BO} called as the forward breakover voltage, the SCR will turn ON and current will be limited by only the applied voltage V_{BO} called as the forward breakover voltage, the SCR will turn ON and current will be limited by only the applied voltage.

V_{BOO} : THE FORWARD BREAK OVER VOLTAGE

This is the voltage to be applied between anode and cathode to turn the SCR ON, when $I_G = 0$.

B.O : Breakover O : Zero gate current.

As the gate current is increased, the forward break over voltage decreases. If the value of I_G is very large $\approx$ 50 mA, the SCR will turn ON immediately when same voltage is applied between anode and cathode.

Usually, some voltage is applied between the anode and cathode of the SCR, and it is turned ON by a pulse of current in the gate.

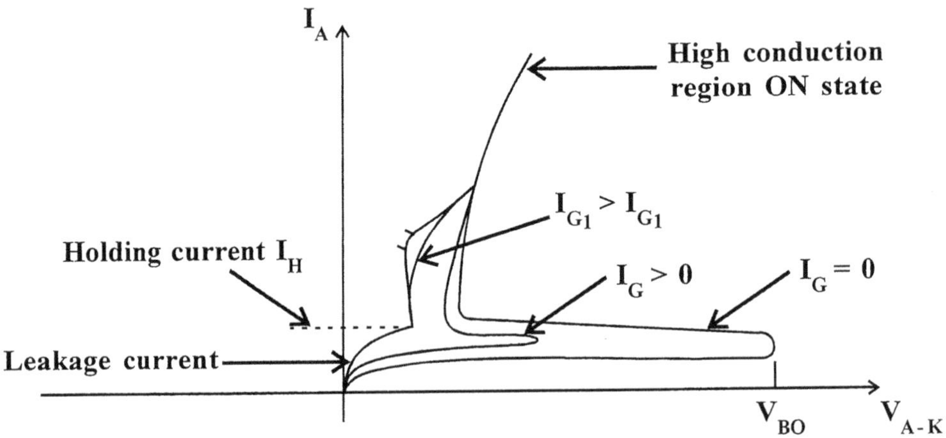

Fig. 8.32 SCR Characteristics

Holding Current I_H

Once SCR is ON, a minimum amount of current is required to flow to keep the device ON. If the current is lowered below I_H, by increasing external circuit resistance, the SCR will switch off.

Once the SCR is turned ON, the gate looses control over the device. The gate can not be used to switch off the device. If V_{A-K} is reduced to zero (as automatically happens in rectifying application) or if the current is reduced ***below*** I_H***, the device will turn off.***

I_H : Holding Current

Minimum Anode Current required to keep SCR in the conducting state. If I_A goes below I_H, SCR is turned off (To hold in conducting state).

Latching Current I_L

Minimum value of I_A to be reached to keep SCR in the ON state even after the removal of gate trigger signal. (To latch on to the conducting state.)

8.15.3 Analysis of the Operation of SCR

The SCR can be regarded as two transistor connected together.

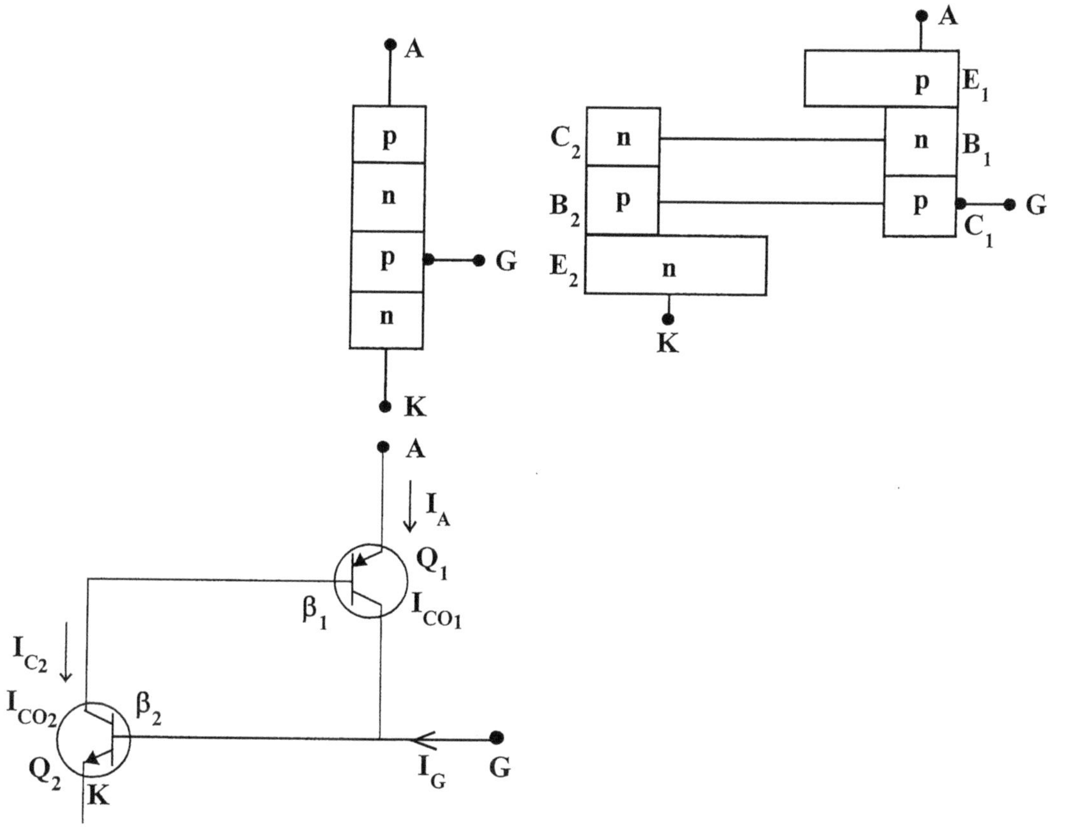

Fig. 8.33 Two transistor analogy of SCR.

The SCR can be regarded as two transistor p-n-p and n-p-n. The Base of p-n-p transistor is connected to the collector of the n-p-n transistor. The base of n-p-n transistor is connected to the collector of p-n-p transistor. The emitter of n-p-n transistor is the cathode. The emitter of p-n-p transistor is the anode.

Suppose a positive voltage is applied to the anode (p-type) since there is no base injection for both the transistors, the transistors are off. The only current is the leakage current. I_{CO1} and I_{CO2} of the two transistors. Therefore when SCR is not turned ON, the current that results is the leakage current ($I_{CO1} + I_{CO2}$).

In a transistor, $I_C = \beta I_B + (\beta + 1) I_{CO}$ ....(8.15.1)

$= \beta I_B + \beta I_{CO} + I_{CO}$ ....(8.15.2)

$= \beta (I_B + I_{CO}) + I_{CO}$ ....(8.15.3)

Consider Transistor Q_1

The base current for the transistor Q_1, I_{B1} is the collector current I_{C2} of Q_2.

$$I_C = \beta (I_B + I_{CO}) + I_{CO}$$

$$I_{B1} = I_{C2}$$

$$\therefore \quad I_{C1} = \beta_1 (I_{C2} + I_{CO1}) + I_{CO1} \quad \text{..........(8.15.4)}$$

Consider Transistor Q_2

The base current of transistor Q_2 is the collector current I_{C1} of Q_1.

$$I_{C2} = \beta_2 (I_{C1} + I_{CO2}) + I_{CO2} \quad \text{..........(8.15.5)}$$

Anode current is $I_{E2} = I_{B2} + I_{C2}$

But $I_{B2} = I_{C1}$

$\therefore$ $I_A = I_{C1} + I_{C2}$

I_A can be written in the forms,

$$\frac{(1+\beta_1)(1+\beta_2)(I_{CO1}+I_{CO2})}{1-\beta_1 . \beta_2}$$

The two transistor will have wide base regions. So β will be small. The value of β depends upon the value of I_E. When I_E is very small, $\beta \simeq 0$.

Therefore when $\beta_1 = \beta_2 = 0$.

$$I_A = \frac{(I_{CO1}+I_{CO2})(1+0)(1+0)}{1-0.0}$$

$$\simeq I_{CO1} + I_{CO2}$$

When a negative voltage is applied from gate to cathode, to inject holes into the base of Q_1, emitter base junction of Q_1 is forward biased. So I_{C1} increases. But this is the base current for Q_2. So transistor Q_2 is tuned ON. This increases collector and emitter currents. The I_{C2} of Q_2 now becomes trigger current and so increase I_{C1}. So the action is one of internal regeneration feed back , until both the transistor are driven into saturation. Once the SCR is turned ON, removal of gate current will not stop conduction because already the transistor Q_2 is turned ON and it provides the base injection for Q_1.

SCR can be represented as a pnp and npn transistors connected together, as shown in Fig. 8.34.

$$I_{B1} = I_{C2}$$
$$I_{C1} = I_{B2}$$
$$\text{Anode current} = I_A = I_{E1}$$

General expression for I_C in term of β, I_B and I_{CO} is

$$I_C = \beta I_B + (\beta + 1) I_{CO}$$

Since for transistor Q_1,

$$I_{C1} = \beta_1 I_{B1} + (\beta_1 + 1) I_{CO1} \qquad \text{..........(8.15.6)}$$

Since for transistor Q_1,

$$I_{C2} = \beta_2 I_{B2} + (\beta_2 + 1) I_{CO2}$$

But $\quad I_{B2} = I_{C1}$

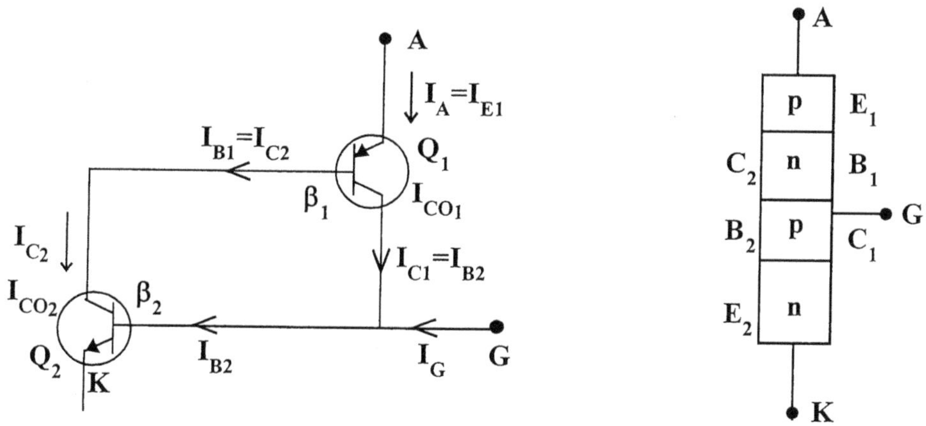

Fig. 8.34 Two transistor analogy of SCR.

$$\therefore \quad I_{C2} = \beta_2 I_{C1} + (\beta_2 + 1) I_{Co2} \qquad \text{..........(8.15.7)}$$

$$I_A = I_{E1} = I_{C1} + I_{B1} \qquad \text{But } I_{C1} \text{ is given by eq. (8.15.6)}$$
$$= \beta_1 \cdot I_{B1} + (\beta_1 + 1) I_{CO1} + I_{B1}$$

$$\therefore \quad I_{E1} = I_A = (\beta_1 + 1)(I_{B1} + I_{CO1}) \qquad \text{..........(8.15.8)}$$

Since we must get an expression for I_{B1}.

Substitute eq. (8.15.6) in (8.15.7).

$$\therefore \quad I_{C2} = \beta_2 [\beta_1 I_{B1} + (\beta_1 + 1) I_{CO1}] + (\beta_2 + 1) I_{CO2}$$

But $\quad I_{C2} = I_{B1}$

$$\therefore \quad I_{B1} = \beta_2 \beta_1 I_{B1} + \beta_2 (1 + \beta_1) I_{CO1} + (\beta_2 + 1) I_{CO2}$$

or $$(1-\beta_1\beta_2)\,I_{B1} = \beta_2(1+\beta_1)\,I_{CO1} + (1+\beta_2)\,I_{CO2}$$

or $$I_{B1} = \frac{\beta_2(1+\beta_1)I_{CO1} + (1+\beta_2)I_{CO2}}{1-\beta_1\beta_2} \qquad \text{..........(8.15.9)}$$

Substitute the value of I_{B1} from eq. (8.15.9) in (8.15.8).

$$\therefore \qquad I_{E1} = I_A = (1+\beta_1)\left[\frac{\beta_2(1+\beta_1)I_{CO1} + (1+\beta_2)I_{CO2}}{1-\beta_1\beta_2} + I_{CO1}\right]$$

$$= (1+\beta_1)\left\{\frac{\beta_2 I_{CO1} + \beta_1\beta_2 I_{CO1} + I_{CO2} + \beta_2 I_{CO2} + I_{CO1} - \beta_1\beta_2 I_{CO1}}{1-\beta_1\beta_2}\right\}$$

$$= (1+\beta_1)\left\{\frac{I_{CO1}(1+\beta_2) + I_{CO2}(1+\beta_2)}{1-\beta_1\beta_2}\right\}$$

$$I_A = \frac{(1+\beta_1)(1+\beta_2)(I_{CO1}+I_{CO2})}{1-\beta_1\beta_2}$$

8.15.4 Anode to Cathode Voltage - Current Characteristics

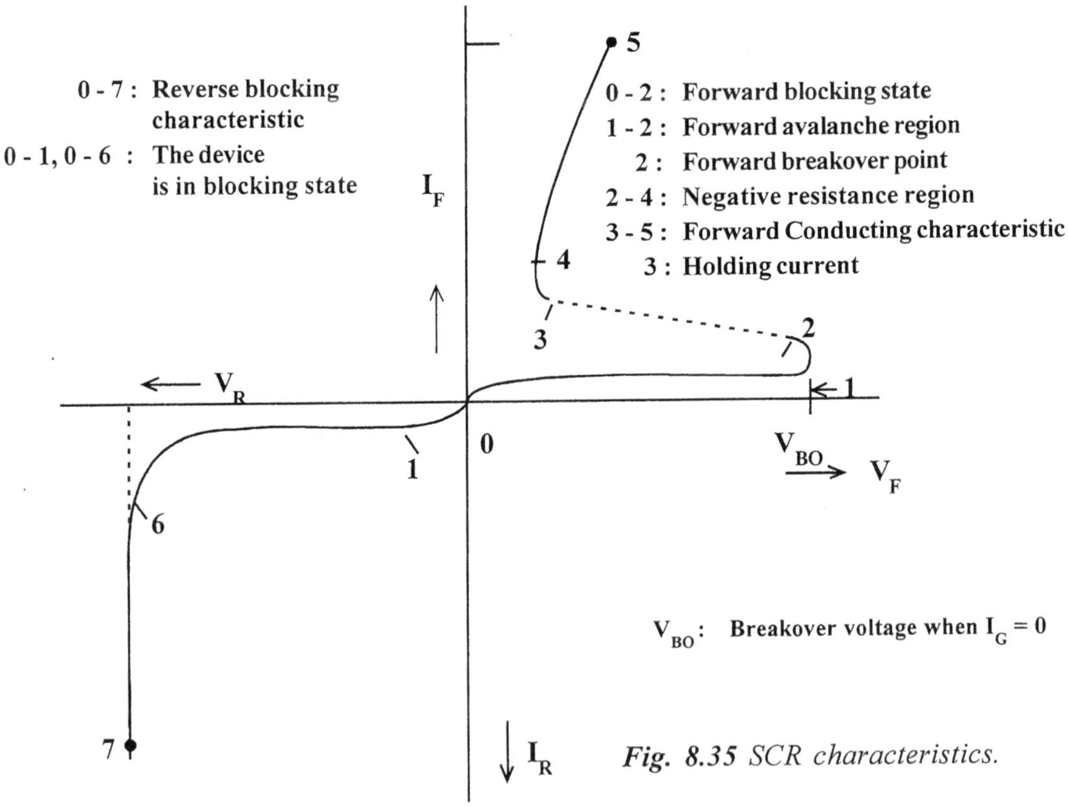

Fig. 8.35 SCR characteristics.

8.16 DIAC

Definition: The name DIAC indicates two (Di) – terminal parallel inverse combination of semiconductor layers that permits triggering in two directions.

Construction and Operation: The semiconductor layers of the diac can be basically arranged as shown in Fig. 8.36.

- Here it is important to remember that neither terminal is referred to as cathode.
- Here they are two anodes referred as anode 1 and anode 2.
- If anode 1 is positive with respect to anode 2 then the semiconductor layers forms p_1 n_2 p_2 and n_3 particularly.
- In similar way if anode 2 is positive with respect to anode 1 then the semiconductor layers forms p_2 n_2 p_1 and n_1.

8.16.1 Diagram for Basic Construction and Symbol

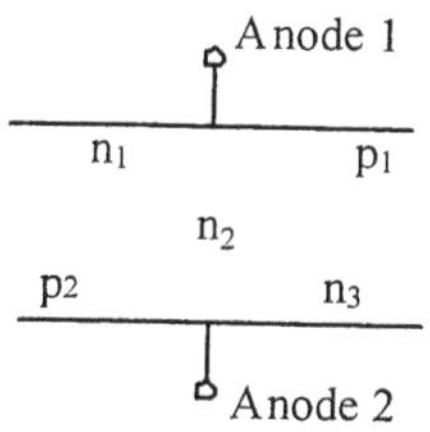

Fig. 8.36 Basic construction

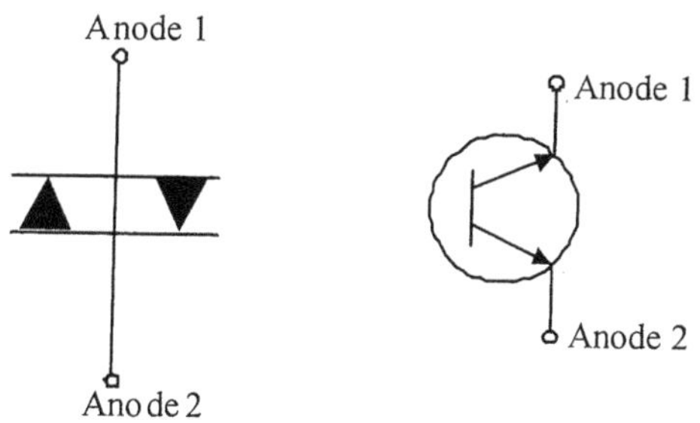

Fig. 8.37 Symbols

8.16.2 Characteristics of DIAC

The special characteristics of this device is it offers break over voltage in two directions this indicates possibility of an 'on' condition in either direction this is the main advantage in AC applications.

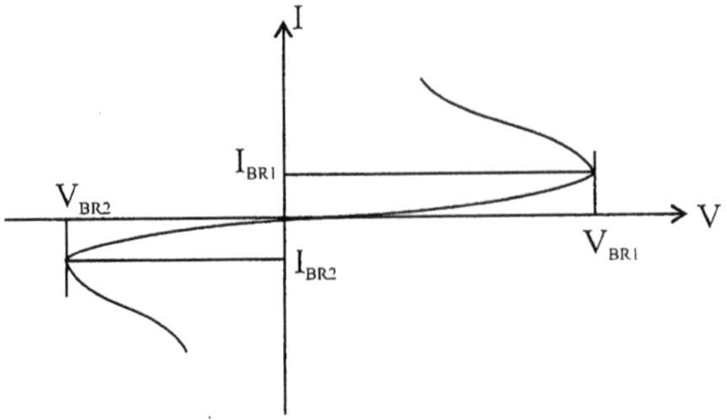

Fig. 8.38 Characteristics

Here the breakdown voltages are related by the following equation since here DIAC has two breakdown voltage in two directions which are mention in specification sheet is given by,

$$V_{BR_1} = V_{BR_2} \pm 0.1\, V_{BR_2} \qquad \text{....(8.16.1)}$$

- The breakdown voltages are very close in magnitude that means approximately equal in magnitude but varies from 28 V (min) to 42 V (max).
- Here the current levels $\left(I_{BR_1} \text{and } I_{BR_2}\right)$ are also approximately equal in magnitude for each device whose value is 200 μA.

8.17 TRIAC

TRIAC is formed from DIAC for which gate terminal is connected which is used for controlling the device in either direction for turning on.

- In other way, the action of the device in either direction can be controlled by gate current.

Symbol:

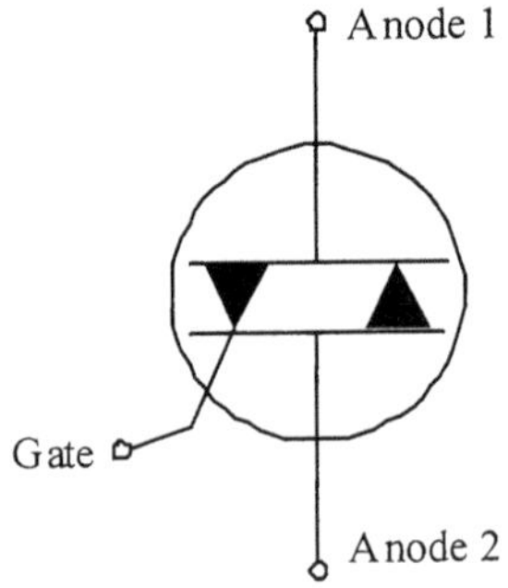

Fig. 8.39

8.17.1 Construction

The semiconductor layers of the diac can be basically arranged as shown in Fig. 8.40.

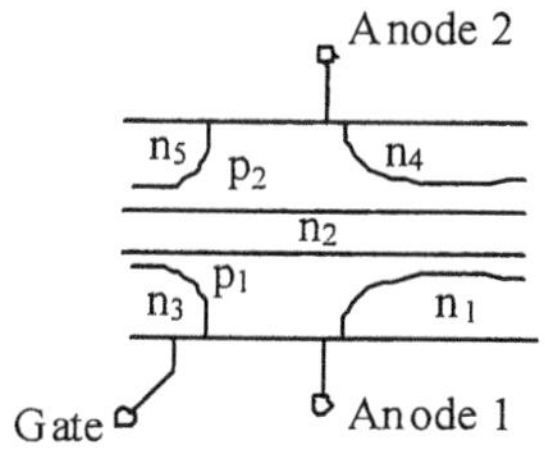

Fig. 8.40

- Here they are two anodes and gate terminal.

***Operation*:** Here for every possible direction of conduction, input signal is applied to the gate terminal can control the combination of semiconductor layers.

8.17.2 Characteristics

Here in TRIAC the characteristics are different from DIAC. When we observe characteristics from Fig. 8.41 the first and third quadrants are different from DIAC.

- Mainly the holding current not exist in each direction in the characteristics of DIAC.

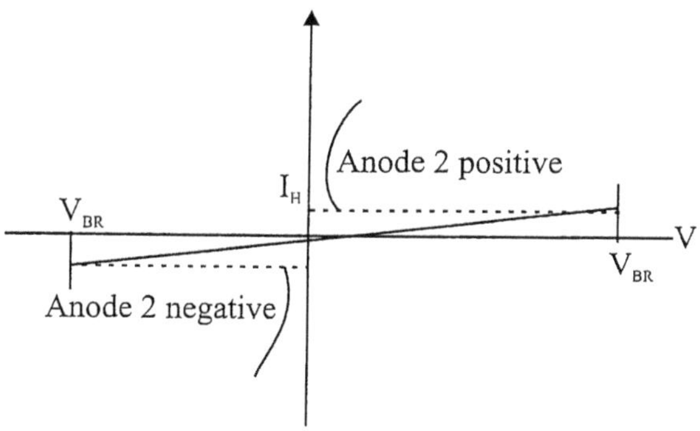

Fig. 8.41 *Characteristics*

SUMMARY

- JFET is UNIPOLAR Device (Unipolar only one type of carriers either holes or electrons)
- JFET device has Higher input resistance compared to BJT and Lower input resistance compared to MOSFET.
- The disadvantage of JFET amplifier circuits is Smaller Gain - Bandwidth product compared to BJT amplifier circuits.
- I_{DSS} = Satuaration value of Drain current when gate is shorted to source.
- For zero drift current in the case of JFET, $0.007|I_D| = 0.0022\ g_m$
- For zero drift current, in the case of JFETs, $|V_p| - |V_{GS}| = 0.63V$
- Expression for g_m for JFET interms of g_{mO} is, $g_m = \left(1 - \dfrac{V_{GS}}{V_P}\right)$
- The two types of n-channel or p-channel MOSFETs are
  1. Depletion type
  2. Enchancement type
- In JFET terminology, $B\ V_{DGO}$ parameter means the value of breakdown voltage V_{DG} at breakdown when the source is left open i.e., $I_S = 0$.
- The parameters $B\ V_{GSS}$ stands for breakdown voltage V_{GS} when drain is shorted to source.

OBJECTIVE TYPE QUESTIONS

1. In JFET recombination noise is less because it is .... device.
2. The disadvantage of JFET amplifier circuit is ....
3. The D, G, S terminals of JFET are similar to .... terminals of BJT respectively.
4. The voltage V_{DS} at which I_D tends to level off, in JFET is called ....
5. The voltage V_{GS} at which I_D becomes zero in the transfer characteristic of JFET is called ....
6. The range of $r_{DS\,(ON)}$ for JFET is ....
7. For low electric fields E_x of the order of 10^3V/cm, $\mu \alpha$ ....
8. Expression for V_{GS} in terms of V_p, ...
9. Expression for I_{DX} in terms of I_{DSS} in I_{DS} ....
10. I_{DSS} in defined as ....
11. JFET can be used as .... resistor.
12. Relation between μ, r_d and g_m for JFET is ....
13. The square law device is ....

ESSAY TYPE QUESTIONS

1. Obtain the expression for the pinch off voltage V_P in the case of n-channel JFET.
2. Deduce the condition for JFET biasing for zero drift current.
3. Draw the structure and explain the static Drain and Gate characteristics of n-channel JFET. Repeat the same for p-channel JFET.
4. What are the applications of JFET and MOSFET devices?

MULTIPLE CHOICE QUESTIONS

1. Compared to BJT, JFET is

(a) Less noisy (b) more noisy

(c) Same noisy (d) can't be said

2. At pinch off voltage in a JFET device,

(a) Channel width is zero and I_D becomes zero

(b) Channel width is zero, but I_D is not zero

(c) V_{GS} becomes zero

(d) None of these

3. For high eletric field strengths $\varepsilon > 10^4$ V/cm, mobility of carriers is proportional to

(a) ε_x (b) $\frac{1}{(\varepsilon_x)^2}$ (c) $\frac{1}{\varepsilon_x}$ (d) None of these

4. The pinch off voltage Vp in the case of n-channel JFET is proportional to

(a) N_D^2 (b) $\frac{1}{N_D}$ (c) $\frac{1}{\sqrt{N_D}}$ (d) N_D

5. Expression for I_D in the case of JFET is,

(a) $I_{DSS}\left(1-\frac{V_{gs}}{V_p}\right)$ (b) $I_{DSS}\left(1-\frac{V_{gs}}{V_p}\right)^2$

(c) $I_{DSS}\left(1+\frac{V_{gs}}{V_p}\right)^2$ (d) $I_{DSS}\left(\frac{V_{gs}}{V_p}-1\right)^2$

6. The condition to be satisfied for zero drift current is,

(a) $0.007\,|I_D| = 0.0022$ gm (b) $0.07\,|I_D| = 0.022$ gm

(c) $0.7\,|I_D| = 0.0022$ gm (d) $0.007\,|I_D| = 0.022$ gm

7. The relation between μ, r_d and g_m for JFET is,

(a) μ = rd/gm (b) rd = $\mu.^2$gm (c) μ = rd . gm (d) μ = rd^2.gm

8. Typical values of μ, r_d and g_m for JFET are,

(a) μ = 5, rd = 100 Ω, gm = 100 m (b) μ = 1000, rd = 10 Ω, gm = 0.1 m

(b) μ = 6, rd = 1MΩ, gm = 10 K (d) μ = 7, rd = 1MΩ, gm = 0.5 mA/V

9. The resistor which is connected in series with source resistance Rs to reduce distortion in JFET amplifier circuits is called

(a) Swamping resistor (b) swinging resistor

(c) bias resistor (d) distortion control resistor

10. The symbol shown 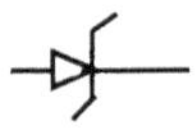 **is that of a**

(a) LED (b) Schottky diode

(c) Varactor diode (d) Gunn diode

11. The symbol shown is  **is that of a .....**

(a) DIAC (b) GTO (c) SCR (d) TRIAC

12. DIAC is a ..... layer device

(a) 4 layer (b) 3 layer (c) 6 layer (d) Two layers

CONCEPTUAL QUESTIONS (Interview Questions)

1. Why tunnel diode device exhibits negative resistance characteristic?

Ans : The current flowing through a device is proportional to the number of free electrons. The higher the number, the more the current. If current through a device 'I' decreases as the voltage across it is being increased, the device exhibits negative resistance.

In the case of a tunnel diode, there are two conditions to be satisfied for tunnelling to take place.

(i) There must be empty energy states on the P-side corresponding to filled energy states on the n-side.

(ii) The barrier width at the junction must be very narrow (~ $3A^0$), corresponding to doping concentration of 10^{19} carries/cm^3.

When forward voltage is applied to the tuneel diode, because of the energy gained from the electric field, electrons with higher energy levels tunnel through the barrier, to occupy the equivalent empty energy states on P-side. As the forward voltage is further increased, electrons with lesser energy also acquire sufficient energy from the increased electric field and will be able to tunnel through the barrier. But as the empty energy levels on the P-side get filled, after reaching the peak current I_P, even though forward voltage is being increased, the number of electrons reaching P-side decreases and hence tunnelling current decreases. So the device exhibits negative resistance region, as current is decreasing with increasing voltage. When the voltage is further increased, tunnelling phenomenon ceases, as the empty energy states on P - side would have been filled and normal forward current of the diode only flows.

www.ingramcontent.com/pod-product-compliance
Ingram Content Group UK Ltd.
Pitfield, Milton Keynes, MK11 3LW, UK
UKHW061537190726
13845UKWH00006B/130

9 789352 301508